# PHYSICAL TECHNICS;

OR PRACTICAL INSTRUCTIONS FOR

# MAKING EXPERIMENTS IN PHYSICS

AND THE CONSTRUCTION OF

# PHYSICAL APPARATUS

WITH THE MOST LIMITED MEANS.

BY DR. J. FRICK,

DIRECTOR OF THE HIGH SCHOOL IN FREIBURG, AND PROFESSOR OF PHYSICS IN THE LYCEUM.

TRANSLATED BY

JOHN D. EASTER, PH.D.,

PROFESSOR OF NATURAL PHILOSOPHY AND CHEMISTRY IN THE UNIVERSITY OF GEORGIA.



PHILADELPHIA:

J. B. LIPPINCOTT & CO.

1861.

# PREFACE

TO THE

# SECOND EDITION.

---

It is impossible to acquire thorough scientific knowledge by the study of books alone, without personal observation and research. But a thorough acquaintance with physical science has become necessary to success in so many pursuits that the chemist, the physician, and the practical mechanic, not less than the professed physicist, must know how to make physical experiments. The intrinsic interest of the subject also attracts many who have no positive necessity for the study. The amateur confines himself usually to some special branch of the science, but the teacher must have a much more general acquaintance with the art of experimenting: for the intellectual and material importance of physical science is daily becoming more evident, though its practical utility is that which will first gain it a place in the school. In order that instruction in natural science may accomplish that which is expected of it, (whatever may be the method employed,) it must be supported by experiment. The time when chalk and black-board were all that were required, is, fortunately for our schools, long past.

To assist instruction by experiment, requires something more than the expenditure of money in the purchase of apparatus: the teacher must know how to make experiments. Many teachers, however, have had no opportunity, after the completion of their studies, to acquire the art of manipulation, and must learn it after the apparatus has been placed in their hands as teachers; and how often is the division of studies made to depend upon which teacher can occupy two to four extra hours a week! Even if the first selection of apparatus be left to the teacher, the means, usually sparing enough, will often, in spite of his best endeavors, be injudiciously expended, though hardly sufficient to procure the most indispensable articles; while no provision at all is made for the future development of the instructor himself. The teacher can only overcome this difficulty by undertaking himself to construct such apparatus as does not require a particular arrangement of the workshop, special dexterity, and accurate execution; or, at least, to have them made under his direction by such mechanics as are accessible. This is possible only when the apparatus is reduced to the simplest form consistent with its object.

Many others are placed more or less in the same circumstances as the teacher, and the amateur in physics will take pleasure in constructing apparatus for himself.

It is the object of this book on the one hand, to furnish an introduction to physical experimentation, to describe all the particulars requisite to success, to call attention to those points which must be considered in the purchase and use of apparatus; and on the other hand, to give instructions for the construction of apparatus in the cheapest and most effective way.

Much of what is here treated of is scattered through the text-books on physics; but there is much more which would be out of place in a text-book, and is preserved only by tradition among physicists. It follows, from what has been said, that the book cannot treat of experiments intended to advance the science, but of such as are suited to the demonstration of known laws, and must even exclude those which illustrate laws too recondite for general instruction. The theoretical importance of the phenomena of fluorescence, and the tension produced on wires by electrical induction, have induced me to mention them, although they have not yet found their way into text-books.

It was no part of my design to make a complete collection of experiments or apparatus; and those experiments which require no particular apparatus or care have been passed over for the most part in silence. Of several experiments for the illustration of the same laws, the preference is given to those which are most easily performed, most striking, and cheapest.

In regard to the making of apparatus, I have described only such as one with some mechanical skill, or the aid of a good mechanic, may make for himself. Such things as air-pumps, telescopes, microscopes, etc., must always be bought from the dealers. If one does not object to the trouble of overseeing, many a thing can be made by a handy mechanic which is in no way contemptible; for this reason, many pieces of apparatus have been drawn in section. The money spared in this way may be expended in the purchase of other articles which cannot be made at home, and thus more accomplished with little means.

In this edition I have, more frequent than in the former, in connection with the simplest apparatus, also described more complex ones, which answer the purpose better. The choice must depend upon the time and the means, as well as upon the scope of the instruction.

It is true that, in beginning, the amateur instrument-maker, with all the pains that he may take, will not make very elegant apparatus; but it will at least be useful, while one often buys apparatus which he could just as well make himself, or which was not properly examined, or well packed before sending.

All costly articles should be purchased from makers of known character.

I have not designed to explain all the mechanical operations necessary, and have assumed technical terms as known; for otherwise, the description of the apparatus would have been too prolix. It was necessary, however, to the object in view, that directions should be given for such common operations as soldering, varnishing, glass-working, etc. I have been induced, partly by experience, and partly by the wish of experienced friends, to collect the description of these several processes, and make of them a separate division of the book. This gave the opportunity to describe several operations more minutely than was done in the first edition. Such operations as are only necessary in a special experiment have been left in their place, as they would be sought for nowhere else.

The size of the apparatus is always given, either in the text or in the figures, as I have learned by experience how greatly one may err in this respect. In these measures, the line is assumed as equal to 3 millimeters.

Although I had in general avoided the designation of apparatus by the names of the inventor, except when this was in common use, some errors of this kind crept into the first edition. I have done my best to correct these errors in this edition, and repeat my former declaration, that I did not feel bound to go into investigations of the right of priority. I am glad to receive corrections, when made in a proper spirit.

In the arrangement of subjects, I have followed Professor Müller's text-book of physics. Many figures have also been borrowed from the same book, with the consent of the author. The desire to make the book as cheap as possible, induced me not to repeat the description of such apparatus as, for example, the air-pump, whose construction is described in every text-book of physics; but, assuming their construction as known, to allude only to the precautions necessary in using or constructing them. Everything relating to chemistry has been omitted, because there are already excellent hand-books on this subject.

My attempt at an introduction to physical experimentation has been received with more favor than, from its imperfections, I had any reason to expect; as is best shown by a new edition being required in so short a time. I have used every opportunity, in the mean time, to supply the defects of the former edition, both by clearer descriptions of processes and by more definite statements of number and measure.

In perfecting the present edition, I have been under obligation to many friends and acquaintances for the description of simple apparatus; and especially to Professor Varrentrapp, of Brunswick. I would here express my thanks to these friends, and also to the publishers, who have enriched this edition with many additional figures, and substituted many new ones for others which were wanting in precision and accuracy. This is a matter of great importance, for the design of the book requires the most careful delineation of the objects described.

May this edition be as kindly received as the last, and contribute its quota to the advancement of science.

THE AUTHOR.

FREIBURG, IN THE BREISGAU,
*January*, 1856.

# TABLE OF CONTENTS.

## PART FIRST.

MANAGEMENT OF APPARATUS IN GENERAL, AND DIRECTIONS FOR SOME OF THE MOST COMMON OPERATIONS.

### CHAPTER I.

OF THE ARRANGEMENT OF THE PLACE AND THE MANAGEMENT OF APPARATUS IN GENERAL.



### CHAPTER II.

ON WORKING GLASS.

## CHAPTER III.

### WORKING IN METALS.



## CHAPTER IV.

### VARIOUS OTHER OPERATIONS.



---

# PART SECOND.

### PHYSICAL EXPERIMENTATION.

## CHAPTER I.

### EXPERIMENTS ON THE EQUILIBRIUM OF FORCES.

#### (*a.*) Solids.

### (b.) Liquids.



### (c.) Gases.

## CHAPTER II.

### EXPERIMENTS ON MOTION.

#### (a.) Experiments on various kinds of motion.



#### (b.) Experiments on hydrodynamics.



#### (c.) Experiments on the motion of gases.



## CHAPTER III.

### EXPERIMENTS ON ACOUSTICS.

## CHAPTER IV.

### EXPERIMENTS ON LIGHT.

#### (a.) On the transmission and intensity of light.

## CHAPTER V.

### EXPERIMENTS ON MAGNETISM.



## CHAPTER VI.

### EXPERIMENTS ON ELECTRICITY.

## CHAPTER VII.

### EXPERIMENTS ON HEAT.

#### (a.) On expansion.



#### (b.) On changes in the physical state.



#### (c.) On the diffusion of vapors through gases.



#### (d.) On the steam-engine.



#### (e.) On specific heat.



#### On the diffusion of heat.

# PHYSICAL TECHNICS.

## PART FIRST.

### PHYSICAL MANIPULATION.

**Management of Apparatus in general, and Directions for some of the most frequent Operations.**

## CHAPTER I.

### OF THE ARRANGEMENT OF THE PLACE AND THE MANAGEMENT OF APPARATUS IN GENERAL.

[1] **The Room.** Convenient arrangement and careful preservation of the apparatus facilitate its use, and contribute to its durability so much, that they are of equal importance to the teacher and the amateur But reasonable desires can seldom be gratified in this matter; the teacher is seldom able to make the desired arrangements or changes. If, however, it be in his power to control the arrangement of his lecture-room, an apartment should be especially appropriated to the purpose of physical experimentation, in which the seats should be arranged in rising succession, with the foremost nearly on a level with the table, so that even those in the first row may look downward upon it. Much time is saved in this way, by enabling the lecturer properly to prepare each experiment beforehand, not to mention the time which is often lost by having to set the apparatus in its place during the lecture. It is very advantageous to have an adjoining room for the preservation of the apparatus. This should in no case be too far from the lecture-room, and least of all in another story. The laboratory must be near the apparatus-room. This room should have the sun on at least one side, and be well provided with shelves and drawers.

[2] Most pieces of apparatus can be best kept in glass cases from twelve to fourteen inches in depth, made in two parts, a lower one about two feet and a half or three feet high, with wooden doors, and an upper with glazed doors; these should move on hinges, not slide. By means of a small ladder of two or three steps, access can be had to the top shelves,

on which may be placed such light articles as are only used occasionally. Larger pieces of apparatus which cannot be kept in these cases, such as air-pumps, electrical machines, etc., should be protected from dust by suitable covers of pasteboard, muslin, silk, or other material. In this room as little work as possible should be done.

[3] In the workshop there should be a wide, firm table, with drawers, in the middle of the room; and, by a window, a heavy work-bench, of three-inch oak, provided with a vice. Under another window should be placed a turning-lathe, with its apparatus. A forge is also indispensable. It should be so situated that the operator shall not stand in his own light. The bellows can be arranged overhead. It need not be large—from two to four square feet are enough for all purposes. In case the operator is obliged to make it himself, the necessary directions are given under the head of glass-blowing. The tube which conducts the blast to the furnace may be made of tin, from one to two inches wide, terminating in a cast-iron nozzle, with an aperture half an inch wide, in order to concentrate the blast. The nozzle projects half an inch into the furnace, and inclines at a small angle toward the hearth.

The blast-pipe is usually furnished with a stop-cock, in order to con-

*Fig.* 1

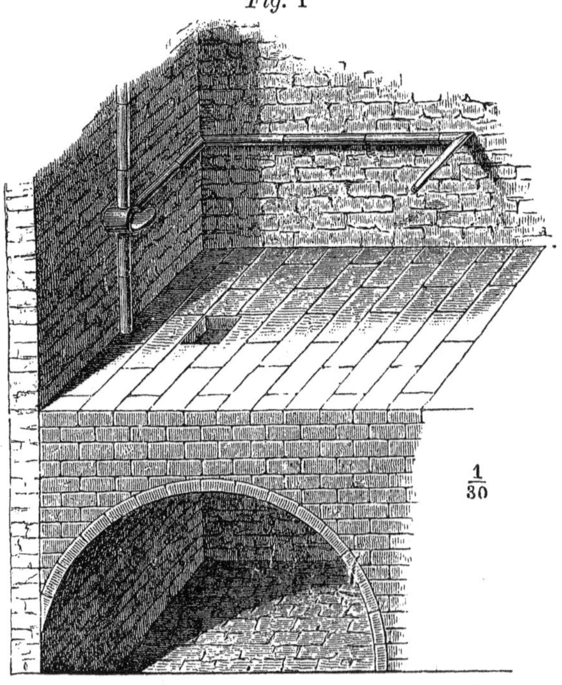

trol the blast or shut it off suddenly. It is well to have a tube leading from the side of the stop-cock, as in fig. 1, the parts of which are only stuck into each other, so that it may be fixed in any direction and brought to bear upon a hand furnace. The tap of the stop-cock is made cylindrical, as is shown in figs. 2 and 3, and has three openings, so that the blast may be entirely cut off, or thrown into either tube at pleasure. The hearth of the forge need not be more than two inches deep; its depth is easily increased by laying a few fire-bricks around it. A valve should be provided to close the chimney in the winter, when the forge is not in use. Near the forge, a blast-furnace should be built in the broad hearth. A glass case should be provided for the necessary chemicals; the tools, however, may be stuck in racks nailed against the walls or in the recesses of the windows. The smaller and more delicate articles may be preserved in wooden boxes or drawers.

*Fig.* 2.

*Fig.* 3.

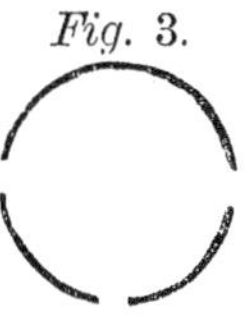

[4] The introduction of gas into a laboratory is a great convenience; it is cheaper than spirits of wine, cleaner and more convenient than any other fuel, and gives a hotter flame. Two burners should be placed over the hearth, one for illumination, the other for heating; fig. 4. For the latter purpose, a ring with three arms, figs. 5 and 6, is fixed on the burner, to support a conical tube of sheet-iron or copper, fig. 7, which is covered at the top with coarse wire gauze. The gas is mixed in the tube with atmospheric air, and burns above the gauze with a bluish flame, without smoke. When a higher and concentrated heat is required, this tube may be surrounded by a similar one half an inch higher, resting in the niches, fig. 6, which brings an

*Fig.* 4.

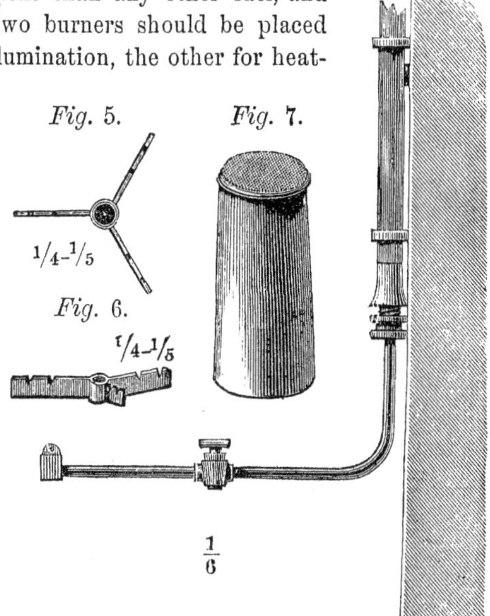

*Fig.* 5. *Fig.* 6. *Fig.* 7.

additional current of air to the flame. The flame is completely controlled by the stop-cock. The branch in fig. 4 can be fitted with a joint, so as to swing round, if desirable. Almost any form of burner can be used. The wire gauze must be two inches above the top of the burner. Burners with numerous apertures are preferable, because they produce a more thorough mixture of the gas with air. Simple or jointed brackets are placed over the work-benches. They should be made with two burners, so that one may be used for illumination, and a flexible tube attached to the other; or a stop-cock, with a short tube bent downward, may be soldered into the pipe, as in fig. 8, for the purpose of attaching a vulcanized rubber hose. If a thin India-rubber tube be attached directly to the burner, it will bend over and stop the flow of gas. For boiling, or igniting, on the work-bench, the wire gauze may be set over the burner, *b*, fig. 8, or the burner, fig. 9, which is connected

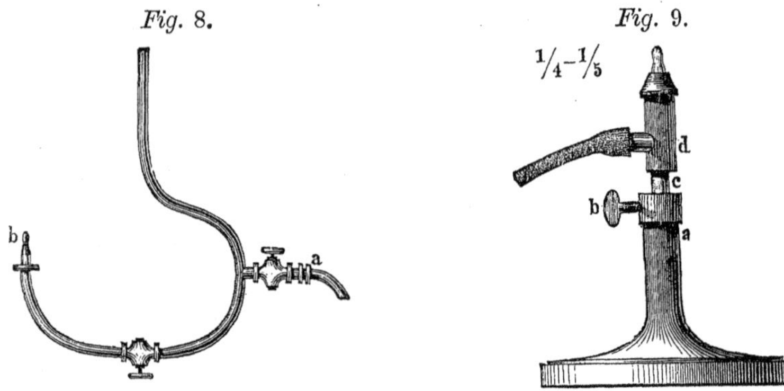

*Fig.* 8. *Fig.* 9.

by a flexible tube, with the stop-cock, *a*, fig. 8. This burner consists of a wooden or metal stand, *a*, in which the rod *c* may be fixed at any height by a screw. This rod supports a gas tube, *d*, on which the burner is screwed.

Gas apparatus for some special purposes will be described hereafter. The arrangement described here involves so little additional expense, when gas is once introduced into the building, that it bears no proportion at all to the convenience attained by it. These requirements will, perhaps, seem too great for a high school or academy, but they can hardly be moderated, if instruction is to be imparted successfully, and not confined to text-books. No provision for instruction in chemistry is here contemplated, for crowding a workshop and laboratory together is attended with much injury to the iron and steel tools.

[5] If the teacher be disposed to devote time to the construction of

such apparatus as he can make, with the instructions here given, or have them constructed under his supervision by skillful workmen, he must be provided with the necessary tools, which are not very numerous. The most of them he can construct himself as occasion requires; and all large operations had better be left to a mechanic. The necessary skill is easily acquired, with a little care and attention. The directions given in this book apply, with the exception of soldering, mostly to operations for which no experienced workman can be found in small towns, and at which the experimenter must often try his hand without ever having seen them performed. The following list may be considered as comprising the necessary outfit:—

1. An iron vice, of about 20 pounds weight.
2. A set of files and a hand-vice.
3. A pair of bellows, with water trough, smith's tongs, coal shovel, and tongs.
4. A horned anvil of from 25 to 30 pounds.
5. Several hammers of from two pounds to one-quarter of a pound.
6. A pair of nippers.
7. Two flat and two round pliers.
8. A pair of tinner's shears.
9. A small lathe.
10. A chisel and a gouge for turning wood.
11. Various augers and bits for the lathe.
12. Brace, with an assortment of bits.
13. A screw-plate for cutting screws of various degrees of fineness, up to two or three lines.
14. A saw for cutting metal.
15. A pair of callipers.
16. A milling wheel.
17. A screw wrench.
18. A jack and a smoothing plane.
19. A couple of chisels and rasps.
20. A panel saw.
21. A few screw clamps of different sizes.
22. A glue pot.
23. A hatchet.
24. A grindstone and whetstone.
25. A charcoal furnace.
26. A spirit-lamp with rather thick wick.
27. A blow-pipe.
28. A soldering iron.

29. A glass-blower's table.
30. A pair of strong scissors.
31. A small iron mortar.
32. A porcelain mortar.
33. An iron bevel.
34. A glazier's diamond, and a diamond for writing on glass.

The cost of all these articles may amount to $100; the lathe alone being worth, say $30. The price of each will of course vary with the perfection of the workmanship, but the first cost cannot be less than this. The most costly article, the lathe, is at the same time about the most indispensable. If the experimenter be not familiar with the use of this instrument, the advice of some experienced person should be taken in its selection, with due reference to the place where it is to be used. When the means allow, it should have cast-iron cheeks and rests. The spindle must be arranged for attaching screw guides, and have on the cheek end an external and internal screw. Two drums will be required, one about eight inches in diameter for turning iron, the other three or four, for turning wood and brass. The fly-wheel should not be less than 50 pounds weight, and two and a half feet in diameter. Besides the tools, a stock of various kinds of wood, such as beech, walnut, cherry, and poplar, must be laid in some dry airy place. Neglect of this causes many difficulties from the shrinking of green wood.

With the above-mentioned apparatus, the operator will be able, by degrees, to construct the remaining tools for himself. At first, all will go on slowly, as to make the smallest or simplest article may require the construction of a special tool; but this difficulty and delay will gradually lessen, and much practical skill be obtained for further operations. The experimenter must learn to do much with small means, and even, as Franklin says, "saw with a gimlet and bore with a saw."

[6] If, after all, the above tools cannot be obtained, the following will answer a great many purposes:—

1. A small bench-vice, with anvil.
2. A hand-vice.
3. A small anvil with a horn.
4. A pair of nippers.
5. Flat and round pliers.
6. Two hammers.
7. A couple of files and rasps.
8. A few gimlets.

These may cost some $25 in all, and are really as few as can be got along with.

[7] The purchase of apparatus depends so much on the means at command and the extent of the instruction proposed, that no general rules can be given for it. If the teacher be inexperienced, he should not be hasty in making his purchases, but rather accumulate apparatus gradually as it is needed in the course of his instruction; he will thus learn best how and where to order apparatus.

Under the supposition that it is intended to construct most of the apparatus for one's self, the following should be immediately ordered from an instrument-maker, even though the instruction to be imparted be of the most elementary character:—

| | |
|---|---|
| 1. Common balance, carrying from five to ten pounds - - - | $25.00 |
| 2. An areometer - - - - - - - - - - - - - - - - - - | 1.00 |
| 3. Siphon barometer, with stopper - - - - - - - - - - | 20.00 |
| 4. Air-pump, with its apparatus - - - - - - - - $10.00 to | 300.00 |
| 5. Models of suction and forcing pumps, with glass tube - - | 12.00 |
| 6. Concave, convex, and plane glass mirrors - - - - - - | 3.00 |
| 7. Prism, of flint glass, if possible - - - - - - - - - - | 1.25 |
| 8. Large convex lens, with several of smaller size; also a concave lens - - - - - - - - - - - - - - - - - - - - | 6.00 |
| 9. Terrestrial achromatic telescope magnifying 12 to 20 times | 30.00 |
| 10. Compound achromatic microscope - - - - - - $18.00 to | 50.00 |
| 11. Thermometer, the scale on glass - - - - - - - - - | 2.00 |
| 12. Water hammer - - - - - - - - - - - - - - - - | 3.00 |
| 13. Horseshoe magnet, carrying 18 to 20 pounds - - - - | 3.00 |
| 14. Magnetic needle with agate cap - - - - - - - - - | 1.00 |
| 15. Electrical machine - - - - - - - - - - - - $20.00 to | 200.00 |
| 16. Six carbon and zinc elements - - - - - - - - - - | 15.00 |
| Add for tools, and pay to workmen for what is given out, and for materials - - - - - - - - - - - - - - - - - - - - | 170.00 |
| | $339.25 |

Thus for an original outlay of from $400 to $800, and an annual addition of $50 for at least the first few years, the experimenter will be in possession of apparatus by means of which he will be able to give very thorough school instruction in physics.

A more complete list of apparatus will be found at the end of the book.

[8] Each of these instruments, as far as possible, should be taken to pieces, in order that its construction may be accurately studied, and manipulation be thus more readily understood. Some too may have imperfections, or at least uncertainties in use, only to be avoided by special precautions. Should this be the case with any, it would be well to attach in writing, to each, the method by which the necessary rectification is to be made.

Likewise, all appurtenances, such as wires, hooks, etc., should be laid beside the apparatus to which they belong. When an inventory of stock is kept, each piece should be numbered, and opposite this number in the list, should be stated the name, use, cost, etc. This will be useful to the present incumbent, and still more so to his successor.

[9] Whenever any piece of apparatus has been used, it should be carefully cleaned and returned to its appropriate place; all articles should, as far as possible, be arranged in systematic order.

Glassware is best cleaned with alcohol and blotting-paper, and dried with a clean linen cloth. Brass should be rubbed, in the direction in which it was polished, with linen and fine Tripoli, or prepared chalk, and dilute alcohol. Brass articles are generally varnished, in which case nothing more can be done than to rub them with clean blotting-paper and fine linen, and this only in the direction of the streak. Articles in frequent use should never be varnished, as varnish soon rubs off in spots, which looks badly. Whenever this is the case, it will be better to remove the varnish entirely by means of alcohol.

Iron apparatus is rarely varnished; it may be cleaned by rubbing with blotting-paper and a little olive oil, with which the surface should always be lightly greased. If harder rubbing is necessary, add to the oil finely levigated emery, or else take emery paper. If the latter cannot be obtained, it may readily be made, by coating stout writing-paper with strong glue, and then upon this sifting emery through fine gauze. The sheet is then to be folded, so as to bring the coated surfaces together, and several placed in a pile between two boards, and dried under a moderate pressure; the loose emery may afterwards be shaken off.

[10] Standards of measure are an important part of physical apparatus, and the experimenter should be provided with those in common use, so that descriptions of foreign instruments may be understood, without the labor of reducing them to the more familiar unit. Obtain, if possible, a slender stick of apple-tree wood, which has been cut a long time; fashion it into a bar, half an inch square, and a meter in length. The ends must be cut exactly square, which may be done with a file. Lay off on this bar the length of a yard, by an official standard, noting the temperature, which, however, has little influence on wood, and divide it into inches. In this way a scale can be obtained whose accuracy may be relied on, at least within known limits, which is not the case with those purchased, even when officially stamped. The lines or the other subdivision of the inch may be marked off by means of a paper scale. For ordinary use, a tape line or wooden measure, whose accuracy has been tested, is good enough.

[11] On the other three sides of the bar mark three other scales, their length having first been ascertained by calculation, and then subdivided. The graduation must be regulated and controlled as much as possible by calculation of subdivisions, so as to make the various scales harmonize with each other. A meter, divided into centimeters, should by all means be among the number.

*Fig.* 10.

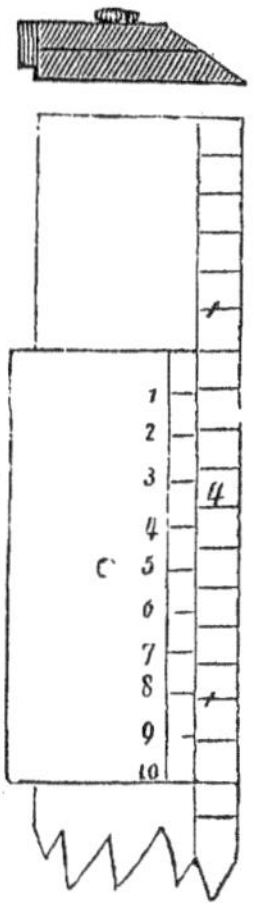

The Vernier is indispensable in making small measurements, and its use should therefore be clearly explained to a class. As those in actual use are generally small, and therefore unsuited to illustration, it will be better to construct one on a large scale, from four to six feet long, fig. 10. If the wood be light colored, and the divisions well blackened, the use of the Vernier may be explained to a whole class at once.

[12] **Measures of capacity.**—More accurate measures than those in daily use will not often be required, and when necessary, it will be best to employ the corresponding weights. If, however, they be really required, then they should be made of metal, and cylindrical, preserving the proper legal proportion between diameter and height. The exact gauging of such measures presents many difficulties, when they are required to be filled to the brim. If this be not the case, and it never is so in physical experiments, one can graduate any glass vessel for himself, by pouring into it the proper weight of water. A glass tube of one-half to one inch diameter, which has been adjusted and graduated into cubic centimeters by successive additions of one gramme of water, will be found very convenient in many experiments. The divisions may be made with the diamond, and should correspond with the horizontal surface of the water, and not its elevated margin. They may also be etched in, according to the directions given in the following chapter.

## CHAPTER II.

### ON GLASS WORKING.

[13] As it is rare to find in small towns any one who can work glass, this generally devolves upon the physicist; it is therefore described more minutely than other mechanical operations.

[14] Emery is a very useful material in working glass; it is, therefore, well to describe the mode of preparing it. Commercial emery is not generally well assorted; the finest dust is mixed with coarse grains, from which it must be separated by levigation. The commercial emery is stirred briskly, with ten times its weight of water, without giving it a rotary motion, and the water instantly poured off from the sediment into another vessel; from this vessel, after standing from three to four minutes, it is poured into a third, in which the emery is allowed to subside—or, after standing from five to ten minutes, it is poured into another vessel. Three or four sorts of emery, varying in fineness, are obtained in this way, the coarsest of which is often adulterated with sand, and the various grades are dried and preserved separately.

[15] **Grinding and boring glass.**—The occasions for grinding glass in physical experimentation are very numerous. The rough grinding is done on a flat cast-iron plate, with quartz sand or coarse emery and water; and the article is ground smooth on an old piece of plate glass with very fine emery and water. The plates which are used for this purpose, especially the glass plates, do not last long; they soon become concave. They may, however, be ground even again upon each other; very good cast-iron plates may be had for this purpose from the machine-shops, but they should be reserved for the finishing touches. In case the glass plate of the air-pump should be broken, a piece of mirror glass, of the proper thickness, should be cut round, the edges ground off, and the hole bored through the center, by a copper coin fixed to a piece of wood in the lathe, with emery and oil. With a little patience, an angular plate of glass may be rounded on a common grindstone. Holes may be bored in glass with ease and rapidity by means of an iron, or, better still, a copper ring half a line thick, secured to a wooden chuck so as to be concentric with the spindle of the lathe: a piece of cork, corresponding in size to the inner diameter of the ring, is cemented to the glass

as a guide. The lathe must be set in rapid motion, and a thin paste of emery and oil constantly applied; a round piece of glass will thus be cut out. Lenses are cut out of thick glass in the same way.

When the glass is nearly cut through, a smooth piece of hard wood must be pressed against it, opposite the ring, and as soon as the lathe cuts through in a single spot, the application of emery should cease; at least, when the plate is thin. It is very difficult to avoid splintering the edges of such holes. The best mode of preventing it, is by boring from both sides toward the middle. Holes in glass may be enlarged by means of a conical block of wood smeared with emery and water. When the piece of glass is of such shape that it can be fixed on the lathe, a circular groove may be cut out with a graver, moistened with turpentine, while the lathe is turned slowly. Holes from one to two lines in diameter may be bored with a copper drill, and those of less size with a three-cornered pointed steel drill, moistened with oil of turpentine. The watchmaker's drill-bow is very convenient for this purpose. The pressure must be very gentle when the point of the drill comes through. Holes may be bored by hand, with almost any hard tool moistened with oil of turpentine, nearly as fast as on a lathe. A dull surface is given to a glass plate most readily by grinding it on another glass plate with emery and water; if the plates be even, the finest emery may be used at once. Grinding crystals, for optical purposes, will be described in the chapter on optics.

[16] **Grinding glass stoppers.**—Although bottles with glass stoppers may be had almost anywhere, there is seldom care enough taken to render them air-tight. If the stopper needs but little alteration, it should be ground by hand, with emery and water; if it be very irregular, it should be fixed to the lathe, and coarse emery and quartz sand used at first. In both cases, the stopper must have not only a rotary, but a longitudinal motion. It is seldom worth while to take much trouble with a stopper that fits very badly, but it sometimes may be necessary, and may be done most easily by making a copper plug the size of the stopper, and grinding out the neck of the bottle with it. When a stopper sticks tight, too much force should not be applied to it; it is better to heat it gently over a spirit-lamp, and then try to turn the stopper: if not successful the first time, let the bottle cool, and heat it again more highly. Care must be taken to hold the bottle so that the contents shall not be spilled in case it breaks. A drawer should be appropriated for preserving all odd stoppers, which will come into play for many purposes.

[17] **Cutting glass.**—It is often necessary to cut off glass tubes, rods, or cylindrical vessels. Tubes not over $\frac{1}{4}$ to $\frac{3}{8}$ inch thick may be broken easily, by making a notch in one side with a three-cornered file,

and then placing both thumb-nails opposite the notch, and bending the tube slowly outward. Thicker tubes must be cut all around with a file dipped in spirits of turpentine; or a red-hot iron may be drawn around the place, and water dropped on the heated surface. An iron ring is best adapted to this purpose, and should be held by a second person, who also holds water in readiness. A crack may be led in any direction, by placing a hot iron on the glass just before it, and drawing it in the required direction. The following method is better for cylindrical vessels: Tie a flattened roll of paper on each side of the spot where the glass is to be cut off, leaving a free space between them less than a line in width, forming a channel between the two folds. Take a fine, tightly twisted cord, long enough to go around the vessel, and leave two or three feet over, after being wound around the hand once or twice at each end; two persons now hold the vessel firmly against the table with one hand, and with the other draw the cord back and forth in the channel between the papers. The glass soon becomes so highly heated that the cord takes fire and burns off; water is then thrown on the heated spot. The glass generally cracks off very evenly, and the more so, the narrower the channel in which the cord moved. Instead of the folds of paper, Mohr uses the wooden instrument, fig. 11. It is screwed to the table, and each operator holds one end of the glass, while the cord is drawn backward and forward through the slit in the wood.

*Fig.* 11.

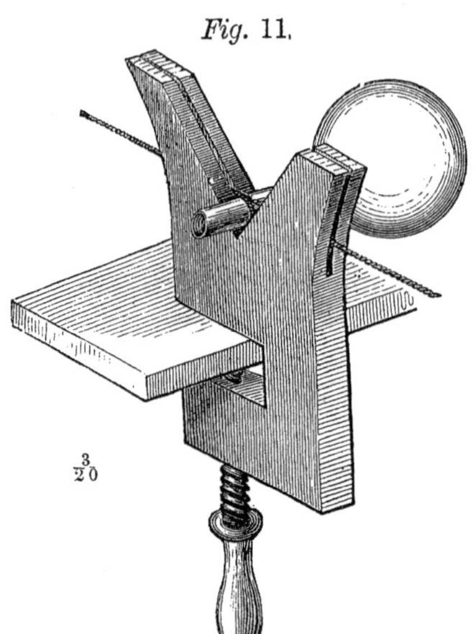


When the glass has already a crack, or one can be started from an edge, the heat can be applied more conveniently with little cylinders of charcoal. These coals are prepared in the following way: 60 grains of gum tragacanth are dissolved in sufficient water to make four liquid ounces of mucilage; 30 grains pulverized gum benzoin are then dissolved in the smallest possible quantity of spirits of wine; the two solutions are then mixed in a mortar, with enough pulverized and sifted beechwood charcoal to form a plastic mass—a little softer than pill mass is generally made. Cylinders, from

$\frac{1}{4}$ of an inch thick and 3 or 4 inches long, are rolled out of this mass, under very gentle pressure, and dried slowly. These cylinders, when ignited, continue to burn of themselves with a pointed end; the coal is used, by holding it at the end of the crack, and drawing it slowly, without pressure, in the desired direction. It is well to mark the direction which the crack is to take, with chalk or ink; the coal must be frequently turned and blown, as the cooling action of the glass extinguishes it where they are in contact. When the crack approaches within a line of the starting-point, it will follow the coal no longer, and the rest must be broken off.

When the glass is not already cracked, a crack may generally be made by notching the edge with a file, and then holding the coal, kept in lively combustion by blowing upon it, against the notch. A thin cylinder of glass may thus be cut into a spiral ribbon, which is tolerably elastic. In the same way pieces of any shape may be cut out of flat glass. The inequalities which are left on the edge of the glass may be removed with the copper emery-wheel, or by grinding on sandstone; the smaller ones can be removed with a file, moistened with resinous oil of turpentine, or a solution of camphor in turpentine. Little pieces may easily be broken off with a pair of flat nippers, and round plates may be cut out of window glass in this way.

[18] **Glass-blowing.**—Scarcely any operation is more frequently necessary than blowing glass; and skillful workmen are only to be found in large towns. Even when a glass-blower is quite near at hand, it is very inconvenient to call in his assistance on every slight occasion, and some dexterity in this art is therefore indispensable.

[19] **Construction of the blast-table.**—It is desirable to have if possible, a separate table for this purpose; but if this be not possible, the whole apparatus may be arranged in a box, which can be placed on any low table during use, and afterwards removed. The bellows should be made to rise vertically, so as to contain more wind, and both the upper and lower parts have a single fold, which may be made of wood. Neither part requires more than two inches play; it is better to increase the blast by increasing the surface, which need not in any case exceed $1\frac{1}{2}$ or 2 square feet. More weight is required, in proportion, to obtain the same pressure with a greater surface; but bellows which rise vertically, *i.e.* have no hinge in the upper part, have an unsteady motion when they rise high, especially when the cheeks are made entirely of leather; they require then to work in guides, which is very inconvenient.

The following directions are given, in case it should be necessary to construct such a bellows for one's self. The middle board, fig. 12, is made of half-inch stuff, from $1\frac{1}{2}$ to 2 feet long, and one foot wide. It has two square openings for the valves, two inches long, one inch wide, and one inch apart. It is better to have two valves than one of double the size.

*Fig.* 12.

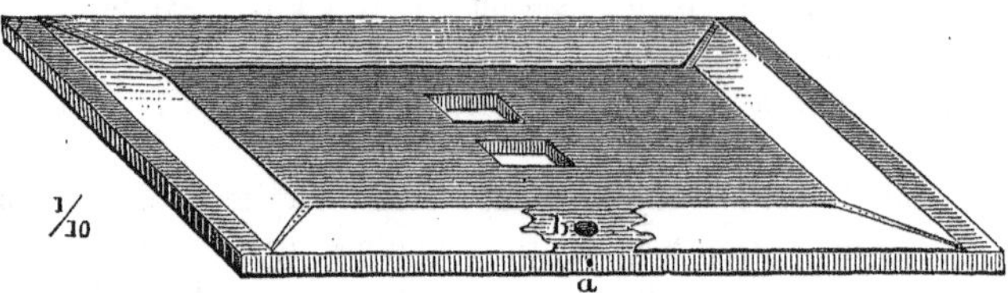


In one of the long sides, a hole, *a*, one-quarter inch wide and one inch deep, is made, connecting with a wide opening, *b*, in the top of the board, to receive the pipe. The board is covered on both sides with stout paper or old parchment. The valves are made of wood, a little thicker than the combined thickness of the cheeks when folded together, so that they will be pressed down by the top board, when the bellows is not in use. Instead of a single board, the valves may be made of strips, one-half an inch wide, tacked close together, upon the valve leather; all warping is thus prevented. The valves extend half an inch beyond the opening, all around, except on the hinge side, and are faced with a piece of sheepskin, glued on with the hair side outward. They must be pressed between boards, and left to dry, so that the leather may remain smooth. The leather projects on one side, to serve as a hinge. Fig. 13 shows such a valve, made of

*Fig.* 13.

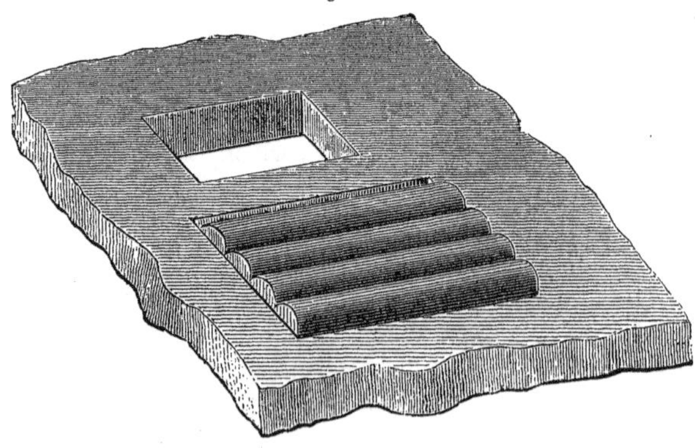

slips, in position. The valves are prevented from rising high enough to fall over, backward, by a narrow strip of leather, which must not be long enough to fall under the valve. The cheeks are best made of thin boards, covered with paper or pasteboard, and so adjusted that, when closed, the corners do not come quite together, as is seen in fig. 12. They are connected by pieces of sheepskin, glued on by the flesh side; the corner strips must be wide enough to allow the cheeks to open freely. The cheeks are connected in the same way, with the middle and upper boards, and it is well to connect the two latter by strips of leather, to prevent the cheeks from expanding to quite their full extent.

The bottom board is made a little smaller than the top board, to afford room for the hinge. The cheeks of the lower part are best made of strips of good buff leather, two inches wide, nailed to the boards, and joined at the edges by gluing on narrow strips of the same The lower valves are exactly similar to those already described. If the bellows can be had from an organ-builder, it is better not to attempt to make one at home. Fig. 14 shows the bellows expanded.

*Fig.* 14.

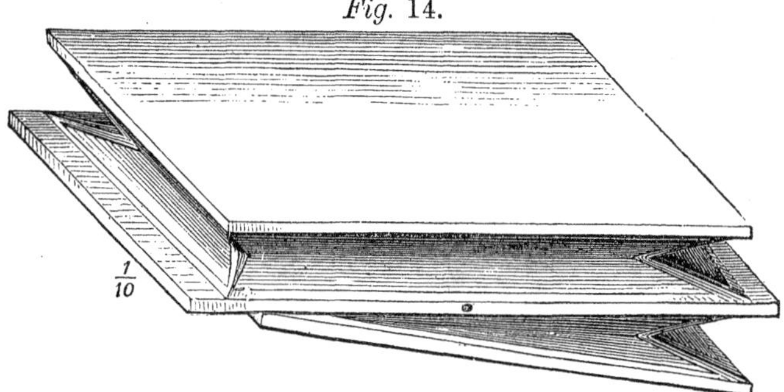


*Fig.* 15.

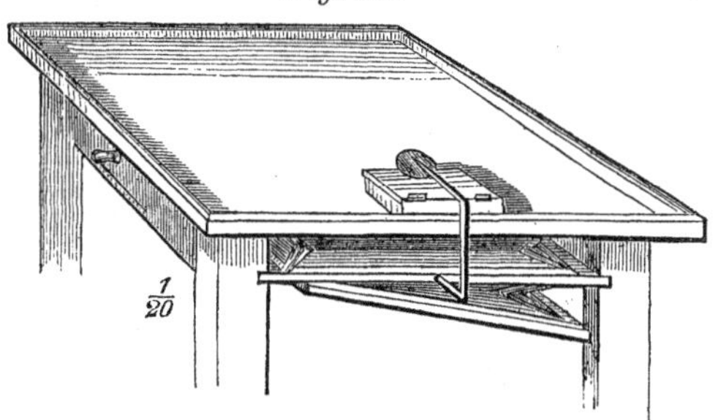


The table should be three feet high, and the bellows hung immediately beneath the top, so that there will be space enough for the operator's feet, below it. The top has a ledge projecting one-quarter of an inch all around, fig. 15. The bellows is worked either by a lever fastened to the table and connected by a wire with a stirrup, or by a cord fastened to a projection on the lower board, and passing upward over a pulley, to the stirrup wire. The bellows is slid into a groove cut on the inner side of the casing, and secured by a screw. The top of the table must be screwed fast to the casing, as it has to replace one of the sides. The conical end of the wind pipe is stuck into the hole in the middle board, so that it can be inclined to the table at any angle. The length of the pipe depends on the height of the lamp. It is very convenient to have a wooden rod projecting about an inch above the table, upon which one of the two slides, fig. 16, is fastened by a binding screw, in order to hold the nozzle in a fixed position. The table may be made a little wider than the bellows, to admit a small drawer to hold the necessary instruments.

*Fig.* 16.

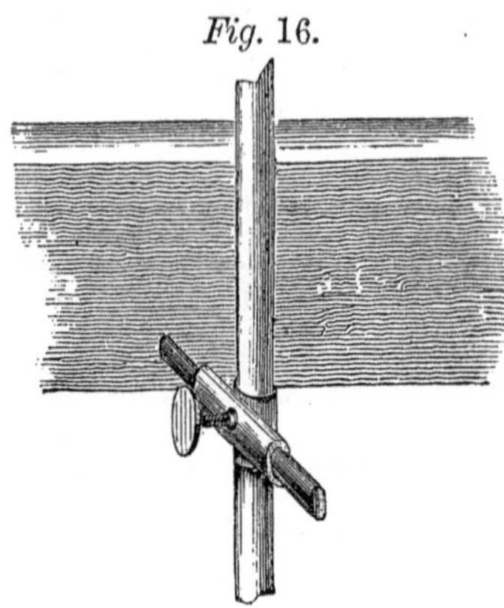

The pressure required for glass-blowing, with a small exit pipe, is about $1\frac{1}{4}$ ounces to the square inch. An old iron plate is the most convenient weight. For larger operations, where the exit pipe is wider, the pressure must be increased; in this case, the bellows is always kept full, and any required pressure applied by the foot.

[20] **The Lamp.**—This is made of tinned iron, narrow in the direction of the blast, and the wick set near the end. It is best to fit it with a tube, in which a solid wick can be screwed up and down, figs. 17 and 18. The tube must of course be slit, to admit the oil. The movement of the wick is effected very simply, as is shown in figs. 19 and 20, where the

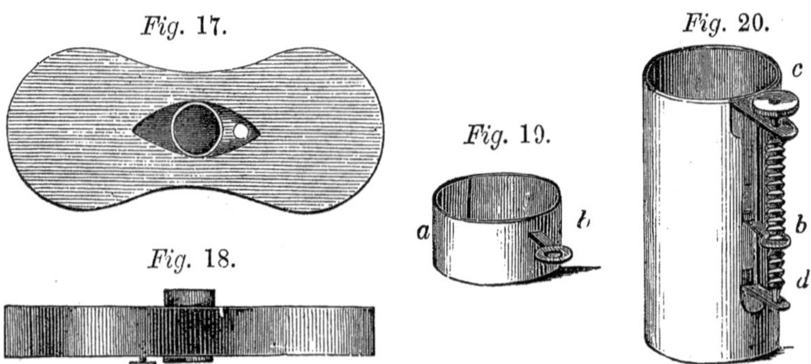


*Fig.* 17. *Fig.* 18. *Fig.* 19. *Fig.* 20.

wick is inserted in the collar, *a*, the stem of which passes out through the slit in the tube, fig. 20. Through this stem, which is of tin, the deep threaded screw, *c*, *d*, passes. Two strips of tin, soldered to the tube, furnish the points of support for the screw, and an excavation is made in the table, to receive the part of the tube which projects below the lamp. This admits of a longer wick, without allowing the oil to sink so low, as it must in a deep lamp. If a little hole be made in the side of the head of the screw, it can be easily and rapidly turned by means of a pin or wire, when it grows hot.

A simple arrangement of the lamp is shown in figs. 21 and 22, where the wick is merely laid in a tube, and is prevented from burning too far back by the plate, fig. 23, which is pushed forward on the top of the lamp, toward the wick.

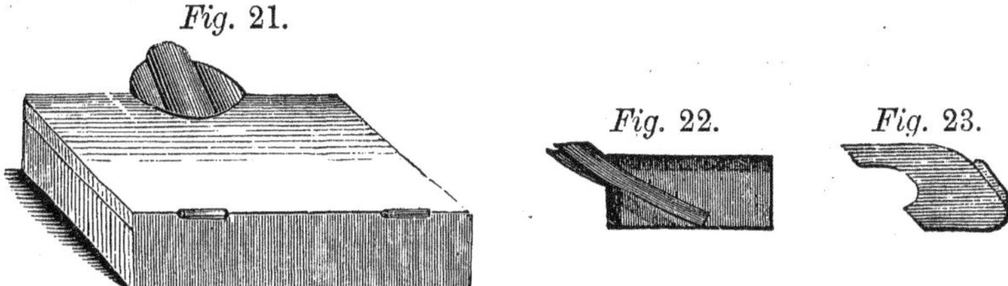

*Fig.* 21. *Fig.* 22. *Fig.* 23.

It is well to place the lamp on a sheet of tin about a foot square, turned up at the edges; it saves the table from being burned, as one is often compelled to lay down hot pieces of glass.

[21] Oil is used as the combustible, and cannot well be replaced by anything else, when a hot flame is required. The wick is made of cotton, loosely twisted together. For the cylindrical wicks, take the wicks woven for argand burners, and fill them with loosely twisted cotton.

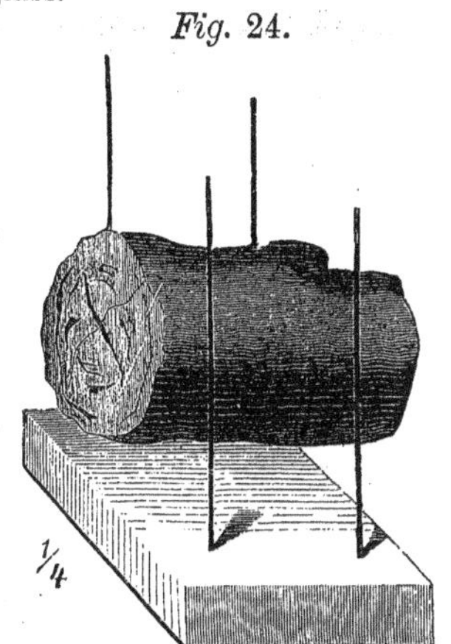


*Fig.* 24.

When a very hot fire is needed, it is a good plan to fasten a round piece of beech-wood coal, between four wires, stuck in a board, fig. 24, with the crown toward the flame; the radiant heat of the burning coal increases the effect very greatly. For the same reason, it is not good to turn the wick next the operator, and let the flame pass over the lamp, apart from the inconvenience of such a position.

Oil, when exposed for some time to the air, becomes resinous, which makes its use very inconvenient for the physicist, who often does not require his lamp for months together. In this case, the spirit-lamp, with a thick round wick, is preferable. If the wick be made about $\frac{1}{2}$ inch thick, it affords sufficient heat to bend large glass tubes, blow thermometer bulbs, etc., especially with the aid of the piece of charcoal just mentioned.

A lamp with double draught, like the argand lamps, is more effective than the common spirit-lamp. The nozzle is fitted by a cork to the inner tube, and the flame driven up vertically. The space between the cylinders should be large enough to receive several cylindrical wicks, fitted one within the other.

[22] The nozzle must be $1\frac{1}{2}$ to 2 lines wide in the clear, and contracted toward the end, where it may be closed by a small plate, in the middle of which an aperture is bored. This should be 1 to $1\frac{1}{2}$ millimeters in diameter, and when a strong blast is required, may be as wide as 3 millimeters, fig. 25. When it is required to heat only a small spot, a smaller nozzle, with an aperture $\frac{1}{2}$ to 1 millimeter wide, fig. 26, is necessary, and is stuck on the larger one. Both should be hard soldered. [Nozzles made of stout glass tube, drawn out fine at the end, answer very well, only they are liable to break, and must be frequently renewed.—*Trans.*] The wind pipe may be of lead. It is convenient, but not necessary, to have a stop-cock in the tube, by which the strength of the blast can be regulated at pleasure.

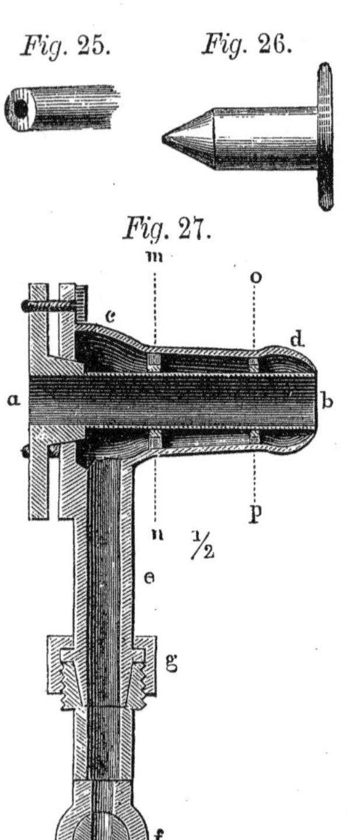


As the whole flame of an oil-lamp is never impelled before the blast, but a part always burns upward, Mohr leads the air first, by a bent copper tube, through this part of the flame, so as to blow hot air into the flame, and thus obtains a much higher heat.

[23] Illuminating gas is the most convenient combustible for glass blowing, as the flame may be completely controlled by a stop-cock, never smokes, and requires no trimming. The only change necessary, is in the burner. A burner with an annular flame is best. Fig. 27 shows a vertical section of such a burner, half the natural size, and fig. 28, a section through the lines *m n*, *o p*. It consists of two parts: a tube, *a b*, screwed at *a* into an outer tube which is connected with the gas pipe; *m n* and *o p* are two disks to support the pipe *a b*, serra-

*Fig.* 28.

ted on their edges, so that the gas may flow out uniformly all around the burner, as in the argand burners. The outer tube is enlarged at *c*, and fits tightly over the disks *m n* and *o p*, so that the gas can only pass through the openings. The space between the sides of the two tubes, at the mouth, need not be wider than from $\frac{1}{4}$ to $\frac{1}{3}$ of a millimeter. It should be made still less at first, and enlarged by grinding the ends of the tubes, until the opening is large enough to deliver, under the diminished pressure in the daytime, as much gas as the air from the bellows can completely burn, when the treadle is worked slowly. The flame is hottest when it burns blue. The stop-cock serves to regulate the flow of gas. The burner is connected with the gas pipe by the screw-cap *g*, which allows it to be inclined to the table at any required angle. The most convenient direction of the flame for most operators, is a little inclined upward, but many blow with a vertical flame.

*Fig.* 29.

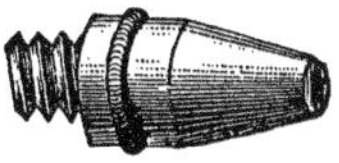

The wind is led through a lead pipe, upon which brass nozzles of various widths can be screwed, fig. 29. Wide apertures must be from 1 to 2 centimeters behind the aperture for the gas, but the finer nozzles for small pointed flames should be pushed up nearly to the end of the tube *a b*.

The burner rests not directly on the table, but on a board about an inch thick, which projects over the edge of the table.* Figs. 30, 31, and 32 show the burner, and the connections in plan; and fig. 33, the

*Fig.* 30. *Fig.* 31. *Fig.* 32.

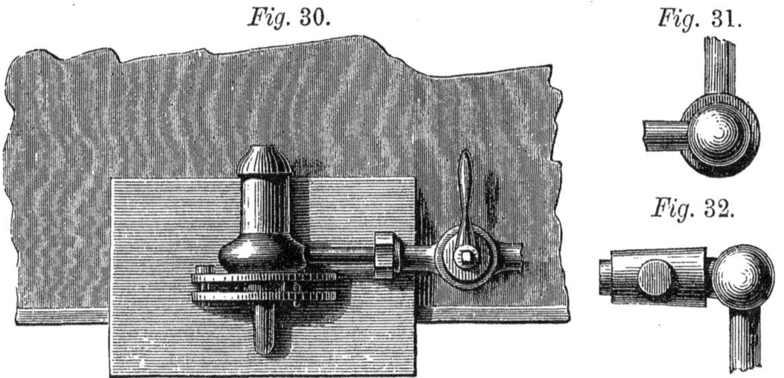

* [A very neat and convenient blast-lamp, on this principle, is made by Ritchie, of Boston. The tubes slide upon a rod, on which they can be fixed at any height. The burner has also a horizontal and vertical motion, and the air passes through a bulb, in which it may be heated by a spirit-lamp. The above lamp is also made after a modified pattern, so as to bring into service simultaneously an assemblage of jets.—*Trans.*]

whole blow-pipe table, with the gas pipes. [If the connections be made with elastic tubing, which is much cheaper than the joints described, the

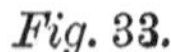

*Fig.* 33.

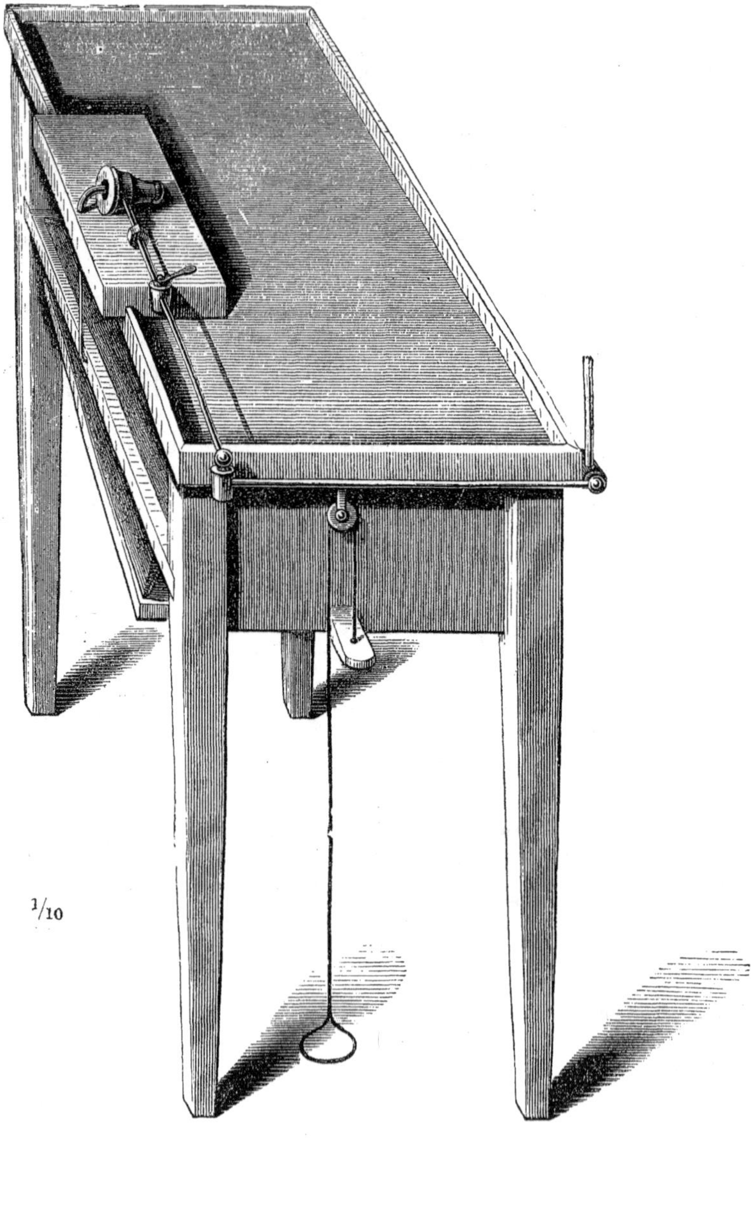

1/10

table may be moved from its place.—*Trans.*] The wind pipe goes from the middle board of the bellows to the block which supports the burner, and the little lead tubes which carry the nozzles are inserted in this board.

[24] **The Eolipile.**—If one is not disposed to incur the expense of a regular blast apparatus, the eolipile, shown in fig. 34, in half the natural size, will answer many purposes, and can be made at small expense by any ordinary coppersmith. It consists of a cup with double sides; the exterior cylinder, which is closed air tight, is filled with weak spirits of wine, (40 to 50 per cent.,) through the opening *a*, which is closed with a cork; the inner cylinder is filled up to the level of the jet with strong spirits. The latter is ignited, which causes the spirits within to boil violently, and the vapor, escaping through the tube *b*, gives a very hot flame. The inner cylinder may be closed by a tightly fitting cover.

*Fig.* 34.

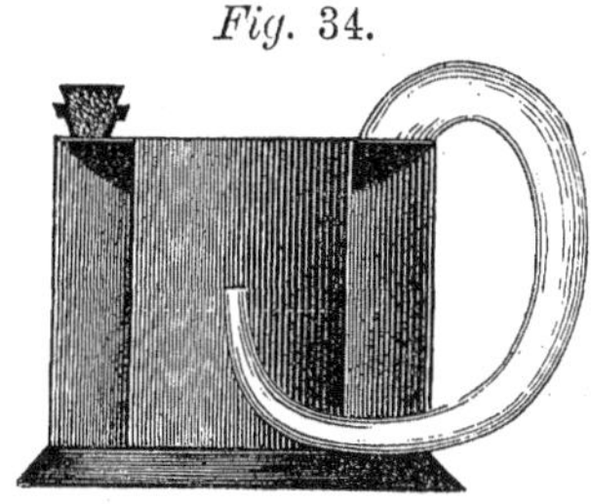

[25] The soldering lamp of the tinners is more convenient, because the parts can be separated and kept in better order, and the flame is horizontal. Fig. 35 represents one, $\frac{1}{4}$ or $\frac{1}{5}$ the natural size. The sheet-iron cylinder M N has an iron ring soldered to the bottom, to mark the proper position of the spirit-lamp; near the top is a row of small holes. On the top of the cylinder is set a strong cup made of hammered metal, which has an opening for spirits, closed by a screw or cork, and a bent tube. A slit is made in the back of the cylinder for the reception of the tube, and a sufficient space left open in the front for the exit of the flame. The end of the tube, which has an aperture 1 millimeter wide, reaches nearly to the flame of the lamp, half way between the wick and the bottom of the cup *b*. This cup is shown in fig. 36. If the opening be closed with a cork, it will serve as a safty-valve. I have, however, never heard of the explosion of such a lamp, though they are often in the hands of very careless workmen. The lamp is filled in the same way as the last.

*Fig.* 35.

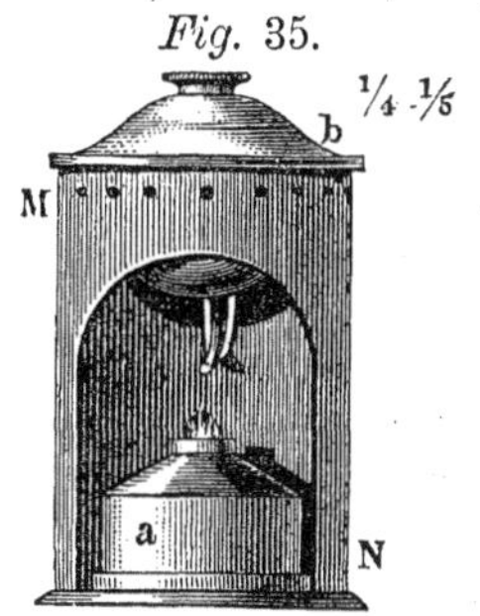


*Fig.* 36.

[26] The only other apparatus necessary for glass blowing is a pair

of sharp scissors for trimming the wick, a three-cornered English file for cutting glass tubes, a pair of flat pliers with narrow jaws, and some small iron rods.

[27] The mode of cutting off glass tubes and rods has already been described in § 17.

[28] The glass should be in the form of tubes, and care should be taken to select such as are of uniform thickness, free from striæ and warts. Glass containing lead should be rejected if possible, for though easily fused, it is apt to blacken from the reduction of the lead. The French soda glass is especially adapted to glass blowing, being easily fusible, and free from lead. Suitable glass may be obtained from almost any dealer in chemical apparatus. It may be observed in general, that all sorts of glass become, by continued heating, crystalline and less fusible.

[29] The first requisite in glass blowing, is to obtain a good flame, and the experimenter may be assured that he can do nothing until this is accomplished. As this depends upon the movement of the nozzle, a hair's-breadth either way, it is, like the whole art of glass blowing, a matter only to be obtained by practice; but the following directions may serve as a guide. The wick should be raised, so as to give a flame 3 or 4 inches high, and separated with the scissors into two parts, so as to leave a passage for the air through the middle. It must then be trimmed perfectly flat, and all projecting, rugged threads carefully removed. The nozzle is inserted a little below the level of the wick, between the two parts; it must project a short distance into the flame for a quiet, pointed flame, and be drawn back about a line, when a large roaring flame is required.

[Having once obtained a good flame, the greatest care should be taken not to change, in the slightest degree, the relative position of the nozzle and the lamp.—*Trans.*] The flame should be directed rather a little upward than downward; it should burn with a faint blue light, deposit no soot, and be over two inches in length. When large objects are to be heated uniformly, a broad, roaring flame is required, which is obtained by placing the jet a little behind the wick.

The glass must be perfectly free from moisture; breathing into the tubes before they are heated must therefore be carefully avoided. The condensed moisture, coming into contact with a highly heated spot, is sure to crack the glass. The glass should be held about $\frac{2}{3}$ the length of the flame from the wick; large and thick pieces are apt to snap if suddenly heated; they should therefore be first warmed in the flame without blowing at all, and then very gradually heated where the soot is thickest. During the heating, the glass must be constantly turned, that it may be heated on all sides alike; some practice and care are necessary when the glass is held with both hands, to turn both ends with equal rapidity, otherwise it will be twisted into a spiral form, as soon as it begins to soften. When the operation is finished, the glass must not be withdrawn from the flame too suddenly, else it is apt to crack, especially when it is of irregular thickness. [It is a good plan to allow it to become thickly covered with soot, which protects it from the cooling action of the air.—*Trans.*] When two different kinds of glass have been fused together, the work generally cracks, in spite of all precaution; it is therefore best to keep the different sorts separate.

[30] Only those operations can be described here which occur frequently and are the basis of all others. The making of special apparatus, such as areometers, etc., should be left to the professional glass-blower. The following operations find place here: (*a*) widening and contracting the end of tubes; (*b*) drawing out and thickening glass; (*c*) sealing the ends of tubes; (*d*) bending tubes; (*e*) blowing bulbs; (*f*) joining tubes. Some special operations will be mentioned in their proper places.

(*a*) Glass tubes, when cut off, have a sharp edge, which is apt to cut the finger, and is easily broken. The extreme end is therefore heated with a pointed flame, till the edge fuses round, by which also the opening is apt to contract a little, which is an advantage when the tube is to be passed through a cork. This operation can be performed on thin tubes, over the flame of the spirit-lamp, with the mouth blow-pipe. If the size of the opening is required to remain unchanged, or to be enlarged, it must be strongly heated while turning rapidly, and flared outward with an iron rod, while still in the flame; it is then allowed to shrink again to the required size, turning rapidly all the while.

(*b*) *Drawing out, narrowing, and thickening glass tubes.* Glass tubes are drawn out thin, by holding them horizontally with both hands, and heating a considerable breadth while turning rapidly; the glass is then removed from the flame, and drawn out suddenly.

When the glass is very thin, it should be exposed to a very high heat, and allowed to shrink, by which the bore becomes smaller, and the glass thicker; it may then be drawn out to a fine point, and still be tolerably strong. To thicken the glass at any one spot, without diminishing the bore of the tube, the two ends must be gently pressed together while the glass is soft, and the tube blown out to its original size. This requires a high heat, and the tube must be turned constantly.

(*c*) To close the end of a tube, it is heated, and the sides pressed together with a glass rod; the rod is then heated also, and stuck to the end of the tube, and the excess of glass drawn out in the flame to a fine thread. It may be necessary to repeat this two or three times, before the excess of glass is removed. The end is then blown out round, before the glass is thickened too much, and then reheated to thicken it, and blown out to the proper size. [A knob of thickened glass, once formed on the end of the tube, can only be removed by drawing it off with a glass rod.—*Trans.*] When the tube is to be cut off and then closed, draw it out to a point at the proper place, and treat it as just described. Capillary tubes can be closed by simply heating the end.

(*d*) *Bending tubes* is an operation of constant occurrence. Thin tubes may be bent in the flame of a spirit-lamp. A well-bent tube must have both limbs in the same plane, and be neither wrinkled on the concave, nor flattened on the convex side. This is best avoided by bending the tube in a gradual curve, instead of a sharp angle. The inside of the curve should be kept hottest. Any flattened places may be blown out again. To bend large tubes, fill them with sand, and heat them over a charcoal fire.

(*e*) *Blowing bulbs.* Before blowing a bulb on the end of a tube, the glass must be thickened at the spot. It is easier to do this before the tube is cut off, and with wide tubes it must always be done first: the tube is then drawn out, and the end closed. The glass must be removed from the flame before pressing the ends together to thicken it, because this cannot be done while turning the tube, and if it were held still, it would be heated unequally. If the glass be too highly heated, it is apt to wrinkle; it is therefore better to repeat the operation several times, having thickened the glass for a short distance, to go over it again and again, until a pear-shaped accumulation of glass is formed. A good bulb can never be blown on glass which is once wrinkled, as seen in fig. 37. Narrow tubes may be closed by simply heating them highly, and blowing gently into them from time to time, to prevent their closing up too far from the end. Very narrow tubes are sometimes closed at both ends, and warmed slightly throughout their whole length, that the expansion of the inclosed air may keep them from closing during the process of thickening.

*Fig.* 37.

When the glass is thick enough, bring it to a white heat while turning rapidly; then take the end in the mouth, hold the tube downward and blow gently into it, continuing the rotation. In this way the mass of glass will be more uniformly distributed. Heat the glass again and blow it out to the required size, gently at first, and with more force as the glass cools. If the operator blows too forcibly at first, the bulb is apt to swell out so large as to become very thin, and burst. The film of glass formed in this way often exhibits Newton's rings beautifully. This precaution will probably be superfluous in the beginning, for the learner will not obtain heat enough to blow the bulb even to the required size.

To blow the large bulbs mentioned in the experiments with the air-pump, [see fig. 167,] take a tube $1\frac{1}{2}$ to 3 lines in diameter, and, after thickening, draw it out at both ends, as in fig. 38. Then close one end and blow out the bulb very thin. After cooling, hold one side in the flame to flatten it, and then draw out the tube very fine close to the bulb, and cut it off. When the bulb is quite cool, introduce this point into the pointed flame, and seal it quickly without warming the bulb, so as to expand the inclosed air. Glass bombs are managed in the same way, except that they are left round, and after cutting them off, they are held in a pair of tongs made of thin wire, fig. 39, heated pretty highly, but not enough to soften the glass, and the fine point then sealed.

*Fig.* 38.

*Fig.* 39.

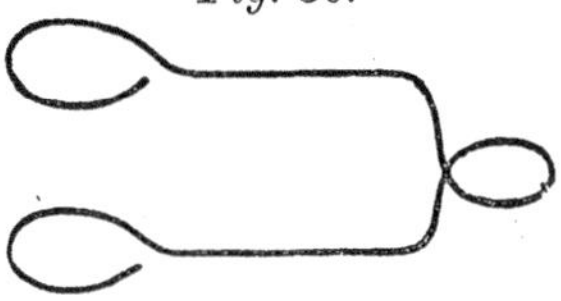

For bulbs with a hook, such as are used in experiments on specific gravity, [see fig. 140,] the glass must be made thicker, to give them strength. The neck between the bulb and the tube must not be drawn out at once, but first be allowed to shrink and thicken, as it is to form the hook. After the neck is cut off, it is heated and bent into a hook, with an iron rod or a pair of small pliers.

(*f*) *To join two tubes*, the ends of both must be made of the same size by widening the narrow one, or contracting the wider. Their edges must be even, and fit together well. Hold the edges together, as shown in fig. 40, and heat them with a pointed flame. When they are united all around, heat the joint highly and blow gently in at one end to swell out the joint a little. Then heat it again until the swelling sinks in; blow it out again, and repeat this process until the joint is smooth, and both pieces well fused into each other. Contract it finally to the proper size, and draw it out, or thicken it if necessary. Without this repeated heating and blowing out, the tubes are apt to crack at the joint upon cooling.

*Fig.* 40.

(*g*) *Opening bulbs and tubes.* It is often necessary to make an open ing in the side of a bulb or tube, as in fig. 41. To effect this, heat a thermometer tube and stick it to the glass, then heat the spot moderately and draw it out to a point. Break off this point and round the edges with a pointed flame. Regulate the size of the opening with an iron rod, and strengthen the edge, if necessary, by laying a thread of glass around it. This may be done by making a slender pointed rod red hot, attaching the point to the edge of the glass, heating the part next the point, and gradually drawing it out to a thread, which is laid around the edge. It may afterwards be more thoroughly united with the edge. Instead of drawing out the bulb with a rod, the spot may be heated with a pointed flame, and a small knob blown out upon it; repeat this several times, and at length blow it out so suddenly as to burst it.

*Fig.* 41.

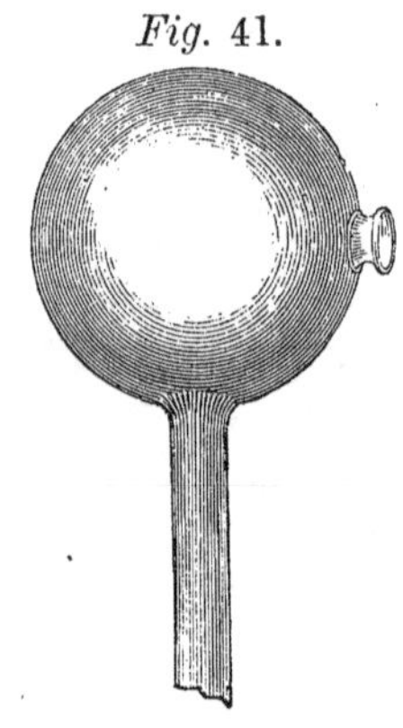

If the bulb or tube is to have only a fine opening in the side, the spot may be heated with the pointed flame, and burst at once, and the opening subsequently modified by heating it. If merely an opening is needed, without an elevated margin, the hole may be bored as described in § 15, or with a diamond, if the glass be thick enough.

[31] **Cutting glass** with a diamond requires only a little practice in judging at what distance from the rule the crack will be made, which depends on the setting of the diamond. A clean cut is accompanied by no scratching noise, and makes no glass dust.

[32] **Etching glass.**—The experimenter has frequent occasion to etch scales, etc. on glass. They cannot be done neatly with a diamond, and the glass is, moreover, apt to break at the scratch. Glass is prepared for etching by being cleaned with lye, and then with pure water, and dried with a linen cloth. It is then covered with a thin coat of wax, melted and applied to the heated tube with a feather. Especial care must be taken to leave no part uncovered. The varnish used by the copper-plate engravers is a better coating. It is made of 2 parts white wax, 1 part mastic, $\frac{1}{2}$ part asphaltum, and $\frac{1}{2}$ part turpentine, kept melted for about half an hour, in a clean earthen pot, to allow the impurities to subside, and the upper part then poured into water. A ball of the composition, as large as an egg, is wrapped in a silk rag, and the heated tube rubbed with it. When the varnish is cold, the scale is scratched through it with a steel needle. The graduation may be effected on a graduating

machine, or, in the absence of this, the operator may make a scale on a strip of paper, pasted on wood. With a stick, 2 to 3 feet long, he then makes a sort of dividers, by driving two needles through the opposite ends, fastens the scale and the glass to be graduated at a suitable distance from each other in a straight line, and, setting the point of one needle on one of the divisions, scratches through the varnish with the other needle; and so proceeds from one division to another, taking care to lay bare the glass at every stroke. The figures are easily added. Powdered fluor-spar is then laid in a leaden trough of proper length, and moistened with an equal weight of oil of vitriol. The graduated glass is laid immediately on it, or suspended at a distance of about 3 inches above it. The vapors are confined by paper laid over the apparatus. The apparatus must be placed under a chimney with a good draught, so that the vapors may be carried away from the operator, *for they are very injurious.* Although the gas will be evolved at the ordinary temperature of the air, it takes place very slowly; it is therefore better to warm the trough gently. The scale is allowed to remain in the white vapors until the glistening divisions appear dull, (about three minutes,) after which the coating is removed by warming the glass, and wiping it with paper and a little spirits of turpentine.

## CHAPTER III.

### WORKING IN METALS.

[33] **Copper and brass.**—These metals need special treatment only when they are required to be very soft or very hard. In the first case, they have only to be heated to redness. Cooling them in water, instead of hardening them as it does steel, only makes them softer; it loosens the scale of oxyde moreover, especially from copper, and gives a bright metallic surface.

These metals are hardened by hammering, and very good springs may be made of them in this way. Wire may be hardened by drawing it a few times through the drawing-plate. The experimenter will have so frequent occasion to draw wire for various purposes, that it will be advisable to obtain such a plate.

[34] **Drawing wire.**—The process of drawing wire is very simple. The drawing-plate is held in the vice between plates of copper, and about an inch of the end of the wire is filed small enough to pass through the next smaller hole. The wire is greased, and the projecting end seized with the pliers and drawn slowly through. After passing through several holes, the wire becomes hard, and must be annealed before drawing it finer.

[35] **Iron and steel.**—Steel is, like glass, a more or less elastic body according to its treatment. To make steel as hard as glass, heat it to bright redness if it be German steel, and plunge it suddenly into very cold water: cast-steel should be heated only to a cherry red, and suddenly cooled. The scale of magnetic oxyde which forms in this process ought to drop off and leave the metal bright; but the steel may still be hard, even if this does not take place. Long pieces of steel are apt to warp in this process, and must be ground afterwards, to bring them to proper shape. The surest, though not an infallible, means to prevent warping, is to immerse the piece suddenly, with its longest dimension perpendicular to the water. It is difficult to harden large, and especially long pieces, because it requires a very large fire to heat them uniformly throughout. Thick pieces are apt to crack or break off in hardening. This is best obviated by plunging them in water at 133° F., which makes them hard enough.

It is seldom necessary to make steel as hard as glass, so as to be brittle, but it is not in our power to temper it properly at once. The metal baths which are prescribed for this purpose may certainly be dispensed with. On the other hand, however, steel which has been made very hard may easily be softened to any required degree, by cleaning its surface, and heating it until it assumes successively yellow, purple, blue, and gray tints. Care must be taken to heat it uniformly; pieces which are to be made equally hard throughout must, therefore, not be held in the tongs, but heated on a tray. Tools of all kinds, for cutting metal, should only be allowed to become yellow; instruments for cutting wood, which have a thin edge, purple red; springs which have not much motion, blue; very large springs, gray. Springs which are required to have much motion, are burnt with grease, *i.e.* they are rubbed with grease, and heated until it burns, after which they are, like all the rest, plunged into cold water.

Drills, turning chisels, etc. are less liable to break when hardened only near the edge. When the tool is so shaped that it cannot be ground, the edge should be cooled first, and the heat allowed to diffuse itself from the other part until the edge has the right temper; the whole tool is then plunged into water.

Cast-steel loses its value when heated beyond a bright cherry red, and cannot be frequently exposed, even to a much lower temperature, without injury. Steel which has been spoiled in this way may be regenerated by heating it to dark redness, and cooling it in a mixture of turpentine, tallow, and fish oil, melted together. Repeated cooling in boiling water is said to have the same effect: thick pieces must always be treated repeatedly. Steel regenerated in this way is hardened as usual.

Steel and iron have often to be made very soft, which is done by heating and allowing them to cool slowly under a covering of charcoal. Iron may be coated with clay, and heated. It becomes extremely soft by being heated in any vessel, under a cover of hammer scales.

[36] **Tool making.**—Making tools is certainly not the business either of the teacher or of the amateur, and it is always better to buy such things, if possible, than to make them. But among the most indispensable tools are some which it is often impossible to obtain, and the experimenter must therefore make them himself, or send frequently to the cutler, and often take to him large pieces of apparatus. Some general directions for their manufacture will therefore be in place here.

[37] **Screw tools.**—A plate for cutting screws has already been mentioned in § 5, as one of the most necessary tools. Some taps for cutting internal screws generally accompany it, but they are usually too

short, and of little use. It is therefore better to make at once a number of taps to suit the thread of the plate. The small end of each should be a little less in size than the large end of the next smaller one. The taps ought not to be too tapering; they work better and are not so apt to break off when nearly of the same thickness throughout. The internal screws can also be cut more cylindrical. Slender taps should not be made too long; six times the diameter is a good length, and in this distance the diameter should increase by the depth of a thread. It is common to make the first third more tapering than the rest. At the thicker end the thread is cut away and the head made square, to afford a better hold for the hand-vice.

When the thread of the tap is nearly cut, remove the roughness with a smooth file, and then finish cutting the thread. File away the thread on four sides, except near the point, to give the tap a better hold at the start. Temper them straw yellow, and blue the head separately.

To cut an internal screw clean to the bottom, when it does not go through the nut, requires a cylindrical tap with few turns. Fig. 42 shows a common screw tap.

*Fig.* 42.

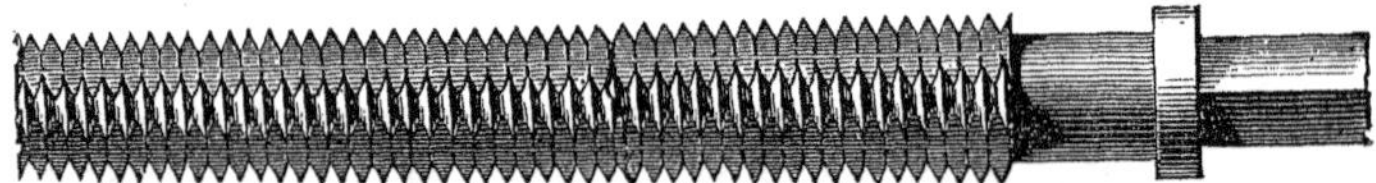

After a screw-plate is bought, the first thing done with it should be to make taps designed to cut new dies when the old ones are worn out, and used for no other purpose. These taps ought, properly, to exceed in thickness the screws cut by the plate by twice the depth of a thread, but this is not rigidly adhered to, for the proportion between the diameter of a screw and the depth and thickness of the thread may vary within wide limits, without rendering the screw useless. A screw is, however, generally useless when the depth of the thread is not at least equal to its thickness, and when the ratio of the external diameter to the internal is less than $\sqrt{2} : 1$; for in this case the thread cannot be filed away on four sides of the tap, and consequently it cannot be used. In this case, the depth of the thread is about $\frac{1}{7}$ of the diameter of the screw, and, except in special cases, it is not safe to go under $\frac{1}{8}$, where the screw has any considerable strain to bear. On large machines, screws will be found where the ratio is $\frac{1}{5}$ of the external diameter; for such cases, the threads of the tap must be cut off, as shown in figs. 43 and 44.

*Fig.* 43.

The reserve taps are made cylindrical, and the threads are not filed away at the sides, but four spiral grooves are cut in them, as seen in fig.

*Fig.* 44.

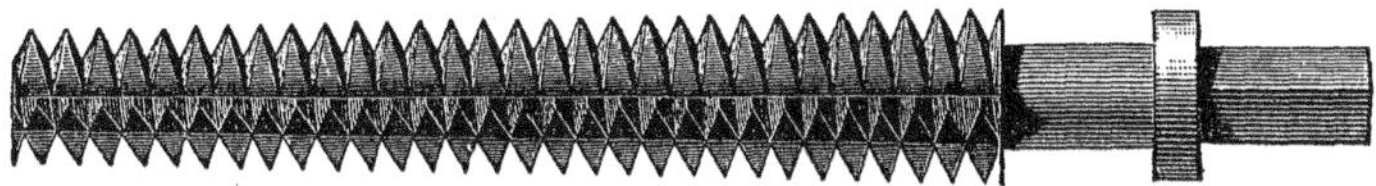

45. They are turned off on two points, so that they can be put in the lathe again.

*Fig.* 45.

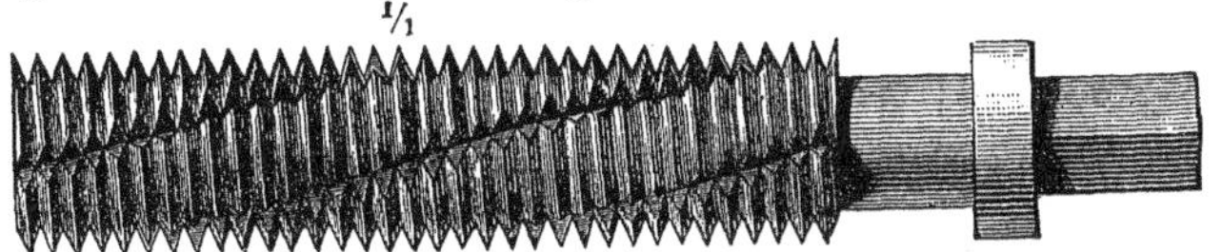


This standard or reserve tap serves to make two other screw tools, which it is easy to buy; but it is very convenient to have them with the same thread as the plate. These are the chasers, one for cutting external, and the other for internal screws. Figs. 46 and 47 show a pair of these tools, which are easily made, if the reserve taps can be used for cutting them on the lathe. Such tools are very convenient, because with them, threads can be cut rapidly, without removing the work from the lathe, and without regard to the thickness of the cylinder. The necessary skill is easily acquired, even with a lathe not furnished with guides, if the precaution is taken to use a steel support instead of one mounted with wood, which is otherwise better for turning metals.

*Fig.* 46. *Fig.* 47.

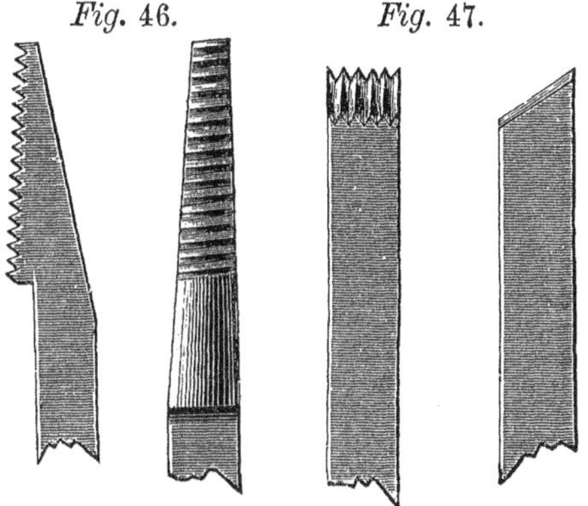

[38] **Cutting external and internal Screws** is a simple operation. For the former, care must be taken to give the cylinder barely the required diameter, for the screw enlarges a little, and to hold the plate at right angles to the axis of the cylinder. If this be a little thicker than the hole in the plate, it does no great harm. The plate must be turned slowly at first, and pressed on steadily until one turn is completed, else it may cut a false thread. The internal screws must be bored

out far enough to admit 3 or 4 turns of the tap, which is then screwed into the hole with a moderate pressure. In both cases, it is necessary to diminish the friction in the part of the screw already formed by turning backward and forward. In cutting the external screw, care is necessary not to twist too forcibly, else a slender screw might be twisted off. In forming the internal screw, it is necessary to bore at last from the opposite side in order to insure uniformity in the thread; but this is not necessary if the thread is turned off for a sufficient distance from the top of the tap and the nut is not too thick.

In cutting screws on iron and copper, oil must be plentifully employed. Brass is cut dry.

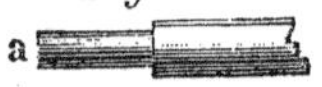

Screws with broad brass heads are often needed. For the heads of these, a square piece may be cut out of a brass plate and filed round, or a disk cut off from a cylinder. The head is then pierced, and the shank turned or filed down so as to fit tightly in the hole, driven in with a hammer and soldered with tin. The portion [*a*, fig. 48] projecting beyond the head serves to fix it in the lathe, or as a hold for the vice in cutting the screw; it is at last filed off, and the head either finished with a file or on the lathe. The heads are generally milled, which is done with a milling tool.

*Fig.* 48.

a

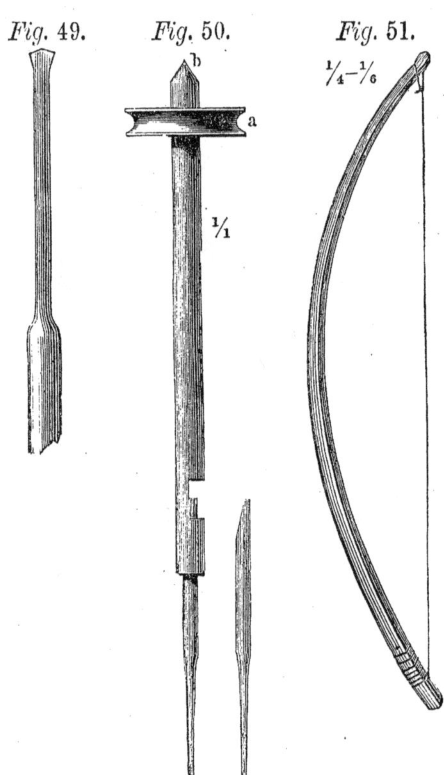


*Fig.* 49. *Fig.* 50. *Fig.* 51.

[39] **Metal drills** are made of cast-steel, which can be bought of any size. The smaller ones are made of round steel, and hammered broad at the end. The edge runs out to a point, so as to be set exactly on any desired spot. The other end is squared to fit into the handle or the chuck of the lathe. Fig. 49 shows a drill, with about the proper slope for the edge. The smaller drills, for holes of 1 millimeter and under,

are not used in the lathe, but set in a handle, such as is shown in fig. 50, and turned by means of a whalebone bow, fig. 51, the cord of which makes one turn around the wheel *a*. The point *b* is set in a depression in the vice, the piece to be bored pressed against the drill with the left hand, and the bow moved up and down without pressure. The hand must be protected from being wounded, by a piece of wood held opposite the point of the drill. This makes the edge of the hole smoother than when the drill is allowed to break through. Large drills are made of round or square steel hammered broad at the end. The edge is beveled in opposite directions on each side of the center, as seen in fig. 52, so as

*Fig.* 52. *Fig.* 53.

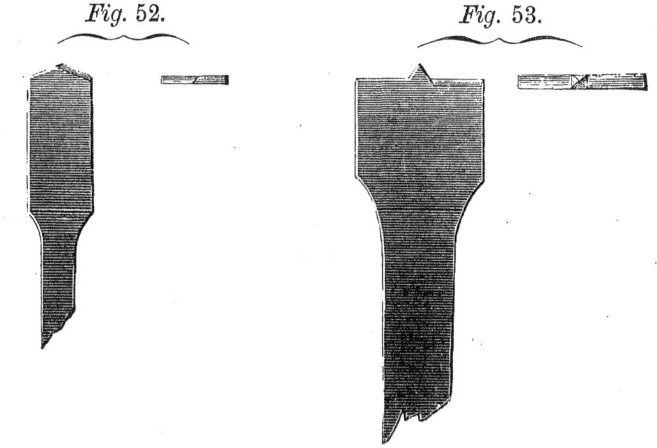

to have an oblique edge instead of a point. The hole must be started for them with a conical drill. Drills of still larger size are usually made in shape of fig. 53. The edge is straight, and has a short, square point in the middle. Such drills are used chiefly for boring shallow holes.

*Fig.* 54.

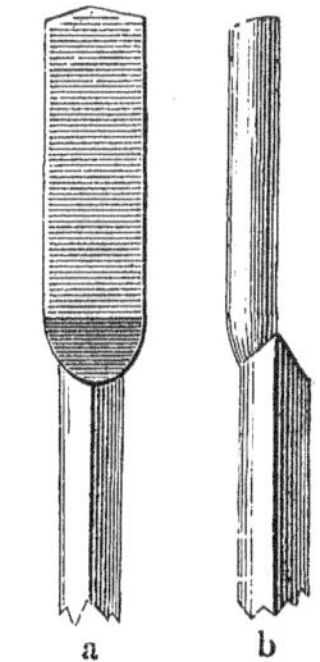


In boring on the lathe, the drill may either be fixed to the chuck, and the object to be bored pressed against it, or the converse. The latter makes rounder and more accurate holes. Cast-iron and brass are usually bored dry; copper moistened with oil.

For deep holes, the so-called cannon drills are preferable to all others. They are made by hammering out of a piece of round steel a shank of sufficient length, turning the rest of the cylinder accurately on a lathe, and then filing away one-half of it, and sharpening the end, as shown in fig. 54. The point is not made in the middle of the edge.

4

Larger drills are forged half round, and then turned. These drills cut very clean and straight, but can only be used on the lathe with the piece to be bored revolving. The point not being in the middle, it is necessary to start the hole, as seen in fig. 55, so that the drill fits into it accurately and the point does not cut at first. For drills less than a line wide, this is not necessary, and the very small ones to be used by hand are shaped approximately with a file.

*Fig.* 55.

**The Reamer,** fig. 56, is an almost indispensable instrument. Its conical head is cut into concentric grooves, and the square end must fit into the lathe and the drill handle.

*Fig.* 56.

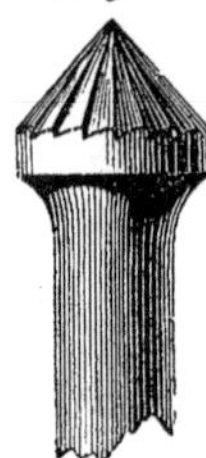

[40] Other tools are easily made when needed. The center drill is a useful tool; it is a cylinder of steel 3 inches long and $\frac{1}{3}$ inch thick, sharpened to a point at an angle of 60°. It is used for marking the point where a hole is to be bored, and making a cavity in the ends of pieces to be turned, by which they are set in the lathe. Turning tools for metal are easy to make. One end of a piece of cast-steel is tapered, to set it into a wooden handle, and the other filed to the desired shape. Chisels for turning brass are made with an obtuse edge, *s*, fig. 57; while such as are intended to be used for iron, have a sharper edge, fashioned like *s*, fig. 58.

*Fig.* 57. *Fig.* 58.

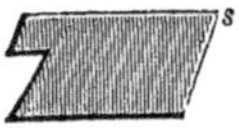

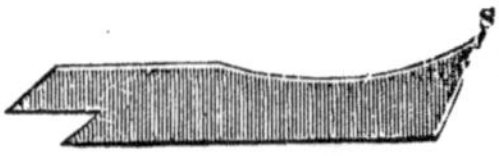

[41] **Soldering.**—We distinguish hard and soft solder; to the latter belong the metals used for soldering which fuse below a red heat, such as tin, tinner's solder, (tin and lead in varying proportions,) and the same with the addition of bismuth. Hard solder includes silver solder, (silver alloyed with brass,) which may be had of various degrees of fusibility from the silversmith, silver coin, (9 silver to 1 copper,) and various fusible alloys of zinc and copper, and lastly copper itself. The last can only be used for soldering iron.

*Fig.* 59.

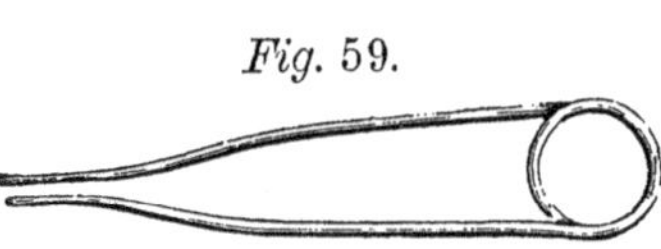

All joints which are to be soldered, must fit as smoothly as possible, and both surfaces be scraped clean; they must also be held in close contact, either by wrapping them with fine binding wire, or holding them with a pair of forceps made of strong wire, fig. 59.

If a piece is to be soldered several times, making it necessary to heat the old solder again, a softer solder must be used each time.

[42] **Soft Soldering.**—A soldering iron is used for soft soldering. For our purposes, a tolerably thick, pointed iron set in a wooden handle is sufficient. It is used in soldering pieces which are too large to be heated entirely, or in any case where only local heat is required. The iron should be tinned before it is used. This is done by moistening it with the soldering fluid, heating it until a piece of tin melts on it and wiping it off with paper. In using the iron, it must be made very hot, wiped clean, and a little solder taken up on the point of it. The joint must be strewn with a little powdered rosin.

Soldering without the iron, by heating the whole metal, is a very simple operation. The joint is moistened with the soldering fluid, and heated with a gas or spirit flame, until the solder laid on it melts. The solder draws quickly into the joint and spreads through it. The excess, if any, is scraped away. Any soldering fluid remaining must be wiped off with a wet cloth, otherwise it will oxydize the metal.

The best soldering fluid is a solution of sal ammoniac, or a mixture of chlorides of zinc and ammonium made by dissolving 32 parts of zinc in the necessary quantity of muriatic acid, adding 22 parts of sal ammoniac, and evaporating the solution to dryness. The salt is dissolved in water, and filtered. With this fluid it is hardly necessary to clean the surfaces to be soldered.

In soldering brass and copper, the solder flows through as soon as it melts, and the color of the metal shows when this heat is attained. Iron requires to be heated higher before the solder will adhere. When objects are to be soldered, therefore, which will not bear a high heat, the iron must first be tinned separately in the way described for the soldering iron.

[43] **For hard soldering** small articles, silver solder or silver coin is best. The latter is hammered to a thin plate and cut into pieces $\frac{1}{8}$ inch broad and $\frac{1}{4}$ inch long. The cost is inconsiderable; a quarter of a dollar goes a long way; and metals soldered with silver can be hammered and bent without the joint breaking, which is not always the case with brass solder.

When the joints are well fitted together, cleaned, and fastened, they are moistened with water, some borax strewn over them, the bits of solder laid on the joint and more borax sprinkled over them, or the bits of solder are mixed up in a paste with borax and a little water and applied so. The metal is heated slowly until the borax ceases to puff up, taking care that the silver remains in its place. It is often difficult to retain the silver on small articles, when ordinary borax is used; in this case, powered calcined borax is used without water.

According to the size of the object, the flame of a spirit-lamp is either

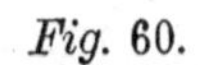

*Fig.* 60.

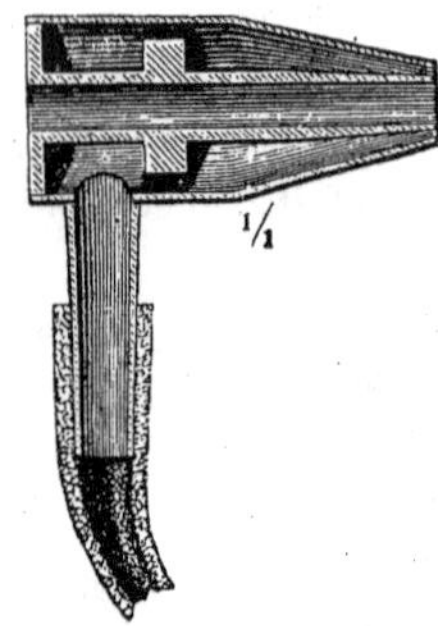


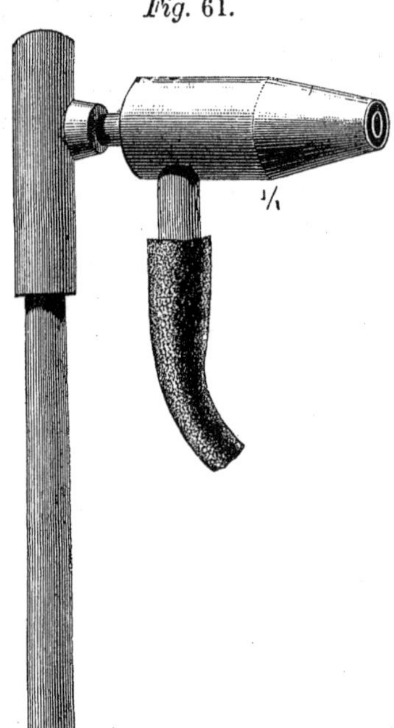


*Fig.* 61.

directed immediately upon it with a blow-pipe, or it is laid on a piece of charcoal, surrounded on three sides with coal and covered with another piece. The flame of the lamp is then directed upon it until the solder melts and adheres, after which the article, if it contain no iron, is immediately cooled in water. Practice in the use of the blow-pipe is of course indispensable.

A gas blast-lamp is very convenient for soldering. A burner is used for this purpose like the one described for glass blowing, but smaller; it is shown of natural size in figs. 60 and 61. The opening for the escape of the gas must only be a fine slit. The burner is connected with any gas pipe by means of a flexible hose. The air pipe is stuck tightly into the burner, and reaches to the mouth of it. With this burner, sufficient heat may be obtained to fuse a bead on copper wire 1 to $1\frac{1}{2}$ millimeters thick, without the aid of charcoal; or the flame may be reduced to a mere point, and made to bear on the smallest spot. The objects to be heated may be laid on the work-bench, and the flame brought to bear on them in any direction.

When a greater heat is required than can be obtained in this way, the articles to be heated must be laid on well ignited coals in the forge or a charcoal furnace. The joint to be soldered must always remain visible. Use the blast of the forge gently, and fan the coals briskly. Large articles are usually brazed. In soldering brass, care must be taken that the metal itself does not partially fuse. This is not so much to be feared in the case of copper and iron.

# CHAPTER IV.

## VARIOUS OTHER OPERATIONS.

[44] **Gluing.**—Common glue only adheres to fresh surfaces, which have not been glued before.

Isinglass cement is worthy of special notice, being suitable for cementing glass, and for all other purposes where a strong cement is required. It is prepared by cutting Russian isinglass to shreds, and soaking 4 to 8 parts of isinglass in 100 parts of water, for 24 hours. The water is then gently boiled for half an hour, being filled up as it evaporates. It is filtered hot. When more is prepared than is used, the thickened mass may be dried on paper, and boiled as before, when needed for use.

The teacher has, frequently, in preparing diagrams, etc., to paint upon paper which is almost always water spotted. This difficulty may be remedied by sizing the paper with thin glue-water, containing a little alum. A solution of 2 parts of isinglass (or 4 parts of the so-called gelatine) and 4 parts of alum, in 100 parts of water, is sufficiently thick. Stretch the paper, and apply the size with a brush or a small sponge. The surface should not appear glossy after the size is dry.

[45] **Cementing.**—Cementing with sealing-wax or shellac is a common operation. The former is preferable, when the color is not objectionable, as being less brittle than pure shellac. The pieces to be cemented must be warmed so that the cement will melt upon them. When both parts are sufficiently coated, they are warmed again to make the wax cohere. If they are put together when very hot, a crack may be caused by their unequal expansion. This precaution is not always necessary, but should never be omitted when glass tubes are to be cemented into close-fitting metal sockets. The pieces must, of course, be held in place until the cement hardens, which may, without injury, be hastened by the application of water. Any excess of cement is afterwards removed with a knife. The closer the pieces fit together, and the less cement there is between the joint, the stronger it will be.

It is often necessary to cement metallic articles to a wooden chuck, to fix them in the lathe. Common sealing-wax is used for this purpose, and dropped, burning, on the wood. The article is easily centered while the wax is still warm. It is well to be cautious in turning, to avoid heating the metal so as to make it necessary to cement it afresh.

To fasten cloth to metal, it is soaked in a dilute solution of galls, squeezed out, and pressed, still moist, upon the warm metal, which has been coated with glue.

The joints of gas and steam apparatus, held together by screws, must often be made steam tight. The machinist's red-lead cement is the best for this purpose. It is made by incorporating red-lead and litharge thoroughly, with linseed oil, until they form a soft plastic mass. It should be kept under water. This cement is readily obtained from the machine-shops or gas-fitter's. Putty answers the same purpose, if too much chalk is not used in it, instead of white-lead. If the parts are not to be very highly heated, a cement made of melted caoutchouc and red-lead or white-lead is very serviceable. Parts which are not to be heated may be made close by laying strips of vulcanized rubber between them.

[46] **Varnishing.**—Physical apparatus should generally be varnished, both for its better preservation and for ornament. Varnish must, however, not be applied to everything, especially not to parts which are to be frequently handled. The varnish is soon rubbed off from such places, and they look worse than if they had been kept clean, without varnishing.

*Making the varnish.*—Take 1 part of finely-broken shellac to 4 parts of strong alcohol (85 to 95 per cent.) and let it stand in a warm place, shaking it often, until the solution is complete, which will require about 24 hours. The addition of $\frac{1}{5}$ to $\frac{1}{10}$ part of mastic renders the varnish less brittle. This addition is more necessary to bleached than to crude shellac. It is well to take a little more resin, to make the varnish as thick as possible.

To filter the varnish, which is often necessary, let it stand quietly, in a warm place, until the insoluble and slimy matters have subsided, then pour off the clear liquid, and filter the rest, hot, through coarse filter paper. The slimy mass filters very slowly, and must be covered with a glass plate, to retain the heat. The varnish should be kept in a wide-mouthed bottle, with the brush stuck through the cork, or else the brush must be washed out with alcohol after using it. A broad camels'-hair brush is used for applying the varnish, or a piece of fine sponge fastened in a tin holder, like a brush, and cut off so as to project only a few lines beyond the tin.

Colored shellac varnish, especially a reddish sort called gold varnish, is needed for some purposes. The coloring resins may be dissolved at once with the shellac, but, to avoid having so many varnish pots standing about, it is better to dissolve and filter the coloring matters separately, and mix them with the varnish as required. For our purposes, a solution of gamboge and another of dragon's blood are all that are needed.

To make varnish of sealing-wax, it is only necessary to lay an excess of sealing-wax in alcohol, and, before using it, to shake it up. Bad sealing-wax makes bad varnish, and good wax is expensive; it is therefore better to add vermilion directly to a solution of shellac, with a good deal of mastic. It is well to add also some carbonate of magnesia, which does not injure the color much, and gives the varnish more body.

Black varnish is also needed sometimes; this is easily made by mixing fine lampblack with a solution of shellac. A solution of 1 part asphaltum, in 3 parts spirits of turpentine, furnishes a good black coating, especially for objects which are to be heated. The tinners varnish stove-pipes with this, but they add linseed oil and red-lead to it.

*Varnishing metals.*—In varnishing metals, the varnish must either be applied immediately after turning or filing, or else they must be polished with pumice-stone and water, and then with emery and oil, taking care to make all the streaks parallel. This makes neater work; but after polishing, the articles must be washed with soapsuds, and wiped with a clean linen cloth. The articles to be varnished should never be touched with the bare hand, before applying the varnish. All metals should be heated over a coal fire, or on a heated sheet of iron, until the hand can hardly be borne upon them, and the varnish then applied.

The brush should be dipped into the varnish, not too deeply, and the strokes kept parallel to the streaks of the metal; if the streak is made on the lathe, the varnish must also be applied on the lathe. In varnishing large articles, pressure on the brush must always be increased somewhat, toward the end of the stroke, in order to apply the varnish evenly. The pieces must not be touched before they are quite cold. Several coats of shellac varnish may be applied, cold, to metals, but the varnish does not attain the same hardness or luster, and it is therefore applied so only when it is desirable to have a very thick coat of varnish, as, *e.g.*, on electrical machines, or where the object cannot be warmed again.

Iron is seldom varnished, and only with bleached shellac; unbleached shellac may be used for brass and copper; it gives the brass a darker color; but bleached, with the addition of the coloring matters mentioned, is preferable. For white metals, such as lead, either bleached shellac is used, or unbleached, with the addition of resins to give it a gold color.

The same varnish is used for wood, but it is applied oftener and in thinner coats. When elegance is not an object, but only protection against moisture, or good insulation, unfiltered solution of shellac may be employed. A second coat must not be applied before the first is dry.

It saves the varnish, to coat the wood first with thin glue. To make the varnish look well, the wood must be rubbed, after the first coat, with fine emery paper. When the varnish has been laid on thickly, it may be rubbed with Tripoli on a woolen rag, and polished with linen and oil. The excess of oil must be removed by strewing the surface with meal and wiping it with a clean linen cloth.

White wood may be stained nut brown by a dilute solution of asphaltum: the addition of dragon's blood gives a mahogany color. Staining with nitric acid does not answer for physical purposes.

Paper is varnished in the same way as wood, but the paper must first be sized, to prevent the varnish from sinking in and staining it.

Glass is varnished warm, when only a single transparent coat is required; but when several coats are needed, they must be applied cold.

[47] **Cork.**—Good soft corks are necessary for various purposes, and nothing is more difficult to obtain. By picking over several bales of fine corks, a few really good ones may be found. Champagne corks, which have not been pierced with the corkscrew, are serviceable for many purposes.

To fit a cork well into an opening, it should first be cut nearly to the size with a sharp knife, and then finished with a file.

*Cork-borers.*—A cheap, but not very durable, cork-borer is made of a tin tube about 3 inches long, thrust through a large tube, for a handle, fig. 62. The size of the tubes is so graduated as to fit into each other in sets of 8 or 10, as shown in fig. 63. The lower edges are filed sharp. The piece of cork cut out remains in the cylinder, and is pushed out by the next smaller one. A better style of cork-borers is made of similar tubes of brass, fitting into each other, but they are more expensive. In order to make a quite smooth, clean hole in a cork, it is better to take a cylinder smaller than the hole required, and enlarge it with a fine rat-tail file. An assortment of these should be kept in the drawer with the corks, and used for no other purpose.

*Fig.* 62. *Fig.* 63.

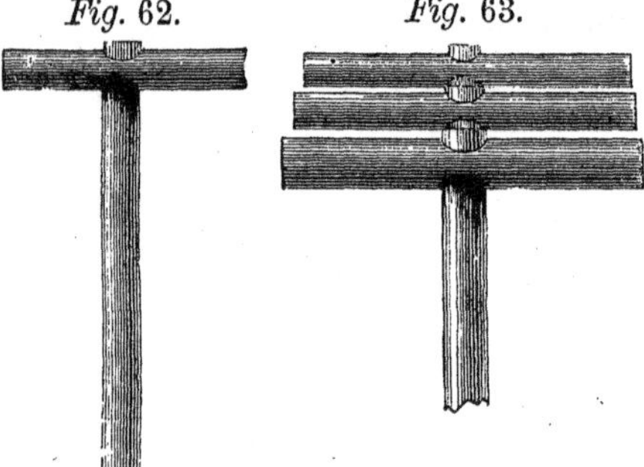

Thin slips of cork may be used as washers, to make an air-tight joint. Supports of all kinds may be made with corks and glass rods, stuck on inverted funnels.

[48] **India-rubber.**—Tubes of vulcanized rubber are almost always to be had now, but in case of necessity they may easily be made from sheet rubber. For this purpose a strip is cut of the proper width, folded around a corresponding cylinder, and the fresh edges pressed together and rubbed a few times with the finger-nail. It may sometimes be necessary to warm them slightly. Care must be taken not to touch the fresh surfaces with the fingers. The cylinder is then withdrawn, and the seam more firmly united by pressure between the fingers. The tube may be strengthened by laying another thickness of india-rubber around it, with the seam on the opposite side. Such tubes must be tied around the glass tubes, to be united. Two tubes may be united by laying around the joint a strip of india-rubber, about an inch wide, warmed, and bound fast with silk; the folds stick together of themselves.

# PART SECOND.

## PHYSICAL EXPERIMENTATION.

---

## CHAPTER I.

### EXPERIMENTS ON THE EQUILIBRIUM OF FORCES.

### (*a.*) SOLIDS.

[49] **General remarks.**—In experiments on this subject, it is necessary to have a number of weights with hooks, and of a unit not too small, so that any quantity from 1 to 100 may be expressed in terms of this unit. Any one can readily make such weights of lead, by casting this metal into paper tubes, and cutting off proportional lengths. The handles may be made of small brass rings, with iron screws to them, which may be had at any hardware-store: the final adjustment of the different weights may be made by means of a knife and coarse file. If brass be preferred as the material, it would perhaps be better to purchase the weights ready made. Copper coins are also very convenient for the purpose, as they are easily combined in any desired number. They may be held together by a string or wire passed through a hole in the center, or by being inclosed in a wire frame or a small tin box. Small metal boxes filled with shot may also be employed. In all cases, the quantity should be stamped on each weight, in terms of the assumed unit.

Besides the weights, several scale-pans, whose weight is accurately determined and marked on them, will be found very convenient. Their weight should, if possible, be integral parts of the standard unit.

[50] **The parallelogram of forces.**—Among the various arrangements for the statical demonstration of this law, the one presented in fig. 64 is especially recommended on account of the simplicity of its construction and the readiness with which the forces may be measured by the length of the lines. To the block *A* is fastened the rectangular bar *a b*, which is pierced with holes at equal distances of about a centimeter or an inch. The point of rotation *c* should form the commencement of the series. Upon the bar moves the slide *d e*, whose length corresponds to a whole number of units of the graduation; it is held in its place by a pin, and the numbers are marked on the bar in such a man-

ner that the one seen at the upper edge *d* of the slide indicates the distance *c F.* The thickness of *d e* is such that its anterior face is parallel to that of *A,* and stands the thickness of one of the bars farther back. The thin strips, *F i, F k, m n, l o,* are pierced with holes at the same distance apart, as in *a b*; they are likewise numbered from their point of rotation, and may be combined into a quadrangular figure by the insertion of flat-headed

*Fig.* 64. *Fig.* 65.

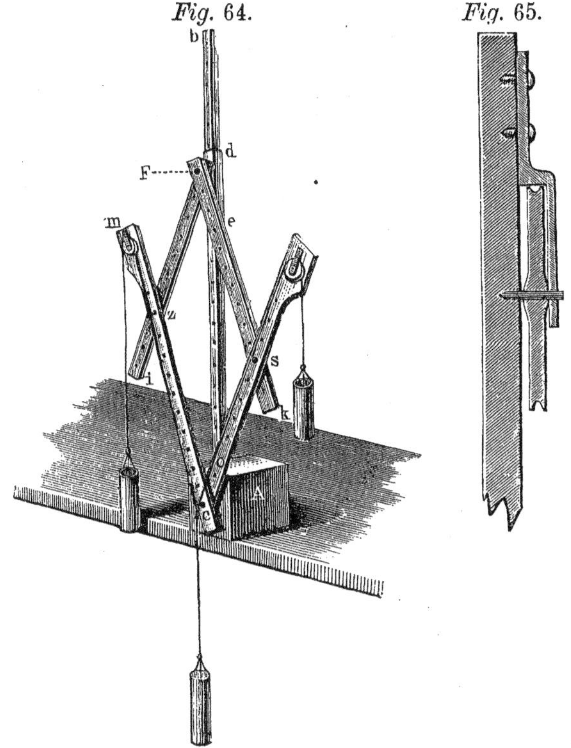


wooden or iron pins, as at *z s.* It is better, though not necessary, for these pins to have screws on the opposite extremity. The strips *m n l o* likewise carry pulleys, one side of which coincides with a line connecting the holes. In the present and similar cases, the pulleys (best of boxwood) should have a slight elevation toward the center, as shown in the one represented of natural size in fig. 65. It is held between the strip on one side and a bit of bent brass plate, screwed to the wood, on the other. The brass must fit sufficiently close to the pulley to prevent the string from slipping off.

In trying the experiment, take three silk strings, and tie them to a small brass ring; pass two of them over the pulleys, allowing the third to hang down. Construct a parallelogram of any size out of the four strips, which will then have its diagonal in *a b*. If weights be now hung to the ends of the three strings, corresponding to the distances *c z*, *c s*, *c F*, equilibrium will take place, and the point of union of the strings will correspond to the point *c*, returning to it whenever drawn away. The small boxes of tin filled with coins will be found very convenient in this experiment. The block *A* may be fastened to a stand, or simply placed on the edge of a table; in the latter case, however, it must be heavily loaded.

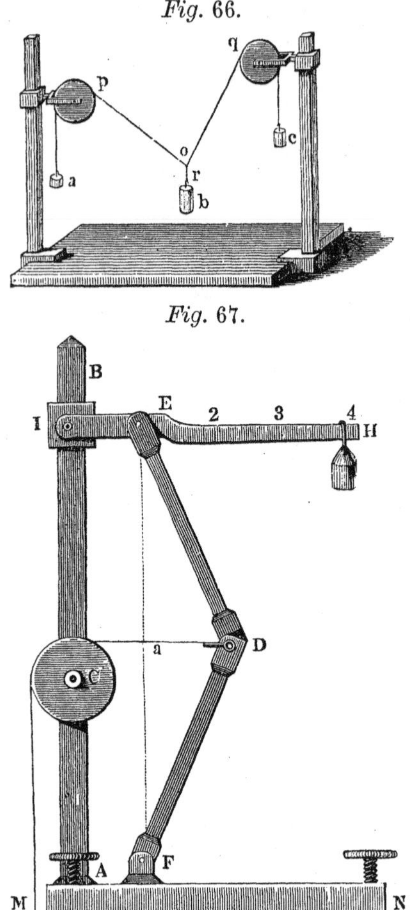


*Fig.* 66.

*Fig.* 67.

If any other weight than that corresponding to the diagonal be used for the middle force, the knot will not come to *c*; the slide may then be moved until this is the case, and the weight will again correspond to the length of the diagonal.

Without such a machine, two pulleys may be fixed to a table in any way, as in fig. 66, and any weights attached to the cord, and then the angle, which, according to the law, the forces would make with each other, sought by construction. If this figure, drawn tolerably large, be held behind the cords, the angle which they make will always correspond to that of the figure. The correspondence will not be exactly owing to the unavoidable friction, but the difference will fall within the same limits on both sides.

[51] **The knee-press.**—The application of the parallelogram of forces, in the so-called knee-press, may be readily illustrated by means of the apparatus represented in figs. 67 and 68. To the board *M N* the upright pillar *A B* is fastened by a screw. Up and

down this moves the pulley *C* and the lever *I H*, capable of being fixed by slides and clamp screws. Thus, in any position of the knee, the lever *I H* and the string *a* of the pulley may be made to assume a horizontal position. The string is attached to the axis at *D* by means of a stirrup. If the bars *E D D F* move on their joints with sufficient freedom and

*Fig.* 68.

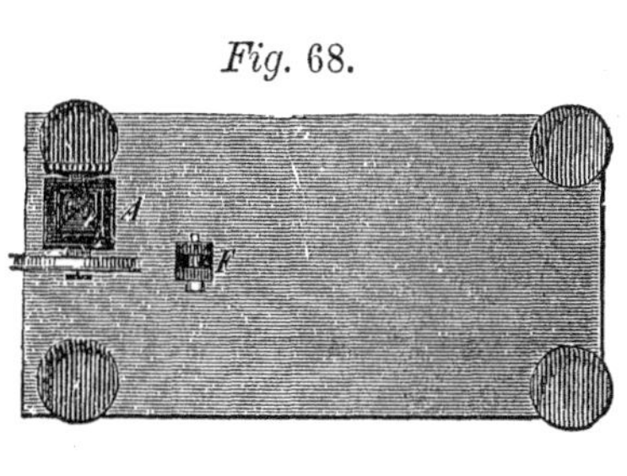

*Fig.* 69.

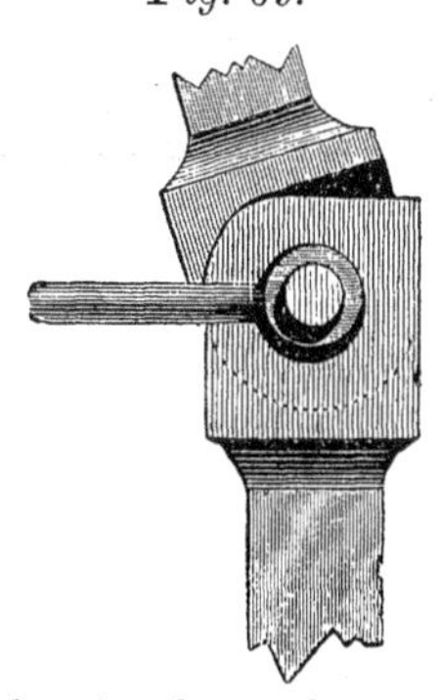

precision, we can readily calculate from the weights in the scale-pan *e*, and the proportions of the lines *a D* and *D E*, the lateral force *D E*, and from the leverage, the weight required at *H*, since in the experiment, *D E* will act nearly perpendicularly on *I H*. One of the joints must be so constructed that the bars *D E D F* will be unable to pass beyond a vertical position. All the bars should be made of iron, the extremities somewhat stronger for the joints, the central joint being constructed as shown in fig. 69, to prevent the two bars from going past the straight line. The holes and pins, as well as all the joints, must be made very smooth, to give the apparatus the proper freedom of action.

The compound action of two forces may likewise be shown by the simultaneous impact of two ivory balls upon a third, all of the same size.

[52] **The composition of uniform forces** may be illustrated by the apparatus shown in figs. 70 and 71. *M N* is an upright board, with a foot *o p*. The board *R S* is fastened perpendicularly to *M N*, and held in place by means of two brackets, *a b*. This board is slit longitudi-

*Fig.* 70. *Fig.* 71.

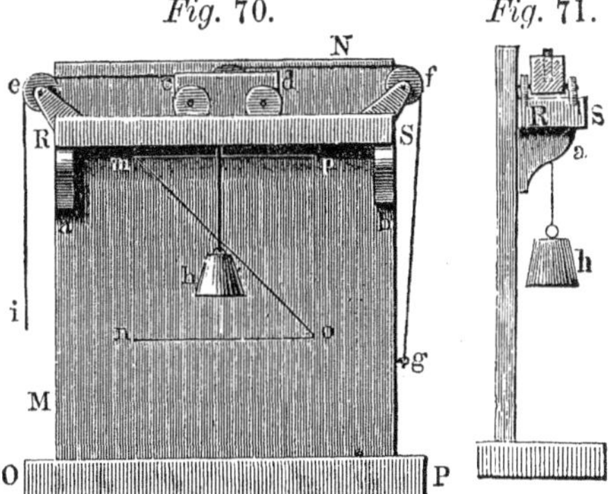

nally along the middle, and has two rabbets on the upper side, parallel with the slit. To a block of wood *c d* four wheels are attached, running in the rabbets of *R S*. In addition to the four wheels, there is a pulley in the block, projecting slightly above the upper side; over this and the pulley *f* is passed a string, fastened at *g*, and carrying the weight *h*. If the block *c d* be now drawn along the board *R S* by means of the string *c* passing over the pulley *e*, the weight *h* will immediately ascend along the diagonal *o m* of the parallelogram *m n o p*.

[53] **The inclined plane.**—An arrangement to exhibit the laws

*Fig.* 72.

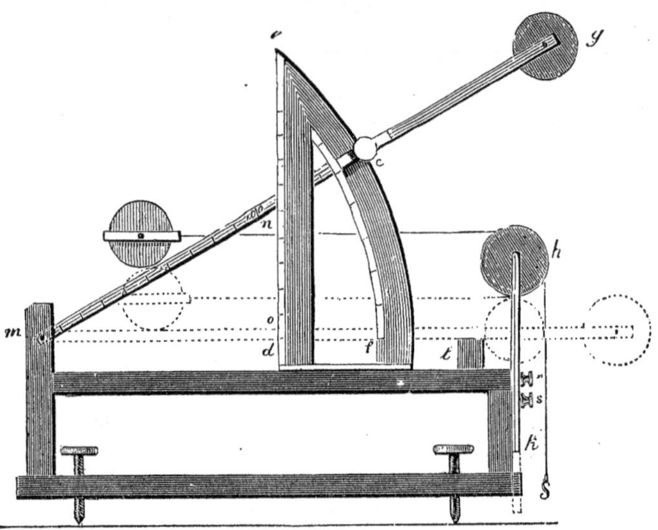


*Fig.* 73.

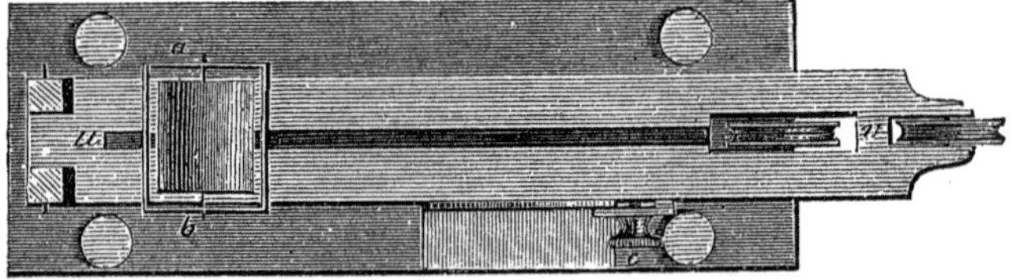


of the inclined plane conveniently and with sufficient exactness, must be arranged so that the force may be made to act parallel, either to the plane or to its base. The inclined plane should be of metal, and capable of being fixed by a clamp screw at any angle. The cylinder running on the plane must also be of metal, having for pivots the steel points of two screws fixed in the brass frame *a b*, figs. 72 and 73. The piece *d e f* has on its side $d\,e$ a graduation in fractions of the base; $e\,f$ likewise is graduated either to degrees or in fractions of the length of the plane. The latter is the more difficult to apply, and in general it is better to calculate the counterpoise required from the known weight of the cylinder, and the triangle $m\,n\,o$. If the counterweight acts over the pulley $g$, it will hold the cylinder in equilibrium at any point on the plane. When it acts, however, over $h$, equilibrium will only exist when the string is horizontal. This latter pulley is carried by the brass strip $i\,k$, fig. 74, and may be fixed at any position by screws, $r\,s$, which, however, will not affect the result. In this case, the inclined plane has a slit, through which the string can pass. The pulleys are made of wood, with slender axles; $t$ serves to support the plane in a horizontal position, for which the slit is widened toward $g$, to pass over the pulley $h$.

*Fig.* 74.

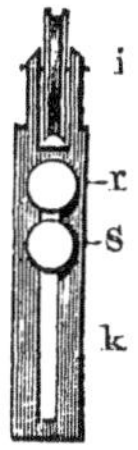

i
r
s
k

[54] A simpler apparatus, though quite as instructive, may be made of wood, as shown in figs. 75 and 76, $\frac{1}{6}$ of the real size.

*Fig.* 75.

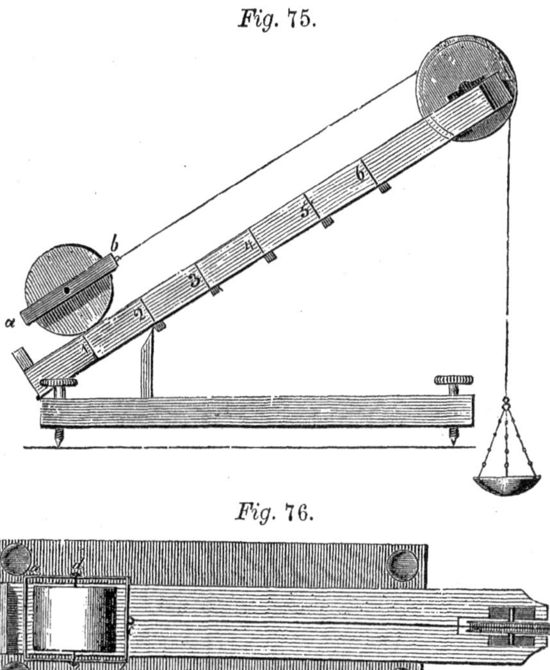


*Fig.* 76.

The two boards may be connected by a hinge, the lower one having four adjusting screws. The inclined plane has a slit in front, to receive the pulley, whose axis rests on the sides of the slit, and is held down by two plates. The pulley must be as large as possible, without touching the table, when the two boards are shut on each other. To adjust the apparatus, we may make use of a pin, which catches against cleats or holes, on the under side of the inclined plane, made at distances apart equal to the height of the pin.

The load is best made of metal; it may, however, be of old wood saturated with hot varnish; its diameter, as shown in the figures, will depend on the height of the pulley. When of wood, it should have metal shafts, and be set in a frame of sheet-brass. The weight of the roller and fractions of the same are marked upon it.

[55] **The screw.**—The theory of the screw is best illustrated by means of a wooden cylinder about two inches in diameter, on which is wrapped a right angled triangle of paper, as *a b c* in fig. 77, one cathetus of which is equal to the height of the cylinder, the other to its circumference. The hypothenuse has a well-defined black margin, and represents, when wrapped around the cylinder, one turn of a screw.

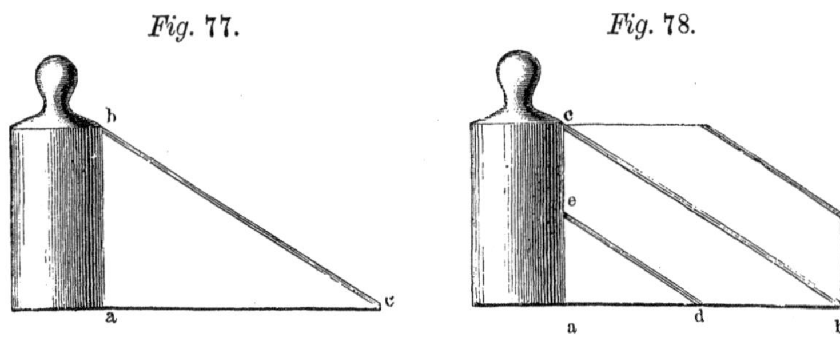

*Fig.* 77. *Fig.* 78.

In a similar way, the double-threaded screw may be illustrated, by having a second black line parallel to the hypothenuse, as in fig. 78, where *a b* is equal to the circumference of the cylinder, and *a e* half the same.

In order really to show the action of screws by weights, it would be necessary to choose such as have a wide thread, and even then the friction would be too great. It would be well, however, to have a few models of flat and sharp threaded single and double screws, and also a model of an endless screw, although they should not be used with weights. The models should not be too small.

If it be desired to illustrate the action of a screw experimentally, the following apparatus, designed by Professor J. Müller, is very convenient. The base *M N*, fig. 79, supports a massive, well-turned cylinder *A* of hard wood, over which fits a hollow cylinder *B*, designed to support the weight

*C*. The latter cylinder is made of very hard and dry wood; the lower margin is cut in the form of a screw, which can be made pretty accurate by first pasting around it a slip of paper, as in fig. 79. Into the solid cylinder, six rollers *a a*... are screwed, so that all may support the screw together. These rollers move on steel bearings. Fig. 80 shows one of natural size. The arm *D* also has a roller, over which the cord with counterweight passes. The screw should have a high angle, and, for con-

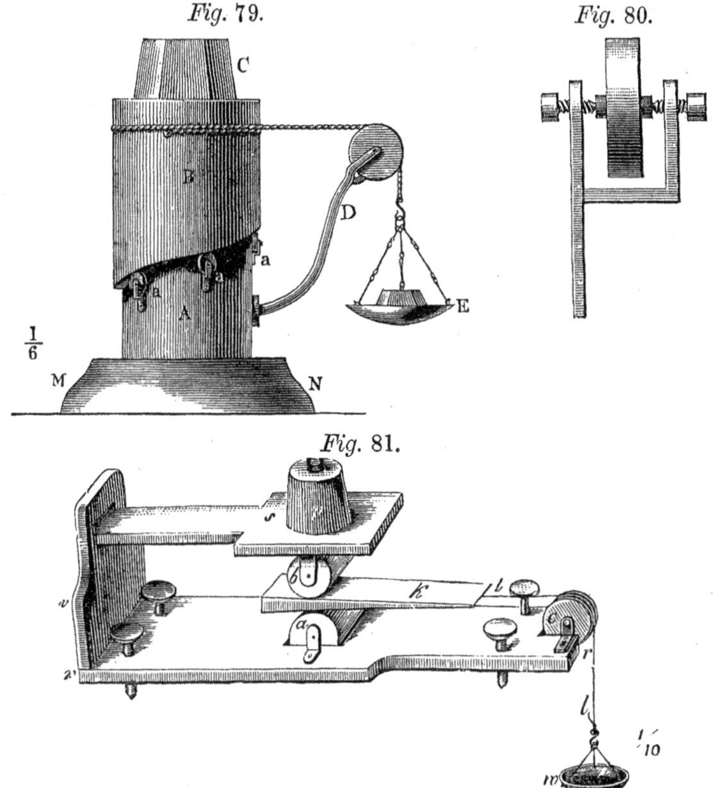


*Fig.* 79. *Fig.* 80. *Fig.* 81.

venience in experimenting, should bear a simple ratio to the circumference of the cylinder. The use of the apparatus needs no explanation.

For further illustration, an Archimedes' screw with a glass cylinder is convenient.

[56] **The wedge.**—The simple arrangement of fig. 81 is intended to illustrate the theory of the wedge. It consists of a wooden base *r r* set horizontally on a table by means of four adjusting screws. In this base and hung between cheeks of brass is sunk the cylinder *a*; it also

carries the pulley *c*, which projects in such a manner as to give free play to the string *l l* to which the scale-pan *w* is attached. In the upright *v* is hinged the board *s*, with its cylinder *b*, which can be pressed against the cylinder *a* by a weight. Between the two is situated the wedge *k* attached to the string *l l*. There should be several of these wedges of the same length, but of different bases.

The entire apparatus is very simple, although the cylinders require nice work. They are made of boxwood and turned on their pins after these have been inserted. The two must rotate perfectly parallel to each other.

The mode of using the apparatus will be evident to every one. The pressure of the board *s* with the cylinder *b* is ascertained by attaching it by a string to one scale of a balance, and putting weights into the other until the board rises on its joint to a horizontal position. This weight is marked on the board, and in each experiment is to be added to *p*. It will be best so to regulate this weight that it shall amount to a whole number of the assumed units.

*Fig.* 82. *Fig.* 83.

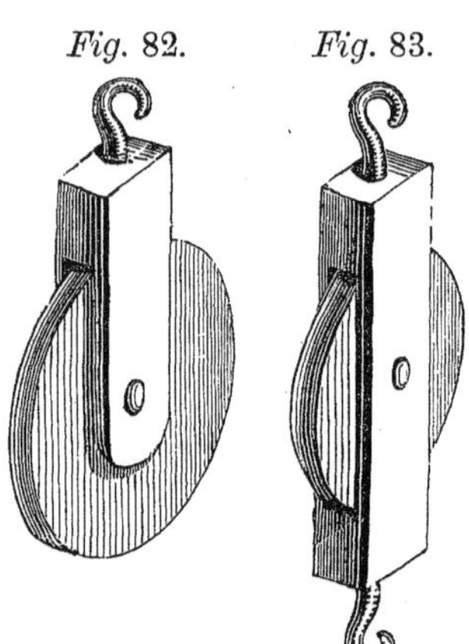

[57] **The pulley.**—In addition to the common form, it will be necessary to illustrate the movable pulley for the case where the two cords are not parallel, as this affords a good opportunity for remarks on the parallelogram of forces. The experimenter should have a number of single pulleys with movable hooks, about the size and shape of those in figs 82 and 83, so as to admit of various combinations. A frame with various sets of pulleys is shown in fig. 84.

The pulleys themselves must move freely enough to require their weight to be taken into account in experimenting, this being determined by means of a scale-pan, as in the simplest combination of fig. 84. They must be a little thicker toward the center so as to diminish the friction against the frames as much as possible, and the space at the top must be so small as to prevent the string from slipping off. To explain the working of a movable pulley when the two strings are not parallel, as in fig. 85, the same arrangement may be used as for the parallelogram in fig. 66; in this case we pass a single string over the two pulleys of the frame, and upon this place a movable pulley, as in fig. 82. By means of the apparatus, it will be easy to explain the advantages and disadvantages of the different combinations.

[58] **Construction of the pulleys.**—If made of brass, pieces as nearly circular as possible are cut out of brass plate with the saw or cold chisel, following up with the file. A hole is next to be bored in the center of each, and a wire filed slightly tapering passed through so as to fit tightly. This should be at least a line in diameter and be soldered

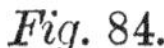

*Fig.* 84.

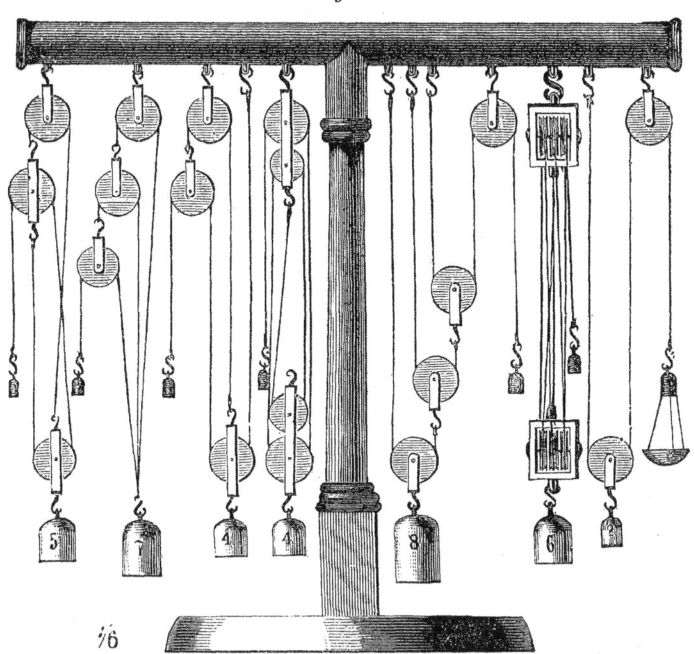


with tin. The pulley should then be put on the lathe and turned at the same time with the axis, care being taken that the pulley is left very slightly thicker at the axis than the circumference. The axis should be left at least half a line in thickness, and the groove must be deep enough to receive the entire thread without any part of it projecting beyond the edge, fig. 86. The pulley is then to be varnished while still on the lathe.

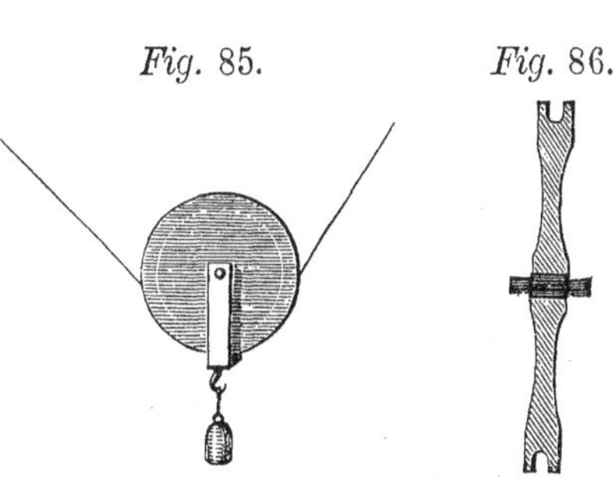

*Fig.* 85. *Fig.* 86.

The cheeks, fig. 87, the hook, with its neck, fig. 88, and the intermediate piece, fig. 89, are then made, and the latter pierced with a hole to receive the hook. The different parts are then combined by means of a hand-vice, which is applied over the upper half of the intermediate piece, allowing the pulley to move freely and to come so near to the intermediate piece as to prevent the string from slipping off. In this situation bore a hole *a*, fig. 90, entirely through the head, and so as to encroach slightly on the one previously bored in the intermediate piece, insert the neck of the hook, and put in a rivet before removing the hand-vice. A second hole may then be bored also passing through the neck of the hook, or any other part of the head, and another rivet inserted. Should the hook be too tight to turn freely, a little oil and working will ease it. The entire block is then to be cleaned and varnished.

The intermediate piece may also be soldered to one of the cheeks, or both made in one piece. The latter course is not to be recommended, and it will at any rate be necessary to use a rivet to have a movable hook.

In the case of wooden pulleys, the frame or block is made of a piece of bent brass plate, the pulley bored through and the axis riveted, or otherwise fastened in the frame. The hook is likewise riveted into the frame before it is bent. If the operator cannot make such pulleys, or time be wanting for the purpose, very good temporary ones can be constructed by getting some wooden pulleys already pierced from the turner, and bending a round iron wire into the shape shown in fig. 91. The part

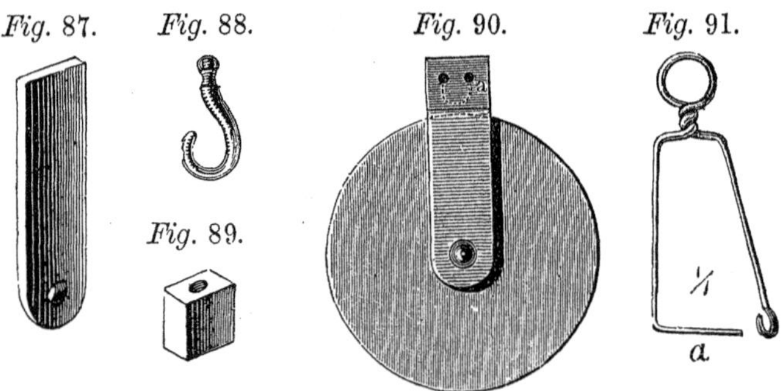


*Fig.* 87. *Fig.* 88. *Fig.* 89. *Fig.* 90. *Fig.* 91.

*a* passes through the hole of the pulley, and in this way we shall have a tolerably good arrangement, and at a very trifling cost, even for a considerable number.

To make the pulleys turn very easily, which is necessary for many purposes, they must run on pointed steel screws passing through the frame. A hole is bored in each end of the axis, which is enlarged conically, as shown on an enlarged scale in fig. 92. The pulley must be centered on this conical hole. The points of the screws which run loosely in the conical depression must be sharper cones, so that only the obtuse edges of the depressions will play on their polished surfaces. The screws are made as hard as glass, and so adjusted as just to allow the pulley to turn freely without pressure or oscillation. For this purpose a second nut is necessary, as seen in fig. 92.

*Fig.* 92.

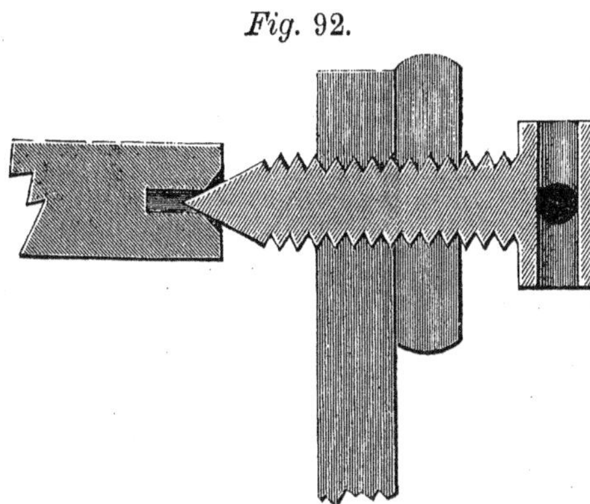

[59] **The lever.**—However simple the theory of the lever, the teacher will find considerable practical difficulty in explaining it to the entire comprehension of his scholars. It will, therefore, be necessary to present experiments in their most varied applications.

The apparatus, figs. 93 and 94, is well calculated for illustrating the

*Fig.* 93. *Fig.* 94.

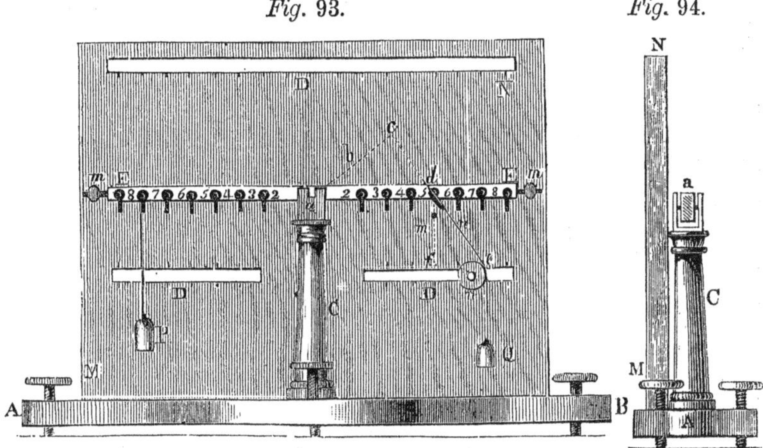


laws of the lever; it is constructed entirely of hard wood, with the exception of some small pieces. The post *C* is fixed in the middle of the base *B A*, which stands on leveling screws, and behind it the upright board *M N*. *M N* has three slits, *D D D*, in which to fasten pulleys. A brass plate *a* is screwed to the top of the post *C*; its cheeks are notched, to receive the axis of the lever. This is a wooden bar turning on a steel axis perpendicular to its length; a piece of stout knitting-needle will answer for this purpose. The axis should be driven in a little above the center of gravity. Along the bar, at distances of an inch, are next to be inserted short pins of rather thinner wire, all to be in the longitudinal axis of the bar and fitted in as truly perpendicular to its vertical plane as possible. To each pin is to be attached a stirrup of brass wire, as shown in fig. 95. These stirrups are made of hard, drawn wire, so as to have some elasticity; they should be all of equal length and capable of being detached, so as to be replaced with the bend above the bar. The pins, with their stirrups, are numbered from the center. The brass balls *m m* moving on close-threaded screws at each end of the bar serve to balance it. To bring forces to act obliquely upon the lever, we make use of a pulley *n* placed in one of the two slits, *D D*, and with its plane parallel to *M N*. This is shown of natural size in fig. 96; it is made of boxwood or brass, and has a steel axis turned at the same time with the pin *R R*, on which a screw is cut, leaving a space as long as the thickness of the pulley. This is embraced between *R* and the convex head *S*, so as to have an easy motion. Over this pulley is passed a string attached at one end to any stirrup and to a weight *Q* at the other. We can then ascertain by direct measurement the distance *a c* of the direction *d c*, in which this weight now acts on the lever; *a c* is, however, inconvenient to measure. If the upper edge of the slit *D* is distant from the middle of the lever, when in equilibrium, by a whole number of inches, (about three inches in fig. 93,) and provided with a graduation corresponding with that of the lever, the triangles *a c d* and *d e f* will be similar, and from *d f*, *d e*, and *a d*, we can find the distance *a c*; *d e* is easily measured. If, in the present case, $df = 3$, $ad = 5$, $de = 3\frac{1}{2}$, then $ac = \frac{ad \times df}{de} = \frac{30}{7.}$ If now a direct acting weight be attached to the left, at a distance, $7 = \frac{49}{7}$ the weights would be inversely as the distances, or $Q : P = 49 : 30$, a proportion which, like any other, is easily constructed by means of coins.

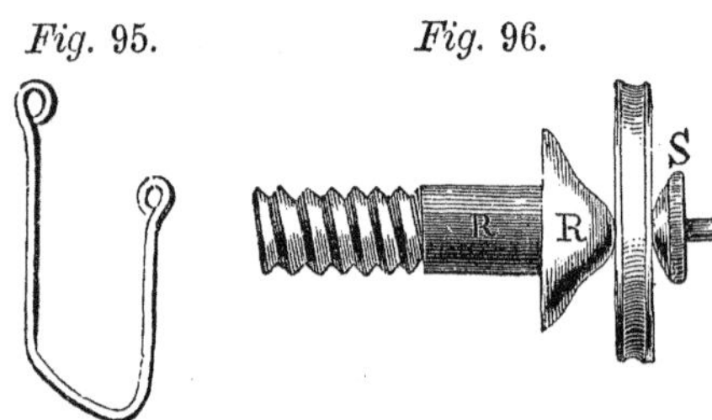

*Fig.* 95. *Fig.* 96.

To make a lever of the second or third class, a pulley is placed in the upper slit, and one of the stirrups turned upward; the attached weight acts over the pulley, and the other weight acts directly at the required distance. To adjust the upper pulley with sufficient exactness for the purpose, its slit has a graduation corresponding to that of the lever.

The axis of the lever itself may be held in a stirrup, and the action of the two lateral forces balanced by a middle force for the purpose of showing what the fulcrum at *a* has to perform, and in general to exhibit the law of equilibrium of three forces, whose centers of action are connected by a straight line.

A simpler and very useful arrangement for illustrating the laws of the lever, is given in fig. 97. It needs no further description. Each of the levers may be used alone or in combination with others. Smooth holes are bored in the levers and the stand, and pieces of smooth wire serve as fulcrums and supports for the brass stirrups which support the little hooks. To use them as one-armed levers, they are balanced by the weights *a a*. To use a single one as a one-armed lever, the upper bar is passed through both supports, and pulleys attached to it for the weights which are to act upward.

*Fig.* 97.

The arrangement in fig. 98 is still simpler. The lever *A B* turns on a steel wire, and similar wires support the stirrups of the weights where they are required. The pulley *C* causes the weights to act upward.

If satisfied with the simplest arrangement, we may attach the lever to a hook, by means of a somewhat longer stirrup of brass wire; in this case, however, only forces acting directly can be employed. We may also furnish any steelyard with a graduation on its short arm corresponding to that on the long, and thus explain the law of the lever, though less comprehensively.

[60] **The wheel and axle.**—An iron axis is passed through a suitable piece of hard wood and both turned off together, so that the

*Fig.* 98.

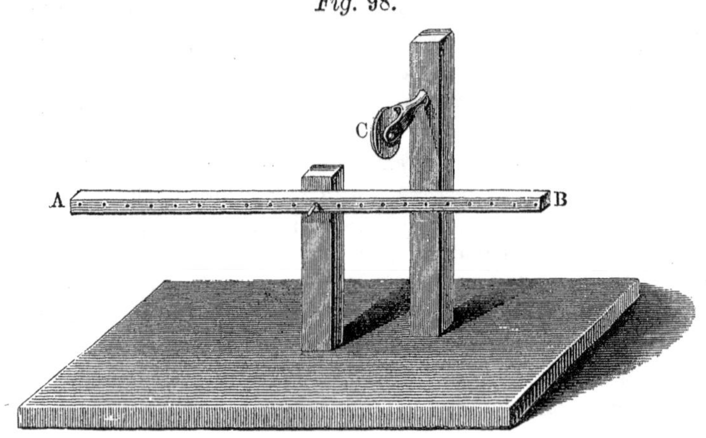


*Fig.* 99.

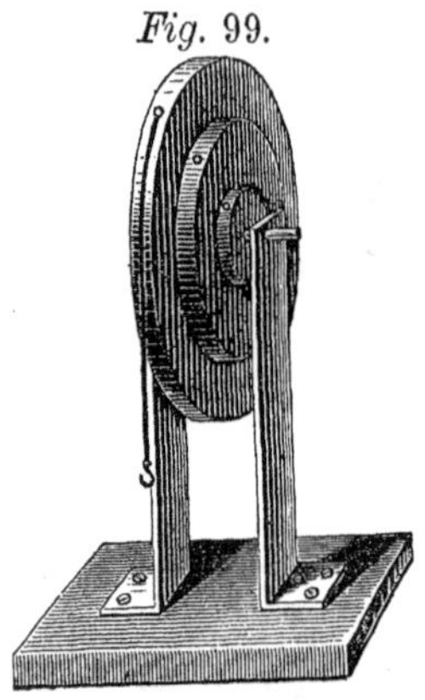

wood may have a number of successive disks, with diameters proportioned to each other, as 1, 2, 3, etc., fig. 99. Each disk has two pins, diametrically opposite to each other, to attach threads. These disks are supported by two equal strips of brass, which are screwed to a board.

For the further elucidation of this subject, it would be well to have a simple wooden axle with iron gudgeons, on which may be placed a winch or a wheel of the same radius, to explain the action of the winch. These gudgeons must also be turned off with the axes, and rest in ground brass boxes. The action of weights in different positions of the crank may readily be shown.

After having thus explained the knee, the pulley, the lever, wheel and axle, and crank, we may next proceed to the consideration of some compound machines.

For this purpose, models will likewise be needed, although the experiments will be no longer conducted with weights. Among these the common carter's jack occupies a prominent place. Instead of constructing a model of this, it will be well to procure a good jack of the smallest size, and either cut through the plate which covers the wheels, or, if this cannot be done, have the plate arranged to screw on, instead of being riveted. A jack of this kind may be used for other purposes, and costs no more than a model.

*Fig.* 100.

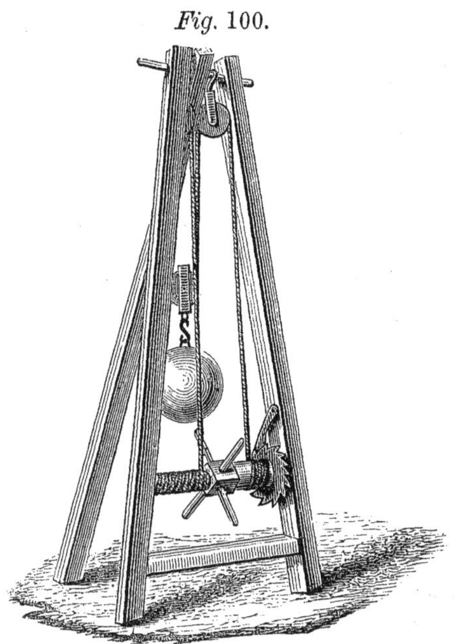

The tripod with a windlass, shown in fig. 100, is easily constructed in model, and is the better suited for illustration of a compound machine, as the tripod is so frequently used with or without the windlass. It depends on the design of the instruction, how much should be done in the way of collecting models; but it is desirable to have, so far as possible, real working models of tolerable size for the sake of exhibiting the actual operation of machines.

[61] **Center of gravity.**—For illustrating the law of the center of gravity in surfaces, a triangle may be made of very even wood, mahogany or ebony, about a line in thickness, and the center of gravity found on it by construction. If a hole be bored from the other side more than half way through, the rest pierced with a fine needle, so that it can be suspended by a thread, as in fig. 101, it will be so nearly balanced that a little wax will easily adjust it. It should be made of uniform thickness. In the same way the center of gravity of a square or polygon may be constructed by combining the centers of gravity of the separate triangles which compose them.

*Fig.* 101.

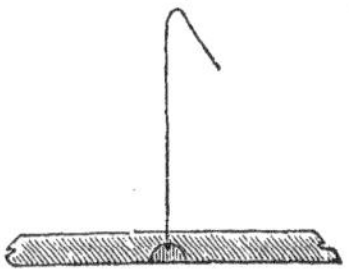

[62] The equilibrium of suspended bodies is most stable. When the mass of a body can be so distributed as to be suspended without appearing to be so, various curious experiments may be exhibited. Thus on the point of a pyramid, *A*, fig. 102, is a slightly concave metal plate. A piece of wood *B*, with a bent wire *M N* passed through it, having a lead ball at each end and provided with a steel pivot, is set in this concavity. The whole system will be suspended when the two balls lie below the pivot. For the block *B* may be substituted a small figure made of light

material resting on one toe, and holding the wire *M N* in its hands like a rope-dancer. Such an apparatus will always maintain its equilibrium, however much it may oscillate or rotate.

[63] **The cone which runs up hill.**—Make a double cone of wood, fig. 103, of 5 to 10 inches in length and 2 to 4 in thickness, with a knob at each end. Next procure two bars, like *m n o p*, fig. 104, beveled

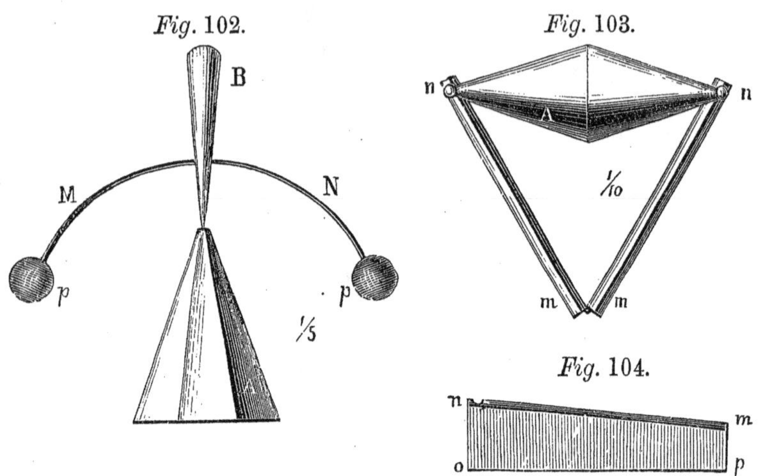

*Fig.* 102. *Fig.* 103. *Fig.* 104.

at the upper edge *m n*, which inclines to *o p* in such a manner that the axis of the double cone, when laid on the highest part at *n*, fig. 103, lies somewhat lower than when the common base rests on the lowest point at *m*. The two boards are united at *m* by a hinge, and separated enough to let the apices of the cones rest in the notch at *n*. On placing the cone at *m*, it will run toward *n*, and come to rest in the notch.

[64] **The Chinese tumbler.**—This may be bought at any of the toy-shops. When the stair is erected, the tumbler is to be bent together with its back downward, as if resting on the hands and feet; place it, however, only on its hands, and it will turn several somersets before reaching the bottom. If the toy does not work well, it should not be bought at any price, for its adjustment will involve much time and trouble.

[65] **Stable equilibrium.**—The apparatus shown in figs. 105 and 106 is very well adapted to illustrate the laws of stable equilibrium. The board *M N* with adjusting screws carries a square iron bar *B*, on which the slide *a b* with the pulley *c* moves up and down. This can be fixed at any height by means of a screw *d*. In the middle, directly across the board, is fixed the thin brass strip *m m*. Make an accurately perpendicu-

lar prism of white boxwood, and bore a hole from below, half the length of the prism and on one side of its vertical axis, having its center in the median line of the prism. Fill the hole with melted lead, and after cooling, fix it more firmly in the cavity by hammering.

*Fig.* 105.

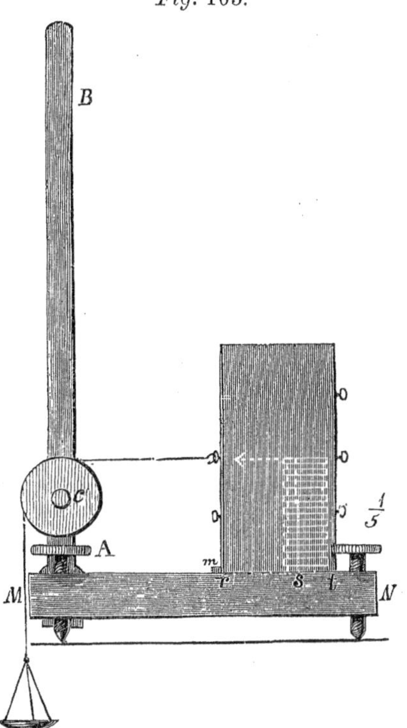


The center of gravity of the body must now be ascertained by applying the sharp edge of a ruler. If wood of uniform texture has been selected, and the hole for the lead made directly opposite the middle of the side nearest to it, the center of gravity will lie in the section passing vertically through this middle line. This line and the one directly opposite are now divided into four parts, and in each division is inserted a small hook, to which is fastened a thread with a scale-pan at one end and passing over a pulley. The pulley must be so arranged that the string shall be horizontal.

The distance of the line of direction of the center of gravity from that edge over which the prism will be thrown, the distance of the point of attachment from the same edge, and the entire weight of the body furnish the elements from which we can calculate the weight which must be attached to the string in order to overthrow the body in any of the four positions in which it can be placed, with the string attached to either of the six hooks. The weight of the body is marked on it, and the distances *r s*, *s t* measured and marked, assuming the interval between the hooks as unity. The weight calculated for each hook should be

*Fig.* 106.

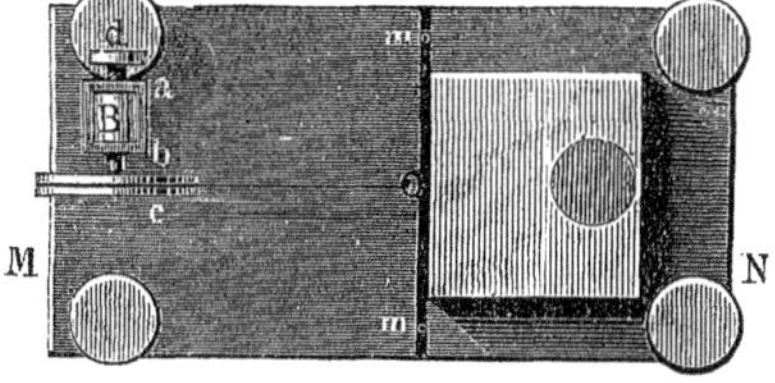

marked beside it. The stand of fig. 67 will serve for this arrangement, after removing the lever and knee.

[66] **The balance.**—The dependence of the delicacy of a balance on the position of its center of gravity can be shown, as suggested by Greiss, very simply as represented in fig. 107. Two half knitting-needles and one whole one are thrust through a cork at right angles to each other, and the apparatus supported on two wineglasses, so that the whole needle represents the beam of the balance, and one of the halves the tongue. By pushing the latter up or down, the center of gravity can be altered at pleasure, and its effect on the equilibrium thus demonstrated.

*Fig.* 107.

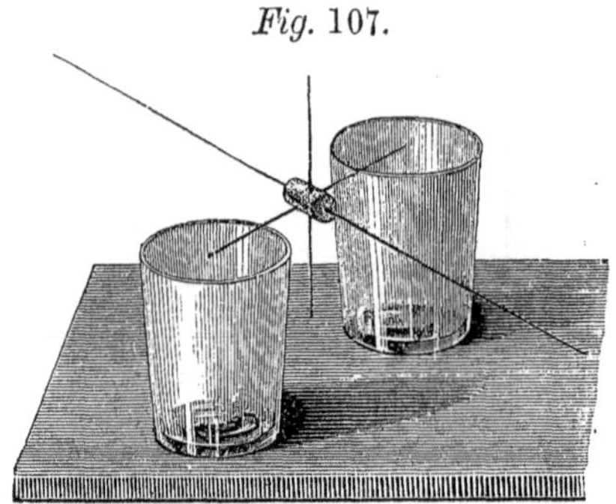

[67] **Management of the balance.**—Two balances are needed; a delicate one, for accurate experiments, (as in specific gravity,) and a good common instrument, for miscellaneous purposes. In case of necessity, the latter may be used for specific gravity experiments, provided they are as sensitive as the better kinds used in the shops.

The characteristics of a good balance are well known. As the outlay is considerable, the fine balance should be bought of a well-known maker, as it will be no advantage to have any but a first-rate instrument. It should carry about 500 grammes, and turn with at least $\frac{1}{100000}$; that is, with the load of a hectogram it should turn with a milligram. The base must be provided with screws for horizontal adjustment, since the suspension in the so-called shears, as in the common balance, is not admissible here. Wedges, indeed, may be driven underneath, or the stand may be placed upon a board, with adjusting screws, which will be found available for other purposes; but either mode is a troublesome one.

This balance should have an extra pan, for the determination of specific gravities.

It is essential to the good preservation of a balance, that, when not in use, the beam should be raised from its bearings by a gentle, steady movement. It will be found very convenient to have the case for the balance glazed, and it should be so arranged that the pans can be reached by opening a door in the case. The equality of the arms may be ascertained by loading the pans nearly to their limit, and balancing and then transposing the pans, which should still balance each other. This may not occur, however, with mathematical accuracy in very delicate balances, and therefore when very accurate determinations are to be made, recourse must be had to the method of double weighing. It will be readily understood that in such a balance, the edges must be parallel with each other, sharp, and in the same plane.

The balance must have a firm resting-place in a good light where it can remain undisturbed. If it become ncessary to move it, the knife edges should be lifted from the bearings, for these parts are easily injured by shaking.

[68] In *weighing*, it is more expeditious to try the weights in regular succession, beginning at the larger, and omitting none, for one is very apt to underestimate the weight, and have to go back again. The balance should not be raised to its full height until equilibrium is so nearly secured, that neither scale touches the bottom. It is only by the most cautious handling that such an instrument can be kept in order.

It will not be necessary to wait for the balance to come entirely to rest, as the index moves along a graduated arc, and a comparison of the amount of motion to either side of the zero will indicate the fact even better than when all is in equilibrium. This, however, soon occurs in specific gravity experiments, when the body to be weighed is suspended in water, on account of the greater resistance of this element.

[69] For the *common balance*, one should be selected carrying 5 to 10 lbs., and of sufficient delicacy to turn, when not loaded heavily, with $\frac{1}{1000}$, (1 decigram;) in this way the finer instrument may be much relieved. It should be suspended in some convenient accessible place, where there should be two hooks, one to support the scales when in use, the other lower down, and allowing the pans to rest on a base. There should also be a point of suspension for this balance over the experimenting table, as it will often be needed here. If this balance be the only one procured, and with the above-mentioned degree of sensibility it will serve for purposes of demonstration, it will be necessary to have a pair of flat scale-pans to be used among other purposes for specific gravity determinations. The instrument must be subjected to all the tests to insure accuracy, as this will frequently be found wanting. Besides seeing to the equality of the arms, the three knife edges must be tested to see if they are in the same straight line. If the construction of the balance will allow it, this latter point may be ascertained by means of a fine, smooth thread attached to a bow; if this test cannot be applied, then recourse must be had to weighing. The sensibility must not decrease

too much, by increasing the load, nor become extreme when loaded to the highest. In the former case, the central knife edge is too high, or rather, the line connecting the outer knife edges is below the center of gravity; in the latter, it passes above the central edge. Thus a balance which when empty turns with a weight of 1 to 2 centigrammes, and when loaded with 3 kilogrammes in each scale-pan turns with 2 or 3 decigrams, may well satisfy the operator as to this point. It is well known that the sensibility of the balance is indicated not by the amount which sets it in motion when at rest, but by that which affects its oscillation when already in motion. If a balance does not come up to all these requirements, it should not be purchased. If, however, an inaccurate one has been already procured, all its parts should be carefully scrutinized. If the bearings and knife edges are not hard enough, (the file should not cut them,) they should be knocked out with a wooden or copper hammer and rehardened; or else they should be replaced entirely by better steel, ground and polished anew. Further hints on this subject will be given when treating of the pendulum. Should the balance still indicate a want of sensibility, its center of gravity lying too low, it will be necessary either to remove some matter from beneath the point of suspension, or else add something to the tongue. Finally, if the knife edges are not properly adjusted, or the arms of unequal length, and the knife edges not movable, something might be done by means of the hammer; but this is a very hazardous business with which we should have as little to do as possible. It will in most cases of this kind be found best to purchase a new one.

[70.] **Weights.**—It will be advisable for the experimenter to have weights of several different denominations, as troy, avoirdupois, and apothecaries' weights, together with those of the new French standard. The latter should be obtained from some reliable workman; for making decimal weights is a very tedious operation without a standard weight. Even with the indispensable precaution of laying each weight in its proper division in the case as soon as it is taken from the scale-pan, one of the smaller weights will occasionally be lost in some way. It is well therefore to make duplicates of these out of thin silver or platina plate. The cost of such plate is inconsiderable. A convenient mode of constructing these is to cut out a small rectangular piece and regulate it carefully to the weight of a grain or decigram. Then with the dividers lay off aliquot parts of the length and cut it with a sharp knife and a metallic ruler into bits of 5, 4, 3, 2 and 1 milligrammes. The accuracy of the several pieces may then be tested by the balance. Chemists often possess scales sensitive to $\frac{1}{10}$ of a milligramme, and access to such should be made use of for purposes of comparison and rectification.

[71] **The Steelyard.**—A steelyard will be necessary only to explain the mode of its use; it is to be tested before purchasing by trial with known weights.

*Fig.* 108.

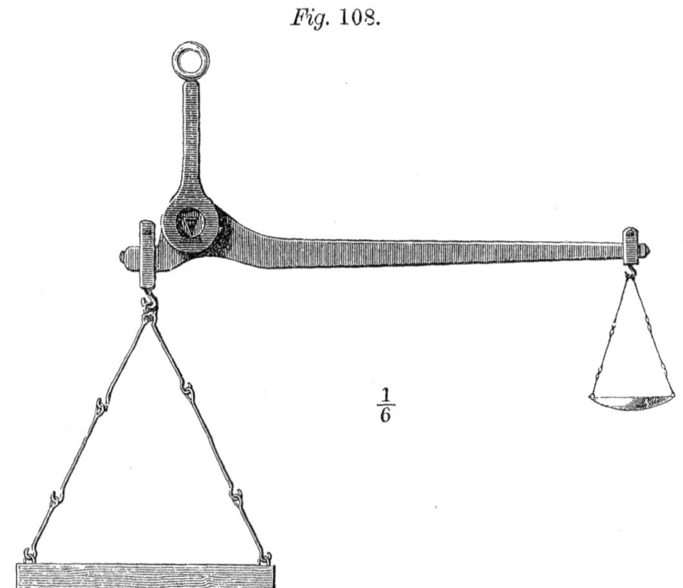


[72] **Counter scales or platform scales.**—The construction of these scales can be readily explained, and should by no means be omitted on account of the almost universal use into which they have come. The main point in testing them is to ascertain whether the decimal relation is maintained with different weights when placed on any given part of the bridge. Their sensibility rarely exceeds $\frac{1}{10000}$. A convenient form for demonstration will be found in one carrying about 50 lbs., the bridge or platform of which consists of two pieces hinged to a frame, which, when closed, entirely fill up the frame, and when opened and turned back, expose the lower lever to view.

[73] **Elasticity.**—The elasticity of ivory is shown by dropping a small ball on a marble slab. If the slab be breathed upon, the ball will make a spot the size of which is proportioned to the height of the fall. The spot is seen best in an oblique light. Blackening the slab with lamp-black makes the spot more distinct. The slab should be leveled by wedges to avoid having the ball rebound obliquely. It will be best to catch the ball as it rises in the air.

Any piece of marble slab will answer for the preceding experiment, after it has been dressed square. It will be best to have a wooden case with projecting edges made to fit the marble. The case is laid down, bottom upward, and the marble laid in it. The remaining space is filled with prepared plaster of Paris mixed to a thin paste, which, when quite hard, should be leveled off, and the wooden bottom screwed on to the frame. Gypsum, or plaster of Paris, (sulphate of lime,) is prepared for this and other purposes by taking it in impalpable powder as ground at plaster-mills, and putting a convenient portion dry into an iron pot over the fire. After a short time, an apparent ebullition will take place in the dry powder owing to the disengagement of the water of crystallization. The mass should be stirred continually until the bubbling ceases, then removed and allowed to cool. Unboiled plaster will set or harden, by mixing with a solution of carbonate of potash, although not so perfectly as when boiled.

*Fig.* 109.

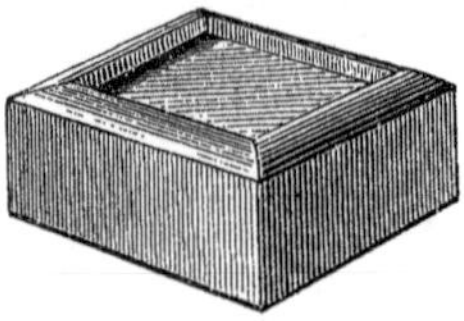

To this experiment we can revert after the laws of falling bodies have been explained; and from the height of the rebound, the elasticity of the ball may be calculated.

The elasticity of well annealed glass is shown by the *glass trumpet*, fig. 110, whose thin anterior wall is bent outward with a rattling sound when blown into. A bunch of glass threads serves a similar purpose, as also a glass tube cut into a spiral. See § 17.

*Fig.* 110. *Fig.* 111. *Fig.* 112.

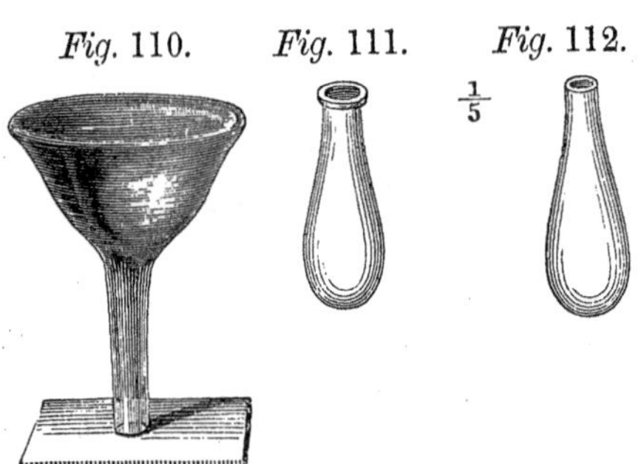


*The Bologna flasks*, fig. 112, are made of unannealed glass, and break to pieces when an angular pebble is dropped into them, although they will bear the stroke of a hammer on the outside. Vessels very similar to these are sold for taking specimens of spirits from casks; they have the same shape, but may be distinguished from the Bologna flasks by having the neck ground even, fig. 111. They are partially annealed, and do not answer at all for this experiment.

The experiment with the *glass tears*, or Prince Rupert's drops, (made by dropping melted glass into water,) is best tried by wrapping them in paper and then breaking off the fine point. In this way the fine sand into which the drop falls will all be retained in the paper, and the danger of injury by the flying about of the particles avoided. These drops, when broken in a tumbler of water, usually break the glass with the violence of the shock.

[74] The *torsion elasticity* of hard wires is shown by fastening a piece 2 or 3 feet long to a weight of several pounds, and suspending it by the other end in a hand-vice. When the weight is twisted out of its place of rest, it will gradually return to it.

[75] **Cohesion.**—Experiments in cohesion are not often undertaken

*Fig.* 113.

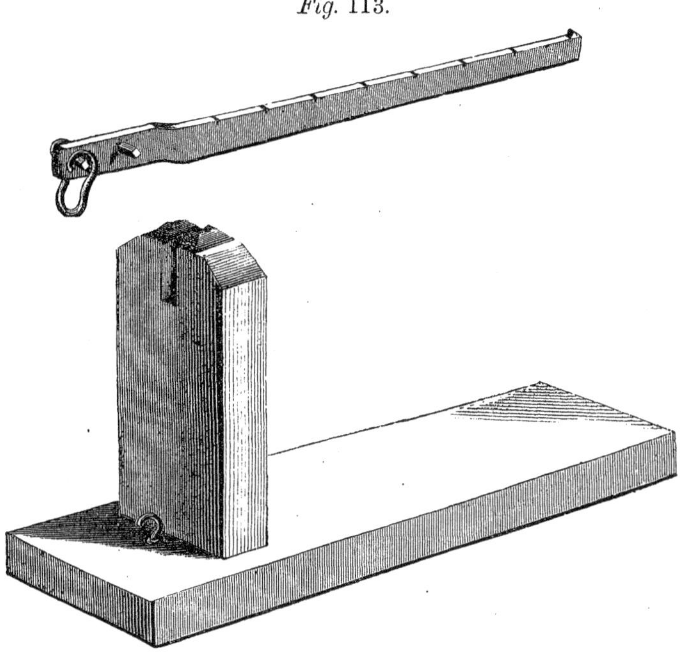

for purposes of instruction, as they consume a great deal of time, and, without constant repetition, lead to no result. The fact of cohesion existing, however, needs no very labored demonstration. An arrangement like that shown in fig. 113 may easily be constructed, (with wooden base,) in which, by means of a sliding weight of some 50 lbs., parallelopipeds of wood may be torn apart. The latter are shown on a larger scale in fig. 114; they should be faced with harder wood at the extremities. The distance between the two holes must be such that after the wood has been fastened between

*Fig.* 114.

the lever and the base by means of bolts, as shown in fig. 115, the former will occupy a horizontal position.

*Fig.* 115.

[76] **Adhesion** is shown (1) by means of ground plates of glass or metal; (2) by pieces of lead pressed together; (3) by Muschenbroeck's cohesion plates; (4) by glass and other plates on mercury and water; (5) by the Scotch waltzer.

1. Plate glass is cut into round or octagonal pieces, and the edges ground off on a grindstone; wooden handles are then attached by means of sealing-wax. The plates must be gently heated until the wax will melt on them. Pieces of broken mirror will answer for this purpose; their cohesion, however, should be tested before going to the trouble of fitting up. Metal plates, after having been turned as true as possible, must be ground down. This must be done with each, singly, upon a perfectly true iron plate or a piece of mirror glass, with sand and water, and finally with fine emery. The plates must then be ground on each other with powdered pumice and oil. These plates require much labor, for even those purchased at the shops are often very defective.

2. Two cylinders of lead, about $\frac{1}{2}$ inch in diameter and length, are smoothed first with the file and then with the penknife; on pressing them together in a vice, they adhere so closely as to be separated with great difficulty.

3. Muschenbroeck's adhesion plates are treated like those mentioned in the first illustration; to unite them they are first heated and then coated with tallow, after which they must be pressed together in a vice and allowed to cool. They will then sustain probably as much as 50 lbs. before separating. See fig. 116.

*Fig.* 116.

*Fig.* 117.

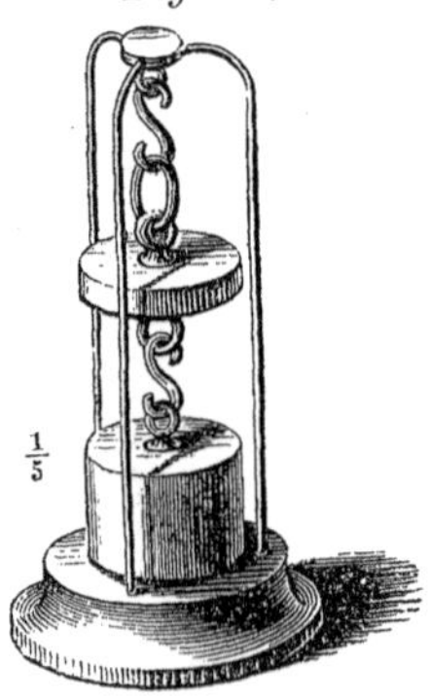


With both kinds of adhesion plates the experiment may be tried under the receiver of an air-pump, to show that atmospheric pressure is not the agent involved in the case. To do this satisfactorily, however, an arrangement similar to that in fig. 117 will be necessary. This consists of a board with three brass rods combined at top, and at their point of union carrying a hook for suspending the plates. The size of the base will depend on that of the receiver and air-pump plate. The weight should not hang high above the board.

4. This experiment is best prepared in the

following manner. To a glass or brass disk is attached a piece of sealing-wax as long as the finger, and to this a string. The scale-pan of a common balance is replaced by this disk, hung so as just to reach the surface of the water in a cup below. The sealing-wax is now slightly softened throughout by very gentle and gradual heat, and the plate allowed to rest on the water; a small weight is placed in the opposite scale-pan. The adhesion of the water holds the plate, and the weight draws the sealing-wax until it becomes perpendicular to the plane of the plate. It will be readily understood that the plate must not be allowed to touch the sides of the cup. As the wax stretches, one or more knots are subsequently made on the string until the balance again has the horizontal position.

In the experiment, the freely suspended plate is first brought into equilibrium by grains of shot, before setting the cup of mercury or water under it. By the gradual addition of weight, the plate resting on the liquid is torn off. The plate must be carefully cleansed, which is best effected by washing with alcohol, and the surface of the mercury in particular must be free from dust. For this purpose it is to be filtered through a paper cone set in a glass funnel. The inferior apex of the cone is cut off so as to allow the mercury to issue in an exceedingly fine stream. The glass plate is first laid obliquely on the mercury and then gradually brought to full coincidence with the surface, to expel all air-bubbles. The same precautions are necessary in using water, with the exception of the filtering.

5. The Scotch waltzer consists of a figure cemented to a convex lens or watch-glass. The glass is placed on a glass plate or porcelain tablet, and a drop of water put between the glass and tablet; on inclining the latter slightly, the glass with the figure turns half around, and by successively inclining the glass to different sides, the figure may be made to keep up a constant and pretty rapid rotation. Considerable practice is necessary to make this experiment successfully.

As it often becomes necessary to pour out of vessels filled with liquid, it may be well to mention here that the liquid is prevented from running down the side of the vessel by first moistening a glass rod with the liquid and then holding it in a vertical position against the side of the vessel to be emptied; the liquid will follow the rod.

## (*b.*) EXPERIMENTS UPON LIQUIDS.

[77] **Uniform transmission of pressure.**—The fundamental law of hydrostatics, that of the uniform transmission of pressure, may

be readily shown by the following apparatus. This illustrates also the application of the law in Bramah's press and saves an expensive model of this instrument. It is shown in plan and section in figs. 118 and 119, of $\frac{1}{5}$ the natural size.

*Fig.* 118.

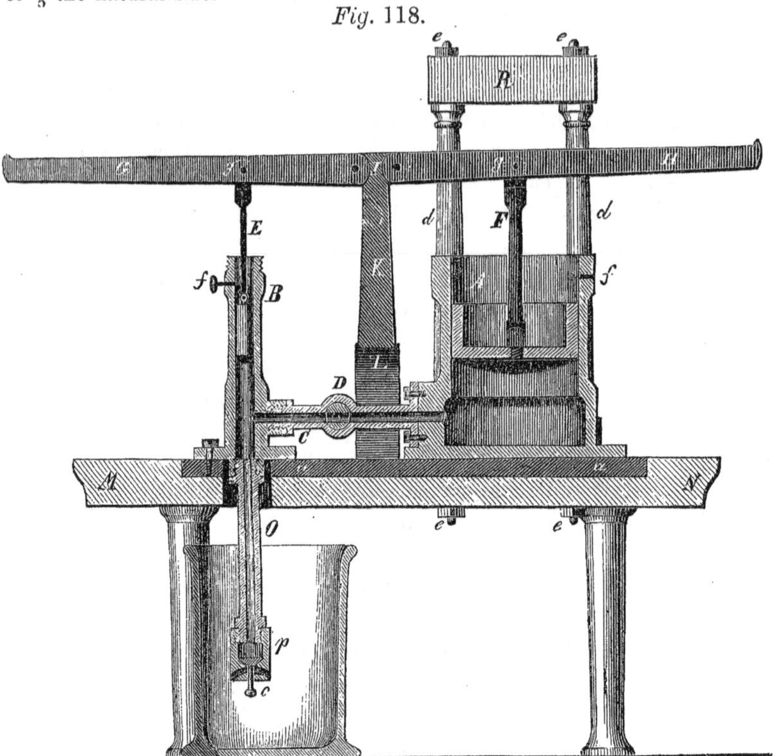


*Fig.* 119.

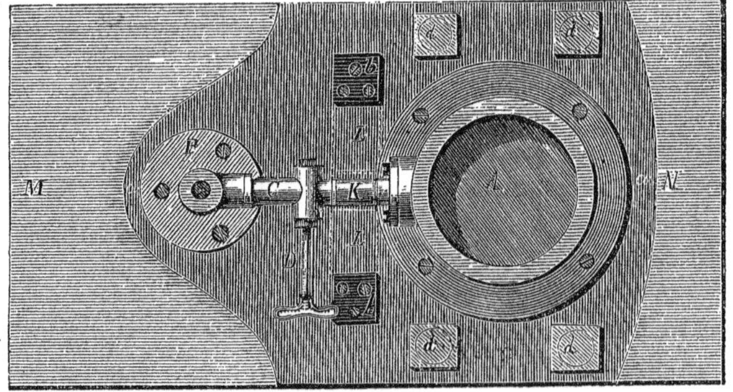

In the table *M N* is inserted the iron plate *a a*, on which are the two cylinders *A B* of gun metal; *A* is exactly 6 times the diameter of *B*, and the two are connected by the tube *C* with its cock *D*. In each cylinder is a metal piston, of which the larger is hollow and open above; these are ground so accurately that the larger, when set into rotation, will turn round several times freely in the cylinder, and yet fit air tight; they must not be greased, but before each experiment the cylinders and pistons must be wiped clean with linen to remove the dust. The pistons are connected with the levers *G H* by the jointed rods *E F*; these levers are of equal length and hinged to the cross-piece *I* carried by the iron column *K*, which again rests on the semicircular iron arch *L* springing over the tube *C*; the arch is fastened down to the base by the two plates *b b*. All the joints are well fitted and ground with emery, so as to have an easy and yet steady movement. Underneath the small cylinder *B* is fixed the suction tube *O* with the valve-piece *p* screwed to it; the valve can be opened from without by the rod *c*. On the base plate are also the four columns *d d d d* carrying the cross-plate *R*; they, as well as *R*, may be of wood; in this case they are bored through and an iron rod introduced, which, by means of the nuts *e e e*, holds the plate *R* to the base plate.

In using the apparatus, a suitable water-vessel is placed beneath the table, the pistons removed, and the cylinders connected by the cock and filled up to the small openings *f f* with water, (*f* on the smaller cylinder can be closed with an iron pin.) In replacing the pistons, the air escapes through *f*, and after the levers are adjusted the valve is raised far enough by hand or by a wire to allow water to escape until both levers are horizontal. Water may be drawn up from the subjacent vessel by raising the piston above the opening *f*. To remove the air entirely, water is pumped up by the smaller cylinder until it appears at *f*, when the lever is rendered horizontal by opening the valve. At the outset, the tare of each lever, with its piston, is estimated by suspension from a common balance at the points *g g*, and the weight calculated which must be laid on the lever of the larger piston to resist of itself the pressure of the smaller. In this tare is included for each lever a hook, as in fig. 120, with a small scale-pan. After restoring equilibrium in this way, each addition of weight to the small piston will require 36 times as much on the large piston, since the two levers have the same proportion. In this way the uniform transmission of pressure by weight can be very readily shown.

*Fig.* 120.

If now the cock *D* be closed and the small piston with its lever removed, the opening *f* of the cylinder *B* closed, and the glass tube

*R R*, fig. 121, with its brass cap *S*, screwed to the small cylinder, then this can be filled to any desired height above the surface of the water in the large cylinder; these heights are marked from 1 to 3 feet on the tube. By calculating the weight of a column of water of the same height and of the diameter of the cylinder *A*, and deducting the tare, we shall know what weight according to the leverage is to be applied to the large piston to hold the water in the small tube in equilibrium after opening the cock *D*. All these calculations are easy, as the diameter of the larger piston is 3 inches, of the smaller 0·5.

*Fig.* 121.

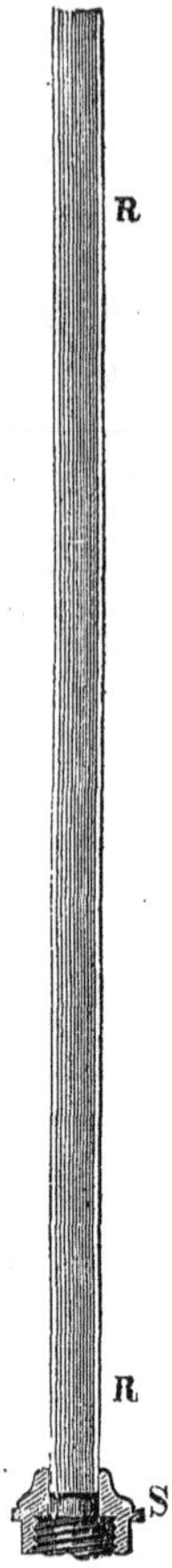

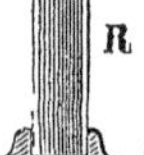


Again: close *D*, and, by raising the valve *C*, let the water run out of the small cylinder to below *f*, replace the piston and lever and remove the lever *H* and the rod *F*; a block of beechwood, fig. 122, which fits well on the sides and bottom of the piston, may then be set on the piston of the large cylinder, and any object to be pressed be introduced between it and the cross-beam *R*. By properly working the cock *D* and the lever *G*, more water can be pumped into the larger cylinder, and the application of the law of uniform transmission of pressure to the hydraulic press be thus demonstrated.

*Fig.* 122.

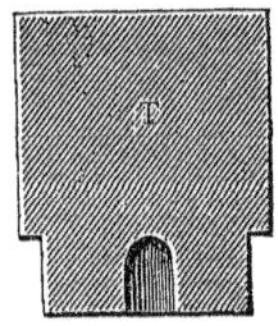

With such a model, a pressure of some hundred weight may be exerted without injury; indeed that of 10,000 lbs. might be applied, but for the danger of bending the iron base plate, by which the pistons might lose their easy working. In the demonstration, the amount of the pressure actually exerted is indifferent, its relation to the pressure in *B* being alone important; hence it will be well as soon as *A* actually exerts a pressure, to replace the hand on the lever *G* by a corresponding weight.

The apparatus is emptied after use, by opening the valve *c*.*

The simplest form of apparatus for this experiment is a strong tin ves-

* Link, instrument-maker to the University of Freiburg, makes this apparatus very handsomely for about 50 florins, ($20;) whereas a model of a hydrostatic press costs from $16 to $80, and cannot be used for general demonstration of the law.

sel with two cylindrical apertures, as seen in fig. 123. These apertures are closed with good corks, the narrower one being long and cylindrical; the other must not be set in more than a line, nor hold too tightly. If the vessel be filled with water, and the smaller cork driven in suddenly, the other will be forced out. Another vessel may have both apertures above, or instead of the second tube, two small holes may be made, in the side or top, and loosely stopped with wax.

*Fig.* 123.

[78] **The anatomical siphon.**—Procure a tin vessel, fig. 124, consisting of two parts, *A* and *B*, the former with a rather stout rim. Into the narrow leg is cemented a glass tube of any length; over the larger is tied a piece of fresh hog's bladder, after being filled with water to the brim. Care must be taken to have no air between the bladder and the water. Should the bladder be sunken in, water must be poured into the other leg, until it is raised up, after which it must be drawn tightly round the edge of the vessel, under several turns of thread, and then tied down securely. On filling the narrow tube with water, the bladder will be stretched, and driven upward, so as to show the muscular fibers crossing each other. By laying a board upon the bladder, a weight can be imposed, corresponding to a column of the height of the water in *B* and the diameter of *A*. On piercing the bladder with a needle, the water will escape, although with a jet not at all corresponding to the height of pressure, the opening being too small.

*Fig.* 124.

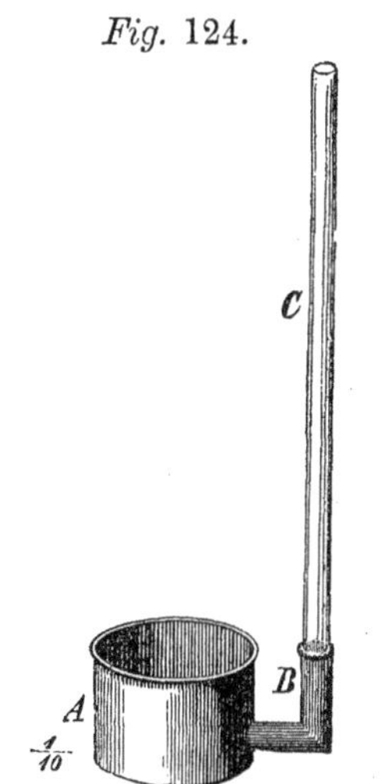


[79] **The hydrostatic bellows** is, however, a much more convenient apparatus. Two circular boards of hard wood are connected by a strip of good, strong leather, 5 inches broad, fig. 125. The

*Fig.* 125.

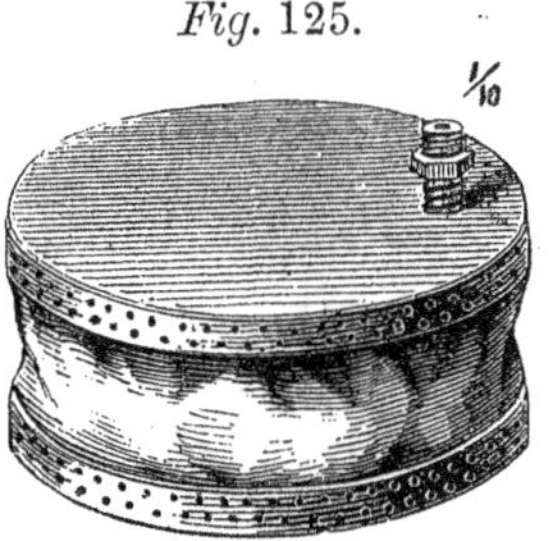

upper board, near the edge, is provided with a conical aperture to receive a brass screw, fig. 126. The boards are first well painted, and the leather stretched tightly over them secured in position by a few tacks. At the seam, which is turned inward, the wood is cut out a little, and in the joint is placed a mixture of wax and lard, with which, well heated, the leather also is saturated. Strips of leather, of width equal to the thickness of the boards, are now bound round them, and over the other leather, by means of a double row of nails, driven close together. The brass screw, fig. 126, is now screwed into the top board. The bellows are filled with water, through the hole in the screw, into which is screwed a glass tube, as represented in fig. 121; any weight may now be placed on the top board, and from the height to which the water rises in the tube, and the diameter of the base, we may calculate the weight. The weight applied must be equal to that of a column of water, with a diameter equal to that of the bottom, and a height equal to the distance which the water rises in the tube.

*Fig.* 126.

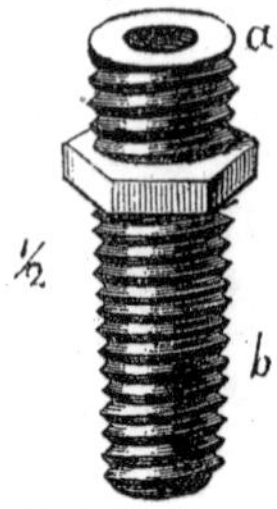

*Fig.* 127.

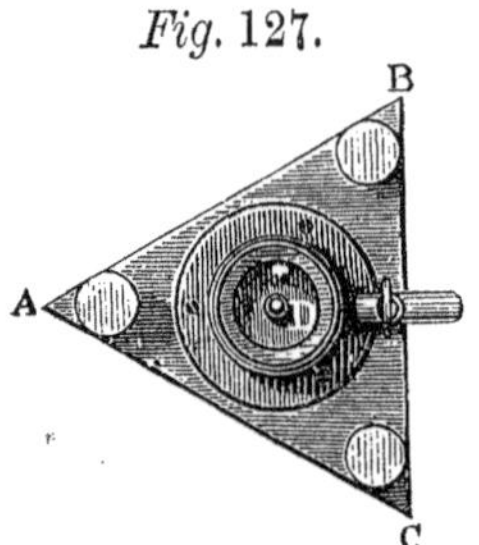


[80] **Pressure upon the base.**—Besides the apparatus of Haldat, that of Pascal may be recommended, as affording the opportunity of measuring the pressure directly by weights. The following is a convenient form for its arrangement:—

*A B C*, fig. 127, is a triangular board, with leveling screws, on which stands the metal cylinder *D*, fig. 128. This is smoothly bored, and contains a piston *E*, fitting water tight, and of such thickness as to go entirely below the cock *F*. In the bottom of the cylinder is an opening for the admission of air, as otherwise the piston would not be movable. The piston has a hook to allow of its being attached by a wire to one side of a pair of scales, while the apparatus stands on the experimenting table. The glass vessels *H H*, figs. 128, 129, and 130, of various forms, but of nearly equal height, may be screwed on the cylinder *D*, by means of the brass fittings, *g g g*. An equal height above the base is marked on each.

*Fig.* 128.

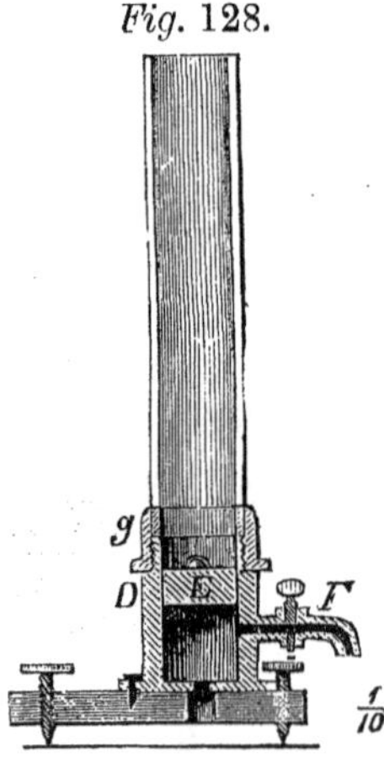

In using the apparatus, the piston *E* is first suspended freely from the balance by its wire for the purpose of taring it; it is then brought back to its place and the cylindrical vessel *H* screwed on, the piston again attached to the balance, and the board *A B C* arranged directly under the hook. The wire must be long enough to allow the balance to stand horizontally, when the piston is above the cock *F*. Water is now poured in up to the mark, and in the opposite scale of the balance enough weight added to preserve equilibrium. If desirable, we may note the weight at which the piston begins to sink, and that at which it commences to rise again, so as to obtain the mean, and to eliminate the very slight friction. The piston should not be greased, but fit tighly without; in any case, oil must be used and not lard. The wire is now unhooked, and the piston allowed to sink below the cock *F*, through which the water escapes. The same weight which was necessary for the cylinder of uniform diameter will also maintain equilibrium when any other vessel, as *H*, is employed. No experiment for estimating friction can be made with these vessels, as the height of the column of fluid varies with the motion of the piston. We may also calculate beforehand the required counterpoise from the diameter of the base and the known depth of the piston below the mark. If the one vessel be actually of equal size with the cylinder, the water in it may be poured into the opposite scale to show that the pressure on the bottom is really equal to the weight of the mass of water which was in the cylinder. In this case, however, we must not overlook the weight of the other pan which is attached in order to hold the water.

*Fig.* 129. *Fig.* 130.

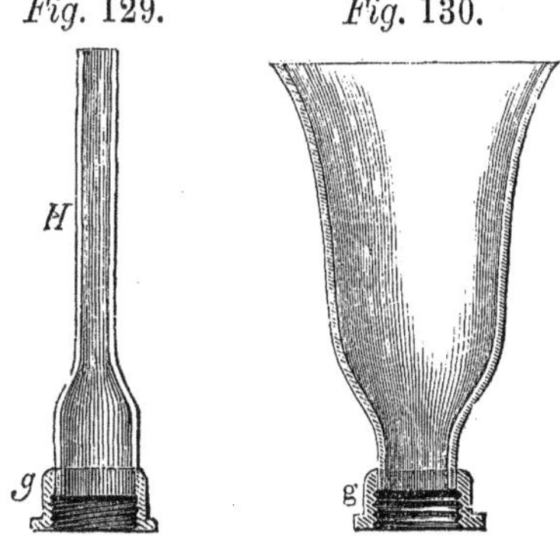


*Fig.* 131.

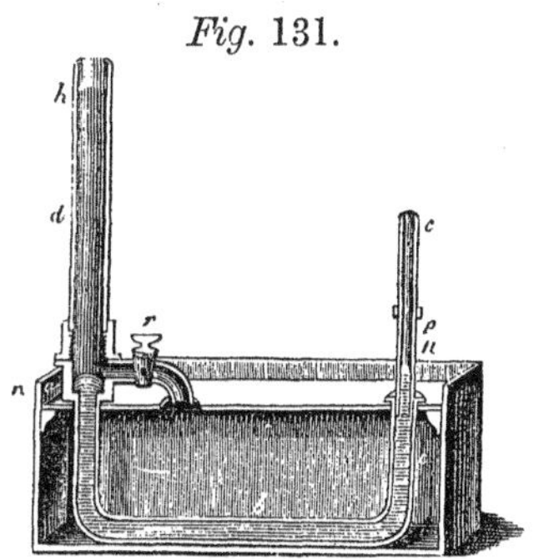

*Haldat's apparatus* depends upon the law of the equilibrium of various liquids in communicating vessels, and therefore only admits of an indirect estimation of the pressure on the bottom. It is represented in fig. 131. In a box of tin is fastened the bent glass tube *a b c*, which, on the one side, projects above the box, and on the other, ends in an iron cap with a stopcock *r*. The tube is filled with mercury nearly up to the cock. On the iron cap can be screwed the cylindrical vessel *d*, which is filled with water, and the height to which the mercury is driven in the other leg indicated by the slide *p*. If other vessels than *d* be screwed on and filled with water up to the same height, the mercury will always rise as before to the point *p*.

Since the vessels which are attached instead of *d* are never high, and the mercury cannot be raised above the cock, it is well to pour a little colored water in the tube *n* as an index.

[81] **The upward pressure of water.**—In this experiment, we may make use of the glass chimney of a lamp, fig. 132. The lower edge is ground even and finished with fine emery. For the bottom we may take a piece of plate glass *t*, which is cut round and then ground even on another plate with emery; a string is cemented to it with sealing-wax. The string is easily torn off from this glass plate; it is therefore better to make it of brass. A rounded brass plate is cemented to a piece of wood turned round and even, and then ground plane. When this is done, a hole is bored through the middle, a screw cut, and a hook inserted. The hook, in place of being screwed in, may be riveted.

*Fig.* 132.

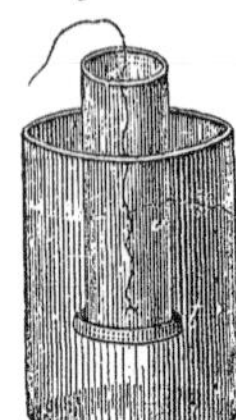

*Fig.* 133.

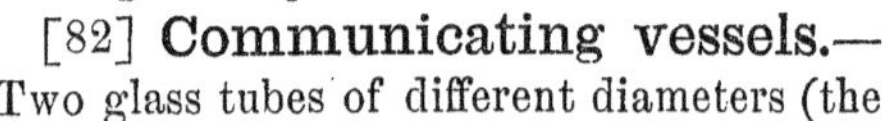

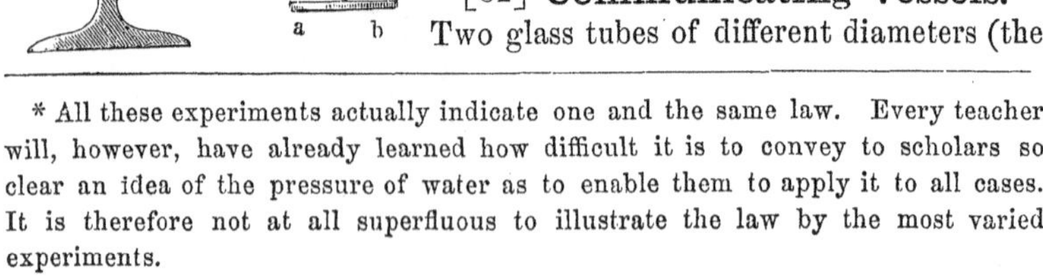


In using the apparatus, no grease must be allowed between the plate and the edge of the glass cylinder, as this would prevent the plate from dropping off from the tube when drawn gradually out of the water; and also when water is poured into it, unless the water was raised higher within than without. To make the apparatus neater and more durable, it should be furnished with brass mountings, as seen in fig. 133, which shows also the mode of making the experiment.*

[82] **Communicating vessels.**—Two glass tubes of different diameters (the

* All these experiments actually indicate one and the same law. Every teacher will, however, have already learned how difficult it is to convey to scholars so clear an idea of the pressure of water as to enable them to apply it to all cases. It is therefore not at all superfluous to illustrate the law by the most varied experiments.

least not more than $\frac{1}{2}$ inch) are cemented into a base of tin, *a b*, fig. 134. To employ mercury and water, an arrangement like that of fig. 135 is used. The tube is about $\frac{1}{4}$ inch in diameter. The mercury is first introduced by means of a funnel into the long leg and then the water. The amount by which the height of mercury in one column exceeds that in the other is to the height of the column of water inversely as the specific gravities.

*Fig.* 134.

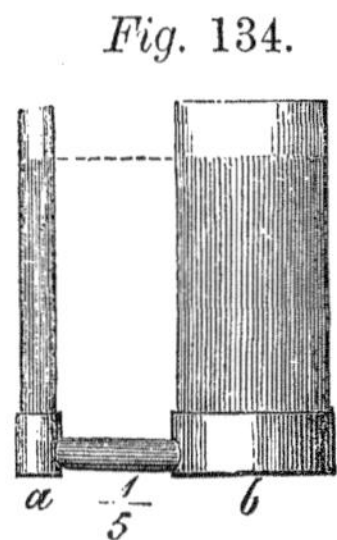


*Fig.* 135. *Fig.* 136.

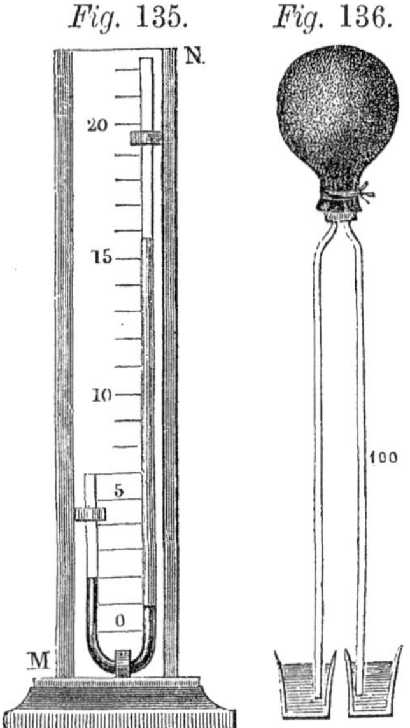


The experiment is also instructive in another form, in which a partial vacuum is created simultaneously in two tubes connected together and standing in different liquids. The liquids will rise to a height inversely proportional to their specific gravities. The best way of making the experiment is, according to Mohr, to pass the ends of the tubes through a cork tied (fig. 136) into the mouth of an india-rubber bottle. The bottle is pressed together with the hand, and, returning by its own elasticity, sucks up the liquids. A rude scale may be marked on the tubes with indian-ink.

[83] **Floating bodies.**—To demonstrate the law, that the center of gravity of bodies immersed in water, when it does not coincide with that of the water displaced, must lie deeper than this, a piece of lead may be cast in a cork, and so adjusted that the specific gravity of the mass shall be equal to that of water. This is easily attained; but unless the cork selected be very free from cavities, it will not long remain so.

To show that the center of gravity of a body floating on water does not necessarily lie deeper than that of the water displaced, but that its position at the same time will not be arbitrary, we may make use of two elliptical rollers, one of them constructed of very light wood, and the other of a cork $\frac{1}{2}$ to 2 inches in diameter. Such rollers are in stable equilibrium only when the short axis of the ellipse is vertical. The place of the metacentrum may also be marked upon the base, if it be deemed expedient to go into such minutiæ.

It may readily be shown by the following experiment, that under certain circumstances the center of gravity of floating bodies may be at rest in two positions. A piece of thick brass wire or a cylinder of lead is inserted in a large cork, as shown of half size in fig. 137, half-way between center and circumference, and near the opposite side a thinner wire. If it be rightly arranged, the cork may float with the larger wire uppermost; but when shaken very roughly, it will turn over. If the thick wire be too heavy, it may be remedied by cutting the cork and filing. For the cork in fig. 137, the large wire weighs 33 grammes, and the smaller 10 grammes.

*Fig.* 137.

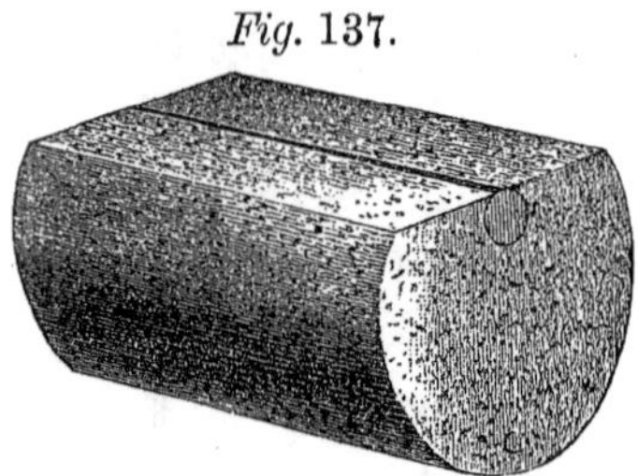

To show that a floating body sinks deep enough to displace an amount of water equal to its own weight, take a tin vessel 2 or 3 inches in diameter, with a lateral outlet near the top; fill this with water until it runs over. A floating body of considerable size is then laid in the water, and the amount displaced caught as it escapes from the tube, and measured or weighed; this weight will be found equal to that of the body as previously ascertained. A piece of wood or a properly loaded glass bulb will be best for the float. Great care is required in this experiment.

[84] **Loss of weight in water.**—The apparatus shown in fig. 138

*Fig.* 138.

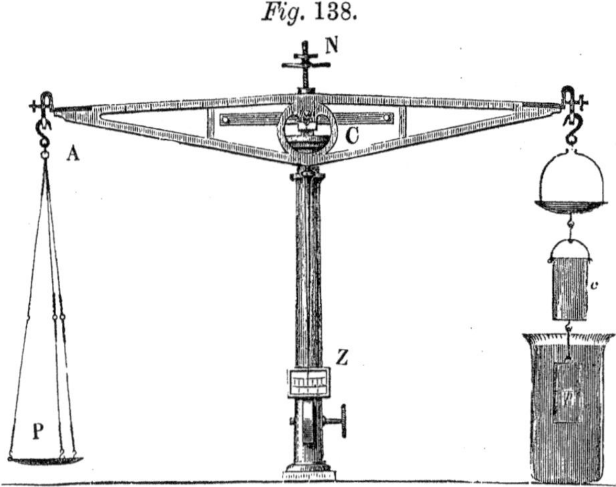


is intended for the demonstration of this law. It consists essentially of a hollow cylinder *c*, which can be suspended from a pan, and has beneath a hook on which can be hung the solid cylinder $p$, fitting exactly in $c$. The two cylinders are now suspended on one end of a balance, after removing the scale-pan, and equilibrium restored by means of weights. A vessel of water is next set underneath, in which the cylinder $p$ is entirely submerged; the equilibrium will again be destroyed, but restored by exactly filling $c$ with water. The vessels may be made by the tinner; but it will be best to construct them of brass, and turn them so that $p$ shall fit accurately in $c$.

A second very instructive experiment may be made with the same apparatus. A glass of water is balanced in one scale after the cylinder $c$ has been suspended from the other. On suspending $p$ in the vessel of water by means of a string held in the hand or attached to a stand, equilibrium will be disturbed, and again restored by filling $c$ with water.

[85] **The Cartesian devil.**—A glass bulb, $\frac{1}{4}$ to $\frac{1}{2}$ inch in diameter, with a little tube terminating in a fine aperture, illustrates the principle of this experiment much better than the ordinary figures made of opaque, colored glass. Both the bulb and figures are filled by gentle warming with a spirit-lamp and subsequent immersion of the aperture in water. The bulb should float near the surface, and any excess of water may be removed by suction. When the bladder tied across the top of the vessel, fig. 139, is pressed upon, the bulb will sink, returning when the pressure is removed. When the apparatus is intended to be permanent, the water is mixed with alcohol, and the bladder replaced by india-rubber.

*Fig.* 139.

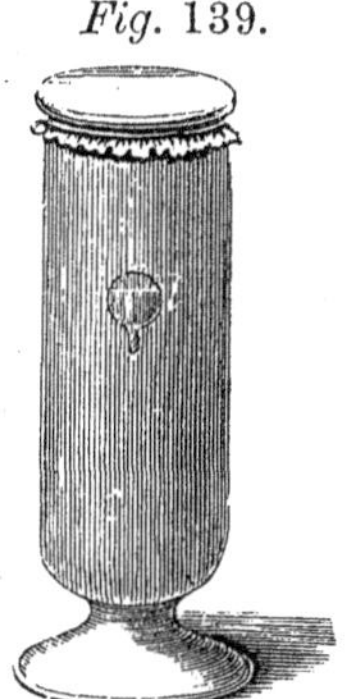

[86] **Determination of specific gravity.** (*a.*) *With the balance.*—The general mode of procedure is well known, and the precautions necessary in weighing will be found under the description of the balance. The finest possible wire must be employed, and, if the weight of the body allow it, a human hair, with the oil removed by lye. In either case, the weight of the wire or hair must be ascertained. An unavoidable error in this method lies in the fact that the loss of weight of the hair in water is ascribed to the body weighed. This error may indeed be avoided by placing the body in question in a glass vessel with ground cover, as is done with powders which are to be determined. Nevertheless, as the sensibility of every balance decreases with the load, the possible error of weighing thus becomes greater, and alone may exceed the slight loss of weight of the body. Thus a greater error will in general be incurred than the one intended to be avoided.

For pulverulent or soft organic substances, a small bottle with ground stopper must be used; for larger masses of soft consistency, a scale-pan suspended by the finest thread, the absolute weight of which $= p$ and its specific gravity $= s$ have been previously ascertained. If the weight of the body be $p'$, and the loss of weight of the whole system $n$, the specific gravity of the body will be $= \frac{n\,s - p}{sp'}$.

Bodies lighter than water are attached to a heavier body whose weight and specific gravity are known; the loss of weight of the heavy body, subtracted from the loss of the whole system, gives the loss of the lighter body.

For liquids, we make use either of a glass vessel with ground cover, or, still better, of one with narrow neck and a thermometer tube ground in as a stopper. In the first case, the edge of the glass and the cover are ground with sand and water on a plane glass plate; when the sand has cut everywhere uniformly, levigated emery is employed, and the two ground on each other. If a stopper is to be ground in, the neck of the glass must first be ground roughly by means of a conical brass tap on a lathe. The glass tube, drawn into a conical shape in the lamp flame, is then fastened in a chuck on the lathe, and ground in the neck of the vessel with sand and emery.

The latter process is much more difficult than the former. In any case the vessel must be as light as possible. The weight of the vessel when empty, and also when filled with water of a given mean temperature, say 60° F., are ascertained by repeated experiments, and marked on a card attached to the apparatus, or else scratched on the glass.

*Fig.* 140.

*Fig.* 141.

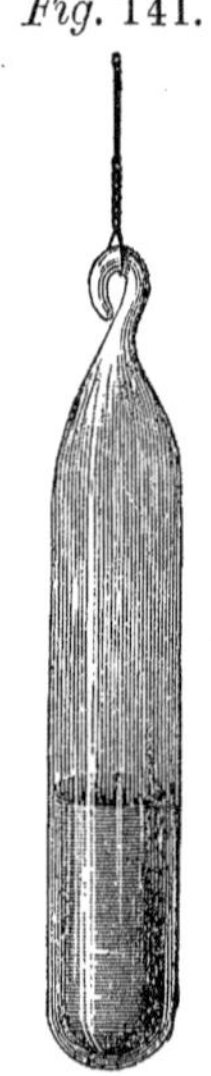

In the case of liquids which evaporate readily, as alcohol, the wide-mouthed vessel is not a convenient one, as the liquid presses out between the vessel and the cover, and evaporates copiously over the large surface; in consequence, small bubbles form on the inner wall of the vessel, and when the weighing is nearly completed, additional weight must be continually taken off. If a glass vessel with narrow neck be not at hand, we may weigh some insoluble body in the liquid, and then in water, thus obtaining very accurate results. In this case, it will be best to use a massive piece of glass, or a glass bulb, an inch in diameter, fig. 140, which, before sealing up, has been loaded with mercury. Instead of a bulb, a glass tube, fig. 141, $1\frac{1}{2}$ to 2 inches long, and $\frac{1}{3}$ of an inch wide, may be employed. It should be suspended by a very fine platina wire, to admit of the determination of acids. The weight of the bulb in air and in water of given temperature is ascertained and marked down, so as to render only a single weighing necessary. It is convenient to have a piece of lead which exactly balances the bulb in the air; when this is immersed in a liquid, the weight placed in the shorter scale-pan gives directly the loss of weight. Under all circumstances, care must be taken that no air-bubbles adhere to the body immersed. Careful cleansing with alcohol, if necessary, and previous wetting in another vessel of water and passing a camel's-hair pencil over the surface, will generally secure the desired end. The method just described has the advantage of requiring very little of the liquid, which is often desirable. Minute quantities may be weighed in a little pipette, like fig. 142, which is filled by suction. A new pipette must be employed for every different liquid.

*Fig.* 142.

[87] (*b.*) *With Nicholson's areometer.*—In the arrangement of this apparatus, it will be found convenient if the sieve in which bodies are placed in weighing can be suspended inverted, in order to admit of bodies which are lighter than water being placed underneath; this may be done in the form represented in fig. 143. When it is intended not to use the areometer for liquids, it is made of thin brass plate, and care must be taken to have the body conical at *a*, otherwise it will sink too suddenly when overweighted. When the instrument has once sunk to the pan *b* so as to wet it, it must be dried; but to avoid this trouble, as well as to prevent the contact of the instrument with the sides of the glass, a wire fork, as in fig. 144, is laid across the top of the vessel so as to embrace the stem.

*Fig.* 143. *Fig.* 144.

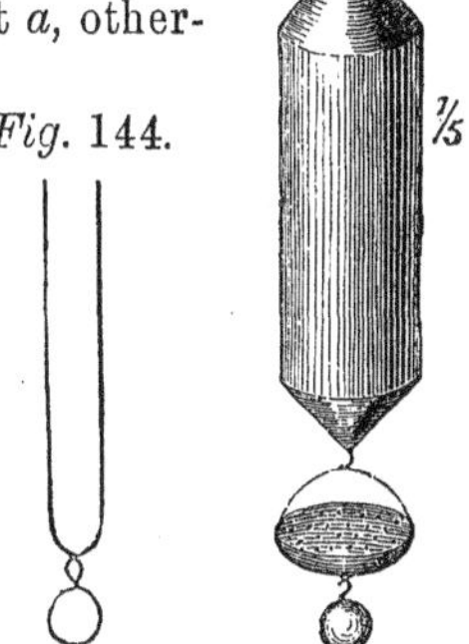


The sensitiveness of the instrument depends upon the delicacy of the stem *c*; to have this therefore as thin as possible, it is constructed of hard, drawn brass wire, screwed into the body, and soldered with soft solder, to make an air-tight joint. The mark in the center of the stem is a fine file scratch, running entirely around.

Body, pan, and sieve are made of equal width, so as to be kept in one paper case. The sieve and bulb are kept in a special division. The proportions of the instrument must be such that, when empty, it shall not sink to the upper cone, and, when sunk to the mark, the limit of stable equilibrium shall be nearly reached, so that the greatest capacity may be obtained in proportion to its dimensions. With the dimensions given in the figure, this will be attained when the brass used does not weigh over 0·5 gramme to the square centimeter; it is reduced somewhat in thickness by finishing with pumice-stone on the lathe. A cheap substitute for the brass instrument may be made of a medicine phial, by filling the cavity of the bottom partially with sealing-wax, and inserting the stem in this; shot are inclosed in the phial sufficient to cause it to float erect though entirely submerged, when weights are laid on the pan. The cork is then pared off smooth and covered with sealing-wax. Fig. 145 shows such an areometer. The substance to be weighed is laid in the cavity of the bottom.

*Fig.* 145.

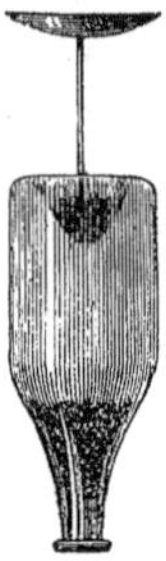

For each instrument, the weight required to sink it to the mark at a given temperature must be accurately ascertained and marked on it; it will then be only necessary for obtaining the weight of the body in air to deduct the additional weight required to bring the stem down to the mark from the above quantity.

All these instruments are more or less inert, *i.e.* with the same weight they remain at different depths in the water. The weight must therefore be so adjusted that the instrument shall oscillate to equal distances above and below the mark. This is easily done by gently raising and sinking it. Special care must be taken to remove air-bubbles from the holes of the sieve, where they are very apt to adhere with great tenacity.

*Fig.* 146.

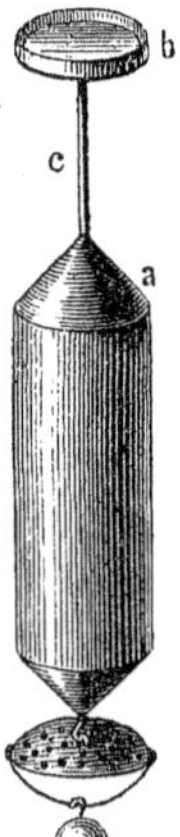

b

c

a

To use the areometer for bodies lighter than water, the sieve is inverted, as in fig. 146, and the body placed under it. To use the instrument for liquids, weight the pan until it sinks in the mark to the test liquid. Its entire weight (when not loaded) and the tare are marked on it.

In all these experiments, observations of temperature, and the consequent corrections, are indispensable. For simple instruction, however, this may be passed over, as the error arising from this omission cannot easily amount to a unit of the second place of decimals, and the coincidence of an experiment with a table of specific gravity is sufficient, if it reach to this place.

[88] (*c.*) *The graduated areometer. The volumeter.*— To explain the theory of this instrument, take a thin glass tube of as

uniform diameter as possible, about 1 centimeter in thickness and from 3 to 4 decimeters in length. Close it below, make the bottom even, and widen the upper end slightly, to take off the sharpness. Place this in water, and load it with shot and some wax, until it swims vertically. Divide the remaining space into two parts, *a b*, fig. 147, and *c d* into 100 equal parts, to be laid off on a slip of paper, and continued over the space *a c*. When the numbers are properly marked, cut away the lowest, $\frac{2}{10}$ or $\frac{3}{10}$, and attach drops of sealing-wax to the four corners of the strip, which must be about half as broad as the circumference of the tube. The paper is now bent over a thermometer tube, and pushed into the tube until the hundredth division comes opposite to *c*; it is then attached, by gently warming the tube opposite to the corners of sealing-wax. A cork, previously coated with sealing-wax, is then placed in the aperture. The loading of the tube is so arranged that the scale sinks to the hundredth mark, when immersed in water at the temperature assumed. To obtain an extended range, it will be necessary to have two instruments, one for heavy, the other for light liquids. In the case of the volumeter for liquids heavier than water, the scale may be made directly; for the lighter, however, it will be necessary to take a liquid whose density, as determined by the balance, is 0·8. The point to which the volumeter sinks will then be 125, and the interval between the water-mark and the other is to be divided into 25 parts, and the remaining space on the same scale. Tolerably wide tubes are necessary in this mode of construction, and the degrees obtained are short; they may be conveniently replaced by narrow tubes, and larger divisions; for this, however, it will be necessary to have a bulb at the lower end. It will be found cheapest to purchase such instruments; they should, nevertheless, be tested before using them. This is done by immersion in a liquid of known specific gravity, and observing—(1) whether it indicates the same specific gravity; (2) whether it sinks to the proper mark in pure water; (3) whether the degrees are all equal. In order to be quite certain that the tube is everywhere of the same thickness, it must be tested in several

*Fig.* 147.

*Fig.* 148.

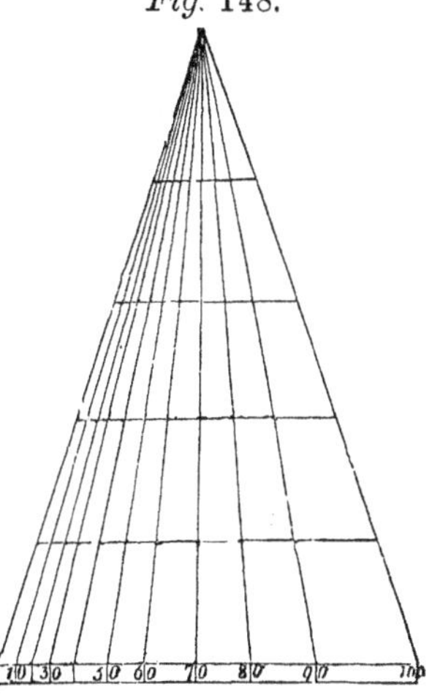

liquids of known specific gravities. However simple, theoretically, the volumeter may be, it is not so in practice, and can never come into extensive use, as it always requires some calculation, even if quite simple, and promises no greater accuracy than the areometer, which is divided directly for specific gravity—only the accuracy of its graduation is more easily controlled. The following method is to be employed for areometers of unequal divisions.

[89] **Testing areometers.**—The graduation is laid down on thick paper, on a somewhat larger scale than could possibly be used for an areometer. On the entire scale is constructed an isosceles triangle, fig. 148, and a straight line drawn from each division to the apex. To prove an areometer scale by this triangle, lay it upon the triangle parallel to the base, and at such a part that the extremities of the scale rest on the two sides of the triangle. If graduated properly, the division of the scale will also correspond to the lines of the triangle. It will be convenient to draw lines on the triangle, parallel to the base, at short intervals. Fig. 148 represents this for the alcoholometer of Tralles.

*Fig.* 149.

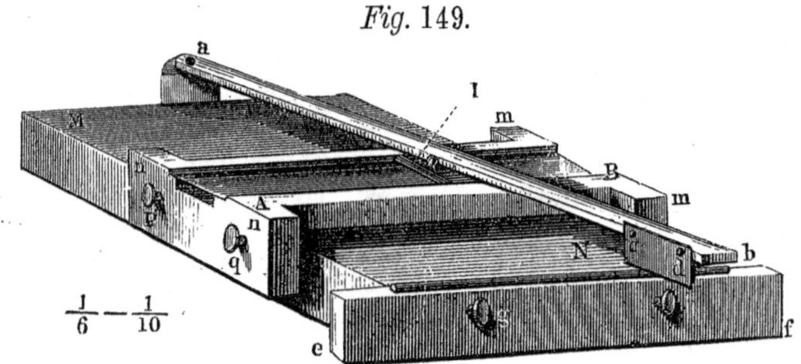


[90] **The graduating apparatus.**—The graduation is effected by instrument-makers on the same principle as that just stated, and it is easy to construct a more or less perfect instrument of this kind. On a solid, stout plate of seasoned wood, $M\ N$, fig. 149, the wooden strip $a\ b$ rotates on the pin $a$, having at the other end $b$ the steel edge $c\ d$ perpendicular to $M\ N$. The strip $e\ f$ is attached to the edge opposite $a$; it has a slit in which a brass or iron bar can be fastened by two screws, $g\ h$. There are several similar strips having along their edge some one of the usual unequal graduations; one is divided into equal parts.

The ruling board $A\ B$ can be slid on the plate $M\ N$, and under the strip $a\ b$; it is kept constantly parallel to $e\ f$ by the two guides $m\ m$, $n\ n$, and may be fixed at any point by the two screws, $p\ q$. Into this board is let the broad iron ruler $l$, which can be moved along in a rabbet by means of the strip *a b*, and a pin riveted to *l*; it must move easily, but always maintain its direction perpendicular to *e f*. Beneath the ruler is a depression in *A B*, in which the scale to be divided is laid. An additional arrangement may be attached to the ruler, for the purpose of scratching or marking off the divisions, when this must be done more accurately and rapidly than by hand.

Now, if such an instrument as the alcoholometer of Tralles is to be graduated, its loading is so arranged that in water it shall sink to *a*, fig. 150, after the paper with its sealing-wax corners has already been slipped into the tube. The approximate length of the tube is ascertained by a previous experiment. Alcohol of a known specific gravity is next taken, and its percentage of pure spirit ascertained from a table; the areometer is placed in it, and the point to which it sinks marked: suppose the spirit to be 80 per cent. and *c* the point. The proper temperature of the liquid is indicated by the zero of the inclosed thermometer, which has a peculiar graduation. The length *a c* is taken off on the paper strip, which is then laid on the ruling board *A B* of the graduating machine at such distance from the brass strip *e f* (*Tralles'*) that when the strip *a b* moves from 0 to 80, the ruler *l* traverses the space between the two marks on the paper. The strip *a b* is then brought back, so that *c d* clicks in the notch marked *O*; it is then carried from division to division, and corresponding marks made on the paper strip by means of the ruler *l*. After the graduation has been completed, the paper is fixed properly in the tube again, the corners melted fast, and the top of the tube fused together without any loss of glass. In ascertaining the arbitrary but tolerably elevated point *c* of the scale, a normal instrument is generally made use of, instead of determining the specific gravity of the alcohol employed by the balance.

*Fig.* 150.

[91] **Tralles' hydrometer,** besides Tralles' scale, usually has a second one, giving percentage of weight of the alcohol; to this the thermometer melted in the wide part of the areometer has reference. It has its zero at 60° F., and is graduated to single degrees. They are so large that when the temperature of the liquid is higher or lower than the normal by *n* degrees, the instrument shall stand higher or lower, inversely, by *n* degrees of Richter's alcoholometer, than it would at the normal temperature. In this case, we first read off the otherwise useless Richter scale, and then see what degree of Tralles' corresponds to that of Richter thus ascertained.

As these instruments are rather long, it will be necessary to have high, narrow cylinders to contain them. A form like that in fig. 151 will be found very convenient for reading off, when the vessel is nearly full. The usual form of the Richter's scale rests upon incorrect assumptions, and, in fact, does not coincide with the true percentage of alcohol; for this reason it is only useful for the above-mentioned correction.

*Fig.* 151.

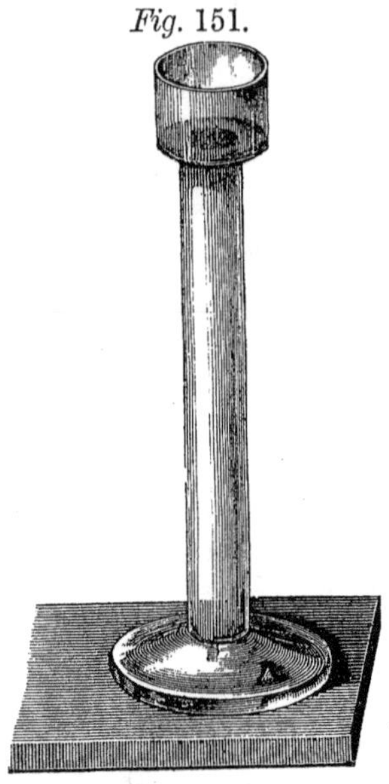

[92] **The areometers of Beck, Beaume, and Cartier** have entirely arbitrary scales, with degrees of equal length. These, by the aid of a table, may, likewise, be used for ascertaining specific gravity; such tables are to be found in most of the works on physics. Their construction is easily intelligible, from what has already been said on the subject, when we know their arbitrary fixed point. They are best tested by means of a liquid of known specific gravity, with the assistance of the tables referred to.

Beaumé's areometers are made in the following way: the one for liquids lighter than water sinks to zero in a solution of 1 part of common salt in 9 parts of water, and to 10 in pure water; the interval is divided into 10 equal parts, and 40 such degrees marked above. The other, for liquids heavier than water, sinks to zero in pure water, and to 15 in a solution of 15 parts of salt in 85 parts of water; the interval is divided into 15 equal parts, and 70 added below. The temperature taken is "the mean temperature of the air." On this account, and the impurity of the salt, these instruments are not very accurate, and scarcely indicate the specific gravity correctly to the second place of decimals. The table annexed is according to Francoeur.

The starting-point in Cartier's areometer is still more uncertain; the degrees are somewhat longer, and the whole scale stands $\frac{1}{4}°$ deeper. According to Francoeur, 16° Cartier = 15° Beaumé + 22.

Beck's areometer is at zero in pure water, and 30 at the specific gravity of 0·85, (water = 1;) the temperature is + 54·5° F.

[93] **Specific gravity areometers.**—The graduation of areometers for immediately indicating specific gravities was formerly very difficult, and for that reason the instrument was not in much favor. As, however, we can obtain the specific gravity from the volumeter scale, fig.

152, when in the equation $y = \frac{100}{x}$, we substitute for $x$ the various readings of the volumeter, so inversely we can from $x = \frac{100}{y}$ find the points of the volumeter scale $a\ b$ corresponding to the specific gravity; that is to say, the points of a scale divided into 80 parts, but bearing numbers from 50 to 130, where the point 50 corresponds to a specific gravity of 2·00, and 100 to 1·0; calculating in this way the divisions of the scale for the specific gravities 1·9, 1·8, etc. to 0·7, we shall find as parts of the scale 52·63, 55·55, 58·81, 62·50, 66·66, 71·43, 76·92, 83·33, 90·90, 100·00, 111·11, 125·00, 142·85. This table must, however, be calculated for the specific gravities to the second place of decimals. The proportional parts are transferred to the normal scale, having given to it such a length that, for gravities of from 2 to 1, the numbers must be carried out, at least to the second place of decimals, and from 1 to 0·7, to the third. This range is always divided between two areometers, constructed as directed for Tralles' alcoholometer.

*Fig.* 152.

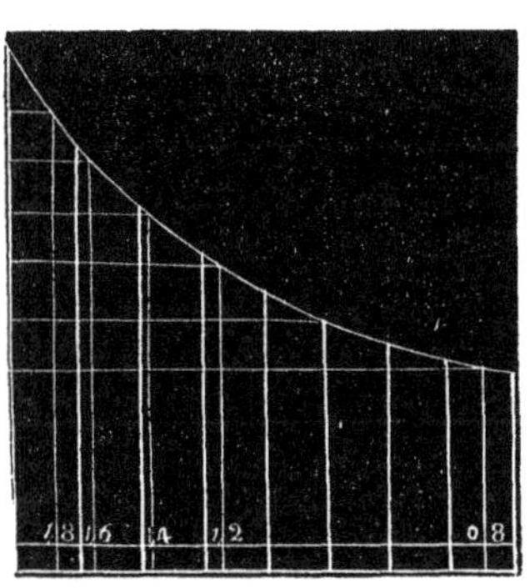

All this presupposes a perfectly cylindrical tube; but as this is never realized, these areometers will not give the specific gravity exactly, unless constructed according to the very elaborate process of Brisson and Schmidt. The same difficulty exists in reference to all other areometer scales, that of the volumeter not excepted. With a little care in the selection, tubes good enough for ordinary use may be procured, and those who require absolute accuracy, will always have recourse to the balance.

The following table contains the distances of the divisions for specific gravity areometers, and those of Tralles. In the first table, the length of the scale, that is, the base of the isosceles triangle, fig. 148, or the length of the strip on $e\ f$, fig. 149, is considered as divided into 5000 parts for liquids heavier than water, and into 4285 for those lighter. The table then indicates at what point of this division any specific gravity occurs. In the latter, the graduation is possible for the third place of decimals; the interval, however, is divided into equal parts only, as no error worth noticing can occur. The table itself shows this, for it contains the divisions to the third place, for the extreme case, specific gravity 0·70 to 0·71.

## SCALE OF 5000 PARTS.

| Specific gravity. | Corresponding division. | Specific gravity. | Corresponding division. | Specific gravity. | Corresponding division. |
|---|---|---|---|---|---|
| 2·00 | 0 | 1·66 | 1024 | 1·32 | 2575 |
| 1·99 | 25 | 1·65 | 1060 | 1·31 | 2633 |
| 1·98 | 50 | 1·64 | 1097 | 1·30 | 2692 |
| 1·97 | 76 | 1·63 | 1134 | 1·29 | 2752 |
| 1·96 | 102 | 1·62 | 1172 | 1·28 | 2812 |
| 1·95 | 128 | 1·61 | 1211 | 1·27 | 2874 |
| 1·94 | 154 | 1·60 | 1250 | 1·26 | 2936 |
| 1·93 | 181 | 1·59 | 1289 | 1·25 | 3000 |
| 1·92 | 208 | 1·58 | 1329 | 1·24 | 3064 |
| 1·91 | 235 | 1·57 | 1369 | 1·23 | 3130 |
| 1·90 | 263 | 1·56 | 1410 | 1·22 | 3196 |
| 1·89 | 290 | 1·55 | 1451 | 1·21 | 3262 |
| 1·88 | 319 | 1·54 | 1493 | 1·20 | 3333 |
| 1·87 | 347 | 1·53 | 1535 | 1·19 | 3403 |
| 1·86 | 376 | 1·52 | 1578 | 1·18 | 3474 |
| 1·85 | 405 | 1·51 | 1622 | 1·17 | 3547 |
| 1·84 | 432 | 1·50 | 1666 | 1·16 | 3620 |
| 1·83 | 464 | 1·49 | 1711 | 1·15 | 3695 |
| 1·82 | 494 | 1·48 | 1756 | 1·14 | 3771 |
| 1·81 | 524 | 1·47 | 1802 | 1·13 | 3849 |
| 1·80 | 555 | 1 46 | 1849 | 1·12 | 3928 |
| 1·79 | 586 | 1·45 | 1896 | 1·11 | 4009 |
| 1·78 | 617 | 1·44 | 1944 | 1·10 | 4090 |
| 1·77 | 649 | 1·43 | 1993 | 1·09 | 4174 |
| 1·76 | 681 | 1·42 | 2042 | 1·08 | 4259 |
| 1·75 | 714 | 1·41 | 2092 | 1·07 | 4345 |
| 1·74 | 747 | 1·40 | 2143 | 1·06 | 4433 |
| 1·73 | 780 | 1·39 | 2194 | 1·05 | 4523 |
| 1·72 | 813 | 1·38 | 2246 | 1·04 | 4615 |
| 1·71 | 847 | 1·37 | 2299 | 1·03 | 4708 |
| 1·70 | 882 | 1·36 | 2352 | 1·02 | 4803 |
| 1·69 | 917 | 1·35 | 2407 | 1·01 | 4900 |
| 1·68 | 952 | 1·34 | 2461 | 1·00 | 5000 |
| 1·67 | 988 | 1·33 | 2518 | ...... | ...... |

## SCALE OF 4285 PARTS

| Specific gravity. | Corresponding division. | Specific gravity. | Corresponding division. | Specific gravity. | Corresponding division. |
|---|---|---|---|---|---|
| 1·00 | 0 | 0·86 | 1628 | 0·73 | 3698 |
| 0·99 | 101 | 0·85 | 1764 | 0·72 | 3888 |
| 0·98 | 204 | 0·84 | 1903 | 0·71 | 4084 |
| 0 97 | 309 | 0·83 | 2048 | 0·709 | 4104 |
| 0 96 | 416 | 0·82 | 2195 | 0·708 | 4124 |
| 0·95 | 526 | 0·81 | 2345 | 0·707 | 4144 |
| 0·94 | 638 | 0·80 | 2500 | 0·706 | 4164 |
| 0·93 | 752 | 0·79 | 2658 | 0·705 | 4184 |
| 0·92 | 869 | 0·78 | 2820 | 0·704 | 4204 |
| 0·91 | 989 | 0·77 | 2989 | 0·703 | 4224 |
| 0·90 | 1111 | 0·76 | 3157 | 0·702 | 4245 |
| 0 89 | 1236 | 0·75 | 3333 | 0·701 | 4265 |
| 0 88 | 1363 | 0·74 | 3513 | 0·700 | 4285 |
| 0·87 | 1493 | ...... | ...... | ...... | ..... |

## TABLE FOR TRALLES' HYDROMETER.

| Degrees. | Division on a scale of 2597 parts. | Specific gravity. Water at +4° C. = 10000. | Degrees. | Division on a scale of 2597 parts. | Specific gravity. Water at +4° C. = 10000. |
|---|---|---|---|---|---|
| 0 | 9 | 9991 | 51 | 735 | 9315 |
| 1 | 24 | 9976 | 52 | 758 | 9295 |
| 2 | 39 | 9961 | 53 | 782 | 9275 |
| 3 | 54 | 9947 | 54 | 806 | 9254 |
| 4 | 68 | 9933 | 55 | 830 | 9234 |
| 5 | 82 | 9919 | 56 | 854 | 9213 |
| 6 | 95 | 9906 | 57 | 879 | 9192 |
| 7 | 108 | 9893 | 58 | 905 | 9170 |
| 8 | 121 | 9881 | 59 | 931 | 9148 |
| 9 | 133 | 9869 | 60 | 957 | 9126 |
| 10 | 145 | 9857 | 61 | 984 | 9104 |
| 11 | 157 | 9845 | 62 | 1011 | 9082 |
| 12 | 169 | 9834 | 63 | 1039 | 9059 |
| 13 | 180 | 9823 | 64 | 1067 | 9036 |
| 14 | 191 | 9812 | 65 | 1096 | 9013 |
| 15 | 202 | 9802 | 66 | 1125 | 8989 |
| 16 | 213 | 9791 | 67 | 1154 | 8965 |
| 17 | 224 | 9781 | 68 | 1184 | 8941 |
| 18 | 235 | 9771 | 69 | 1215 | 8917 |
| 19 | 245 | 9761 | 70 | 1246 | 8892 |
| 20 | 256 | 9751 | 71 | 1278 | 8867 |
| 21 | 266 | 9741 | 72 | 1310 | 8842 |
| 22 | 277 | 9731 | 73 | 1342 | 8817 |
| 23 | 288 | 9720 | 74 | 1375 | 8791 |
| 24 | 299 | 9710 | 75 | 1409 | 8765 |
| 25 | 310 | 9700 | 76 | 1443 | 8739 |
| 26 | 321 | 9689 | 77 | 1478 | 8712 |
| 27 | 332 | 9679 | 78 | 1514 | 8685 |
| 28 | 344 | 9668 | 79 | 1550 | 8658 |
| 29 | 355 | 9657 | 80 | 1587 | 8631 |
| 30 | 367 | 9646 | 81 | 1624 | 8603 |
| 31 | 380 | 9634 | 82 | 1662 | 8575 |
| 32 | 393 | 9622 | 83 | 1701 | 8547 |
| 33 | 407 | 9609 | 84 | 1740 | 8518 |
| 34 | 420 | 9596 | 85 | 1781 | 8488 |
| 35 | 434 | 9583 | 86 | 1823 | 8458 |
| 36 | 449 | 9570 | 87 | 1866 | 8428 |
| 37 | 465 | 9556 | 88 | 1910 | 8397 |
| 38 | 481 | 9541 | 89 | 1955 | 8365 |
| 39 | 498 | 9526 | 90 | 2002 | 8332 |
| 40 | 515 | 9510 | 91 | 2050 | 8299 |
| 41 | 533 | 9494 | 92 | 2099 | 8265 |
| 42 | 551 | 9478 | 93 | 2150 | 8230 |
| 43 | 569 | 9461 | 94 | 2203 | 8194 |
| 44 | 588 | 9444 | 95 | 2259 | 8157 |
| 45 | 608 | 9427 | 96 | 2318 | 8118 |
| 46 | 628 | 9409 | 97 | 2380 | 8077 |
| 47 | 648 | 9391 | 98 | 2447 | 8034 |
| 48 | 669 | 9373 | 99 | 2519 | 7988 |
| 49 | 690 | 9354 | 100 | 2597 | 7939 |
| 50 | 712 | 9335 | ..... | ...... | ...... |

## TABLE OF SPECIFIC GRAVITIES,

According to Beaume's and Beck's Hydrometers, for liquids heavier than water, at the temperature of 54·5° F.

| Degrees. | Beaumé. | Beck. | Degrees. | Beaumé. | Beck. |
|---|---|---|---|---|---|
| 0 | 1·0000 | 1·0000 | 39 | 1·3451 | 1·2977 |
| 1 | 1·0066 | 1·0059 | 40 | 1·3571 | 1·3077 |
| 2 | 1·0133 | 1·0119 | 41 | 1·3694 | 1·3178 |
| 3 | 1·0201 | 1·0180 | 42 | 1·3818 | 1·3281 |
| 4 | 1·0270 | 1·0241 | 43 | 1·3945 | 1·3386 |
| 5 | 1·0340 | 1·0303 | 44 | 1·4074 | 1·3492 |
| 6 | 1·0411 | 1·0336 | 45 | 1·4206 | 1·3600 |
| 7 | 1·0483 | 1·0429 | 46 | 1·4339 | 1·3710 |
| 8 | 1·0556 | 1·0494 | 47 | 1·4476 | 1·3821 |
| 9 | 1·0630 | 1·0559 | 48 | 1·4615 | 1·3934 |
| 10 | 1·0704 | 1·0625 | 49 | 1·4758 | 1·4050 |
| 11 | 1·0780 | 1·0692 | 50 | 1·4902 | 1·4167 |
| 12 | 1·0857 | 1·0759 | 51 | 1·4951 | 1·4286 |
| 13 | 1·0935 | 1·0828 | 52 | 1·5200 | 1·4407 |
| 14 | 1·1014 | 1·0897 | 53 | 1·5353 | 1·4530 |
| 15 | 1·1095 | 1 0968 | 54 | 1·5510 | 1·4655 |
| 16 | 1·1176 | 1·1039 | 55 | 1·5671 | 1·4783 |
| 17 | 1·1259 | 1·1111 | 56 | 1·5833 | 1·4912 |
| 18 | 1·1343 | 1·1184 | 57 | 1·6000 | 1·5044 |
| 19 | 1·1428 | 1·1258 | 58 | 1·6170 | 1·5179 |
| 20 | 1·1515 | 1·1333 | 59 | 1·6344 | 1·5315 |
| 21 | 1·1603 | 1·1409 | 60 | 1·6522 | 1·5454 |
| 22 | 1·1692 | 1·1486 | 61 | 1·6705 | 1·5596 |
| 23 | 1·1783 | 1·1565 | 62 | 1·6889 | 1·5741 |
| 24 | 1·1857 | 1·1644 | 63 | 1·7079 | 1·5888 |
| 25 | 1·1968 | 1·1724 | 64 | 1·7273 | 1·6038 |
| 26 | 1·2063 | 1·1806 | 65 | 1·7471 | 1·6190 |
| 27 | 1·2160 | 1·1888 | 66 | 1·7674 | 1·6346 |
| 28 | 1·2258 | 1·1972 | 67 | 1·7882 | 1·6505 |
| 29 | 1·2358 | 1·2057 | 68 | 1·8095 | 1·6667 |
| 30 | 1·2459 | 1·2143 | 69 | 1·8313 | 1·6832 |
| 31 | 1·2562 | 1·2230 | 70 | 1·8537 | 1·7000 |
| 32 | 1·2667 | 1·2319 | 71 | 1·8765 | 1·7172 |
| 33 | 1·2773 | 1·2409 | 72 | 1·9000 | 1·7347 |
| 34 | 1·2881 | 1·2500 | 73 | 1·9241 | 1·7526 |
| 35 | 1·2992 | 1·2593 | 74 | 1·9487 | 1·7708 |
| 36 | 1·3103 | 1·2687 | 75 | 1·9740 | 1·7895 |
| 37 | 1·3217 | 1·2782 | 76 | 2·0000 | 1·8085 |
| 38 | 1·3333 | 1·2879 | ... | ......... | ......... |

## TABLE OF SPECIFIC GRAVITIES,

**According to Beaume's, Beck's, and Cartier's Hydrometers, for liquids lighter than water, at the temperature of 54·5° F.**

| Degrees. | Beaumé. | Cartier. | Beck. | Degrees. | Beaumé. | Cartier. | Beck. |
|---|---|---|---|---|---|---|---|
| 0 | ......... | ......... | 1·0000 | 31 | 0·8742 | 0·8707 | 0·8457 |
| 1 | ......... | ......... | 0·9941 | 32 | 0·8690 | 0·8652 | 0·8415 |
| 2 | ......... | ......... | 0·9883 | 33 | 0·8639 | 0·8598 | 0·8374 |
| 3 | ......... | ......... | 0·9826 | 34 | 0·8588 | 0·8545 | 0·8333 |
| 4 | ......... | ......... | 0·9770 | 35 | 0·8538 | 0·8491 | 0·8292 |
| 5 | ......... | ......... | 0·9714 | 36 | 0·8488 | 0·8439 | 0·8252 |
| 6 | ......... | ......... | 0·9659 | 37 | 0·8439 | 0·8387 | 0·8212 |
| 7 | ......... | ......... | 0·9604 | 38 | 0·8391 | 0·8336 | 0·8173 |
| 8 | ......... | ......... | 0·9550 | 39 | 0·8343 | 0·8286 | 0·8133 |
| 9 | ......... | ......... | 0·9497 | 40 | 0·8295 | ......... | 0·8095 |
| 10 | 1·0000 | ......... | 0·9444 | 41 | 0·8249 | ......... | 0·8061 |
| 11 | 0·9932 | 1·0000 | 0·9392 | 42 | 0·8202 | ......... | 0·8018 |
| 12 | 0·9865 | 0·9922 | 0·9340 | 43 | 0·8156 | ......... | 0·7981 |
| 13 | 0·9799 | 0·9846 | 0·9289 | 44 | 0·8111 | ......... | 0·7944 |
| 14 | 0·9733 | 0·9764 | 0·9239 | 45 | 0·8066 | ......... | 0·7907 |
| 15 | 0·9669 | 0·9695 | 0·9189 | 46 | 0·8022 | ......... | 0·7871 |
| 16 | 0·9605 | 0·9627 | 0·9139 | 47 | 0·7978 | ......... | 0·7834 |
| 17 | 0·9542 | 0·9560 | 0·9090 | 48 | 0·7935 | ......... | 0·7799 |
| 18 | 0·9480 | 0·9493 | 0·9042 | 49 | 0·7892 | ......... | 0·7763 |
| 19 | 0·9420 | 0·9427 | 0·8994 | 50 | 0·7849 | ......... | 0·7727 |
| 20 | 0·9359 | 0·9363 | 0·8947 | 51 | 0·7807 | ......... | 0·7692 |
| 21 | 0·9300 | 0·9299 | 0·8900 | 52 | 0·7766 | ......... | 0·7658 |
| 22 | 0·9241 | 0·9237 | 0·8854 | 53 | 0·7725 | ......... | 0·7623 |
| 23 | 0·9183 | 0·9175 | 0·8808 | 54 | 0·7684 | ......... | 0·7589 |
| 24 | 0·9125 | 0·9114 | 0·8762 | 55 | 0·7643 | ......... | 0·7556 |
| 25 | 0·9068 | 0·9054 | 0·8717 | 56 | 0·7604 | ......... | 0·7522 |
| 26 | 0·9012 | 0·8994 | 0·8673 | 57 | 0·7565 | ......... | 0·7489 |
| 27 | 0·8957 | 0·8935 | 0·8629 | 58 | 0·7526 | ......... | 0·7456 |
| 28 | 0·8902 | 0·8877 | 0·8585 | 59 | 0·7487 | ......... | 0·7423 |
| 29 | 0·8848 | 0·8820 | 0·8542 | 60 | 0·7449 | ......... | 0·7391 |
| 30 | 0·8795 | 0·8763 | 0·8500 | ... | ......... | ......... | ......... |

[94] **Capillarity.**—For illustrating the fundamental facts of capillarity, bent glass tubes may be employed, as seen in fig. 153, with the wider limb half an inch in diameter in the clear; the capillary tubes united to them should be of different degrees of fineness, from two lines to that of a thermometer tube, and in series of about four for water and as many for mercury. Those belonging to a set may be fastened to a board having a graduation in inches and tenths for each tube. For the watery liquids, a deep blue, as a solution of ammoniacal oxyde of copper or of indigo will be best: whichever be employed, care must be taken to wash out the tubes after using, as otherwise the residuum of the solution, on drying, will choke up the tubes, and greatly interfere with subsequent experiments. The mercurial arrangement should be kept filled. With liquids which wet the glass, the tubes should be inclined toward the smaller limb, so as to wet the glass for some distance up, otherwise the greatest elevation will not be attained.

*Fig.* 153.

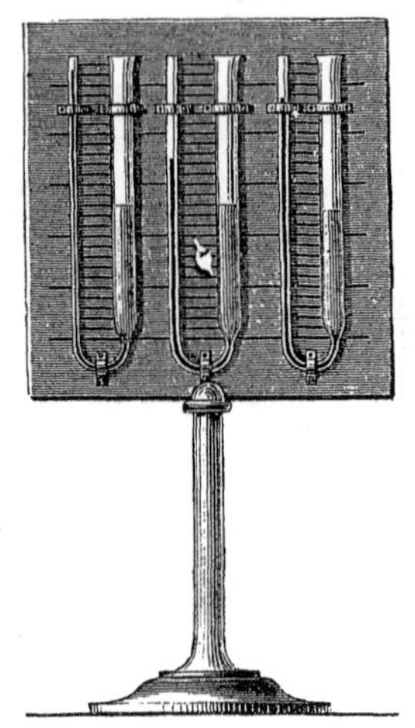

For experimenting with glass plates, take two pairs cut from plate glass, from 3 to 4 inches long, and from 2 to 3 broad, and grind off their edges. Four pieces of thin sheet brass, of uniform thickness, are then laid at the four corners, between one of these pairs, to make parallel plates.

Four other pieces of sheet brass, two of them the thinnest procurable, the others 1 or 2 lines thick, will enable us to give the remaining plates an inclination to each other; each pair may then be bound around several times with waxed thread. The same liquids may be employed for experimenting as recommended for the tubes. After using, the plates must be taken apart, well washed, and combined again, in readiness for the next experiment. Absolute cleanliness is required with these plates, as when only slightly soiled the outlines of the liquid will not be straight in the parallel pair, nor hyperbolic in the inclined. It is best to immerse the plates entirely in the liquid, in order to wet them inside.

Bulbs, from $\frac{1}{4}$ to $\frac{1}{2}$ inch in diameter, made of wax and glass, will conveniently illustrate the experiment of attraction and repulsion of floating bodies. There is no difficulty in performing the experiment in water, but in mercury the glass bulbs do not sink deep enough. This difficulty may be obviated by half filling the bulbs with mercury, and then melting over the aperture. The surface of the mercury, in performing the experiment, must be perfectly clean.

*Fig.* 154. *Fig.* 155. *Fig.* 156.

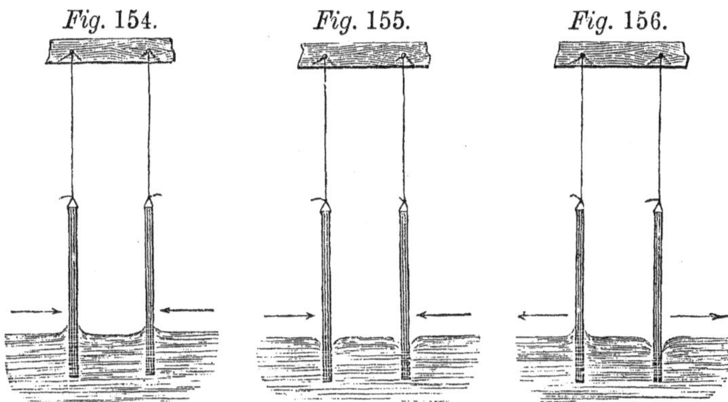

For the experiments on attraction, between immersed plates, take thin plates of glass or mica, attached to threads by means of wax, figs. 154, 155, 156; if they are not designed to be wetted, they need only have a film of grease on the surface.

*Fig.* 157.

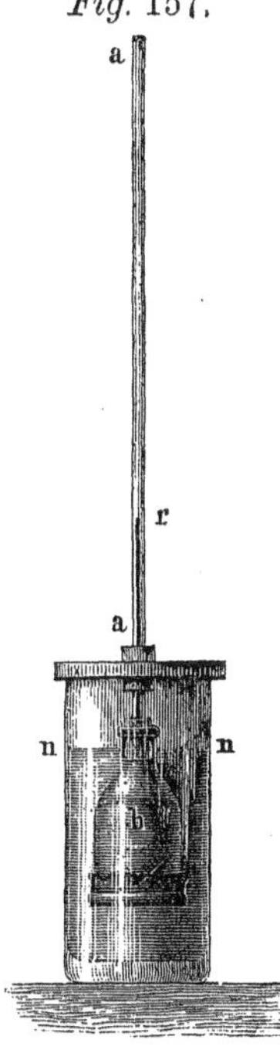


To make pins, needles, etc. float upon water, it is only necessary to draw them a few times between the fingers; in this way they get a thin coating of grease, which protects them from the water. When carefully laid on, they will exhibit the same relation, both toward each other and the sides of the vessel, as the floating balls.

[95] **Endosmosis.**—The inner vessel, *b*, fig. 157, should have a flange at its wider opening, so as to admit of tying on a piece of bladder securely. A vessel of lacquered tin will answer for some experiments, but it is better to procure at once a suitable glass vessel. The narrow tube may be of about the size of that used for barometers, and is inserted into the neck by means of a cork. The inner vessel may be supported by passing the tube through a disk of wood, which covers the exterior vessel. Both vessels should be filled to the same

height, the inner with a concentrated solution of sulphate of copper or alcohol colored with cochineal; the outer with water.

For an experiment with sulphuric acid and water, the clay cells used in the constant galvanic battery will be found very convenient. All that is necessary is to place one in a glass vessel and fill it with one of the liquids, pouring the other round it in the glass.

## (*c.*) EXPERIMENTS ON GASES.

[96] **Torricelli's experiment.**—For the preliminary explanation of this experiment, pour mercury into a rather high glass cylinder, and introduce a glass tube, open at both ends and just wide enough for the effect of capillarity to disappear. Pour water into the cylinder, and its pressure will cause the mercury to rise in the inner tube. As this will be the case with all fluids, the mercury must rise in an empty tube when the external pressure is that of the atmosphere.

Again, take a glass tube, 36 inches in length and about $\frac{1}{6}$ inch in diameter, hermetically sealed at one end. It must be kept perfectly clean, and when not in use the opening should be closed by a cork. The mercury must be perfectly clean: should it leave any impurity in the tube, a new one must be taken, or the sealed end opened and the tube cleaned by wiping it out with pieces of soft blotting-paper on the end of a wire. When barometer tubes are dirty they retain bubbles of air, which rise in the Torricellian vacuum, and thus interfere with the result. The cistern should be wide enough to permit the immersion of the open end, with the finger applied to prevent the escape of the mercury. These tubes should be left full; the tube is inclined until the mercury touches the top, the bottom closed with the finger and the tube inverted; in this position it is hung up by a waxed string tied tightly around it.

The tube is filled through a stiff paper funnel, as directed in § 76, or the mercury must at least be passed through such a funnel, to remove dust and oxydized metal, before being introduced into the tube. The mercury can be cleaned still better by agitation with warm dilute nitric acid, and afterwards with water; it is then dried with blotting-paper. To make the experiment more accurate, the tube and mercury should both be warmed, and the mercury poured in through a long, narrow funnel tube, drawn out to a fine point. When the tube has been filled to within about $\frac{1}{2}$ an inch of the top, it is closed with the finger and inverted several times; the air which occupied the open space, in traversing the tube gathers the minute vesicles which adhered to the sides. The tube is next filled entirely, and inverted in the cistern of mercury, with the finger over the aperture. The experiment should also be made with tubes shorter than 30 inches.

The mercury used in these experiments should be kept for this special purpose; that used for mercurial troughs or for electrical purposes is usually alloyed with other metals.

Another experiment which belongs here is the following: A glass with ground edge is filled full of water, and a sheet of paper laid over it so as to inclose no air bubbles. If a board or a book be now laid on it, the glass may be inverted, and held up in an inverted position without spilling the water. If the glass be set on the table in this position, the paper may be withdrawn, another sheet slipped under it, the glass then slid upon the board again and turned upright. A sheet of good stiff paper, not much larger than the glass, should be used to insure success.

[97] **The barometer.**—Barometers are among the instruments which it is better to buy than to construct for one's self. Besides the art of filling them so as to be perfectly free from air, a good dividing machine is required for the scale.

It will be best to have, for purposes of demonstration, in addition to the cistern barometer, a simple siphon barometer, with the shorter leg contracted near the bend, to admit of being closed up by a rod wrapped with cotton. If possible, it will be best to have a good mountain barometer, to be used for scientific purposes. The characteristics of a good barometer will be found in any text-book of physics.

Should it, however, on account of the fracture of the tube of a valuable barometer, or for any other reason, be necessary to fit up a barometer tube, one should be chosen not too thick in the glass, and carefully cleaned before sealing it up.

The required amount of mercury is then boiled in a porcelain capsule; the tube itself heated along its entire length, and then filled through a funnel, for about two inches, with the hot mercury. The bottom of the tube is now heated over a spirit-lamp until the mercury boils, taking care to turn the tube round and round continually; this is to be kept nearly vertical, and the operation should be performed over a large flat capsule, to prevent the loss of mercury in any possible breakage of the tube. With proper precaution, however, there is little danger of this. The tube must be heated anew along its entire length, and a few additional inches of mercury introduced, the same operation of heating being repeated; the heat should be applied a little below each fresh portion of mercury. When the tube has been filled in this way to within half an inch of the top, hot mercury is poured into the remaining space. A paper cornet, cut off at the bottom, should previously be attached to the top of the cistern, by means of paste and thread, serving as a funnel for the mercury which is poured in, filling the cistern and part of the cornet. The finger is now placed on the filled tube, which is next inverted into the cistern. After removing the funnel or cornet, the excess of mercury is poured out of the cistern; enough, however, being left to have its level correspond with the zero of the attached scale. The tube may be finally fixed in the neck of the cistern by bits of cork, or by an entire cork previously fitted on the tube; this must, however, have a small aperture to admit the air.

In observing the barometer, the mercury must be gently agitated to obtain the proper level, even with a tolerably wide tube. The level of the top of the convexity of the mercury must be observed and the eye held in the same level, to avoid the error of parallax. This is very easily done by holding a piece of mirror behind the tube, and placing the eye so that the top of the column coincides with that of the image. The barometer must hang perpendicularly during the reading.

After purchasing a cistern barometer, it should always be carefully measured, to ascertain whether the parts of the scale which are marked are at the right distance from the level of the mercury; if not, the error should be corrected by adding or taking away mercury.

In removing a barometer, it must be gently inclined until the mercury fills the whole tube, and the cistern then closed if there is any provision for this. This inclination will show whether the tube is free from air, or whether an air bubble remains at the top. For distant transportation it is always completely inverted, and the opening for the admission of air previously closed. In setting it up again, care must be taken that the mercury in the cistern shall unite with that in the tube before the latter begins to move, so that no air shall be introduced into the tube.

*Fig.* 158.

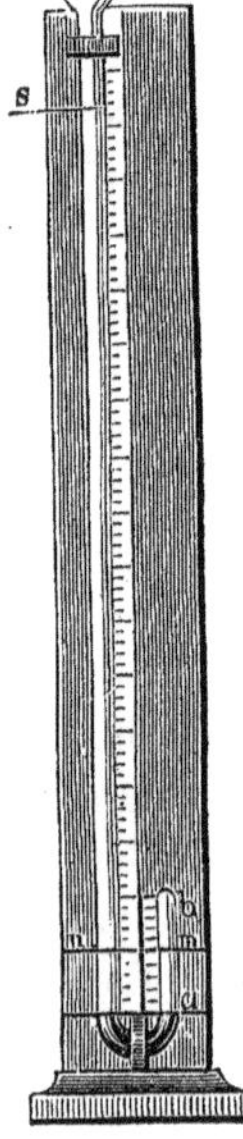


As Huygens's double barometer and wheel barometer are frequently seen, it will be well to take advantage of an opportunity to obtain them cheaply for the purpose of illustration, though they are not used for scientific purposes.

[98] **Mariotte's law.**—To illustrate this law for condensation, we make use of a tube hermetically sealed at one end and bent round, as in fig. 158. This tube is fastened to a board by strips of brass, after a strip of paper, divided into inches and tenths, has been attached; this is to be numbered from below upward, and varnished with white shellac. The tube need not be more than 30 or 36 inches long, as the law can be shown by calculation,

even when the pressure does not amount to entire atmospheres. Longer tubes may, however, be made, if deemed necessary, by uniting several joints by means of iron ferules an inch long.

In experimenting, enough mercury is introduced to fill the bend and stand at the same height in both legs; this may require a continued tilting of the tube one way or other, to let air in or out of the short leg. When accomplished, this height is marked. Mercury is now poured into the open end, up to about $s$; the mercury of the short leg by advancing to $m$ will compress the air to a volume, $b\ m$, and if $B$ be the height of the barometer at the moment, then $a\ b : m\ b = B + n\ s : B$. If the tube be long enough the pressure may be increased by whole atmospheres, and the volume of air reduced successively to $\frac{1}{2}$, $\frac{1}{3}$, $\frac{1}{4}$, etc., for which reason it is convenient to have the original volume a number of inches divisible by 2 and 3.

To demonstrate the same law for rarefaction, we may use a glass tube about 30 inches high and two wide, closed at one end and funnel shaped at the other. Such a tube can readily be obtained at any glass house. It should be cemented into a solid wooden foot, to guard against breakage, or, still better, placed in a frame, as in fig. 159, (in section.) The enlargement above is not absolutely necessary; it is, however, convenient, that the level of the mercury may not change essentially during the experiment. It can easily be made by cutting off the bottom of a bottle and cementing the tube into the neck. Any tube of sufficient width will answer for the experiment, even without being fused together at one end, if it be cemented tightly into the hole in the bottom of the frame, fig. 159. The narrow glass tube used is ground flat on one side and there divided off into inches, the mark being made with indian-ink. In the experiment, the large tube is filled with mercury, and the narrow one inserted so far that only a few inches project; the end is then closed with the finger, and the tube drawn out until the contained air is made again to occupy any given whole number of inches, when the height of the column of mercury in the inner tube is measured. The experiment may also be made with a tube closed at one end, by partially filling it with mercury.

*Fig.* 159.

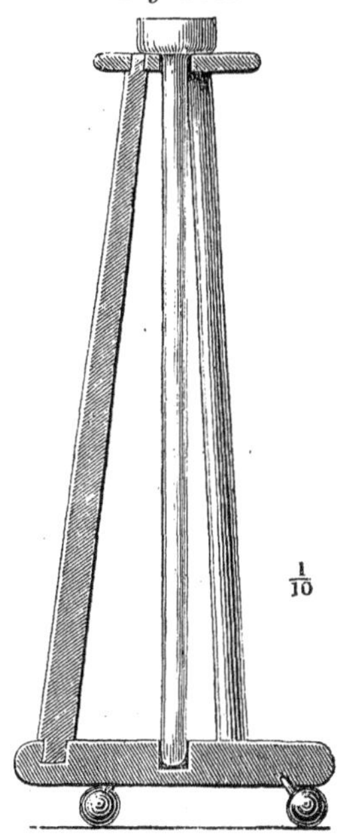

$\frac{1}{10}$

In experiments, both for condensation and rarefaction, the same scale must be used as that of the barometer which may be on hand, as the indications of the latter instrument must be introduced, and complicated reductions will thus be avoided. This should be contracted for when an instrument is ordered from the maker.

[99] **The air-pump.**—This instrument is of the first importance, both to the amateur and to the teacher, and, at the very outset, as good an instrument as the funds will allow should be procured. For the amateur as well as the chemist, a hand-pump, *i.e.* one without lever or guides, will answer most purposes, if well constructed; a few minutes more or less expended on an experiment make no difference. Fig. 160 shows such an air-pump, intended to be screwed on a table; the vessels to be exhausted may be connected by caoutchouc tubes with the arm *a* or *b*; or a glass tube, 30 inches long, reaching down into a vessel of mercury below the table, may be connected to *b*, to indicate the degree of exhaustion. It is a good plan to have such an air-pump furnished with a screw clamp by which it may be attached to the table. When the air-pump has no plate with its stop-cock, special arrangements will be necessary for the experiments to be described. The case is very different in the school, where a number of experiments must usually be gone through in a short time. In this case the instrument should be large, with double cylinders if possible; perfection of action is not so necessary, as experiments before classes rarely require any very great amount of rarefaction.

*Fig.* 160.

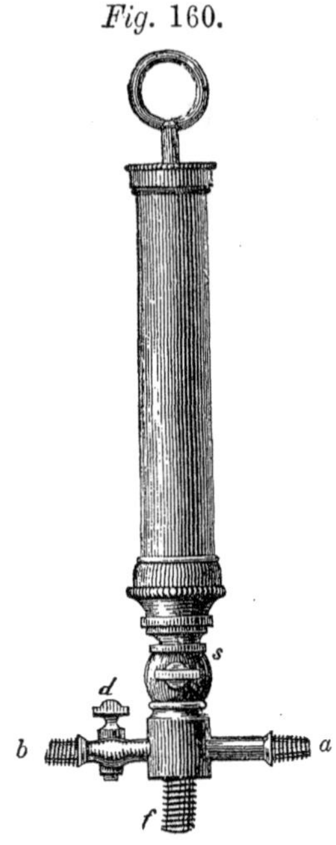


An important consideration in respect to the air-pump is the greater or less facility with which it can be taken to pieces and cleaned.

A new air-pump should always be tested by a good barometer, to see whether it performs all that is claimed for it, and whether it closes tightly. In respect to the first, we remark that the experiment should be tried either with no receiver at all, or else with a very small one. The second point is ascertained by exhausting the air to the proper degree, and then watching the barometer gauge to see whether any air penetrates.

[100] **Management of the air-pump.**—The air-pump, from its nature, requires very careful handling if it is to render good service

for a long time. Above all, cleanliness is necessary, and it should therefore be taken to pieces once a year and carefully cleaned throughout, after which fresh grease should be applied. This can be done most conveniently when the air-pump experiments for the course have been finished.

Experiments on boiling water in vacuo, and on freezing water, by the evaporation of ether, almost necessarily require the cleaning of the air-pump after them; on which account they should be performed immediately after the air-pump experiments proper. The cleaning is best done with blotting-paper, which may be at first moistened with alcohol; afterwards the piston and cocks should be greased afresh. The grease must be applied to the piston, and not to the cylinder: in the latter case the superfluous grease would be pushed to the bottom, and prevent the complete descent of the piston as well as stop up the tubes. For the pistons we may employ hog's lard; for the cocks, tallow, or a mixture of this with an equal quantity of lard. The piston rods and rack should be oiled. It will be well to keep the grease covered in a vessel, and especial care should be taken to have it entirely free from grit.

It is a good plan to pour a stratum of oil $\frac{1}{4}$ to $\frac{1}{2}$ an inch deep above the piston, which acts like the grease in the stuffing-box of a steam-engine. The pistons close better and work easier, but they must fit tight enough to prevent the oil from running down into the cylinders.

Should it be impossible to keep the air-pump in a glass-case, make a cover of pasteboard, to protect it from dust.

A good rule for keeping the air-pump in order is never to attempt moving the piston when the instrument is cold; thus in winter it should be kept in a warm room some time before use.

The edge of the receiver should be rubbed with a tallow candle, or a mixture of lard and beeswax, before being placed on the plate; well ground receivers, however, require no grease.

The motion of the lever must be slow at first, as the air in the narrow channels requires time to diffuse itself in the entire space; the audible hissing of the air, or the continued sinking of the barometer gauge, will indicate the rapidity of exhaustion; as rarefaction proceeds, the movement of the piston may be accelerated. Without this precaution much labor will be lost, and the trouble of making air-pump experiments greatly increased. Scholars, however, or unskillful assistants, do not possess the knack of moving the handle regularly, and irregular action is always very injurious to it.

A receiver should never be removed from the plate of the air-pump when the air within is in the least degree rarefied, without previously letting in the external air. Removal should be effected by sliding the receiver off the edge of the plate, rotating it slightly in the operation.

For every experiment, the smallest possible receiver should be selected. Very small receivers may be readily made by taking glass tumblers, with stout bottoms, and grinding off the edge.

[101] **Minor repairs of the air-pump.**—When the piston becomes loose by long usage, it may for a time be repaired by screwing up the movable disk on the lower face of the piston. The plugs of the stop-cocks have a tightening-screw, by means of which too great freedom of motion may readily be corrected. If, however, neither piston nor cocks can be tightened in the way mentioned, the instrument should be returned to the maker. It is only in case of absolute necessity, or of competent skill, that recourse should be had to fresh grinding of the plug in the hole, by means of oil and fine pumice.

The glass layer of the plate is sometimes loosened from the subjacent metal, in consequence of the unequal expansion of the two substances by heat. In this case, the plate is unscrewed, each part heated separately until the cement flows readily, the glass laid face downward on a table, protected by several layers of paper, and the metal tightly and accurately pressed down upon it, with a twisting, concentric motion. After cooling, the expressed cement is removed with a knife. In case of fresh cement being needed, it may be made of shellac, with $\frac{1}{6}$ of wax and the same quantity of turpentine.

If the air-pump is found to leak in any part covered with leather, this must be replaced with fresh. Chamois leather, saturated with a hot mixture of tallow and lard in equal parts, is best for this purpose. A small stock of such leather should be always kept on hand. If the leak occur where glass is cemented in, the part must be unscrewed, heated, the glass removed and freshly cemented with sealing-wax, care being taken to warm well both glass and metal, (section 45.)

[102] **Experiments with the air-pump.**—(1) A bladder partly filled with air and tied tightly expands when placed under the receiver of the air-pump and the air exhausted, which may be carried so far as to burst the bladder. This bladder, unless preserved under the receiver of the air-pump, is apt to be eaten through by insects. Small india-rubber bags may be used for the same purpose.

(2) The expansion of the air may be shown by a glass matrass, nearly filled with water and inverted in a tumbler of water, which is placed under the receiver, fig. 161. Bubbles of air are at the same time extricated from the water.

(3) The receiver is fastened on the plate by the pressure of the air.

Care is necessary to avoid injuring the air-pump by efforts to loosen the receiver.

(4) *The Magdeburg hemispheres.* The best way of exhibiting this experiment is to attach a weight to the lower hemisphere by a hook, or else to have the lower fastened to a board, and the ring of the upper to the short arm of a lever, at the opposite end of which weights may be attached. In this way the exact force required to pull apart the hemispheres can be estimated. The ground edges of the hemisphere should be coated with tallow. It should first be shown that the hemispheres may be easily separated when filled with air.

*Fig.* 161.

(5) *Bursting of bladders.* In this experiment, take a fresh bladder, well blown out, and tie it tightly over a cylinder or open receiver of metal or glass, 3 or 4 inches in diameter, ground on both edges. The upper end should have a bead around it to fasten the bladder better, fig. 162. Immediately before the experiment, the bladder must be thoroughly dried by a hot fire or stove; without this precaution, there is great danger of a failure in the experiment. A thin glass plate may also be broken, if level enough to fit air tight on the receiver; the hole in the plate must in this case, as well as in many others, be covered in such a way as to prevent the splinters from getting into the pipes, which may be done by laying a piece of felt inside the receiver. The communication with the barometer gauge, in such experiments, must be cut off, to prevent injury from the sudden entrance of air.

*Fig.* 162.

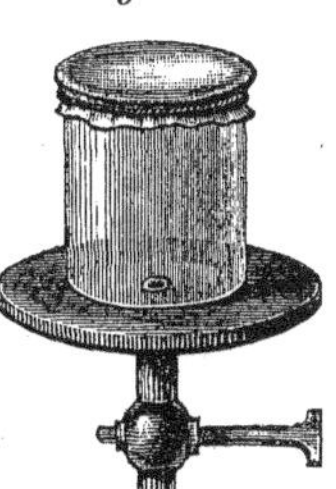

(6) The barometer, placed under the receiver of an air-pump, falls during the process of exhaustion. For this very instructive experiment, make a high, narrow receiver, fig. 163, out of a large glass funnel, and a

long glass tube, hermetically sealed at one end. A ferule of brass or tin, an inch or more long, is then procured, fitting over the ends of the two glass articles when in contact. By means of sealing-wax, a perfectly air-tight joint is made; the funnel must, of course, be ground off at the rim. For the barometer, may be used a Torricellian tube without scale, fig. 164, fitted into a bottle by a cork. All superfluous mercury must be poured out of the bottle before inserting the cork, to avoid the danger of running over when the mercury descends in the tube. Care must, of course, be taken to have the tube so narrow as not to hold too much. Another very good form for this experiment is shown in fig. 165. An air-pump receiver, open above, is fitted with a brass collar with two tubes: in one tube is fitted an ordinary siphon barometer, in the other a double bent tube with a vessel for mercury. As the air is exhausted, the mercury rises in the latter, while it sinks in the barometer. In this case, also, caution is necessary in admitting air.

*Fig.* 163. *Fig.* 164. *Fig.* 165.

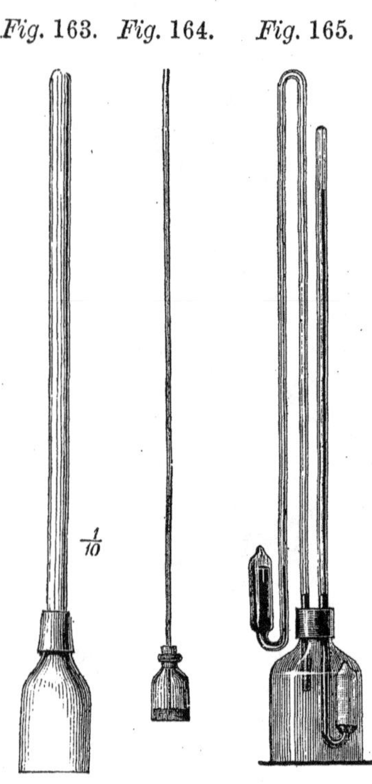


(7) Wood is specifically heavier than water. To show this by means of the air-pump, confine a block at the bottom of a tumbler of water by a weight. Place this under a receiver, exhaust the air, and then let it in again; the pores of the wood will thus be filled with water. The lightest wood will be found to remain at the bottom after removing the incumbent weight.

(8) Sound ceases under an exhausted receiver. The experiment may be tried by means of an alarm-clock placed under a receiver, or by means of a bell suspended from the top. In either case, the experiment will fail if the apparatus touch any of the solid parts of the pump. The alarm must, therefore, be set upon a loose bunch of fine hemp; or the bell suspended by an untwisted thread of the same material.

(9) Water boils at a low temperature, say 90° F., when the atmo-

spheric pressure is removed. To perform the experiment satisfactorily, make use of a small receiver, and a high, narrow vessel of water. With too large a receiver, the working of the pump will not carry off enough of the watery vapor, which of itself would keep the boiling point from sinking, and with a wide water vessel too much vapor will be formed. Ether evaporates still more readily. To show this, put some water into a narrow glass tube, sealed at one end, and drop in a small quantity of ether: invert the tube in a small vessel of water, and place the whole under the air-pump. On exhausting the air, the ether will vaporize and drive out all the water from the tube, to be again condensed on readmission of the air.

(10) To bring water to the freezing point by its evaporation, assisted by sulphuric acid or by the vaporization of ether, the broad, shallow vessel containing it must rest on pieces of cork, to prevent conduction of heat. The ether is best, but the experiment should be performed at the close of the course, as its vapor acts injuriously upon the grease of the air-pump. A rapid exhaustion will be necessary for success. A neat form of the experiment consists in setting a watch-glass filled with ether into another partly filled with water: the two glasses will be frozen together.

Strong acid must be used for the experiment, of which fig. 166 shows the details. The upper watch-glass containing the water is set into a ring, fastened by its stem to a board, on which rests the flat vessel containing the acid. As it is difficult to get acid of sufficient concentration to work satisfactorily, it will perhaps be best to adhere to the ether.

*Fig.* 166. *Fig.* 167.

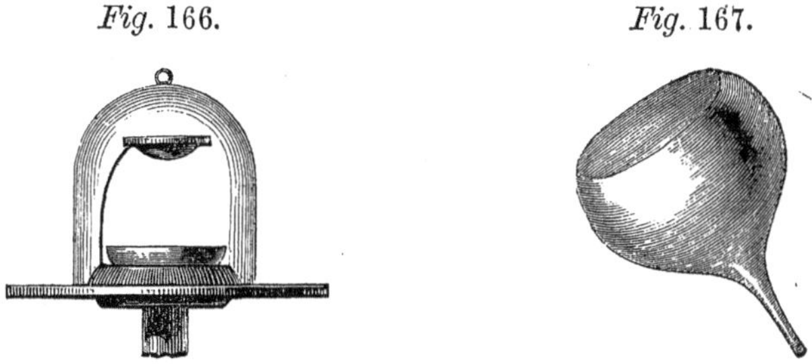

(11) The expansion of confined air will burst a vessel when the outward pressure is removed. For this purpose, we may best use thin, blown bulbs, fig. 167, of an inch in diameter, flattened at the bottom, and the stem sealed after cooling.

(12) To show that a light goes out in rarefied air, place a small

burning wax light in a receiver tall enough not to be affected by the heat. The light will be found to go out with a few strokes.

(13) Steel gives no sparks, and gunpowder burns without explosion in a vacuum. Fig. 168 explains the general features of the arrangement by which this can be readily shown. The lever arrangement *a* is set off by means of a rod sliding through a stuffing-box. Powder may be set on fire by means of a lens within an exhausted receiver; but it will be found to burn away slowly without explosion.

*Fig.* 168.

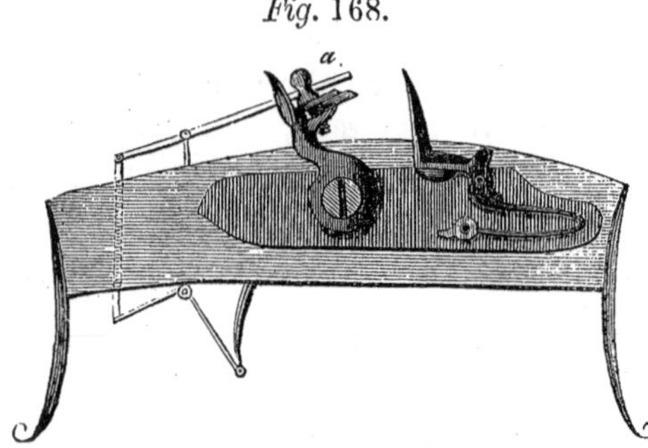


(14) The siphon will not act in a vacuum. For this experiment, construct a siphon out of a thermometer tube, as in fig. 169; the opening at *o* is closed with the finger while the tube is being filled by suction. The experiment can only be made with certainty with mercury. This is poured into a rather high glass vessel, and a lower one set by its side to receive the outflow. The opening in the tube at *o* may readily be made by closing one end and blowing strongly into the other, when the convexity at *o* is softened over a spirit-lamp.

*Fig.* 169. *Fig.* 170.

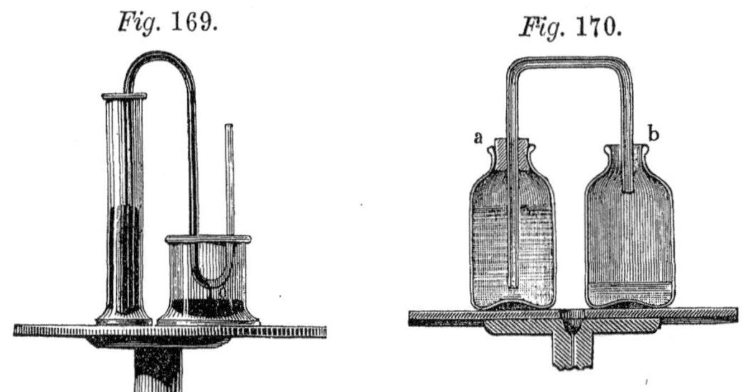


(15) If one leg of a doubly bent glass tube, open at both ends, be inserted, air tight, into a vessel half full of water, *a*, fig. 170, the other end dipping into *b*, and the whole placed under the air-pump, rarefaction outside will cause the air in *a* to expand and force the contained water

over into *b*. If the tube is vertical, a Hiero's fountain of the simplest construction will be obtained; if this latter be brought under the air-pump, care must be taken not to continue the experiment too long, else water will get into the pump, which would be injurious in other experiments.

(16) The following experiment is instructive for the theory of the suction-pump. A large tube, or a high, narrow receiver, *A*, fig. 171, is

*Fig.* 171.

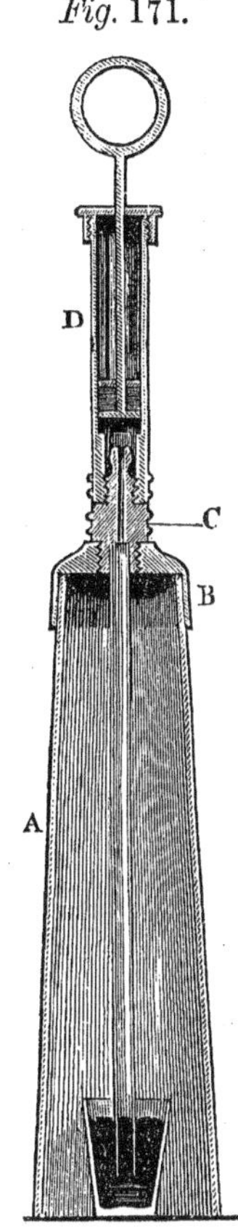


*Fig.* 172. *Fig.* 173. *Fig.* 174. *Fig.* 175.

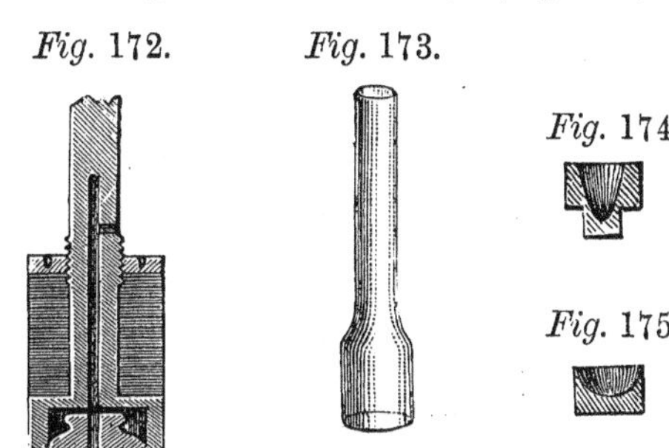

fitted with a brass cap *B*, into which the plug *C* is screwed; this has a tolerably wide aperture below, into which a stout glass tube, nearly as long as the receiver, is cemented; the rest of the plug is perforated with a fine hole. It has a flat head, on which a valve of oiled silk is fastened, similar to that described in No. 25 of this section. On *C* is screwed the small air-pump *D*, the piston of which has a similar valve, shown in natural size in fig. 172. A mercury cup is placed under the receiver, into which the tube dips. When the small pump is worked, the mercury will be drawn up, which will not take place when the air in the receiver is exhausted. One stroke of the pump raises the mercury sufficiently high; the valves may therefore be dispensed with and the pump screwed directly on the mounting *B*.

(17) A very simple apparatus for the quicksilver rain may be made out of a lamp chimney, fig. 173, with the lower end ground off. Into the other end is cemented a piece of walnut wood (shown in section in fig. 174;) a second piece, fig. 175, is fitted into the bottom, to receive the mercury as it falls through, and prevent even a drop from getting into the air-pump.

(18) *The balance manometer.* To one arm of a small balance,

fig. 176, provided with a support, is attached a large, hollow metal ball, soldered air tight, which is balanced by a solid ball on the other side. The balance must have an adjusting screw *a*, as the changes of pressure of the air disturb the equilibrium. In an exhausted receiver the large ball will sink, because it had been buoyed up by the air more than the solid one.

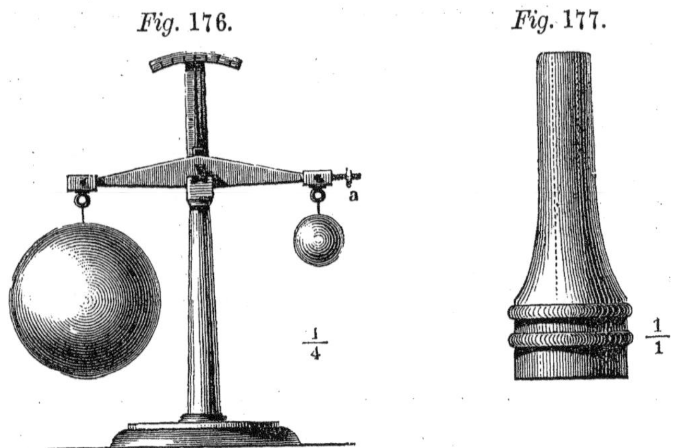

*Fig.* 176. *Fig.* 177.

(19) It frequently happens that air is to be rarefied in a vessel which can neither be screwed to the air-pump nor attached to the plate. For this the caoutchouc tubes, recently invented, are very convenient. To one end is fastened a brass tube, fig. 177, with double-milled head, to facilitate the screwing on to the pump.

(20) *Condensation of air.* If the air-pump is arranged for condensation as well as rarefaction, attention must be paid, at every experiment, to the cocks, for the purpose of ascertaining whether they be properly adjusted or not. If, for instance, the apparatus be adjusted for condensation, and a receiver be placed on the plate for rarefaction, the adhesion between the two will be sufficient to retain the receiver, which, at the second stroke, may be thrown entirely off. The receivers, in which the air is compressed, should be made of glass, $\frac{1}{4}$ of an inch thick, so that it will not be necessary to cover them with wire gauze, which nearly destroys their transparency. It is better to have a separate plate for this purpose, to which the receiver may be screwed down, as in fig. 178.

(21) A manometer of the simplest character may be constructed out of a glass tube closed at one end, and of nearly equal diameter throughout; this is divided into inches and tenths, by means of ink or scratches,

and then inserted into a vessel with mercury, by means of cork or leather. It is not advisable to use the air-pump for condensing to more than two atmospheres.

*Fig.* 178.

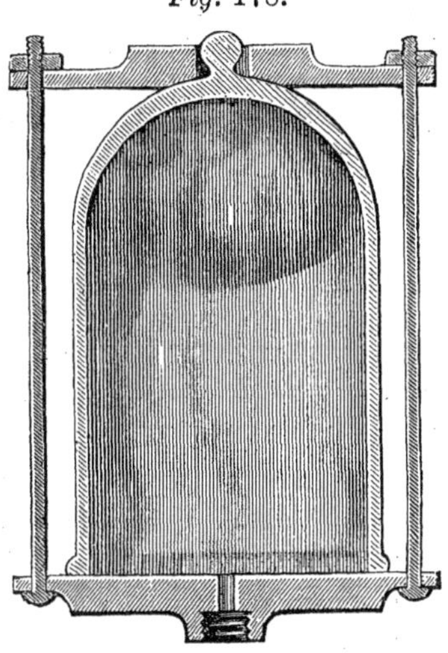

*Fig.* 179.

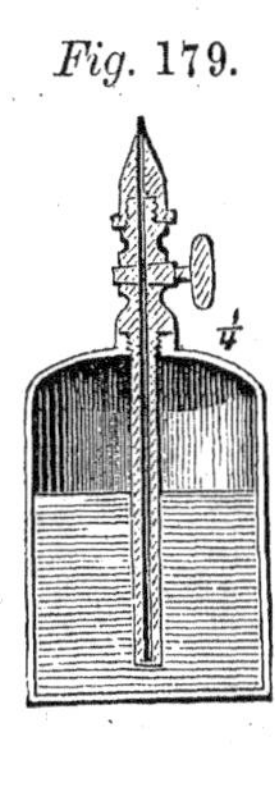

(22) If air be forced into a Hiero's fountain, the escape pipe of which is provided with a stop-cock, as in fig. 179, jets of various kinds may be screwed on after removing it from the air-pump. The metal of which this fountain is made must be at least 1 millimeter thick, and will then bear, with the dimensions assigned, a pressure of two atmospheres, if the seams are hard soldered.

(23) A flattened glass bulb, such as described in No. 8, is crushed in the condenser.

(24) If the inside of the condenser be moistened and the compressed air allowed to escape suddenly, the whole receiver will be filled with vapor, owing to the rapid lowering of the temperature.

(25) Determinations of the specific gravity of gases will scarcely form part of an experimental course. For this it will be sufficient to show the difference in weight between a glass balloon filled with air and the same emptied; and again, when filled with condensed air. A glass tube may be used to exhibit the same experiment. We may, in this place, describe the mode of drying gases, as it will come into play in other experiments. A glass tube, some 3 feet long and $\frac{1}{2}$ to $\frac{3}{4}$ inch in diameter, fig. 180, is filled with chloride of calcium, and closed by corks. The corks are perforated either by a hot wire or else by the cork borer, so as to admit

at each end a glass tube from 1 to 2 lines thick. The stoppers are then cut down to the level of the large tube, and bound with strips of caoutchouc.

*Fig.* 180.

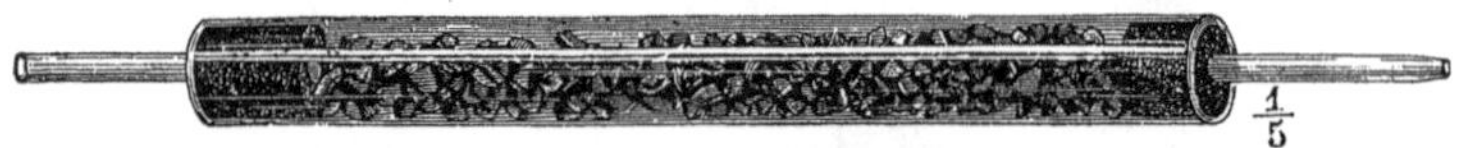


The apparatus thus arranged is connected first with the receiver or bag holding the gas, and then with the exhausted balloon, after enough gas has been forced through the tube to expel all the atmospheric air.

In determining the relative specific gravities of air and water, involving the filling of the balloon with water, especial care must be taken in drying again, which can only be accomplished effectually by repeated admission and exhaustion of dry air. For this reason the experiment is not well adapted for school instruction. The volume of smaller vessels may be determined by means of mercury. If no vessel with stop-cock be at hand, and it be merely desired to show the difference in weight between a vessel full of air and one empty, any thin flask may be made to answer the purpose, by closing the mouth with a good cork, pierced by a hot wire and driven in up to the top. A piece of oiled silk is tied tightly over the stopper and the end of the vessel, and two parallel cuts made in the silk, a short distance on each side of the perforation, fig. 181, and on a larger scale in fig. 182. In this way we obtain a very simple and effective valve, applicable in many other cases. The flask, after being weighed, is placed under the smallest receiver possible, and the air exhausted, after which it is again weighed. A pint flask will answer for the purpose. By the intervention of the contrivance, fig. 183, the flask may be connected by flexible tubes with a hand pump. By inserting the end *b* in the flask, it may be used for condensation. The corks are varnished with sealing-wax.

*Fig.* 181. *Fig.* 182. *Fig.* 183.

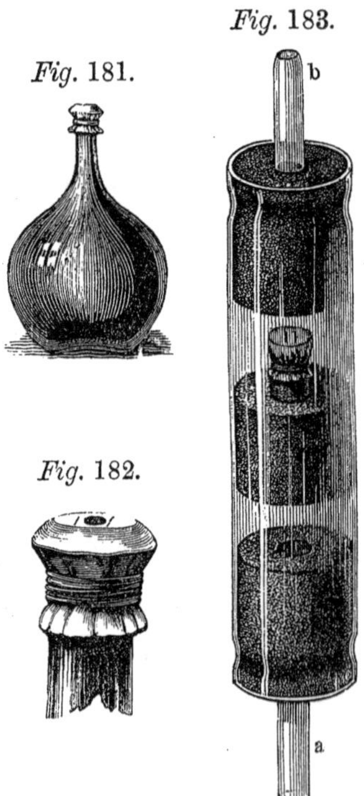

(26) *Fall of bodies in a rarefied medium.* The simplest mode of exhibiting this experiment is by taking a glass tube, from 2 to 3 feet long and from 1 to $1\frac{1}{2}$ inch in diameter, with tight brass caps at both ends, and a stop-cock at one end. A couple of bits of letter-paper and a leaden shot are introduced, and the difference in the time of falling exhibited before exhaustion, by repeatedly inverting the tube. The same is done after exhaustion. A feather will take longer to fall than the paper, but there is more danger of its sticking to the sides of the tube, in consequence of grease or electrical action.

*Fig.* 184.

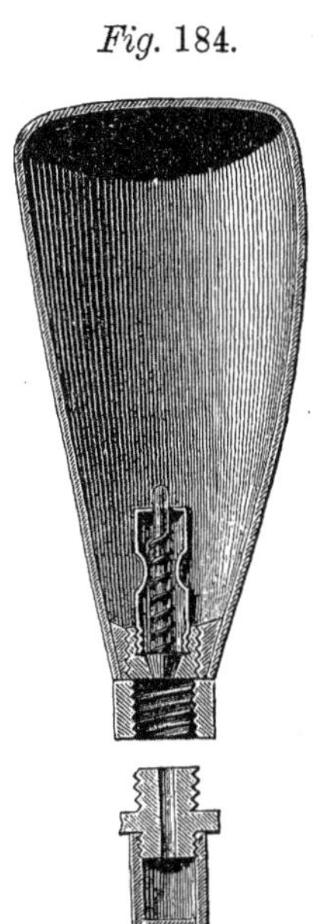

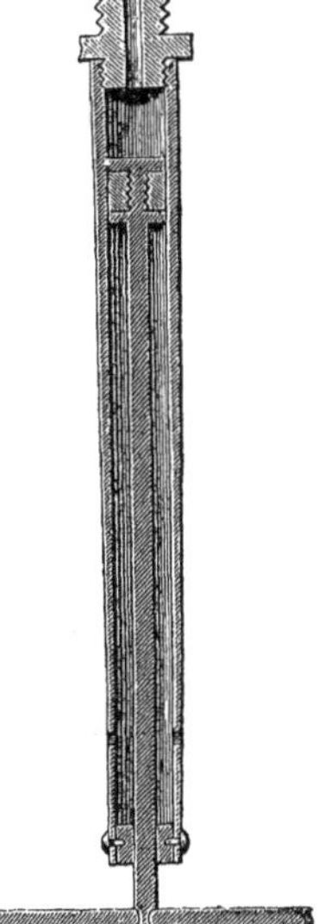

A glass should be selected which is a good insulator, so that it may also be used for electrical experiments. There are numerous arrangements for letting the two bodies fall at the same instant, but they all require the receiver to be exhausted for every descent; in the way described it can be repeated quickly, if not distinct the first time.

[103] **The air-gun.**—The form of this varies so much that a detailed description would be useless. Some care must taken in charging the magazine. When it is doubtful how many strokes of the condenser the magazine will bear without bursting, it is safest to ascertain the relative capacity of the two by filling them with oil, and measure the thickness of the metal. The negative space, *i.e.* the space between the valve of the magazine and the piston, is often so large as to render an overcharge impossible.

In charging the gun, the magazine is screwed on to the condensing apparatus, fig. 184, the cross-bar of the piston rod held firmly under the feet, and the magazine moved rapidly up and down, the piston being drawn out far enough at each stroke to admit the air through the lateral aperture, and then quickly thrust down again. The piston must not move too tightly, for the operation is very fatiguing. Very few air-guns will retain a full charge over 24 hours.

[104] **The siphon.**—A siphon with suction tube, fig. 185, is useful to explain its action. While sucking, the end *b* is closed with the finger, and care is necessary to avoid drawing the liquid into the mouth. Such a siphon may be made by cracking off the bottom of a Cologne bottle, and fitting it up as shown in fig. 186.

*Fig.* 185.

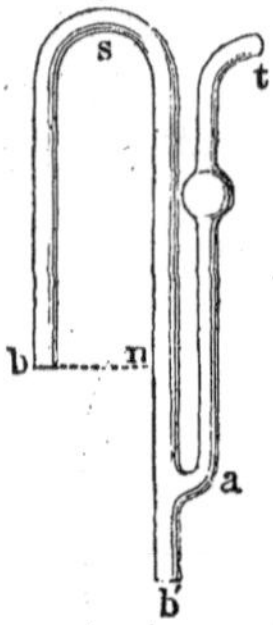

We have already given an experiment with the siphon in No. 14 of the air-pump series, (p. 118.) Another may be shown by taking a tumbler and cementing air tight on it a tin cover, with two short tubes in it, fig. 187. Into one is tightly fitted a bent glass tube, the other being for the introduction of water. If the siphon is set in operation, the flow will continue as long as the aperture *a* is open, but will cease on closing this with the finger or a cork. The cup of Tantalus may readily be constructed by any tinner, after fig.

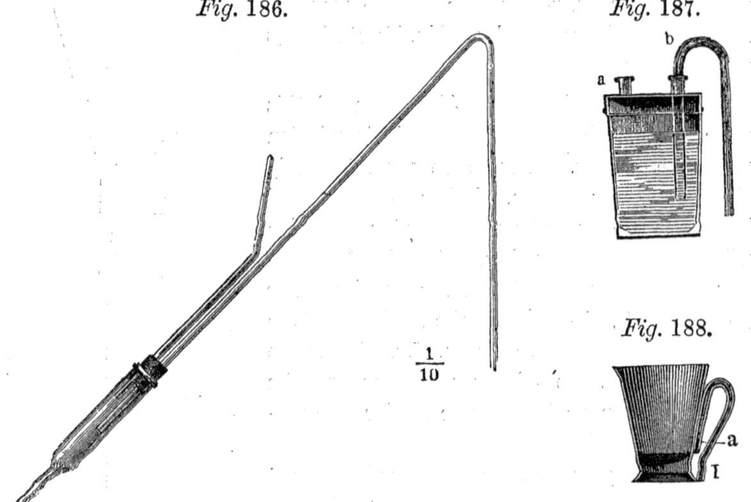

*Fig.* 186. *Fig.* 187. *Fig.* 188.

188. The siphon serves as a handle and opens in the raised bottom of the cup at *a*, while the other end is on a level with the foot. On filling the cup with water, this will immediately run off, and empty the vessel completely. If a glass vessel with an opening in the bottom is at hand, a curved siphon may be fixed in it with a cork, fig. 189; or, as in fig. 190, simply a straight tube with a larger tube closed at one end inverted over it. This tube must be a little longer than the inner one, and not so high as the top of the vessel. The water will flow out as soon as it reaches the top of the siphon. Small bottles with the bottom cut off, and the neck set in a wooden foot, will answer the purpose.

A siphon with an air chamber may be made by taking a Cologne water bottle and cutting off the upper part, just below the neck. Two tubes are inserted through a good tight cork, the shorter being drawn into a

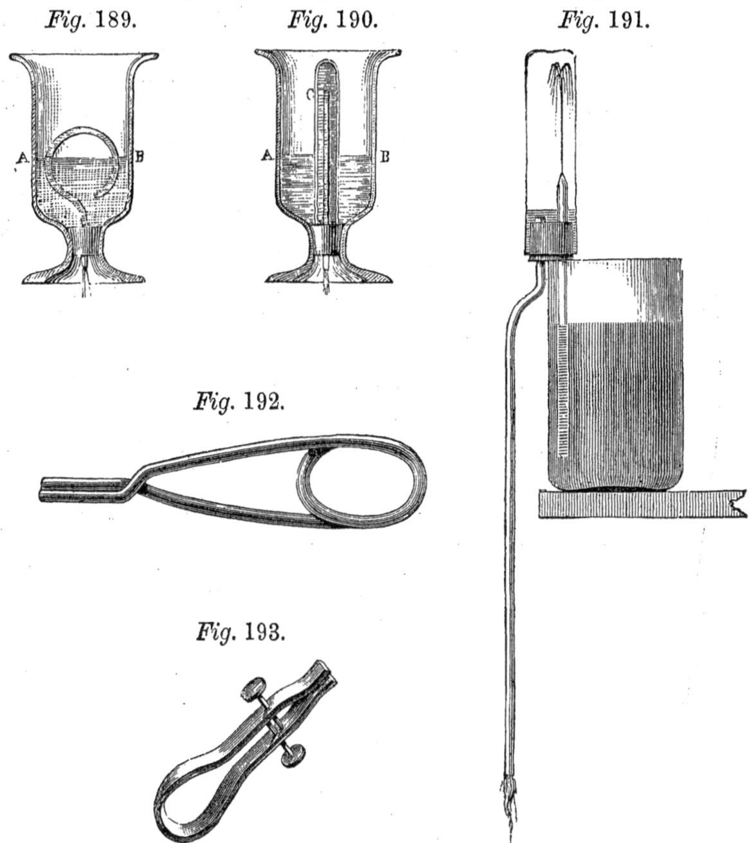


*Fig.* 189. *Fig.* 190. *Fig.* 191. *Fig.* 192. *Fig.* 193.

point and projecting more than the longer, fig. 191. On sucking at the lower end of the long tube until water enough is drawn over to cover the upper end, we will have a continued jet of water within the bottle.

The Wirtemberg siphon consists of a tube bent at an acute angle, with the legs of equal length and the extremities turned up; this may be hung up filled and ready for immediate use.

On account of the general use of the siphon in decanting from large vessels, it is often convenient to attach a piece of india-rubber tube to the end of the longer limb, and close this with a clamp made of bent wire, such as is shown in figs. 192 and 193.

[105] **The pipette** is a tube of metal or glass, about 2 inches wide, fig. 194, which can be closed with the thumb at the top, and ends in a fine point less than a line in diameter.

*Fig.* 194.

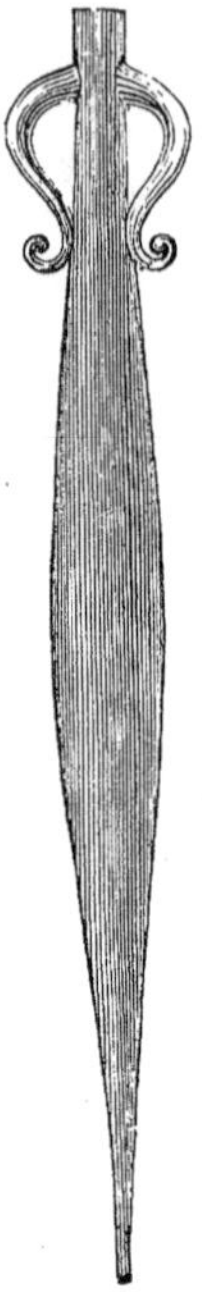

A great variety of amusing toys may be constructed on the same principle: such as the magic funnel, the widow's cruse or oil jug, the vestal's sieve, the magic can, etc., any of which can readily be made by the tinner.

*The magic funnel,* fig. 196, consists of two funnels, set one inside of the other, leaving a small space between, into which the minute opening at *a* leads; this opening is formed by the points of the two funnels, which are soldered air tight together on the upper edge. The handle is partly hollow, and communicates above with the concealed cavity; it has a small opening at *b*. The funnel is filled after closing the nozzle at *c*, and the liquid penetrates into the cavity; as long as the finger is held on the opening at *b*, the liquid will not run out of the interspace, while any desired portion may be let off by covering and uncovering *b*.

*The widow's cruse,* on somewhat the same principle, is readily intelligible from fig. 197.

*The magic can,* figs. 198, 199, has likewise an opening in the handle at *b*; after removing the false bottom *m*, which is only slipped on, it is filled through an aperture in the true bottom *n*, closed by a screw. The spout must, of course, be a small one. If the can, as shown in plan in fig. 199, be

*Fig.* 195.

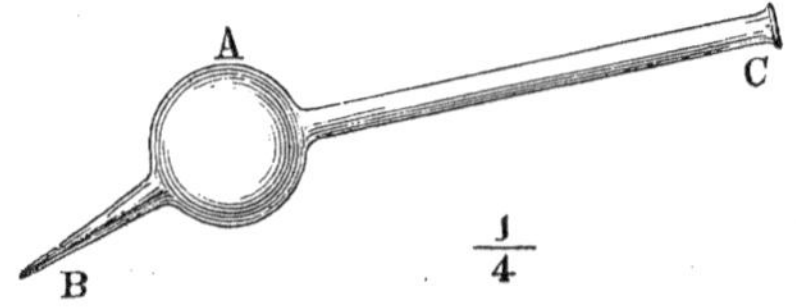


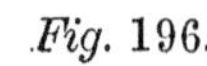

*Fig.* 196.

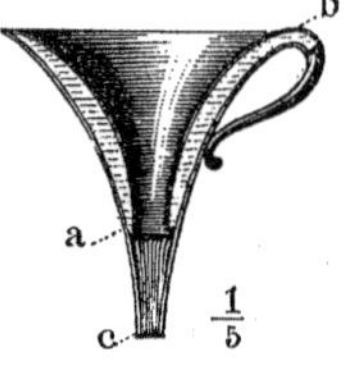


*Fig.* 197.

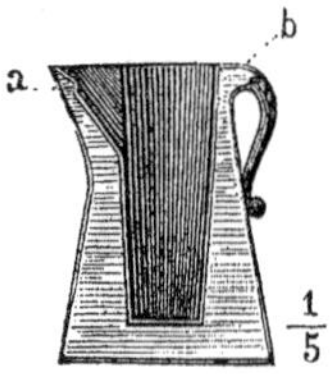

divided by a partition, we can fill in two different liquids through the openings *a a*, and either or both be allowed to flow according to the pressure on the holes *b*, *b*.

*The vestal's sieve,* fig. 200, is a vessel of tin, the bottom of which is minutely perforated. Liquids will be held in as long as the neck remains closed by a cork stopper.

A more practically useful application of this principle

may be made in the construction of a pipette, the simplest form of which consists of a glass tube, from 4 to 5 lines wide, drawn out to a point at

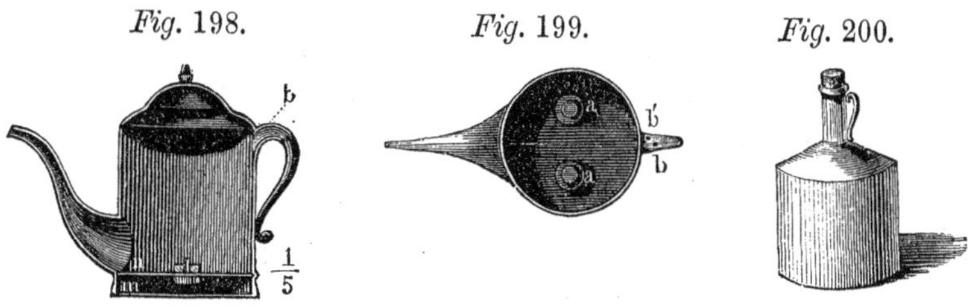


*Fig.* 198. *Fig.* 199. *Fig.* 200.

one end, and at the other so narrowed as to be readily closed with the forefinger. The tube itself may be 5 or 6 inches long, and the narrow extremity from 1 to $1\frac{1}{2}$.

[106] **Hiero's fountain.**—This may be constructed most simply, as in fig. 201. The ends of the tubes are made somewhat conical, so as to fit more tightly into the corks. Considerable skill in glass blowing will be required to make the fountain in one piece, as seen in fig. 202. Fig. 203

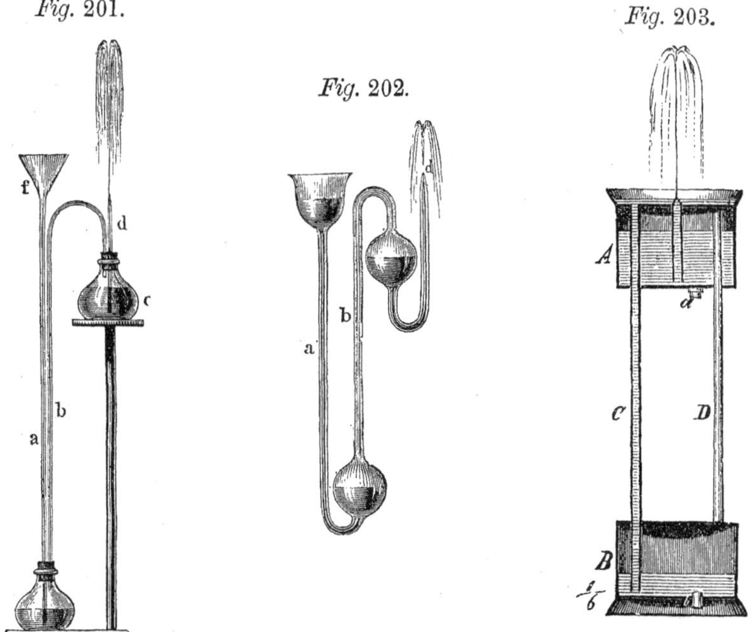


*Fig.* 201. *Fig.* 202. *Fig.* 203.

shows a simple construction in tin of a Hiero's fountain. The two vessels *A* and *B* have each an opening *a b* beneath, closed with corks. *A* is filled with water by the one, and *B* emptied, after the experiment, through the other. In addition to the two tubes *C D*, the two vessels, for greater security, may be connected by means of two similar columns, having no connection with the vessels. In this form of the apparatus the water is again collected in the basin of the upper vessel, and exerts a pressure on the lower.

The magical cask, fig. 204, depends on the same principle as Hiero's fountain. It is divided into two parts, and the partition has an opening above. Wine is poured into *A*, through the opening *a*, to be forced out through the cock *E*, by the pressure of the water column in the tube *C D*; the tube *c* of the cock extends to the bottom of *A*. The compressed air passes through the opening in the partition from *B* into *A*, and, after the experiment, *B* may be emptied through *b*.

*Fig.* 204.

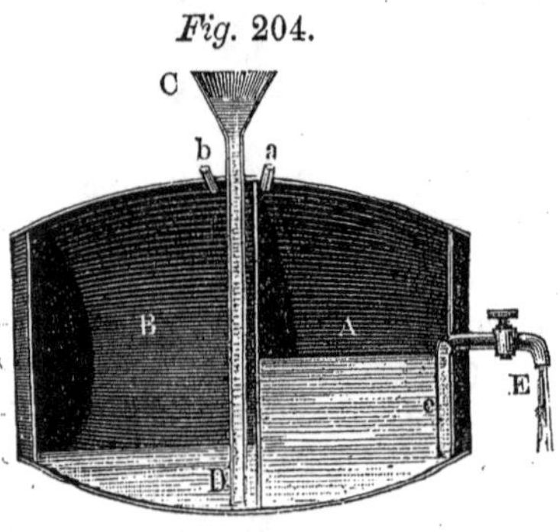


*Fig.* 205.

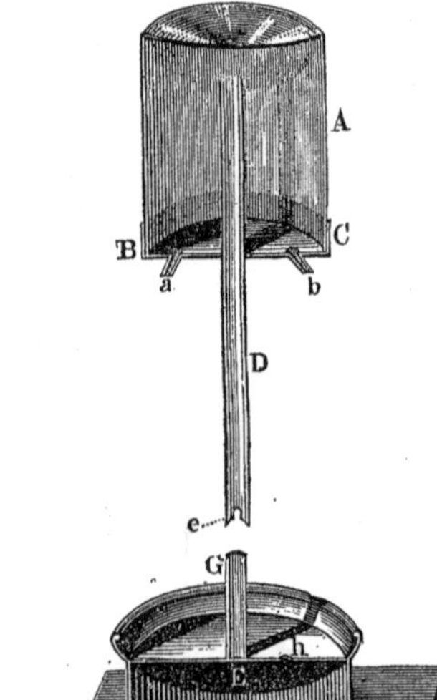


[107] **Pumps.** — Models of suction and forcing pumps had better be purchased ready made, with the cylinders of glass if possible. As no oil can be used with glass cylinders on account of diminishing their transparency, the piston must be wrapped with thread, and consequently soaked in water before use, to make it swell and fit tightly. If the working of the forcing-pump has been explained in connection with the Hiero's ball, the application of the former in the fire-engine will be illustrated quite as well as with a special model.

[108] **The intermitting fountain.**—A simple form of this apparatus may be constructed by applying a tin cap *B C* to a glass vessel *A*, fig. 205, through which

passes the tube $D$, having a small notch in the end; the cap has also the little tubes $a$, $b$. The top of the second vessel $E$ forms a basin, and has in its center the tube $G$, which also has an opening in the bottom of the basin; the tube $D$ is set on the tube $G$, after $A$ has been filled with water through the former. The small opening $h$, leading from the bottom of the basin into the vessel $E$, lets less water run off into $E$ than is discharged by $a$ and $b$ together. A cork is fitted into an opening in the bottom of $E$, to let off the water. As soon as $A$ is set over the tube $G$ the water runs out of $a$ and $b$ and collects in the basin, because $h$ discharges less water than these two openings; the opening of the tube $D$ is thereby closed, air no longer passes through the tube $D$ into $A$, and the flow of water from the latter vessel ceases until more water runs out of the basin, when the notch at $e$ becomes free, and the flow commences again from $a$ $b$.

[109] **The manometer.**—In order to observe the pressure of gases in a state of motion or of rest, open or closed manometers are used, which are also known under the name of wind gauges.

Under open manometers, Welter's safety tube must be mentioned, fig. 206. It serves in the first place, in evolving gases, to guard against the bursting of the vessel, the liquid in the tube being driven up into the funnel by the pressure of the gas. When the evolution of gas ceases, acid can be poured in through the tube. The manometers which are used with gasometers and gas pipes to measure the pressure are made on the same principle, and provided with a scale for measurement. Fig. 207 shows such a pressure gauge where the scale is fixed to the ascending tube. The fluid used is cochineal water mixed with sulphuric acid.

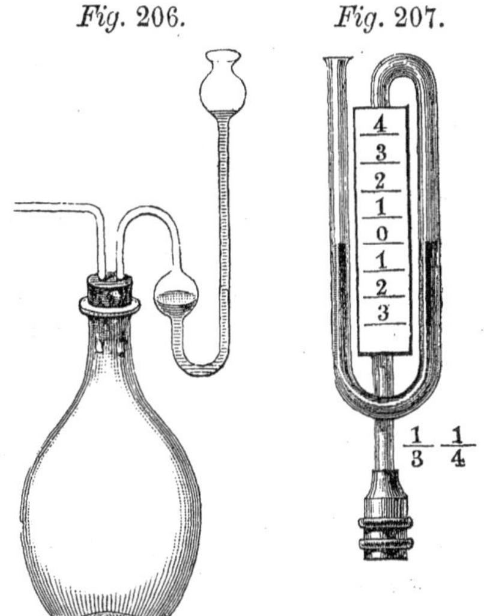

*Fig.* 206. *Fig.* 207.

To measure the pressure of the air in blasts, a simple apparatus of tin plate is used, which is shown in fig. 208 about $\frac{1}{5}$ the real size. $M$ $N$ is a box, of which the tube $a$ $a$ is fitted tightly through a cork, and stuck in an opening in the blowing apparatus. In the short tube $b$ a stout glass tube, from 4 to 5 inches long, is inserted, and divided into quarter inches. Before the trial, water is poured in through *d* until it stands at the zero of *c*. In organs the pressure is not usually over 4 inches, but it must remain at the same

*Fig.* 208. *Fig.* 209.

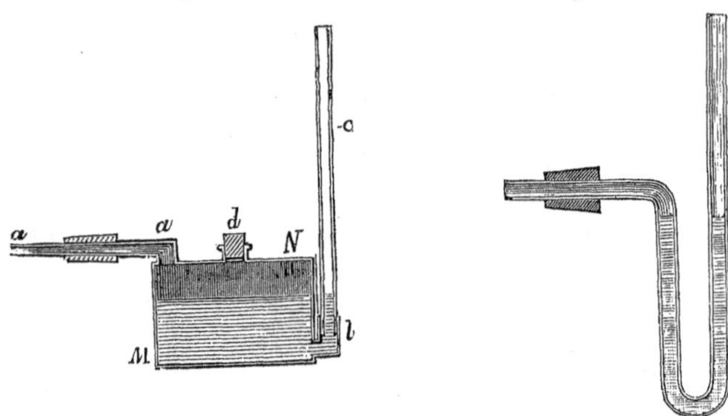


height when all the valves and registers are open. Double bent tubes, fig. 209, may also be used; they have a double reading, like the siphon barometer.

*Fig.* 210.

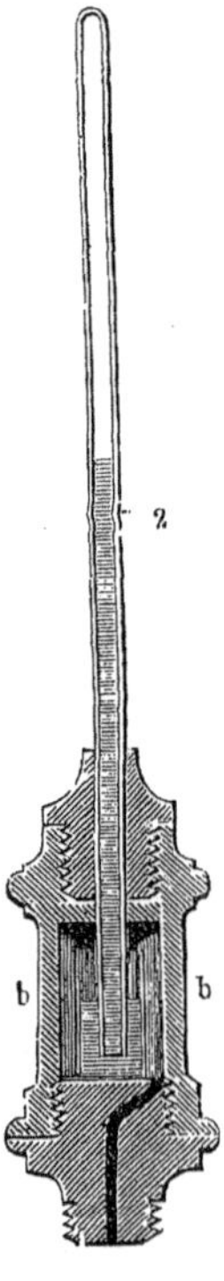

Closed manometers are generally used only for high pressure, which is indicated by the compression of a portion of air inclosed by mercury, according to the law of Mariotte. Fig. 210 represents a manometer of this kind, which may be used for steam-boilers, or with a stop-cock in connection with the condensing air-pump. It consists of a chamber *b b* connecting with the gas through the canal *a*, and inclosing a vessel with mercury. A strong glass tube is screwed down into *b b* by a metallic cap, and reaches below the level of the mercury. The tube is graduated to atmospheres. The pressure of the gas forces up the mercury into the tube, condensing the air before it.

The manometer is also used to indicate the pressure of liquids, *e.g.* in the hydraulic press.

[110] **Döbereiner's lamp.**—The liquid used consists of sulphuric acid, diluted to $\frac{1}{6}$; instead of a piece of zinc, it will be best to use a coiled ribbon of zinc plate, on account of the larger surface. The quantity of liquid must be so great as nearly to reach the cover, fig. 211, when at its highest, so as to have as great a pressure as possible. Very small machines are

not desirable, and care must be taken to have the gas chamber large enough to supply small quantities of hydrogen. For this purpose it will likewise be convenient if the discharge pipe is so fixed as to admit of screwing in another tube, to which glass tubes leading into the pneumatic trough may be attached by means of caoutchouc.

*Fig.* 211.

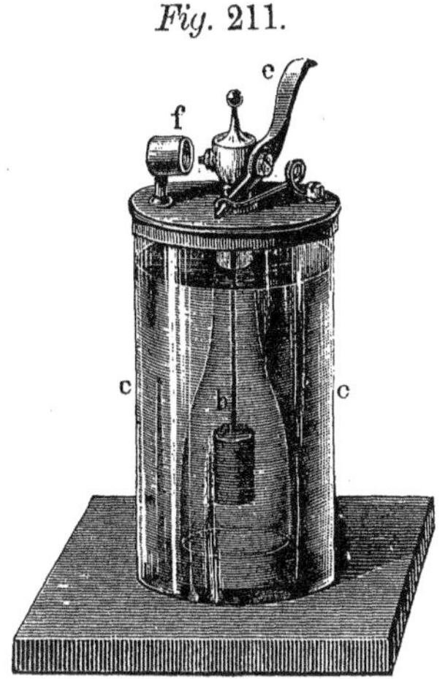


The cock will fit much better if the tap be bored longitudinally for the passage of the gas, as in fig. 212.

Although the cock described is most convenient, there are other forms employed which it is hardly necessary to mention here. The points to be attended to in the purchase of any kind are whether it fits closely without being too tight, and whether, after rapid turning backward and forward, the tap exhibits any traces of friction. After a recent filling, or when the apparatus has not been in use for a long time, an explosive mixture of the hydrogen with air will be found to exist, and on this account care must be taken not to ignite it without having previously let out all the gas (with a piece of paper interposed against the platinum sponge) and getting a fresh supply. After long standing the platina sponge must be heated red hot in the current of gas, previously inflamed, before its properties will be restored.

*Fig.* 212.

While such an arrangement will be found convenient for small quantities of gas, larger gasometers will be required for other purposes; they may be purchased ready made, and are easily constructed. Similar arrangements may be used for other gases. Such an apparatus may be made at little expense according to Varrentrapp, in the manner shown in figs. 213 and 214. For the exterior vessel a candy jar with a tin cover may be taken. The cover has two holes on each side, through which a pin is stuck to hold the cover fast to the rim of the glass. There is a short tube in the center of the cover into which the neck of the inner vessel is fixed by a cork. This vessel is made of a bottle with the bottom cut off. The gas tube *a* is fitted into the neck of the bottle by a good cork; it leads into the vessel *A*, which is shown on a larger scale in fig. 214. The tube *a* passes through the cork nearly to the bottom of the larger tube *B*; a second tube *b* projects a little way below the cork to conduct off the gas. The tube *B* stands in a little conical glass, the bottom of which is filled with mercury, so as to close the tube *a* when the glass

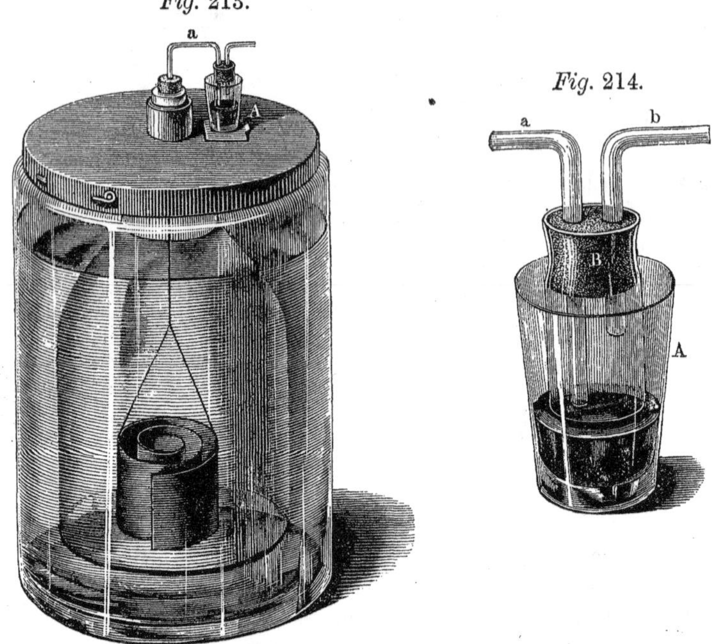


*Fig.* 213. *Fig.* 214.

stands on the top of the reservoir, as in fig. 213. Under *A* there is a hole in the cover, which is closed with a slip of tin; by taking away the tin, the glass *A* sinks into the cover far enough to uncover the end of the gas tube *a*, but not the tube *B*. It is closed by raising the vessel *A* and slipping the tin under it again.

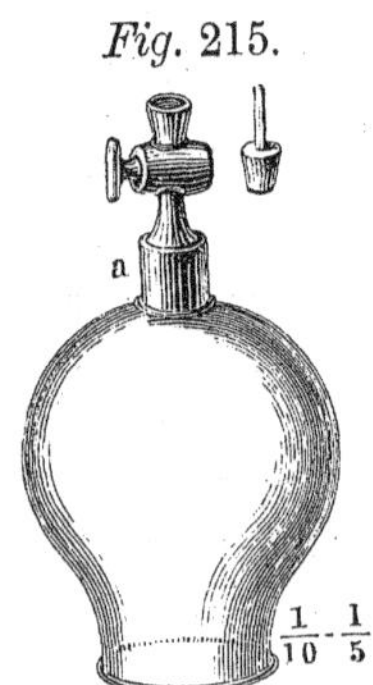


*Fig.* 215.

[111] **The air-balloon.**—Small balloons of gold-beater's skin may be purchased ready made, and answer very well. To fit these up a small quill is inserted and tied around with fine thread, and a cork fitted to the aperture. The hydrogen gas is collected in a glass receiver, with a stop-cock at the top, fig. 215, such as is used for filling bladders, etc. Upon the stop-cock is fitted a cork with a quill inserted, and coated all over with sealing-wax. After collecting the necessary quantity of hydrogen, and allowing it to cool sufficiently, all the air is pressed out of the balloon, which is then set over the quill of the cock, and the receiver

depressed into the water of the pneumatic trough, thus forcing the gas into the balloon. This, when filled, is closed with the stopper. Should the balloon have become wet, it must be blown up, and then allowed to dry. It is not advisable to allow the hydrogen to pass directly from the generator into the balloon, as, owing to the higher temperature, there will be a portion of watery vapor passing over, not to speak of the vapor of sulphuric acid. Nevertheless, the Döbereiner apparatus, mentioned in the preceding paragraph, may be used, when large enough.

Balloons made of collodion are lighter still. Such balloons may be made by shaking up collodion in a flask holding from 6 to 12 ounces, with a short neck of the width of one's finger, until the sides are moistened with it in every part, and then pouring out the excess. After drying, the film of collodion is loosened carefully from the neck of the flask and tied around a small tube. The air within is then carefully exhausted by suction, the film thus gradually detached from the glass and drawn out.

To send up a Montgolfier balloon, it will be necessary to suspend it from a thread carried over two pulleys in the ceiling and counterpoised by a weight. On heating the air within by a spirit-lamp, the balloon may be made to rise: or a piece of sponge, soaked in alcohol, and fixed inside the hoop of the opening, may be set on fire. Without the counterpoise referred to, the volume of the balloon required will be larger than is convenient for a lecture-room.

[112] **Diffusion of gases.**—This experiment is usually performed with carbonic acid gas, poured out of a vessel into a larger one filled with air. A wire is bent into a hook, and on the hook is placed a piece of wax-candle. If the ignited candle be let down into the larger vessel it will be extinguished, but on repeating the experiment soon after, less carbonic acid will be found, and the light will ultimately burn freely. In pouring out the carbonic acid, (which also shows its greater specific gravity,) the two cylinders must be inclined to each other, and the emptying performed slowly; otherwise the carbonic acid will become so mixed with atmospheric air that the light will not be extinguished.

The experiment may be varied, by taking two vessels with equal mouths, filling one with oxygen and the other with hydrogen, inverting the hydrogen vessel over the other, and connecting the mouths of both by a strip of india-rubber. In a short time both will contain explosive gas, which may be shown by igniting a small portion of it. The apparatus fig. 216 is still more convenient for this purpose.

[113] **Absorption of gases.**—1. When powdered quick-lime and sal ammoniac are mixed together and heated in a small retort over a spirit-lamp, ammonia will be liberated, and may be collected over

mercury in a glass tube about a finger's breadth in diameter, and closed at one end. The mercury may be kept in a low tumbler, 2 or 3 pounds being sufficient. If now the tube, when filled with gas and the mouth closed by the finger, be uncovered under water, this will rapidly absorb the ammonia, and rush up into the tube.

*Fig.* 216.

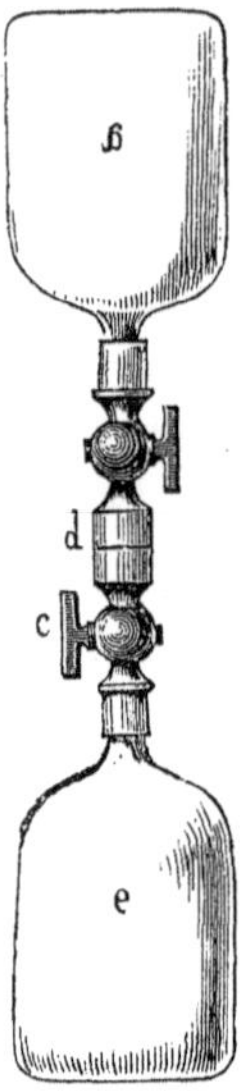

2. Carbonic acid is generated from chalk and sulphuric acid, and collected in a tube as in the last experiment. A bit of recently-heated charcoal is introduced under the mercury into the tube of carbonic acid gas, when it will absorb a great portion of the latter. After the experiment it will be well to pound up the coal, to recover any mercury which may have been absorbed. The gas may be liberated from the coal again by placing the apparatus under the receiver of an air-pump, and exhausting the air.

[114] **Breath images.**—To produce the breath images of Moser several methods may be employed, some of which are as follows:—

1. Cut any desired image out of card paper, lay it upon a glass plate, and breathe upon it. After the deposit of moisture has evaporated, remove the plate and breathe upon the plate again; the part formerly covered by the figure will condense the moisture in a different manner from the rest, and the figure again become visible. The experiment may be often repeated, and at considerable intervals, with equally satisfactory results.

2. On a glass plate well rubbed off with freshly heated Tripoli, or upon a plate cleaned ready for a daguerreotype image, place a seal of metal or stone which has not been recently cleaned off, allowing it to remain for some hours. The plate, on being moistened, will show the image of the stamp. Or, inversely, the stamp or seal may be cleaned, instead of the plate. Even without fresh cleaning, images may sometimes be obtained, although a longer exposure is required. Difference of temperature in the bodies facilitates the formation of the images. Instead of breathing on the metal plate, it may be exposed in the daguerreotype mercurial apparatus to the vapor of mercury, by which means the images will remain permanent. They will also become visible by exposure to the vapor of iodine, or, at least, they appear when the iodized plate is brought out to the light. Coins are less adapted to this experiment than seals or gems, especially when the latter have a simple drawing or writing on a clean surface.

# CHAPTER II.

## EXPERIMENTS ON MOTION.

### (*a*.) EXPERIMENTS ON VARIOUS KINDS OF MOTION.

[115] **Free fall.**—The laws of the free fall of bodies are shown by means of Atwood's machine, the general arrangement of which is shown in fig. 217. In purchasing this machine, attention should be paid to having a seconds pendulum, which shall release the falling weight. This release can, indeed, be effected by hand, with the use of any other pendulum, and the requisite skill may soon be attained by practice, but this is lost again from year to year, and before each new experiment fresh practice is necessary. The wheel must move very easily, and, to secure this, it must run on steel points; the points of the axis and the bearings must be wiped clean after use, to prevent their becoming gummy, and fresh oil must be applied each time before using it. Friction wheels increase the cost of the apparatus very considerably, and, if the work be otherwise good, are unnecessary. When the wheel is a solid disk of uniform thickness, half its weight must be added to the other two weights of the machine, in order to calculate the accelerating weight which must be added: no rule can be given for open wheels, which are lighter, but the accelerating weight must be determined by experiment. It is better, therefore, to choose a solid wheel. To place the scale, the divisions of which should be about an inch in length, perpendicularly, the arresting ring is fixed to the

*Fig.* 217.

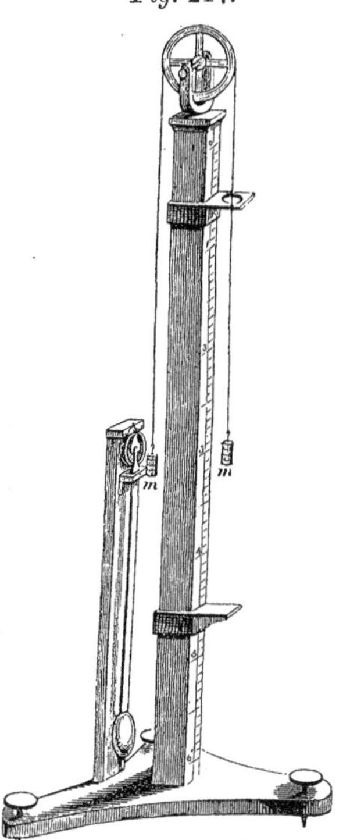

bottom of the scale, and the falling weight used as a plummet; the adjusting screws are then set so that the weight shall hang over the center of the ring. Before proceeding to the actual experiment, the friction must be compensated; this is effected by the disks of tin plate, fig. 218, which are laid upon the falling weight, until a slight impulse with the finger will cause it to traverse uniformly the whole length of the scale, without, however, being able of itself to overcome the friction.

*Fig.* 218.

The weight which is to cause the fall is placed opposite 0 of the scale and the ring which arrests it at 4, 9, 16, etc. The stage is placed at a distance, corresponding to the velocity of the weight, at the point where the accelerating weight is removed; that is, when this weight is arrested at 4—in 2 seconds—the number of divisions at which the stage is placed below the ring is expressed by 4, 2 × 4, 3 × 4; when the ring is at 9 by 6, 2 × 6, 3 × 6, etc., according as the weight is to run 1, 2, or 3 seconds. The stage must be placed so that the top of the descending weight will be opposite the required division when it is arrested, because the bar is removed when this upper surface of the weight passes the level of the ring. For this reason, the length of the weight is always made to correspond to 1 or 2 divisions of the scale. When the experiment is begun at the first division, with 1 second fall, the results are not accurate, because the influence of the disturbing causes is felt too sensibly.

When the ring and stage are properly adjusted, and the load placed at zero, the weights must be brought to rest perfectly, before starting the pendulum, otherwise the experiment will not succeed well. Everything else depends on the special construction of the machine. It is desirable to have the top of the pillar broad enough to support a glass bell over the rollers, or else to have the whole running gear movable, so that it can be packed away in a box.

The laws of falling bodies can also be demonstrated by means of balls running down a long inclined beam, as was done by Galileo. The beam must be perfectly straight, and made of several pieces glued together. The groove is formed by two ledges nailed to the sides. It must be divided into half feet, and have such an inclination that the ball will descend half a foot the first second. The beam must be from 8 to 10 feet long.

[116] **Motion of projectiles.**—The parabolic motion of projectiles can be most readily shown by means of the apparatus represented in fig. 219, in which a ball rolling down the groove *a b* acquires such a velocity as to describe the parabola *b*, 1, 4, *g*. The groove *a b* forms part

of a circle, and is carefully smoothed out with pumice. Care must be taken that the horizontal line *c d*, extending from the end of the groove

*Fig.* 219.

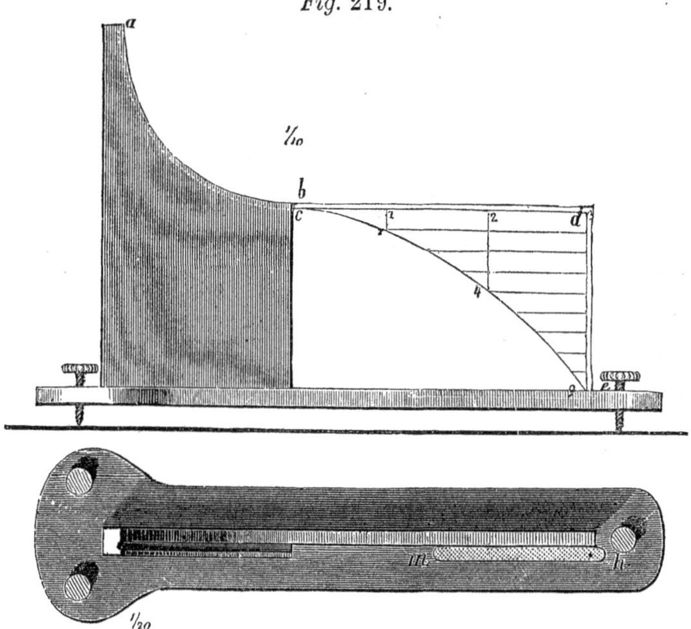


*a b* along the board *b e*, be actually tangent to the curve of the groove. As the theoretical velocity of projection can never be attained, it will be best, before describing the parabola, to let the ball roll down the groove and thus determine the random by experiment, and afterwards to draw the corresponding parabola *c* 1, 4, *g*. The base contains a cavity *m n* filled with sand, to arrest the motion of the ball. Before experimenting, the apparatus must be placed in a vertical position by means of leveling screws. The board *b e* should be made of light-colored wood, so that the well-defined line of curve may be seen at a distance. Any wood may be used if it be painted white or covered with paper.

*Fig.* 220.

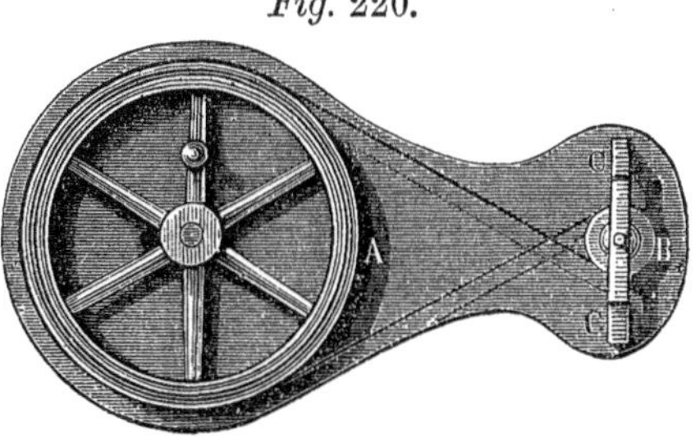


[117] **Rotary motion.**—The laws of central motion are well illustrated by the apparatus in fig. 220. In its essential features it consists

of a large wheel *A A*, connected by an endless cord with a smaller wheel *B*, on the axis of which various appendages may be screwed, and thus set in rapid rotation. The latter has one pivot on the base board, and the other in the cross-piece *c c*; this is even on the top, to allow a circular board to be screwed upon it, for the performance of a certain electrical experiment. The axis of *B*, with its screw, must project above this board, represented in the figure by a dotted line. In constructing the apparatus, care must be taken to have the two wheels in a simple proportion to each other, so that the number of revolutions which the smaller makes in a given time may be easily deduced from those of the larger. For uniform motion, the larger wheel should have considerable weight, so as to admit of the apparatus being used for other experiments than those now referred to, *e.g.* on sound. This can be effected by screwing a cast-iron ring to the under side of the wheel. The ring can be cast at any foundry, from a pattern cut out of $\frac{1}{4}$ inch board, and the few holes necessary for its attachment by screws are easily drilled out. To keep the cord at any required degree of tension, the axis of the driving-wheel may be made to move along the base board *c c*, fig. 221, and secured by a clamp *a*. The

*Fig.* 221.

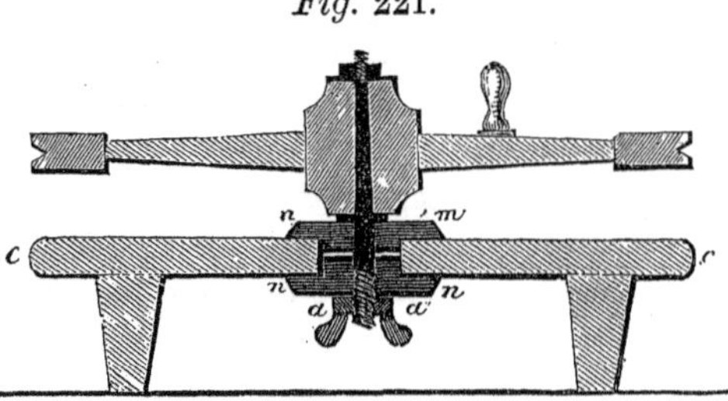


axis must then rest firmly upon the cross-piece *m m*, with a square shoulder passing through *m m* and *n n* and ending in a screw with a nut and washer, which presses both cross-pieces firmly against the base board.

Instead of this mode of fastening the driving-wheel, the sliding piece *B*, figs. 222 and 223, may be inserted into the base board, carrying the fixed

*Fig.* 222.

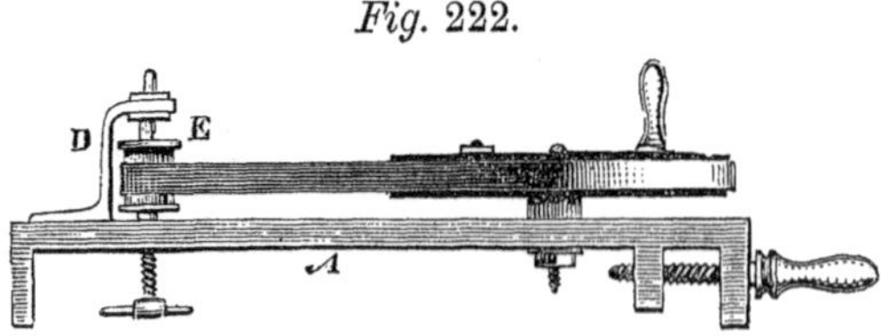


axis of the wheel, and adjustable to the proper tension of the cord by a screw. In this figure, the driving-wheel is of iron. The pulley-wheel runs on a screw below, and the upper end of the axis works in the brace

*Fig.* 223.

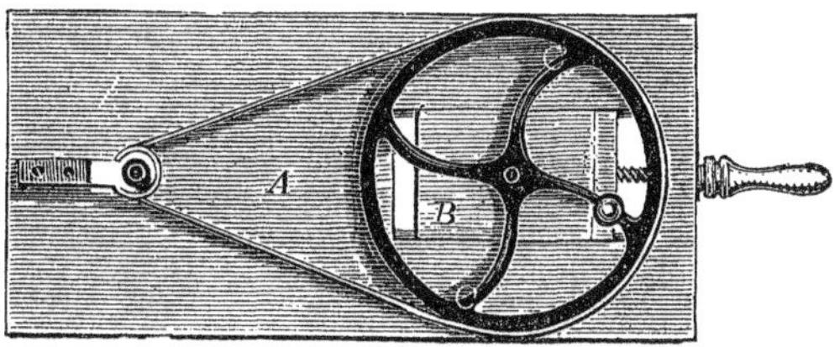


$D$; this arrangement is, however, not suitable for the attachment of the circular board before mentioned.

[118] **Experiments with the whirling machine.**—Screw on to the pulley the apparatus shown in fig. 224, $\frac{1}{10}$ the natural size. It

*Fig.* 224.

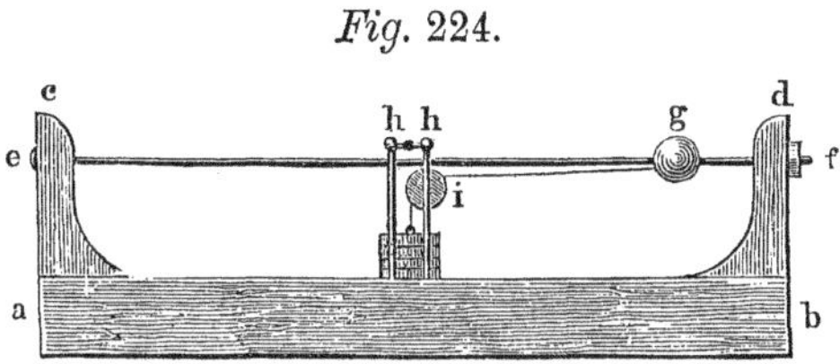


consists of a thick block $a\ b$, loaded above and below with lead, which need not be more than an inch thick, if the driving-wheel is heavy enough. Two uprights $a\ c$ and $b\ d$ are mortised into the ends. A polished steel rod $e\ f$ goes through these uprights, and is fastened by a head at one end, and a screw and nut at the other. It should be 1 to $1\frac{1}{2}$ lines thick. The wooden or ivory ball $g$ slides easily on this rod by a hole through the center. Four brass rods $h\ h$ are fixed exactly in a square at equal distances from the center of rotation, and connected by cross rods with screws passing over the wire $e\ f$. Two of them serve as supports for the pulley $i$, so adjusted that the axis of motion shall be tangent to its circumference. A silken cord passing over this pulley connects the ball $g$ with the little brass platform shown in fig. 225. Its disk $k\ k$ is just large enough to move up and down freely between the rods $h\ h$ without escaping from them. A number of disks, like fig. 226, can also be laid on the platform, and their weight is so arranged that half of the weight of the platform can be assumed as

*Fig.* 225. *Fig.* 226.

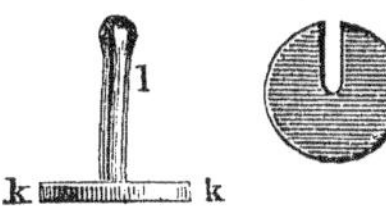

unity. The cord has three loops by which the ball *g* can be placed at distances from the center, which are to each other 1 : 2 : 3.

We may now show in a general way, that with the same position of the ball, the greater the weight to be raised the greater must be the velocity of rotation, and that when the ball is moved from the distance 1 to 3, and the speed of the wheel reduced to one-third, so that the ball moves in its orbit with the same velocity as at the distance 1, the weight will not be raised without accelerating the rotation. By means of a pendulum, the number of revolutions of the driving-wheel in a given time may be ascertained, and from the ratio of this wheel to the pulley the number of revolutions of the ball are known. Computing from this the periods corresponding to the radii 1, 2, 3, it may be shown by calculation that in any two successive experiments the centrifugal forces deduced from the radii and the periods of revolution are proportional to the weights raised. In this experiment, the velocity must be very gradually increased and maintained uniformly, which is much easier with a very heavy driving-wheel. With a little practice, the weight may be held suspended at a little height, and the period of revolution then estimated with a pendulum. The exact moment of raising the weight is easily observed, by pasting a disk of white paper, exactly the same size as the weights, on the board directly beneath them, so as not to be visible until they are raised.

(2) The arrangement fig. 227 is designed to show that, with equal

*Fig.* 227.

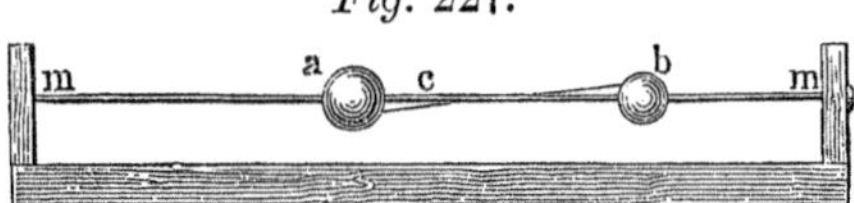


periods of revolution, the centrifugal force is proportional to the weight as well as to the radius. The weights of the two ivory balls *a b* are in the proportion of 2 : 1, and both slide easily on the wire *m m*, which must not be over a line in thickness and very uniform. The balls are connected by a silk cord, and are placed for the experiment at distances from the center inversely proportional to their weight. If this be exactly attained, the balls will retain their position at every velocity.

*Fig.* 228.

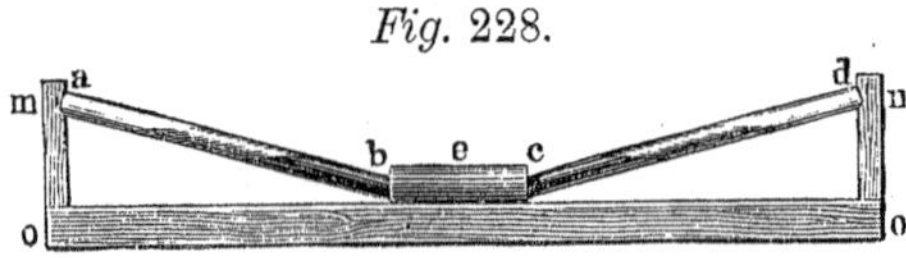


(3) The influence of mass may be shown by the apparatus fig. 228:

*a b*, *c d* are two glass tubes about the thickness of a finger, sealed at one end. They are $\frac{1}{3}$ filled with mercury, $\frac{1}{3}$ with water, and $\frac{1}{3}$ left empty, and the end tightly closed with a good cork. The upper ends are let a little way into the uprights *m n*, and the lower ends held fast by the semi-cylindrical block *e*, which is excavated to receive the corked ends and screwed down to the base board. When the apparatus is set in rapid rotation, the contents of the tubes assume an inverted position, the mercury appearing above and the air below at *b* and *c*. The rapid increase of the centrifugal force with the velocity of rotation may be exhibited very well by slinging a tumbler filled with water, by three cords, as seen in fig. 229, and whirling it rapidly round the head.

The same experiment can be shown more neatly with the apparatus fig. 230, consisting of a globular glass vessel cemented to a wooden foot, by which it can be screwed to the whirling machine. Mercury and colored water are poured into the vessel in quantities not quite sufficient

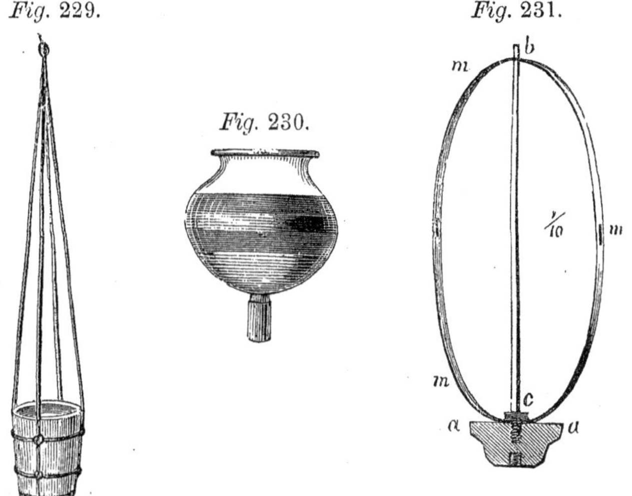

*Fig.* 229. *Fig.* 230. *Fig.* 231.

to fill the bulged portion and leave a cylindrical space, as wide as the mouth, empty. When a rapid rotary motion is communicated to the glass, the mercury and water will form rings around the bulged sides.

(4) The appendage fig. 231 serves to illustrate the flattening of the earth by the centrifugal force. It is easily made as follows: *a a* is a turned block of hard wood, which can be screwed to the axis of the whirling machine; *b c* is a slender square rod, somewhat thicker at *c*, ending in a good screw; *m m m* is a thin brass hoop, pierced with square holes at four opposite points. This hoop is fixed on the rod as shown in

the figure, and the lower end screwed down fast. When rapidly turned, the hoop describes a spheroid. Two hoops fastened on the rods at right angles to each other show the form of the spheroid more distinctly.

The effect of rotary motion may be illustrated very well by means of

*Fig.* 232.

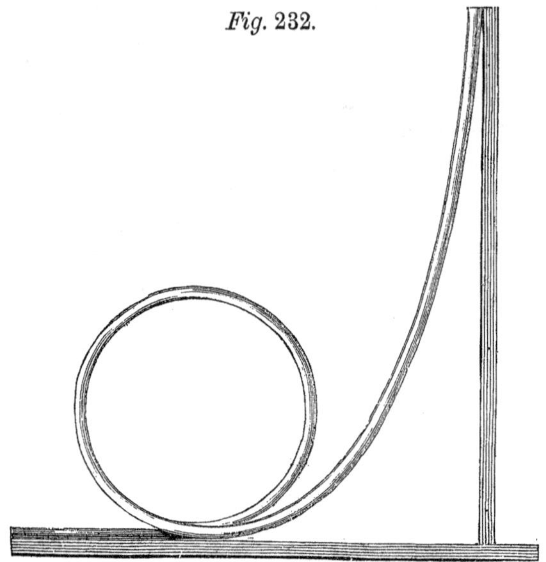

*Fig.* 233.

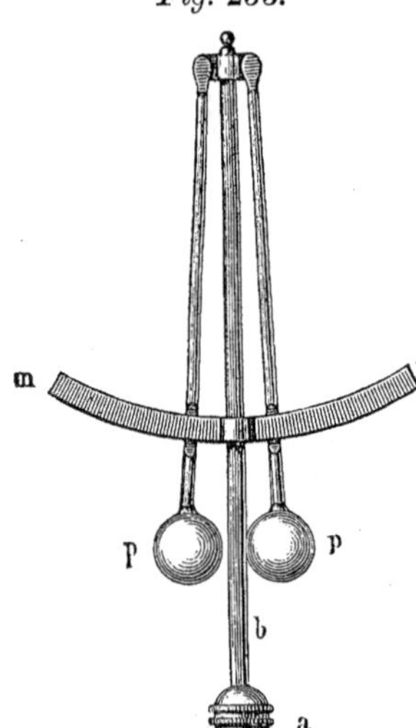


the centrifugal railway, which can be purchased in the toy shops, or made in a very simple way, by fastening a gutter of tin to a board, as shown in fig. 232. The diameter of the ring should not exceed $\frac{1}{10}$ of the height of the fall. A smooth bullet laid in the top of this trough will run through its whole length.

(5) *The centrifugal pendulum.* This arrangement is instructive on account of its use in the steam-engine. An upright rod *b* is fixed into the cap *a*, fig. 233, which screws on the axis of the machine. The two pendulums *p p* work on hinges at the top of this rod, but can only move in the slits of the arc *m n*, shown in section in fig. 234. The more rapidly the machine is turned, the more rapidly the balls diverge.

[119] **The gyroscope.**—This apparatus consists of a heavy metallic ring, revolving on an axis which is hung in gimbals, like the mariner's compass, so as to be free to assume any position. It is also contrived so that the prolongation of the axis may be supported at one end on a steel pivot while the other end moves freely around it. When set in rapid rotation, and one end of the axis thus supported, the ring will revolve around the support in the direction in which the particles at the bottom are moving. The details of its construction vary so much that no general directions for its management can be given. This apparatus illustrates very beautifully the effect of the centrifugal force of each particle in resisting any change of the plane of rotation. A clear comprehension of this principle is necessary in order to understand the cause of the change of seasons.*

*Fig.* 234.

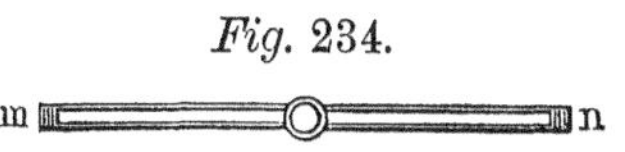


[120] **Foucault's experiment.**—The principle of this experiment, which proves the earth's rotation on its axis from the apparent rotation of the plane of oscillation of a pendulum, may be clearly illustrated by screwing the apparatus fig. 235 on the whirling machine, setting the pendulum in motion, and turning the machine very slowly. As the ball is suspended by a very fine thread, the apparatus may be turned a number of times before the torsion of the thread changes the plane of oscillation. Its application to the rotation of the earth may be shown by setting the little arch on the meridian of a globe, fig. 236, and showing its action at the poles and at the equator.

*Fig.* 235. *Fig.* 236.

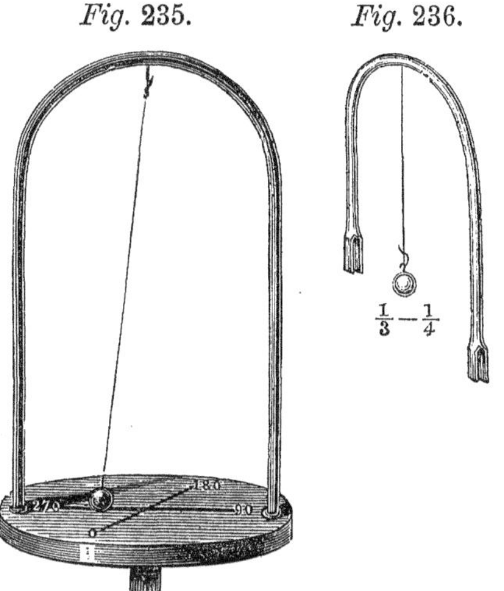


In making Foucault's experiment, the first requisite is a firm, lofty point of suspension, with a height of 20 feet or over; the rest is very easily managed. The experiment succeeds, indeed, with a pendulum from 14 to 15 feet long, so far as to show the apparent revolution of the plane of oscillation from west through north to east; but the amount of

* Very excellent gyroscopes are sold by the Holbrook School Apparatus Manufacturing Co., in Hartford, for $8.—*Trans.*

motion is sometimes more and sometimes less than $15 \times \sin. \varphi$ ($\varphi =$ latitude.)

The pendulum should be a cannon ball, of at least 12 lbs. weight. It should be allowed to float on mercury and the uppermost point marked as the point of suspension, first with chalk and afterwards with the center drill. With a vessel of proper shape, the experiment does not require much mercury. If the point of the drill be blunted in marking the spot, it will not be possible to work the ball properly, and heating will not mend it. The ball must be centered in a lathe on the spot marked, and turned accurately. A hole about $\frac{1}{4}$ of an inch in diameter must then be bored through the center, a piece of wood shaped like fig. 237 inserted in the lower end, and the upper end bored out wider and an internal screw cut in it, to receive a brass plug, as in fig. 238. This plug must be bored

*Fig.* 237.

*Fig.* 238.

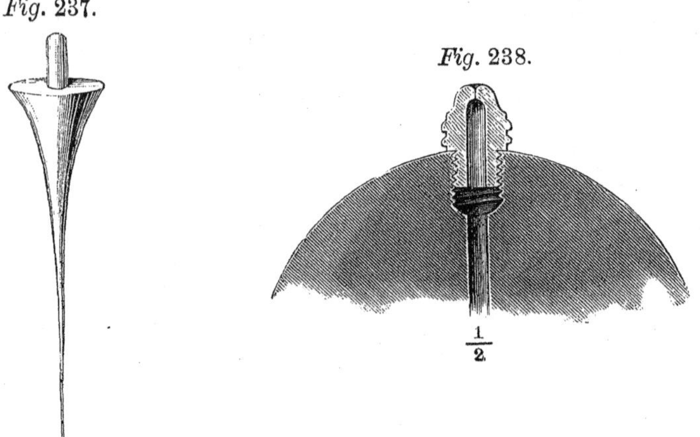


nearly through from the under side, turned off concentric with the ball, and a hole the size of a fine knitting-needle bored quite through in the center. A steel or iron wire, just strong enough to support the ball, is inserted through this hole, and fastened by wrapping it around a bit of stout brass wire resting against the bottom of the plug. The upper point of suspension may be formed by simply boring a fine hole through a stout hook, passing the wire through it, and wrapping the end around the shaft of the hook; but a better mode of suspension is shown in figs. 239 and 240. A polished steel plate with a slightly concave surface is inserted in the upper side of the hook, and the steel stirrup supported on it by the steel pin *b*; the wire passes through the small hole *a* in the stirrup.

Before making the experiment, the ball must be allowed to hang quietly for a considerable time, in order to stretch the wire and destroy the torsion, which may be aided by the hand.

Fasten a string to a nail in the direction in which the pendulum is to move, and observe exactly the point of the ball which lies in a straight

*Fig.* 239.

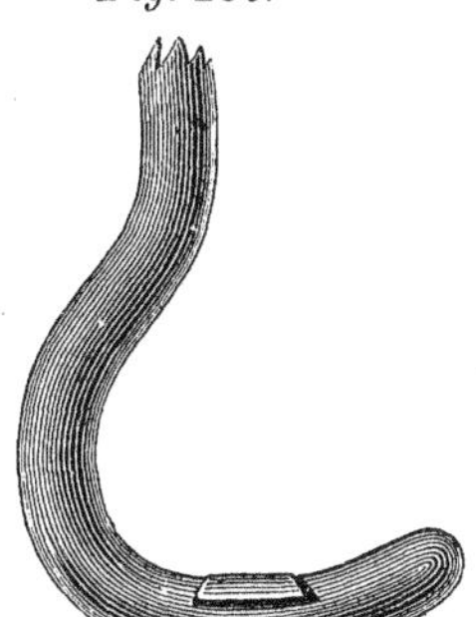

*Fig.* 240.

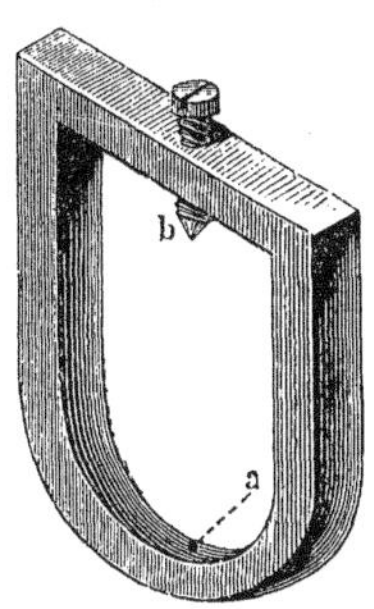


line between this nail and the wire, and mark it with chalk. Make a wide loop on the string, fix this around the middle of the ball, and draw the ball out, as far as its vibrations are to extend, by winding the string around the nail; the arc of vibration should not be very long. Turn the ball in its sling, until the point marked is again in the plane of the wire and the nail.

Describe upon a circular board the angles over which the pendulum will move every quarter of an hour, place the center of this board under the point, fig. 237, before fastening the ball in the sling, and let the point at which the graduation begins be in the line between the wire and the nail.

When the ball is perfectly still, which may be hastened by letting it strike gently against a stick the end of which is held on the floor, (it is better to wait about an hour,) set the ball in motion by burning off the string beyond the loop.

[121] **The pendulum.**—The fall of bodies through chords and arcs on the diameter of a circle may be illustrated by the apparatus figs. 241 and 242. It consists of a wooden ring *A A A*, with a smooth groove *a a a* on the inner side: this ring is fastened by means of a small foot to a base board with adjusting screws, and the foot itself is widened on one side so as to form a small dish to hold sand.

A very shallow dish, or even a flat board with an opening corresponding in depth and breadth with the inner groove, is fixed to the top of the ring opposite the lower dish. One or more inclined planes *m m n n*, also grooved, can be screwed to the sides of the ring.

Now suppose one of the chords *m m n n* be screwed on, and a ball held with one hand in the opening of the upper dish; with the other hand hold a second ball on that part of the chord corresponding to the inner groove, and let them fall simultaneously: they will reach the bottom at

*Fig.* 241. *Fig.* 242.

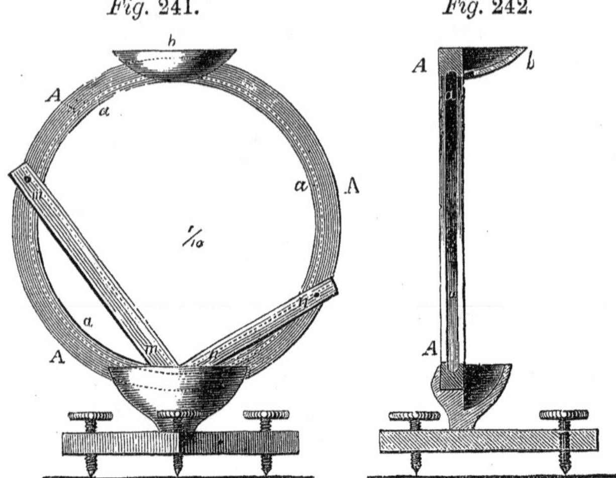


nearly the same instant. In the same way, balls may be allowed to roll down an arc of the circle and its chord at the same time. The isochronic property of the cycloid may be easily illustrated by cutting out a board in this shape with a groove on the inner side.

The apparatus figs. 243 and 244 is very convenient for experiments with the pendulum. Its construction is as follows: the narrow upright *a b*, with an arm *c*, is secured to a stout triangular base, with or without leveling screws. A number of small hooks are screwed into the under side of this arm, to which simple seconds pendulums of cork, lead, wax, stone, etc., are attached; and also one beating half seconds, and another, to which small leaden balls are fastened, at short intervals. The length of the seconds pendulums is measured from the bottom of the hook to the middle of the ball: the proper length is easily attained by making the silk string longer than necessary and tying knots in it. The pendulum, made of

*Fig.* 243. *Fig.* 244.

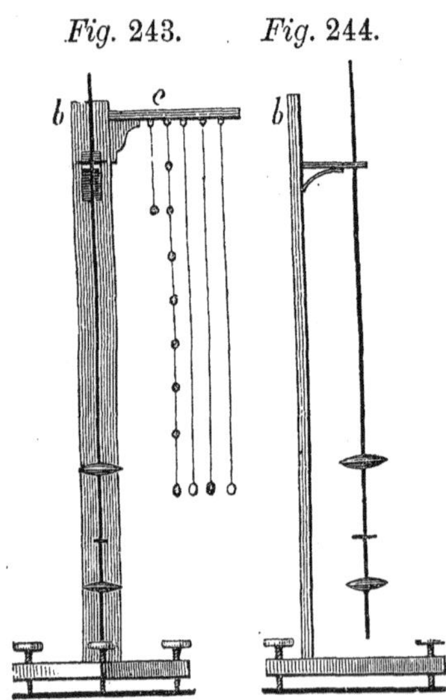

several leaden balls, is formed by screwing two little hooks into opposite sides of each ball and connecting them by cords. By casting the balls around a wire smoothly coated with asphaltum, this may afterwards be drawn out and the labor of boring the holes spared, which is no trifle in lead. This pendulum serves to illustrate the difference between the physical and the mathematical pendulum. Besides the arm *c* the upright has an iron bracket *d*, which is figured on a larger scale in figs. 245 and 246. The arm *c a* is riveted to the plate *b*, which is let into the wood of the stand and fastened with 6 screws. The forked ends *a a* of the arm

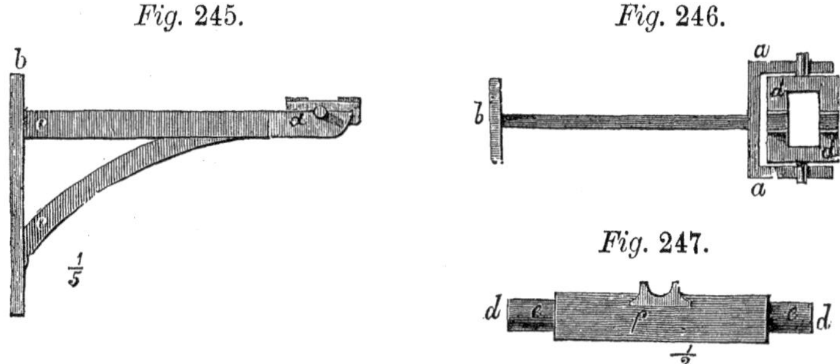


*Fig.* 245. *Fig.* 246. *Fig.* 247.

have two semicircular depressions to serve as bearings for the iron plate *d d*, a front view of which is seen in fig. 247; they are first worked out as accurately as possible with a file and then polished with emery on a cylinder of wood set in a lathe. The plate *d d* is wrought out of one piece, and the axles *e e* turned. Two plates of cast-steel, as hard as glass, and of the form shown in fig. 247, with highly polished cylindrical faces are inserted in opposite sides of *d d*, as bearings for the knife edges of the pendulum. These two plates may be made by laying two pieces of cast-steel together, screwing them up in a vice, and then boring a hole through them so that one half of the hole shall be in each plate; they are then ground together on a cylinder of brass with emery, filed to the proper shape, and then hardened. They are afterwards ground again with fine emery on wood, and polished with jeweler's rouge. The pendulum suspended in this manner is free to move in every direction and can assume at once a perpendicular position, so that the stand does not require to be leveled, thereby saving much trouble.

The knife edges are made of cast-steel, shaped as accurately as possible with the file, and hardened; they are then ground with emery on a glass plate, to give them a straight edge. They may be finally polished with rouge on a linen rag. This can be done more rapidly, however, with a disk of wood 5 or 6 inches in diameter, coated on the edge and surface with felt and fixed on the lathe. This disk is very convenient in making many optical instruments. It is smeared with rouge and water, and turned very rapidly.

The pendulum rods are made of walnut or apple wood, about a centimeter square. Where the knife edges go through, two brass plates, *a a* *b b*, figs. 248 and 249, previously perforated to receive the knife edges, are let into the wood and fastened with rivets going through the rod. A triangular perforation is afterwards cut through the wood with a chisel, and the knife edges driven through with a few gentle taps with a mallet. One pendulum has two such knife edges at the requisite distance to form a simple seconds pendulum, (= 104 centimeters,) the other has only one in the middle for experiments on inertia. The rods are graduated to centimeters.

*Fig.* 248. *Fig.* 249.

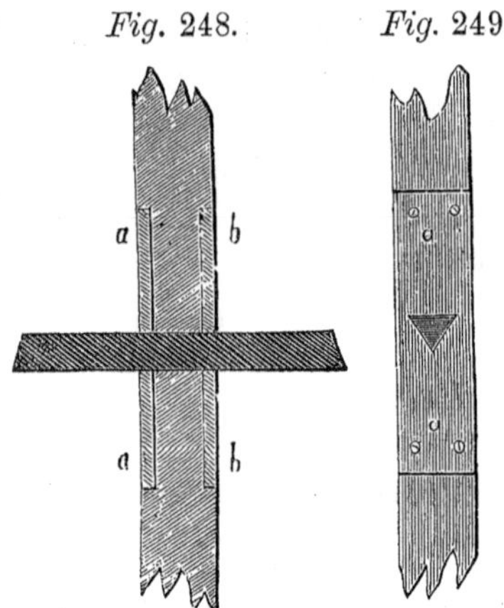


The balls are made of lead cast in wooden moulds coated with graphite. Through the mould is passed a short rectangular bar of smooth iron slightly thicker on one side than on the other, as seen in the cross section, fig. 250; the smaller side is about 1 centimeter in thickness. Points are marked with a punch on the ends of the bar, a a little to one side of the center. This bar is set into the mould so that the points marked will be in the center, and the opening in the weight therefore eccentric. The balls are afterwards turned off on this bar until they have the proper weight, and the bar is then knocked out. The weight of the bar must, of course, be known, and likewise that of the brass piece *b*, figs. 251 and 251 *a*. The latter is set in before inserting the bar, and completely fills the aperture; the thumb-screw *a* pressing against this fastens the weight to the bar in any position. The edge *b* must be a whole number

*Fig.* 250.

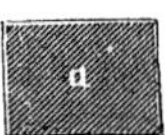


*Fig.* 251. *Fig.* 251 *a.*

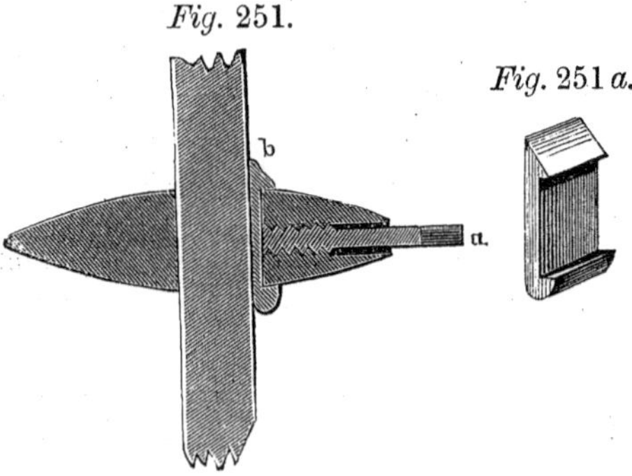

of centimeters from the center of the ball. The ball may be varnished black with shellac and lampblack. The balls may be held in place by a simpler but not so neat arrangement, of two brass slides fastened to the rods by thumb-screws; the hole for the bar is then made square, fig. 252. When the points where the weights should be fixed for the reversion pendulum are calculated, they should be marked on the rod. The weight of the balls is also marked on them.

*Fig.* 252. *Fig.* 253.

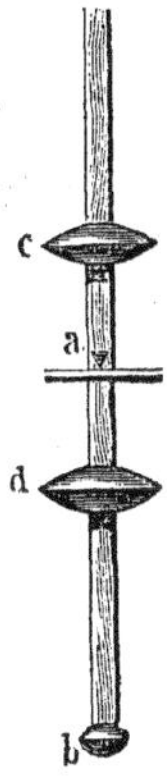

For experiments on inertia, it is necessary to have a small ball of about $\frac{1}{4}$ to $\frac{1}{2}$ lb. weight, 2 of 2 lbs. each, and 2 of $\frac{1}{2}$ lb. To make the experiment, fix the heavy balls at equal distances from the point of suspension, and the lightest one at the end of the rod, as in fig. 253, and observe how many beats the pendulum makes in a minute. The time of the beats will remain unchanged if the $\frac{1}{2}$ lb. weight be fixed at twice the distance from the point of suspension.

To make the experiment with the reversion pendulum, fasten two balls of equal weight to the rod with two knife edges, so that the middle of one will be 80 and of the other 120 centimeters from one of the edges; the time of vibration will then be nearly equal on both edges.

[122] **Pendulums with audible beat.**—A seconds pendulum, constructed as above described, will go an hour or more, but it is inconvenient for many purposes, because its beat is not audible. To make a pendulum which shall beat audibly, fasten to the top of the rod an iron ferule *a*, fig. 254, in which the iron tongue *b c* is soldered. The axis passes through this tongue at *e* and a round pin through the top, upon which the bent hammer is hung through a slit *r*, fig. 255. The strokes will then be audible, but must be counted.

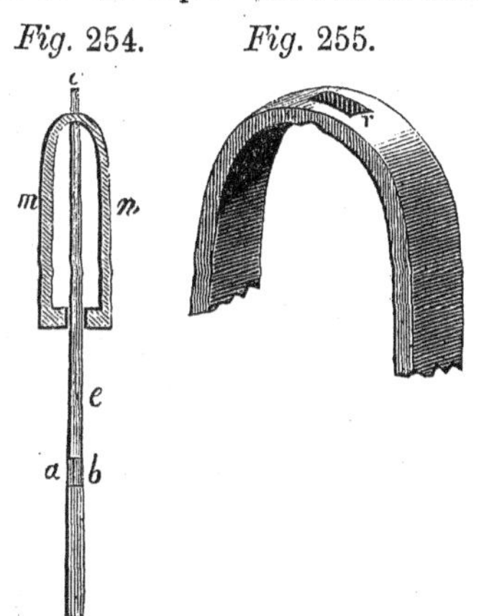

*Fig.* 254. *Fig.* 255.

It is more convenient and advantageous to procure from a clock-maker an arrangement with a ratchet-wheel and anchor escapement, with a hand fixed on the axis to mark the seconds on a dial-plate, as seen in fig. 256. The pendulum rod is then made of iron, so that a leaden ball bored lengthwise or sidewise can be fixed on it by a screw. For the short time during which the experiment will last, no compensation arrangement is necessary. The clock-work is supported on a frame, behind which the weight and the pendulum hang. This frame should not be over 4 or 5 feet high so that the motion of the escapement can be conveniently seen, as this arrangement serves at the same time to illustrate the commonest way in which a pendulum is connected with clock-work. To allow it to run longer, make the wheel around which the cord runs very small, and hang the weight on a movable pulley. The pendulum must be regulated by a simple seconds pendulum or a good watch.

*Fig.* 256.

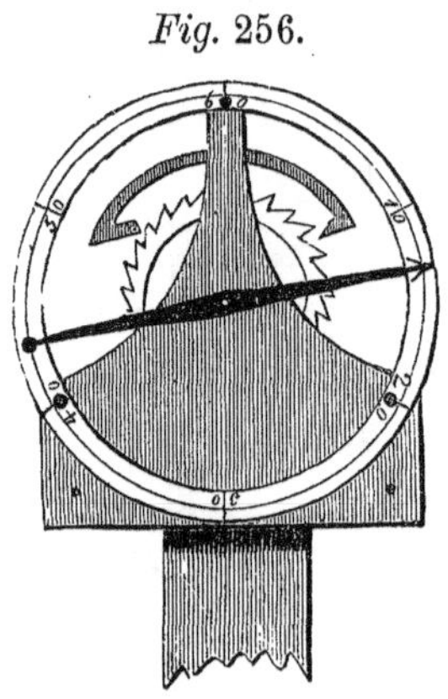

A more complete but more expensive arrangement is shown in figs. 257 and 258, $\frac{1}{4}$ to $\frac{1}{5}$ the natural size. The weight is so arranged that it can be wound up without stopping the motion of the clock. The ball may be made of cast-iron, and should not be less then 10 lbs. in weight.

[123] **Impact.**—For experiments on the laws of impact the inelastic bodies may be balls of unburnt clay of various sizes, into which hooks are fixed while in a soft state. Their weight may be adjusted with a rasp, after drying. But these balls have still some elasticity, and the results will, therefore, not correspond closely with the law. They are suspended by silk strings from a frame, such as is shown in fig. 259, of $\frac{1}{6}$ the natural size. The hooks of the frame pass through long slits, and can be adjusted by means of two nuts, so that the balls shall just touch each other and their centers lie in the same straight line. The arrangement might be made cheaper by fastening the cords with cork plugs. The experiments to illustrate the law are very simple: it is only necessary to see that the balls hang in the same straight line, and not to make the impact too strong, for fear of breaking the balls; but, as already mentioned, the success of the experiment will be indifferent.

Experiments with elastic bodies succeed better. The balls for these are made of ivory or lignum-vitæ; five or six of them of the same size, one twice and another half as large. The same precautions must be observed. The balls will not break, but the results are not accordant when the velocity is too great, owing to the imperfect elasticity of ivory. Special care must be taken that the balls be barely in contact without pressing upon each other, and that their centers be in a straight line.

Experiments on oblique impact against a plane may be made by

*Fig.* 257. *Fig.* 258.

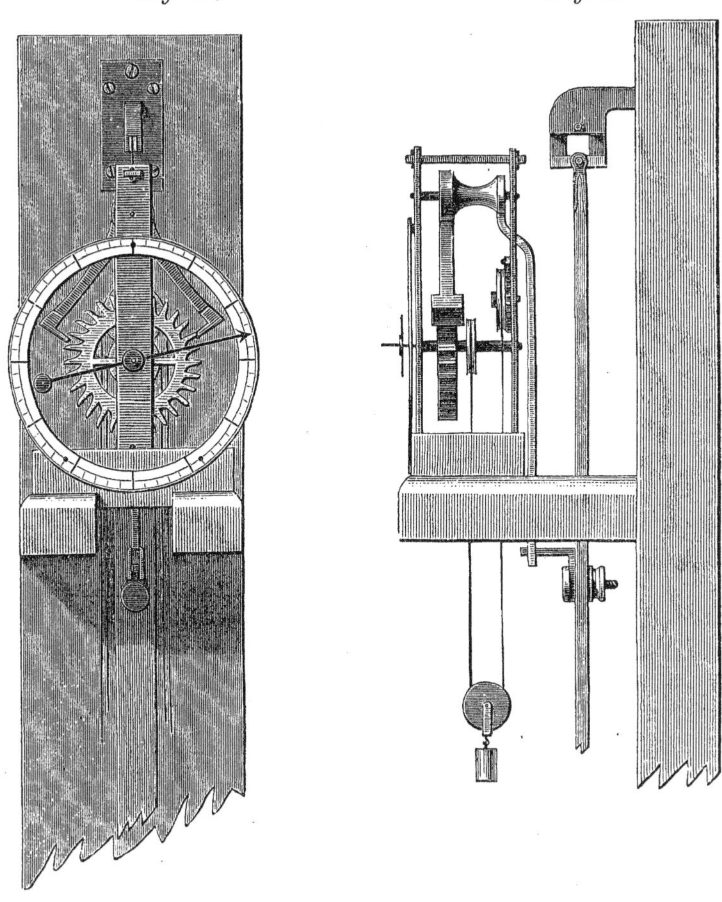

placing the marble slab described in § 73 upright on a table, and suspending a ball by a single string from any kind of a support, fig. 260. Describe the angles of incidence and reflection on the table, and give the ball an impulse in the direction of one of these lines. The less perfect the elasticity of the ball, the less will be the angle of reflection compared with the angle of incidence, especially when the latter is great. In the case of ivory, the ball glides during the compression along the surface of the slab, by which the portion of the velocity which is parallel

*Fig.* 259.

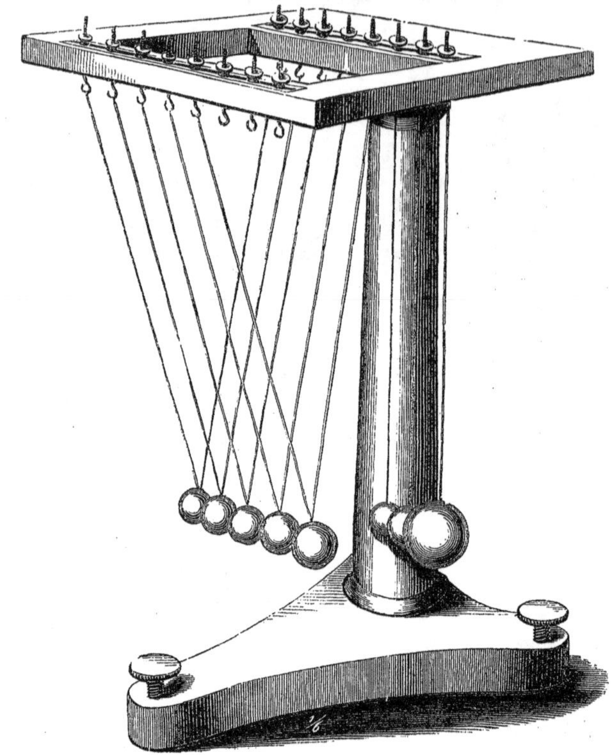

to the slab is diminished, and after several rebounds the ball falls perpendicularly against the surface. The experiment succeeds the first time, however, very well. The law of impact cannot be so well illustrated on a billiard table, because the balls have, besides the gliding motion, which fails entirely when the stroke is feeble, a rotary motion, which sometimes coincides in direction with the gliding one, sometimes opposes, or even makes an angle with it. The imperfect elasticity of ivory is particularly evident in the case of oblique impact between two balls, where, if the

*Fig.* 260.

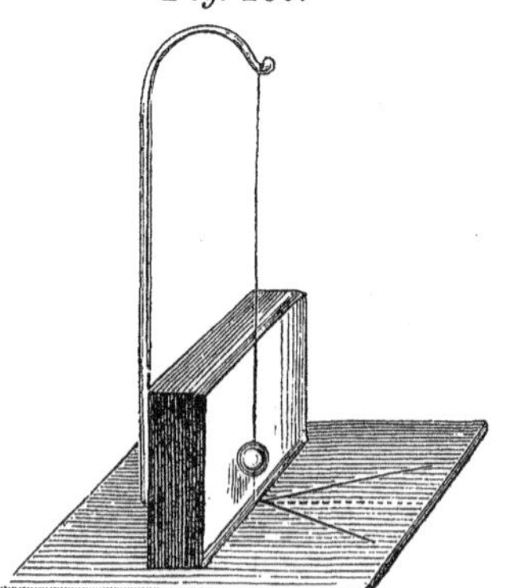

elasticity were perfect, the balls should diverge at right angles after the impact.

[124] **Friction.**—For these experiments the apparatus shown in fig. 75 may be taken when it is only required to show the amount of friction of wood upon wood, and illustrate the principal laws of friction. For this purpose a vertical prism of wood with unequal faces is drawn over the horizontal board by weights, with the fibers sometimes parallel, sometimes perpendicular to those of the board. The prism must have a hook for each position in which it is used, so that the cord passing over the pulley, to which the weight is attached, shall always be horizontal.

The angle of inclination may also be determined at which the prism begins to slide down the inclined plane, and the force acting parallel to the plane calculated from this.

The apparatus seen in fig. 261 may be used to show the difference of friction upon different substances. The wheel *A A* is made of wood saturated with hot oil, or of brass; a band passes over it, having a small scale pan at each end. These are loaded with a considerable weight, and the excess of weight noted which is required to be laid in one pan to turn the wheel when the axle rests successively on the friction wheels *a a* and the bearings *m m m*. Each pair of these bearings is made of a different material: iron, brass, gun-metal, zinc, wood, etc., but all bored alike and equally well polished. The experiment may be made with and without grease. The apparatus is supported on a suitable wooden frame. This is, of course, not sufficient to determine the coefficients of friction, but only to show that the friction varies with the substance used. Friction rollers deserve especial notice, because they are indispensable in hanging large bells.

*Fig.* 261.

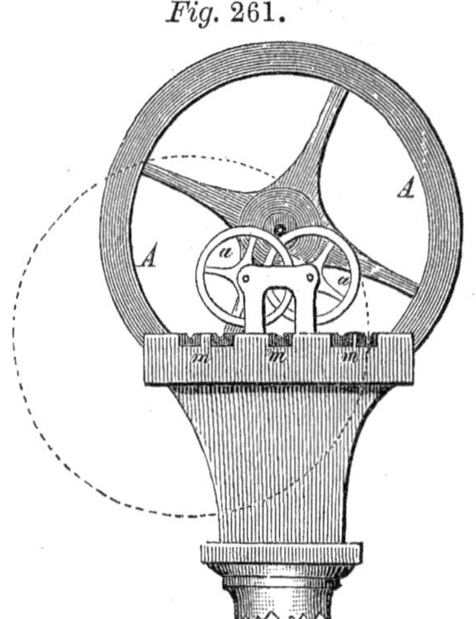

## (b.) EXPERIMENTS ON HYDRODYNAMICS.

[125] **Velocity of efflux.**—Since Torricelli's law holds good only when the velocity of the fluid in the vessel is so small in proportion to that of the stream flowing out that it may be neglected, it requires a tolerably wide vessel, and the velocity of efflux is still reduced 1 per cent. when the diameter of the vessel is 100 times greater than that of the opening. A tin vessel, like that shown in fig. 262, $\frac{1}{10}$ the actual size, is chosen for the experiments and provided with a communicating tube at the side, graduated to inches and tenths, so that the height of the water in the vessel may be conveniently observed. There may be four openings,

*Fig.* 262. *Fig.* 263.

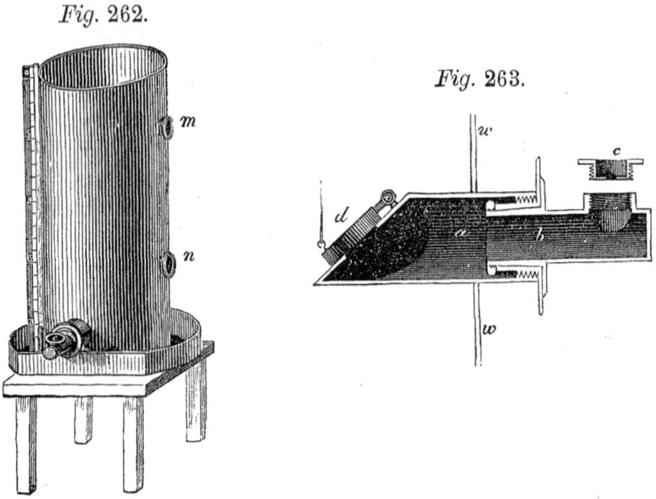


of which $m$ is 4 and $n$ 16, or any whole number of inches from the level to which the vessel is to be filled, as indicated by the scale. All these apertures consist of short tubes, 2 inches in diameter, set in the sides of the vessel $w$ $w$ and closed by a valve like fig. 263, which can be opened by a wire or cord. The real outlet tubes are screwed into these; they have sharp edges and vary in diameter from $\frac{1}{3}$ to 1 line. Into the lowest opening the tube $b$, fig. 263, is set in such a manner that it can be turned on its axis; the outlet tube $c$ is inserted in this, and the inclination of $c$ may be read off on the graduated circle fixed to $b$ and turning on $a$. Water is introduced from a second vessel as fast as it flows out of the first, so as always to maintain the fluid at the same level; but with very small outlet tubes this precaution is not necessary if the stream be allowed to flow but a short time. Instead of the valve shown in fig. 263 the tube may be furnished with a slide outside of the vessel, as in fig. 264, which has the advantage of not disturbing the flow of the water when opened as the valve does. It would be better still to have each outlet tube provided with such a slide close to its mouth, and the tube *a* furnished with a valve which could be closed while changing the tubes. With the arrangement in which the tube is closed some distance from its mouth, it is necessary to open the valve and allow the water to flow out until it flows quietly, and then at a certain moment to begin to collect it, and cease before closing the valve; because after the valve is opened the wide tube *a* must be filled, and emptied after closing it, which gives inaccurate results. But this procedure always makes the room very wet. The revolving tube *c*, fig. 263, serves to compare the vertical height to which the stream is thrown with the height of the column, and the range at different elevations with each other and with the vertical height, and to exhibit the parabolic curve of the stream.

*Fig.* 264.

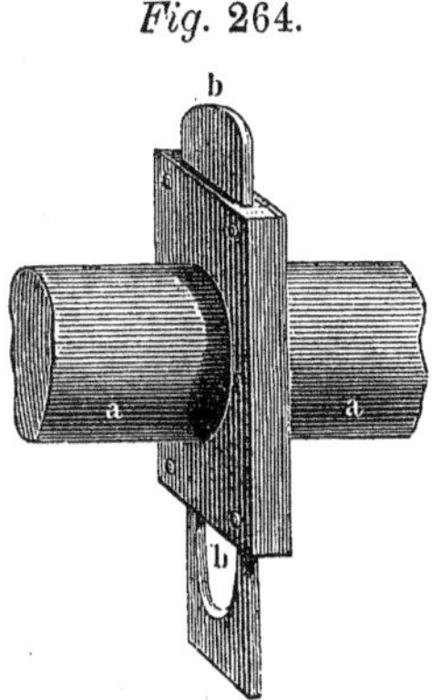


The fourth opening, in the bottom of the vessel, serves for the same experiment as *m* and *n*; the stand has a hole through it for this purpose. The tray around the bottom of the vessel is to catch any water that may leak through the valves.

The quantity of water which has flowed out is most easily ascertained by weighing, and for this reason it is very convenient to have the apparatus graduated by French measure, so that gramme weights can be used, which saves all trouble of reduction.

If one of the outlet tubes be provided with several conical openings instead of one opening in the thin wall, their effect may also be shown. The effect of long cylindrical tubes on the quantity of water delivered may be shown by taking two pieces of the same glass tube, one four times as long as the other, and inserting them successively in the aperture by means of corks.

The duration of the experiments must be estimated by a watch with a second hand, or a seconds pendulum with audible tick. The time must in this, as in all other similar cases, be counted from 0 and not from 1 as is apt to happen.

To experiment on the range, the horizontal range should be chosen. From the height of the aperture above the floor calculate the time required to fall this distance, and from the observed velocity, the distance of projection for this time: mark this on the floor, and the stream will always fall exactly on the spot indicated; whereas the vertical projection will be at most only about 95 per cent. of the height of the column.

In experimenting on the quantity flowing out, the outlet tubes used must not be too small, in order to obtain actually 63 per cent. of the theoretical efflux, *i.e.* of the quantity obtained by multiplying the area of the opening by the calculated velocity. Of course the area of the aperture must be very accurately determined, within at least 0·1 millimeter. If the apparatus be ordered of a mechanic it should be required that the diameter of the openings should be expressible in as small units as possible, *e.g.* in millimeters.

Such a vessel as is here described is rather expensive, but the experiments can be made with Mariotte's vase, to be described directly, or with any vessel of sufficient width and about 6 inches in height. Bore a hole exactly 2 millimeters wide in the side of the vessel, and another a centimeter wide, and solder a short tin tube to it. Close the openings with cork, and do not begin to collect the water until it flows quietly. To change the apertures the vessel must first be emptied. With Mariotte's vase this is not necessary, the air tube need only be closed and the vessel inclined to one side to insert any tube.

[126] **Mariotte's vase.**—A small vase of this kind is an interesting piece of apparatus from theoretical considerations, and is very easily constructed. A stout glass bottle is closed with a good cork, fig. 265. A piece of barometer tube *a*, open at both ends and drawn out a little at the lower end, is heated nearly hot enough to burn the cork and pushed through a hole bored in it a little smaller than the tube, by which a very tight joint will be made. Bore a hole in the side of the bottle with a copper ring about half an inch in diameter, and cement over it a tin tube a little wider than the opening, with a rim fitting accurately to the surface of the glass. This tube may be arranged to receive various escape tubes, or a short glass tube *v* fitted into it with a cork.

*Fig.* 265.

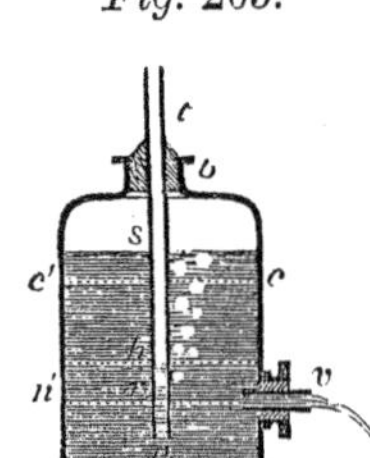

This apparatus is very convenient when a continuous, uniform, and not very large stream of water is required for any purpose. For experiments on efflux proper it must have a capacity of 3 or 4 quarts and allow a vertical pressure (*i.e.* the distance between the lower end of the tube *a* and the level of the escape pipe) of about 12 centimeters. Although the pulsations in the stream caused by the entrance of air bubbles are very perceptible with a lower column of water than this, tolerably correct results will still be obtained. Tubulated bottles suitable for this apparatus may be obtained at small expense.

If the object be only to obtain a uniform stream of water, the bottle does not require any aperture in the side; it is only necessary to bore two holes in the cork, and insert beside the air tube a siphon reaching nearly to the bottom of the bottle.

[127] **The character of a stream of water flowing vertically downward** can be best shown by means of a tin vessel with a small hole in its bottom, uniform pressure not being required in this case. The same is shown, apart from the contraction of the stream, by water or quicksilver flowing through a siphon.

[128] **Barker's mill.**—As the object in instruction is more to show the effect of lateral pressure than to obtain the maximum of effect, it is not necessary to have the arms of the machine bent, and it may be made, in a very simple manner, of tin, as shown in fig. 266. The iron pin *a*, soldered to the bottom of the cylinder *A*, turns in a depression in the iron plate *b*. The bar *c c* runs across the open top of the cylinder and to it is screwed the pin *d*. A strip of tin *C C*, strengthened at the edges and bent twice at right angles, is soldered to the vessel *B*, and serves as a guide for the pin *d* and a support for an easily moving pulley *e*. A string is wound around the pin and passes over the pulley, to the end of which a small scale pan to contain weights is attached. If the experiment is to be continued some time the lower vessel *B* should have an outlet tube. *A* is easily kept full.

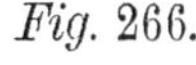
*Fig.* 266.

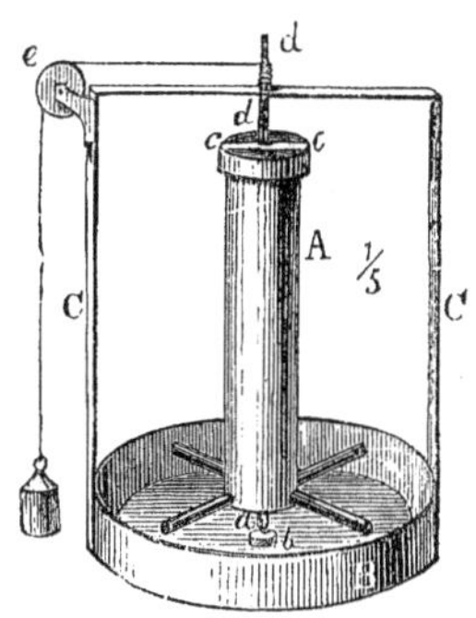


[129] **The water-ram.**—Mohr has devised an apparatus on a small scale, which exhibits the action of this machine very well, and can be constructed without much expense. It is represented in figs. 267 and 268. The cistern *h* is supported on a frame which contains another cistern below, upon the bottom of which the lower parts of fig. 267 rest. The tube *f* which conducts the water downward is made either of glass or lead: the lower end of it is cemented into the valve shown in natural size in fig. 268; the part *g* which passes through the cork must be of glass. After the end of *g* is passed through the cork, a tin tube with a simple flap valve is cemented to it. The tube *p* serves as the jet. The puppet valve, fig. 268, consists of a flat metal plate, with a rod working through

the stirrup *d*; the length of the stroke is regulated by the ball *r*, which screws up and down; the weight may be adjusted by little plates of lead

*Fig.* 267.

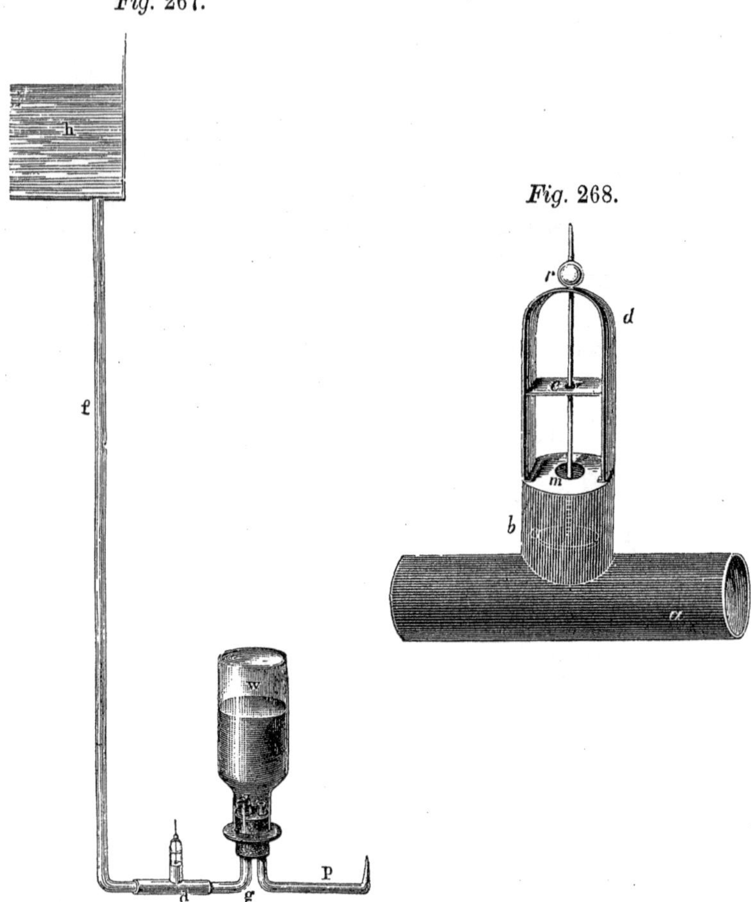


*Fig.* 268.

strung on the rod. The tubes *a* and *b* are made of tin. It requires some patience to regulate the valve, for the apparatus does not always work at first.

## (c.) EXPERIMENTS ON THE MOTION OF GASES.

[130] **The gasometer.**—The form of gasometer which is generally used is shown in fig. 269, $\frac{1}{10}$ the natural size. The cistern $b$ is connected with the vessel below by two solid supports $c$ $c$ and two tubes furnished with stop-cocks, of which $a$ reaches from the bottom of the upper vessel nearly to the bottom of the lower one, but $b$ only enters the arched top of the lower vessel: $d$ is a short tube sloping upward, and capable of being closed with a cork; $e$ is the discharge pipe; and $g$ a gauge to show the height of the water in the vessel. It is well to have the gauge lie in a depression in the side of the vessel made to receive it, where it will be less exposed to injury. To fill the gasometer with water, close $d$, open the stop-cocks at $a$ $b$ and $e$, and pour water in the upper vessel $B$, which will expel the air through $b$ and $e$. When the water begins to flow through $e$, close the stop-cock and expel the rest of the air through $b$. To fill it with gas, close all the stop-cocks, open $d$, and introduce through the opening a gas tube bent slightly upward; the water will flow out beside it as fast as the gas enters. When the vessel $A$ is filled with gas, close $d$. The gas may either be drawn off through $e$, or allowed to rise through $b$ into a bell glass. In either case the vessel $B$ is filled with water and the cock $a$ opened. The gasometer may be cheaply constructed of zinc, and the stop-cocks purchased from a plumber or a gas-fitter. The stop-cock $e$ need not end in a screw; the tubes may be connected with it by india-rubber.

*Fig.* 269.

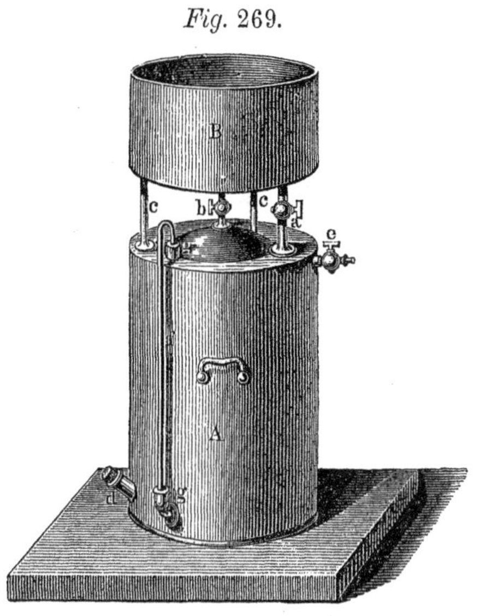


A cheaper gasometer may be made on a small scale like those used at the gas-works. Fig. 270 represents one of this kind made of zinc, with a leaden tube which serves both to introduce and deliver the gas. If made of larger dimensions, so as to have the inner cylinder much over 5 or 6 inches wide, it becomes too heavy when filled to be carried about.

In this case, the lower cylinder should contain an air-tight hollow cylinder, so as to leave only a narrow space between the sides of the two for the

*Fig.* 270. *Fig.* 271.

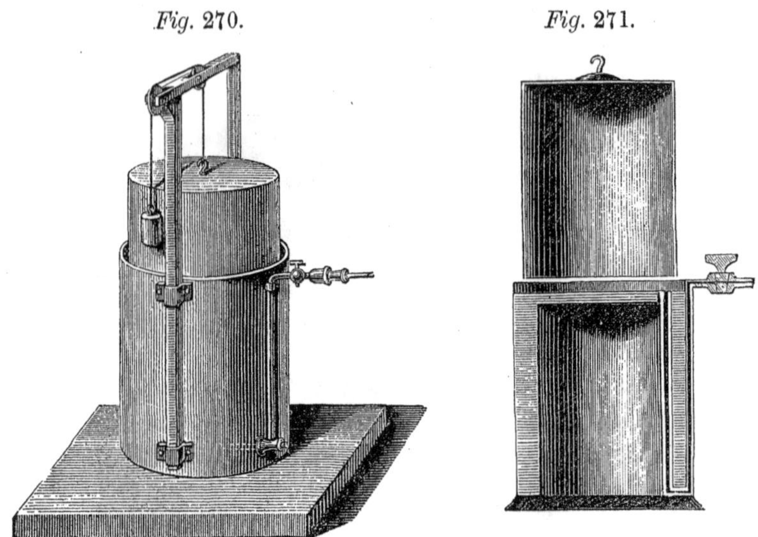

water into which the upper cylinder dips. Fig. 271 represents such an apparatus in section; the conducting pipe being carried up along the side of the inner cylinder.

To fill such a gasometer, open the cock, remove the weights from the movable receiver and let it sink slowly into the water, then connect the gas pipe with the generator by means of a gallows screw or india-rubber tubing. The counterpoise weight should preponderate just enough to overcome the friction.

[131] **Flow of gases through tubes.**—Illuminating gas is the most convenient for these experiments. Fill the gasometer with this, connect with it a very small lead pipe 1 foot in length, open the cock and light the gas at the mouth of the tube, and observe how much burns in a given time. Then replace the tube by one of the same size but much longer, not neglecting the length of the tubes in the gasometer in the comparison, and make the same observation as before. The size of the flame gives a very decided indication of the diminution in the quantity delivered, when the pressure on the gas does not exceed one centimeter of water.

The gasometer may be filled by connecting it by a flexible tube with a gas-burner. A pressure-gauge like fig. 272, but double the size of the drawing, is screwed into the top of the receiver and the requisite weights laid on.

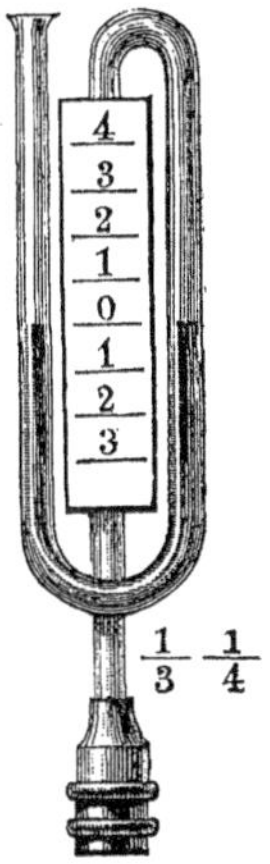

[132] **The experiment of Clement and Desormes.**—Take a circular plate of tin *M M*, fig. 273, from 2 to 4 inches in diameter, with a hole in the center: solder a collar *b* to this, and cement a glass tube *a* into it. Blow into the tube while holding the plate vertically, with a sheet of paper parallel to it and $\frac{1}{2}$ an inch distant: the paper will instantly close over the opening and oscillate backward and forward before it. The apparatus may be varied by passing 3 pins through the disk *M M*, sticking a disk of pasteboard of the same size on these, so as to move freely, and bending them over; the tin disk is then held horizontal, as seen in the figure.

*Fig.* 272.

4
3
2
1
0
1
2
3
$\frac{1}{3}$ $\frac{1}{4}$

The apparatus is still further simplified by thrusting the glass tube through a short cork, cutting it off even with the tube, and then gluing a disk of thin smooth pasteboard, 2 or 3 inches wide, to the cork, and piercing a hole through it. Near the edge stick three smooth pieces of wire, and on these a disk of thick paper of the same size as the pasteboard. The holes in the paper must be tolerably large, and the ends of the wires bent over.

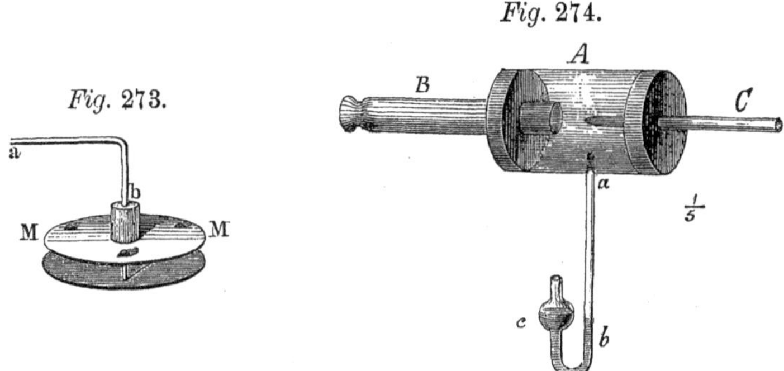

*Fig.* 273.

*Fig.* 274.

The action of the draught in locomotives depends on the same principle, the waste steam being allowed to escape into the chimney, as may be shown by the apparatus fig. 274. *A* is a short, wide glass tube, which may be cut off from a lamp chimney. Corks are cemented into both ends, and through one of them is passed a wide tube *B*, through the other a narrow tube *C*, drawn out to a point. A hole, $1\frac{1}{2}$ or 2 lines in diam-

11

eter, is bored in the side of the tube *A*, into which is cemented a glass tube *a b c*, not over 1 line in diameter, bent like a siphon. The bulb at *c* is convenient but not at all necessary. If the tube *a b c* be partly filled with water, it will rise in the tube by blowing in at the narrow tube *C*, and fall when the blast is in the opposite direction. The most effective position is sought by moving the tubes back and forward, for which reason they should not be cemented.

The experiment may be still further varied, by laying a few burning coals in a funnel-shaped expansion of the tube *a b c*, and blowing through the narrow tube: a good draught will be created over the coals.

# CHAPTER III.

## EXPERIMENTS ON ACOUSTICS.

[133] **Waves in water.**—It is necessary here to recall observations which might often be made, in respect to the simplest phenomena. For this purpose a large tub filled with water is used, into which a drop of water is let fall from a pipette, to show the effect of a single drop, and then several drops, on the same spot, at regular intervals, to form a system of waves. In the same manner two systems of waves are created, whose centers are at some distance apart, to show the interference of the two systems. Finally, the reflection of the waves by the sides of the tub, or by a board placed in it, is exhibited. By placing in the tub a board with a slit an inch wide, the bending of the waves may also be shown.

Müller's stroboscopic disks, and Wheatstone's apparatus, as improved by Eisenlohr, are very well adapted to exhibit these, and all other phenomena of undulation. The latter is rather expensive, but has the advantage that the phenomena can be shown to the whole audience at once. The following apparatus can easily be made anywhere. Fig. 275

*Fig.* 275.

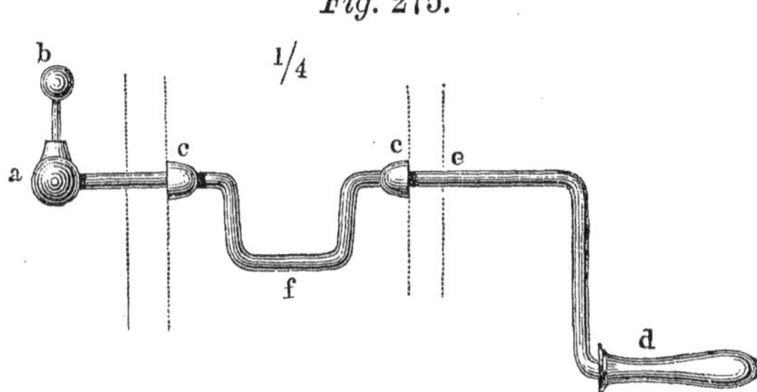


shows an axis made of bent wire, terminating at one end in a screw, on which the wooden ball *a* is screwed, which carries in turn a large glass bead *b* on a slender wire. The other end forms a winch. Two shoulders *c c* soldered to the axis prevent it from slipping about. About twenty such wires, without the shoulders *c c* or the winch, and about half as thick, are fixed in the box *A*, fig. 276, which furnishes a common support for all, the axis fig. 275 being placed in the middle. The sides of the box are made in two pieces, as shown by the dotted line, and the wires are laid in their places before putting the pieces together, the knobs being all on the same side. The cranks *f* of all these wires are now connected by the bar *e*, fig. 276, which consists of two strips united by screws, so that all the axes can be turned by the motion of one winch. The knobs *a* are now turned until the glass beads *b* occupy the positions shown for a succession of twelve in fig. 277.

*Fig.* 276.

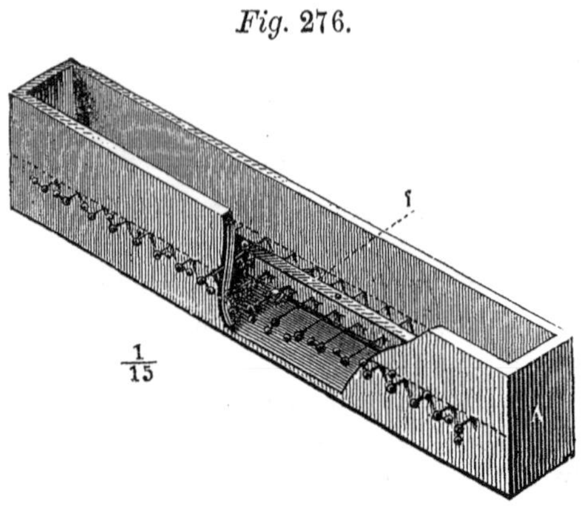

The following is a pretty experiment on the Reflection and Interference of Waves of Liquids: make a trough of wood or pasteboard, 3 or 4 inches long, with elliptical sides. If it be of wood, cut an elliptical hole through a piece of hard wood, and glue a bottom on it.

*Fig.* 277.

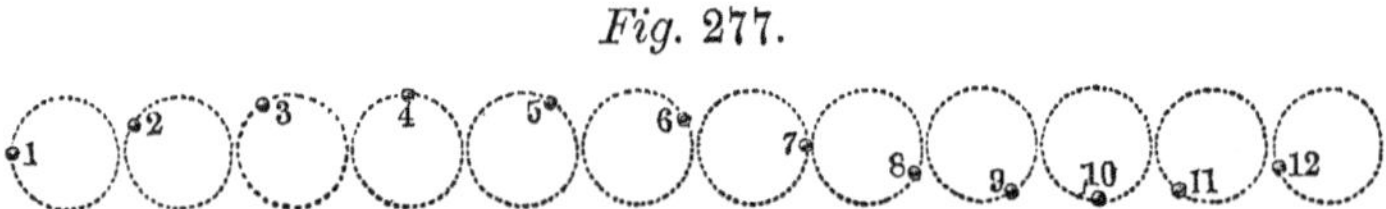


Fill the trough with mercury, and let drops of mercury fall from a pipette on one of the foci: the waves reflected from the sides will interfere with the direct waves and form a little elevation at the other focus. It is well to have the pipette fastened in some way over the focus, in order to let the drops fall with better aim.

[134] **Waves of a rope.**—The rope used for this purpose should be as long as the room allows, and not too thick: for a length of 20 feet, $\frac{1}{4}$ of an inch is thick enough; it must, above all, be very soft, and, therefore, when a suitable old rope cannot be obtained, the new one must be rolled up and beaten with a wooden mallet on a block until it is quite soft. One end of the rope is fastened to the wall, and the other end held slack in the hand. A horizontal jerk produces a wave, varying in length with the extent and rapidity of the jerk, which runs to the other end, is there reflected back to the hand, and this is repeated once or twice with decreasing intensity.

By continuing to move the hand, it is easy to form a succession of waves of such length that they will be aliquot parts of the whole length of the rope, and the reflected waves will then coincide with the succeeding forward waves.

[135] **Nodes of vibration in bars.**—In the holder described in one of the succeeding paragraphs, fix a bar of straight-grained pine wood, about 1 centimeter square and 3 or 4 decimeters long. By strewing the upper side with sand, placing the finger at about one third the length of the bar, and drawing a violin bow in a vertical direction over the bar near the finger, with a motion toward the holder, a node of vibration may be produced.

[136] **Nodes of vibration in strings.**—These are exhibited by the monochord, to be described presently. Care must be taken to make the strips of paper which lie on the nodal points very narrow; they may also be of a different color from the rest. Fig. 278 shows such a rider of the natural size. The degree of tension of the string is indifferent, if it only give a sound, and the bridge set under it form a little elevation, so that the string presses hard upon it. The vibration is best produced by a violin bow, laid on near the bridge, and drawn at right angles to the string.

*Fig.* 278.

[137] **Nodal points on surfaces. Sound-figures.**—In selecting panes of glass for this purpose, choose clear glass of uniform thickness about equal to that of ordinary window glass. When it is thicker it is not so easy to produce complicated figures. The size is indifferent, but it is difficult to obtain complicated figures with small panes, and large ones break easily: 1 to 2 decimeters square is a good size. The sharp edges must be rubbed off on a grindstone. The apparatus fig. 279 is very convenient for holding the plates; it is made of iron or some tough wood. The little knob *a* and the round end of the screw are covered with thick buckskin. It is clamped on a table by the screw *c*. The vibrations should be produced with a violoncello bow, the violin bow is too weak; it is stretched tight and well rubbed with rosin.

*Fig.* 279.

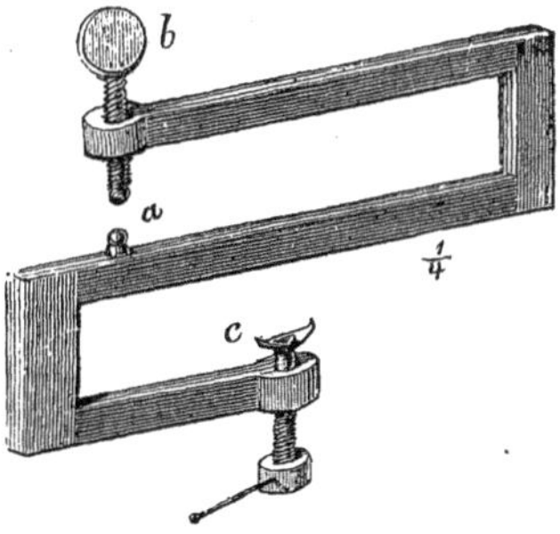

To produce any given sound-figure: clamp the pane of glass at a crossing point of the nodal lines between *a* and *b*, strew the surface with fine sand, hold the tip of the finger on a point of the edge of the glass near the point of attachment, to which a branch of a nodal line is to run, and draw the bow vertically downward across a part of the edge where the glass must have most motion. The stroke must generally be repeated several times before the figure is developed. Writing-sand is best for the purpose. A figure different from that sought will often be produced; this depends upon the variable elasticity of the glass in different directions. If it does not succeed with the three points chosen, try others. For the purposes of instruction it is, however, sufficient to show the perpendicular and the oblique cross on a square plate, and the cross and six-pointed star on a round disk, and these seldom fail. For all of these the plate must be held by the middle, the finger held on one limb of the figure, and the bow drawn across the edge between two limbs—that is, for the vertical cross on a square plate, near one corner. After each experiment, the leather which covers the knobs *a* and *b* must be carefully freed from sand, otherwise a grain of sand may be forced into the glass by the pressure of the screw, and then the glass will break when the bow is drawn across it. But with every precaution, glasses will often be broken in this way; this is an unavoidable difficulty with glass, and is all the more annoying when it is desired, for an experiment to be mentioned presently, to produce always the same pitch. Metal plates would, therefore, be unquestionably superior, if it were not so difficult to procure them of tolerably uniform elasticity. Complicated figures, which belong to high notes, are easily obtained with a flat plate of brass, but the simple ones not so easily. The desired result is most easily obtained in the following way: have a brass plate about 1 to $1\frac{1}{2}$ millimeters thick and 6 or 8 inches on a side, beaten flat with a wooden hammer, file it square, and keep it a uniform red heat for 5 or 10 minutes in a heap of coals briskly fanned. After cooling slowly, polish it first with pumice-stone and water, and afterwards with a piece of beechwood charcoal and oil. If the brass be good, the plate treated in this way will always give the two crosses.

To render the nodal lines of bell-shaped bodies visible, any wide glass cylinder will answer, or even a thin-edged tumbler. Fill it half full of water, and draw a well-rosined violin bow across the edge: holding the glass fast if necessary by two fingers near the bottom.

[138] **Propagation of sound in tubes.**—Tubes of gutta-percha, peculiarly adapted for this purpose, may now be obtained of any length.

[139] **Reflection of sound.**—A concave mirror designed for the reflection of heat may be used in these experiments. A watch held in the focus can be heard at a considerable distance, when everything around is still.

[140] **The theory of the organ pipe** is illustrated most

clearly by means of Professor Müller's wave-disks. They, and the stroboscopic disks in general, should be held 1½ to 2 feet from the mirror. Their construction is explained in § 177, fig. 351.

[141] **Experiments with pipes.**—A whole octave of wooden pipes may easily be had from an organ-builder. A bellows with a wind-chest to fix the pipes into is convenient, but not necessary; if one is purchased, it should be arranged for about 12 pipes, and have keys. Into one is inserted a plug, like fig. 280, the head of which is covered with leather on the sides, but not on the base, for experiments on closed pipes. The depths to which this plug must be pushed in to produce the various notes of a whole octave, and such others as may be required, should be marked on its stem. For the purpose of showing the nodes of vibration, or rather the waves between them, in such a pipe, holes may be bored with a center-bit, about 2 or 3 lines in diameter, and provided with keys. These keys are easily made of a piece of wood, cut as shown in fig. 281. A piece of sheepskin is glued to the bottom of the key and the projecting end *b* of this leather fastened to the pipe, with the flesh side downward; a wire spring *c* serves to press the key against the opening. The holes may also be closed with simple slides, but this requires more accurate workmanship. The simplest way is to close them with the fingers, but then the pipe can be used for no other purpose, and fewer holes must be made. A pipe for experiments of this kind must be comparatively long, otherwise the waves will not form at exactly the spot required by theory, because the vibrations in a pipe, in which the vibrations of the lip are produced by a small current of air are not regular near the end. The pipe must be at least two feet in length. Four holes may be bored in the same pipe, one for the closed and another for the open pipe, in the proper places, and two others at points not corresponding to any belly, for the purpose of showing the effect in this case also. The pipe must, however, be closed, not by a plug, but by a pasteboard cap, so that the length of the tube will not be altered. The hole for the open pipe is made in the middle of the length, that for the closed pipe about ⅓ the length from the bottom.

*Fig.* 280.

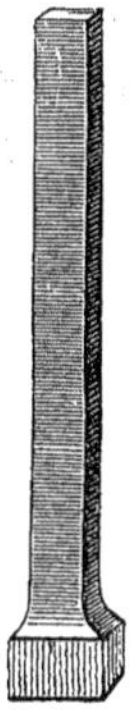

*Fig.* 281.

[142] To show that the pitch is not affected by the *Material of the Pipe*, a mouth-piece may be made as in fig. 282, and pipes of wood, tin, pasteboard, etc., of the same size and length, set on it; they will all give the same pitch but not the same timbre. This difference of sound is still

greater when the lip is made on the piece set on. But as the thickness of the wood at *a* acts as so-called "ears," making the tone deeper, similar pieces must be soldered on when the pipe is made of tin. The introduction of gas into the building makes it very easy to show the influence of the kind of gas blown through a pipe, on the pitch. It is only necessary to connect the pipe with a gas burner, by means of an india-rubber tube, to obtain a uniform tone, about one second higher than usual. With narrow pipes the experiment succeeds when the gas has a pressure only equal to 1 centimeter of water.

*Fig.* 282.

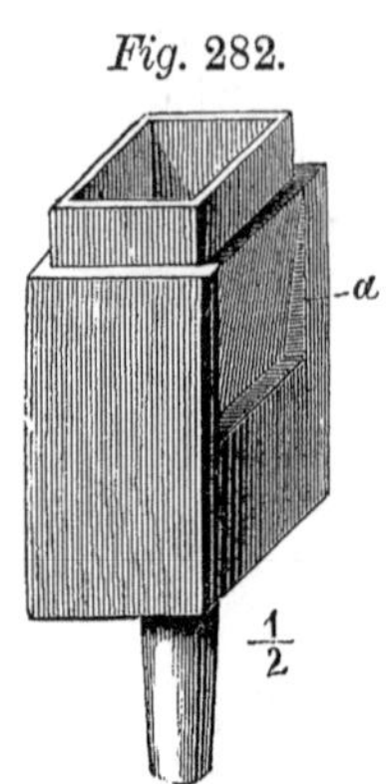


To make Savart's experiment, in which sympathetic notes are produced in a tube by the vicinity of a sounding body, the following arrangement is best adapted: Make two tubes, about 3 or 4 inches in diameter, of pasteboard, so as to slide into one another, the larger of which is provided with a bottom. Mark a scale on the narrower tube, on which the length of the pipe can be read off directly. Select a glass bell giving a rather deep sound, and determine its pitch by comparing it with a piano, so as to be able to calculate the corresponding length of the pipe by assuming the number of vibrations of ā as 440, and dividing the velocity of the propagation of sound by the ascertained number of vibrations, in order to obtain the length of the wave, the fourth part of which is the required length of the closed pipe. The tone of the bell should fall within the minor octave, for with high notes the result is very doubtful. Almost any vessel which is not too narrow can be made to resound more or less with a tuning-fork; it succeeds best with one about 7 inches deep and 2 to 4 inches wide. The effect is far greater with deep tones. When the length of the tubes is arranged, place them upright on the table, hold the bell by the knob, and draw the bow across it so as to produce its deepest tone; this will often be tried in vain, but once obtained, it can be easily repeated. Hold the side of the vibrating bell close to the mouth of the tube.

*Fig.* 283.

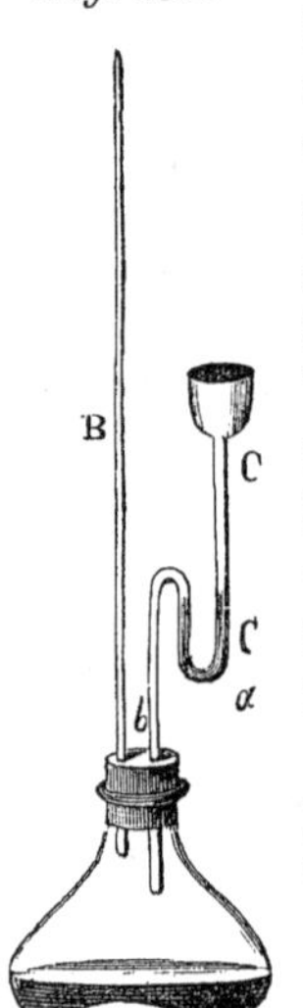


[143] **The chemical harmonica.**—For the experiment use a wide-bottomed flask, as seen in fig. 283, into which a long cork, pierced with two holes, fits tightly. Fit two glass tubes *B C* tightly into these holes, without wax. Draw out the tube *B* to a point, so as to leave only

a fine opening; *C* must be bent into the shape of a safety tube. The funnel on the end is useful to introduce fresh acid into the flask without taking out the cork. Put in zinc, or iron filings, and sulphuric acid diluted with four times its weight of water to evolve hydrogen, and when the evolution is going on briskly let the flask stand for 10 minutes to allow the expulsion of all atmospheric air and explosive mixture before lighting the gas issuing from *B*. For greater safety, the flask should be wrapped with a strong cloth before lighting the gas. The lower end of *B* should be cut off obliquely, so that the drops of condensed water will fall off more readily.

The jet must be of considerable length, and the tubes raised and sunk until the loud, and anything but agreeable, sound is produced. Tubes of various lengths give sounds of different pitch.

[144] **Hopkins' experiment to show the nodes of vibration.**—Fasten a tube, $1\frac{1}{2}$ to 2 inches wide and 2 to 4 feet long, to the perpendicular part of a screw-clamp, and screw it to a table. A glass plate would answer very well for the vibrating plate, but on account of the fragility of glass, a brass plate is preferable, because it is necessary always to produce the same tone. The plate needs no special preparation. Fasten it by the center in the holder, fig. 279, place it immediately under the tube, and endeavor, by holding the finger and drawing the bow at different points, to produce a rather simple sound-figure. Fig. 284 shows the arrangement of the whole apparatus. The reciprocal action of the tube on the plate induces in it much more readily a tone corresponding to the tube; and it will be known at once, from the intensity of the tone, whether the air in the tube is sounding in unison. Having once obtained a figure, such that a division of it corresponding to the edge of the plate is about equal to the diameter of the tube, mark on the plate the spots where the finger was held and the bow drawn, for, on account of the unequal elasticity of the metal, it is seldom that any other points can be selected in place of them. The tube must then be placed at a distance of $\frac{1}{2}$ to 1 line above as large a division of the figure as possible: one near the edge is best. Instead of the little frame usually figured covered with a thin membrane, it is better to take a metal hoop about 1 or 2 lines broad and half as great in diameter as the tube, and cover it with very fine paper in the same way as in stretching paper on a drawing board. This has the advantage that the sand does not slide off continually as it does on the square frame. A difference in tension of the paper according to the pitch is not at all necessary, if it were practicable. Suspend this by three strings uniting in one, and pass the string over the upper edge of the tube which is coated with paper, so that the frame may be kept at any

height by a counterpoise weight. If a special holder for the tube is desired for the sake of convenience, the one figured in fig. 285 is very good; it is made of wood.

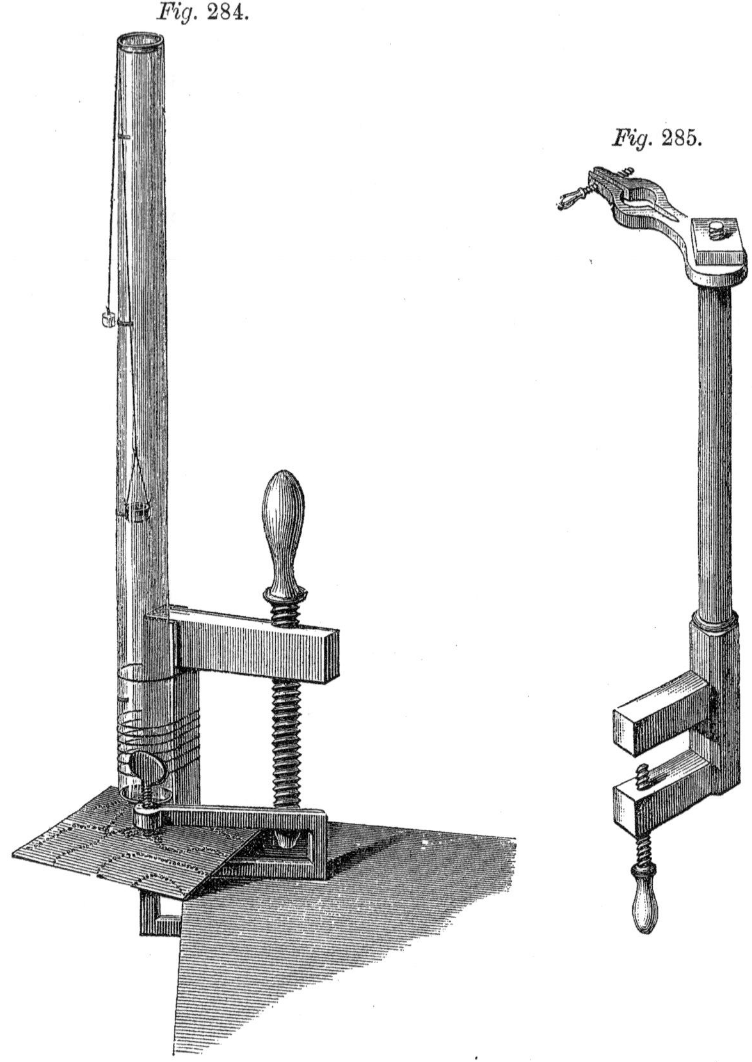

*Fig.* 284.

*Fig.* 285.

Now seek for the position in the tube in which the sand in the frame is least agitated when the bow is drawn across the plate. Having found two such positions approximately, it will be easy to deduce from their appropriate distance and the length of the tube, their true distance apart, since the distance between the nodal points is an aliquot part of the length of the tube, so that the highest and lowest knots must lie at half this interval from the end of the tube. As soon as the correctness of these positions is verified by experiment, mark them, by pasting slips of paper opposite them, so as to be able to exhibit quickly the difference in the motion of the sand according as the frame is on a nodal point or a wave. To show the same thing with a closed pipe, make a pasteboard tube closed at one end, as long in the clear as the interval between the nodes, and just wide enough to fit over the glass tube. Make a little hole in the middle of the bottom, and draw the string through it. If this cap be put on the glass tube so that its lower edge reaches the last nodal point, the whole length of the tube is increased by half a wave length, and the sand in the frame will show the knots still in the same place; the motion of the sand in the waves will even be livelier than before. If the cap be raised or depressed a little, the tone will become weaker and the sand will scarcely move at all.

The length of the tube may be found from the nodal interval, and the pitch deduced approximately from this by taking ā = 440 vibrations, and this compared again with the piano or monochord. Although this is not suitable for class instruction, it is good practice for one who must study these things closely. The same experiment may be repeated with various notes from the plate, in order to attain the requisite dexterity and certainty.

[145] **The siren.**—If the siren has an index, it may be connected with the bellows of the blast-table, by substituting a bent glass tube for the brass one, and connecting this with the tube of the siren. A seconds pendulum may be used to measure the time. The pitch of the siren may be altered and maintained at pleasure, by changing the weights on the bellows. The real siren has two disks revolving one above the other, and the sound is produced by the joint effect of numerous holes, by which means even the low notes become clear. When only one tube blows against the apertures in a wheel, the low notes are indistinct, and are rendered inaudible by the unavoidable hissing. Any siren will give high notes distinctly enough. The whirling machine may be made to serve instead of a real siren. Screw upon the axis of the machine the block *a a*, fig. 286, which has a wooden screw above, over which disks of pasteboard may be slipped, and kept down by a wooden burr *b b*. The pasteboard disks must be very even and perforated with holes about a line in diameter, at regular intervals near the circumference.

*Fig.* 286.

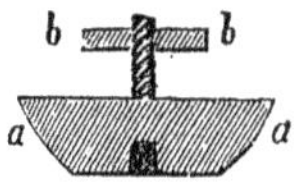

The holes may be made with an ordinary punch. A second and third row of holes may be punched in the same disks, and the intervals should always very slightly exceed the diameter of the holes.

Bend a stout wire, as in fig. 287, and fix the clamp *b b b* on the side of the base board, so that *a* will stand directly over the row of holes; fasten the blow-pipe to this so as to come as near to the disk as possible without touching it. The blast-pipe may be a wide tin tube with a fine hole in the bottom, or a glass tube drawn out to a point. In either case a mouth-piece of glass is fitted to it by a flexible connector. A strong blast not being required, it is generally produced by the mouth, and the wheel

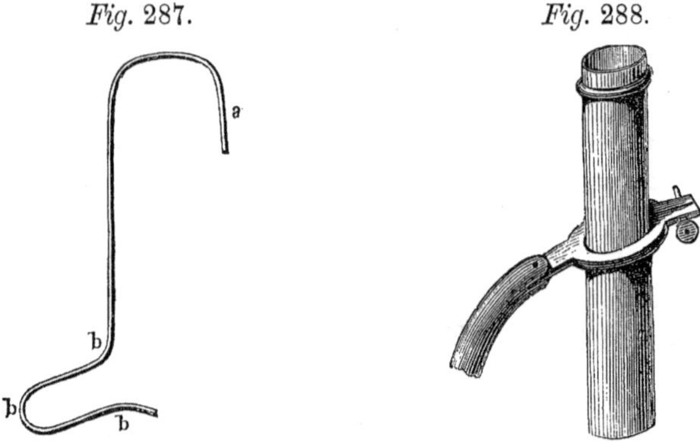

*Fig.* 287. *Fig.* 288.

turned more and more rapidly until the desired pitch is attained. With a heavy driving-wheel the pitch may easily be maintained long enough to observe, with the seconds pendulum, how many revolutions the driving-wheel makes in a second. From the ratio of this wheel to the pulley, and the number of holes in the disks, it is easy to calculate the number of vibrations corresponding to the pitch. The experiment should be made by two persons.

Unless the driving-wheel be heavy, a low note cannot be maintained uniformly, but high ones may even with a light wheel.

The upper part of the bent wire might be finished as in fig. 288.

A wheel with sharp teeth may be fixed concentrically on the axis of the whirling machine, and made to strike against a card held nearly parallel to the surface of the teeth, producing a musical tone.

[146] **The monochord.**—The monochord consists of a rectangular frame *A B C*, fig. 289, made of strong hard wood, with a sounding-board of straight-grained pine free from knots. Two bridges *a a, b b* rest partly on the sounding-board with their perpendicular side toward it.

The sounding-board is coated with white paper. Two steel piano wires are fastened at one end to slanting pins *c c*, and one of them stretched by

*Fig.* 289.

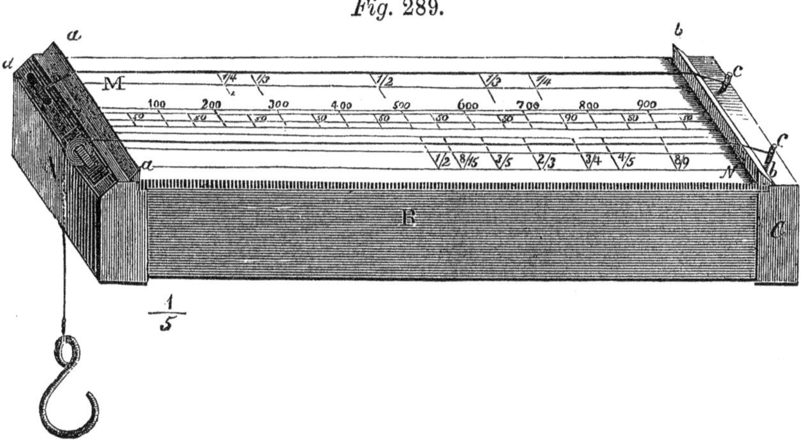


the pin *d*, the other by a weight working over a pulley. The sounding-board must be very even, so that every point of the string will be equally distant from it. In order to restrict the vibration to a given part of the string, it must be held by a bridge which will not alter the tension. This is easily made by taking two pieces of hard wood with sharp projecting edges, covering them on the inner side with leather, and fastening them together by a leather hinge. The lower one must project somewhat and have a mark drawn from the sharp edge perpendicular to the base, fig. 290; this mark serves to place the edge of the bridge directly over a division of the monochord. The upper piece is then shut down and held by the finger, or a small weight. The string is thus held fast without injury, and the rest of the string prevented from vibrating. If both parts are to vibrate, a simple bridge like fig. 291 is set under the string which alters its tension somewhat; but this is of no importance in this case, the only object being to show the nodes of vibration.

*Fig.* 290.

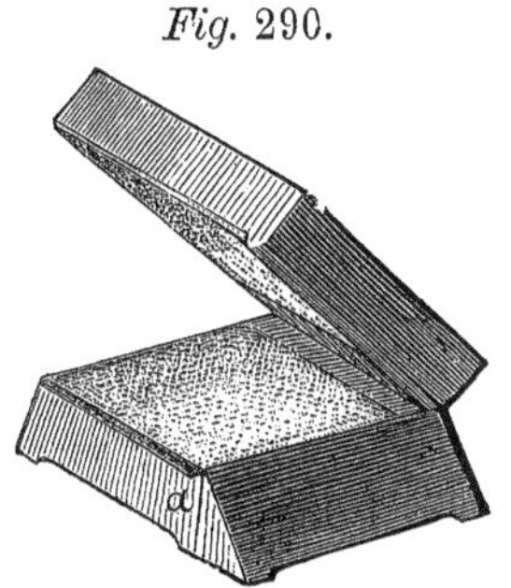


For this purpose a line is drawn under the string, which is stretched on a pin, and divided into thirds, fourths, and fifths. Set the mark *a* of the bridge, fig. 291, over one of these lines, hang little paper riders on the wire, and set it in vibration, with the finger or violin bow.

*Fig.* 291.

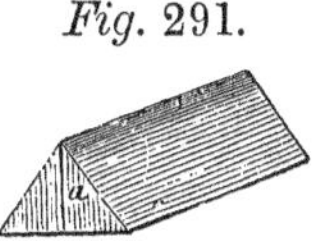

The monochord may be used to show that the number of vibrations is inversely as the length of the string. A line is drawn under the string which is stretched by a weight, and marked with divisions corresponding to the notes of the octave, as shown in fig. 289. The string is drawn tightly enough to give a clear note, with which the other string is made to accord. The string must be slightly raised with the finger, so that the effect of the weight will not be lessened by friction on the bridge. Set the bridge, fig. 291, on the proper division, the segment of the string will give the note required, which may be compared with the key-note given by the other string.

The monochord can seldom be used to show that the number of vibrations is proportional to the square root of the tension, on account of the length of the string, which would make a greater weight necessary than the string could bear. But by taking the weights = 4 : 9, including the weight of the hook, the string will give the fifth, which may be compared with the other string, previously tuned to the same pitch as the string with the weight = 4.

To show the effect of the thickness, the diameters of the strings must be in a simple proportion, and they must be stretched by equal weights. It will be very hard to find such strings.

Besides these experiments, the monochord is very useful to the physicist in studying the musical intervals, and many other points in acoustics. For this purpose the length of the sounding-board should be divided into 1000 equal parts, and marked under the weighted string for the chief notes of the octave, and all the secondary notes for sharps and flats, and likewise the different temperatures. These divisions may be distinguished by lines of different length, and by different colors. Every one who wishes to study this part of natural philosophy thoroughly should therefore make his monochord as complete as possible.

[147] **Longitudinal vibrations.**—These are most easily produced with a glass tube, about 1 centimeter in diameter, held in the middle with two fingers, and one half gently rubbed lengthwise with a damp woolen cloth. As long as the glass tube gives a sound, its moist surface will seem ruffled. A loose cork introduced into the end of the tube will move while the sound continues, and advance gradually toward the middle. If wooden or metallic rods be used, the cloth must be strewn with rosin, or a short glass tube must be cemented to the rod, and the surface rubbed with a wet cloth. By varying the force of the friction, notes of different pitch may be produced, but it does not succeed well with every glass tube. Wooden rods are easily made to sound. By fixing into a block four rods of fir, the longest $1\frac{1}{2}$ meters long and varying

from 6 or 8 to 5 or 6 millimeters in thickness, they may be made to form together an accord, by trimming off the ends of the smaller ones, fig. 292. The upper half of the rods is rubbed with two fingers dipped in powdered rosin.

[148] **The tuning-fork.** — Tuning-forks of different pitch may be purchased at the music stores. Choose such as give a clear tone near ā, on a sounding-board, and maintain it for some time without rising. Two are needed: one tuned accurately to ā, the other so as to produce an interference with the first about 4 times in a second. They are tuned by shortening both legs with a file when the pitch is too low, and filing off the inside of the legs when too high. Both legs must be made as nearly alike as possible. The fork is made to vibrate either by striking one prong gently against a board and resting the handle on a table or a sounding-board, or by drawing a well-rosined bow across the ends of both prongs, or by striking them with a little wooden hammer with a sharp edge covered with leather; the latter mode produces a very clear sound.

*Fig.* 292.

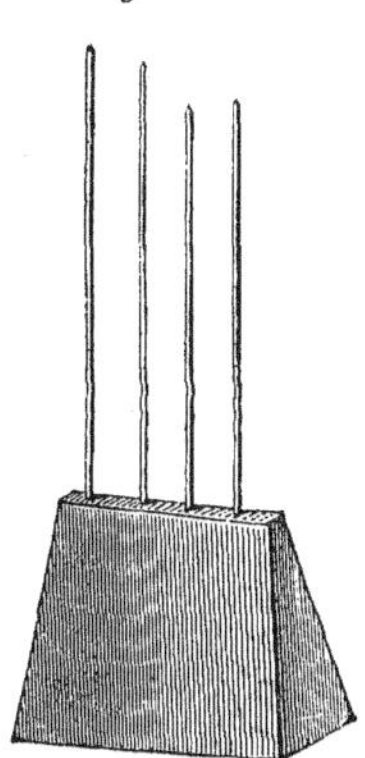

[149] **Interference of the sound-waves.** — To show the interference of the waves of the two prongs of a tuning-fork, hold it while vibrating, horizontally over a vessel 7 inches high and 3 to 4 inches wide, and turn it slowly on its axis: during each revolution the sound will be heard to rise and sink four times. The same observation may be made by simply revolving the fork near the ear.

The pulsations of sound produced by the interference of two tuning-forks set on the same sounding-board, a table, or an empty box, is most distinct when the prongs of both forks vibrate in the same plane. The easiest method of bringing two sonorous bodies into perfect accord is by observing these pulsations. The ear alone is apt to deceive us when the accord is nearly perfect, especially when the timbre of the two sounds is very different.

To produce the interference which the key-note makes with its fifth, take the closed pipe and place the plug so that the pipe will be exactly a fifth lower than another open pipe; mark the position of the plug on its handle for future experiments. By taking the two pipes between the lips and blowing with the same force through both, the next lower octave will be distinctly heard with the other notes. It is, however, not easy to blow with the same force through two pipes without the aid of a bellows.

If a forked tube, like fig. 293, be made of wood or pasteboard, with legs long enough to extend over the screw *b* in fig. 279, to within half a line of the glass plate fastened in the clamp, and the upper end of the tube be covered with thin paper strewn with sand, the sand will be motionless when the two ends of the tube are held over two parts of the glass moving different ways. But if the tube be held over two places which are moving upward and downward together, the sand will be briskly agitated. The latter is the case when the tube is held over two diagonally opposite squares of a pane of glass which gives the simple cross, while the former result is produced by holding it over two adjoining squares.

*Fig.* 293.

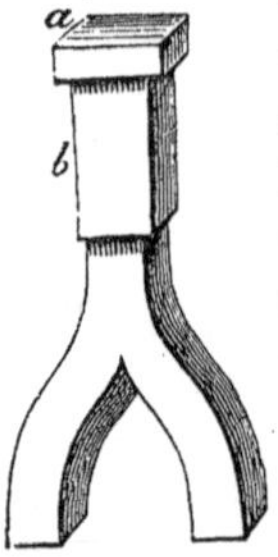

The sand on the paper often forms a sound-figure, which has, however, no relation to the one below. To make this figure clear, the paper must be stretched on a movable cap *a b*, so as to alter the length of the tube as required.

[150] The interference of the waves is still better shown by an experiment of Herschel's, which Nörremberg has repeated in the following simple manner: The tube shown in profile and section in fig. 294 is made

*Fig.* 294.

of wood, and is built into a partition wall, either entirely or so as to leave the greater part of the end *a* free. The sound produced at *a* divides on the sharp edge of the inner partition, and unites again in *b*, after traversing the unequal arms *c* and *d*. If the length of the wave producing the tone be double the difference between the two arms *c* and *d*, the two sets of vibrations will be half a wave apart when they come together, and no sound will be produced at *b*; but the sound will instantly become audible when either arm is closed by a slide fitting accurately in *b*. When the pitch is raised an octave, the two sets of waves will coincide, and the sound will be louder when both arms are open. The two openings *a* and *b* must be in two entirely separate chambers, otherwise the direct wave will interfere with the result.

The sound in *a* is best produced by an open pipe, the length of which is equal to the difference in length between the two arms *c* and *d*: it must be comparatively narrow so that the pitch can easily be raised by a stronger blast; the third octave will then often renew the interference. The axes of the pipe and tube must coincide, and their apertures be together. Fasten the pipe in this position, and connect it with the bellows by flexible tubes and a small stop-cock; it is easy to regulate the blast by means of this stop-cock so that the sound of the pipe shall be steadily maintained. If the bellows act irregularly, let the blast first pass through a large glass bottle, with which the pipe is connected, as seen in fig. 295. With the aid of such a bottle and a long flexible tube, a constant tone may be obtained with the mouth; but the moisture of the breath spoils any pipe after a time.

*Fig.* 295.

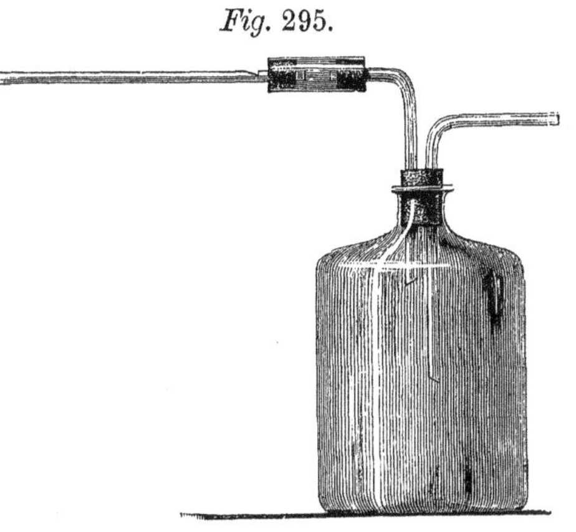

A very steady blast may be maintained, according to Nörremberg, by an arrangement of two large bottles like fig. 296, the upper acting as a Mariotte's vase with a siphon, the lower having besides the pipe a funnel with a stop-cock. If one has no funnel with a stop-cock it is easy to make one with a good cork, as shown in the figure, where the small cork closes the longitudinal opening of the larger more or less. The flow of water, and consequently the pressure of the air, is so regulated by the cock that the pipe will give the tone desired, and the air-tube of Mariotte's vase is set at such a height that the funnel is kept always full. When the upper bottle is emptied, change the corks and their tubes, and take the lower one for the upper, which is the work of a moment.

Patience is needed in making such narrow pipes, for it is not easy to obtain the key-note with them: for this reason they had better be made of glass, and the plug of wood or cork can then be altered until the right

*Fig.* 296.

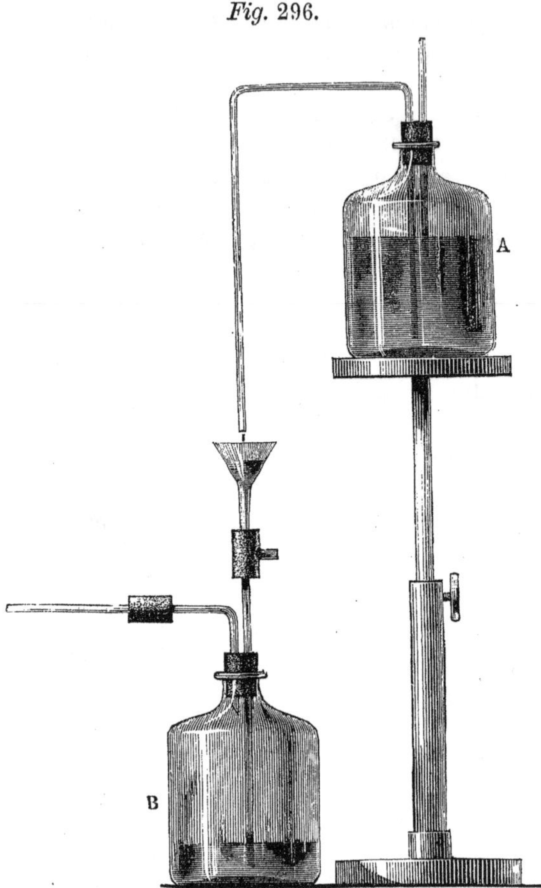


note is obtained. The notch may be cut in the tube with a file dipped in turpentine, and the tube should be cut off in the same way, so as to have exactly the proper length.

[151] **The communication of vibrations** may be shown by simply sounding a clear note, either with the throat or a pipe, near an open piano, the corresponding string of which will sound in unison. The experiment is made much more striking by passing an iron wire, about 2 or 3 millimeters in thickness, through a small hole from one apartment to another remote one, and placing one of its rounded ends on the bridge of

a piano, and the other end on the sounding-board of the monochord. A tune played on the piano will be heard at the other end of the wire, more or less distinctly according to the excellence of the sounding-board, but with a somewhat altered sound.

Two tubes open at both ends and provided with sounding-boards, fig. 297, on which turning-forks of precisely the same pitch are set, show the same phenomenon. If the bow be drawn across one fork, the other will sound in unison, even when at some distance, which may be perceived more distinctly upon stopping the other with the finger.

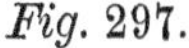

*Fig.* 297.

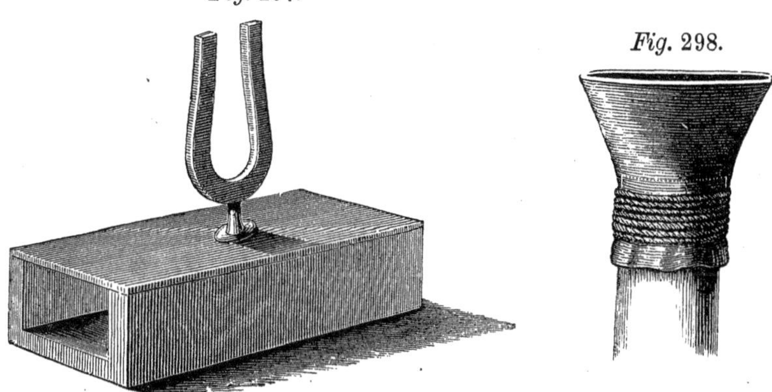

*Fig.* 298.

[152] **The organ of voice.**—The simplest illustration of the general action of the human vocal organ, is furnished by the lips; just as these produce various tones by different degrees of tension, *e.g.* in blowing a trumpet, so the so-called vocal chords act. To illustrate this by a thin membrane, tie a piece of india-rubber tube around a glass tube, fig. 298, and stretch out the sides with the fingers so as to form a simple slit, and blow through it; by varying the tension different tones may be produced.

[153] In exhibiting the structure of the organs of hearing, models of papier-maché and the ear-bones of a calf will be found sufficient for every purpose.

# CHAPTER IV.

## EXPERIMENTS ON LIGHT.

### (*a.*) ON THE TRANSMISSION AND INTENSITY OF LIGHT.

[154] THE general discussion of the nature and distribution of light needs little aid from experiment. A few definitions, such as Reflection, Refraction, Transparency, etc., may be illustrated by examples.

Reflection is best shown by a sunbeam falling on a mirror in a dark room, the direction of the reflected ray being made visible by the illumination of floating particles of dust.

Refraction may be exhibited by laying a silver coin on the bottom of a shallow vessel, fig. 299, so as to be just out of sight, and pouring water into the vessel, when the coin will appear in the direction *i i*.

*Fig.* 299.

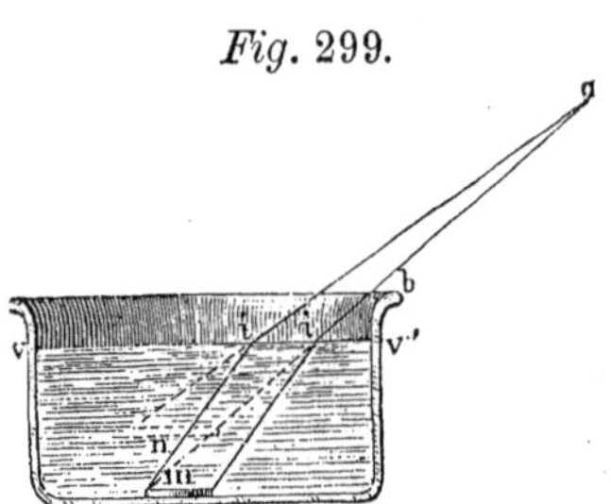

The transparency of thin laminæ may be illustrated by gold. Lay a sheet of the thinnest gold-leaf, about 1 or 2 inches square, on a pane of even glass, cover it with another piece of glass the same size, and paste paper around the edges of the two pieces. To lay the gold on smoothly, breathe on the glass and lay it on a sheet of gold-leaf, and trim the edges with a sharp knife well cleaned on a piece of leather, by simply pressing on the edge rather than drawing it along.

The law of the diminution of light with the distance needs no experimental demonstration. It may be shown approximately by placing four candles in a row on a stand, and comparing their light with that of a single candle of the same kind at half the distance. The comparison may be made either by observing the darkness of the shadows cast on a white wall by an intervening body, or by the photometer.

The candles must all be taken from the same package, and four of them arranged on a stand in a straight line with the photometer, as in fig. 300, so as to measure from the middle point.

[155] **The photometer.**—Among the various means of comparing intensity of the light from two sources, the most convenient are Ritchie's

photometer, and the comparison of the depth of the shadows. Ritchie's photometer consists of a long box, *A B*, *A B*, fig. 301, blackened intern-

*Fig.* 300. *Fig.* 301.

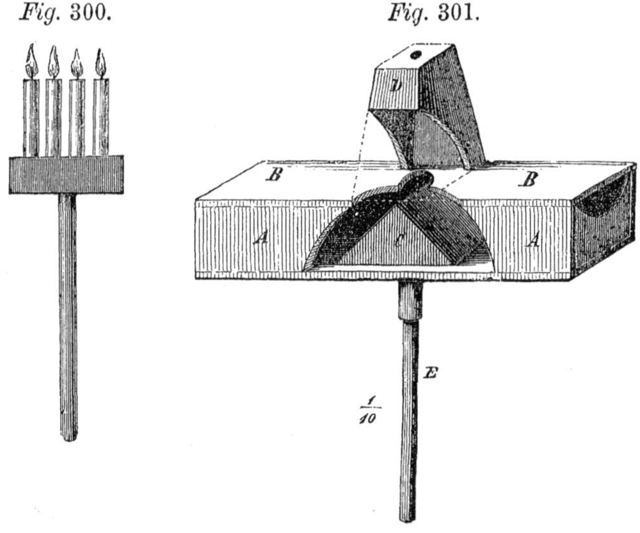


ally. The light, passing through two semicircular openings in the ends, falls upon a rectangular prism of wood *c* covered with white paper, placed exactly in the middle, and illumines both sides of it. A circular opening is made in the top *B*, two-thirds of the width of the box, and so placed that the edge of the prism corresponds to the diameter of the opening. This aperture is covered by a short, blackened tube *D*, to exclude extraneous light, with a small hole for the eye. The apparatus may be fixed by the stem *E* in one of the sockets of the beam described in connection with the concave mirror, or in the little stand, fig. 302. Several such stands and some larger tripods like fig. 303 will be needed for various purposes.

*Fig.* 302. *Fig.* 303.

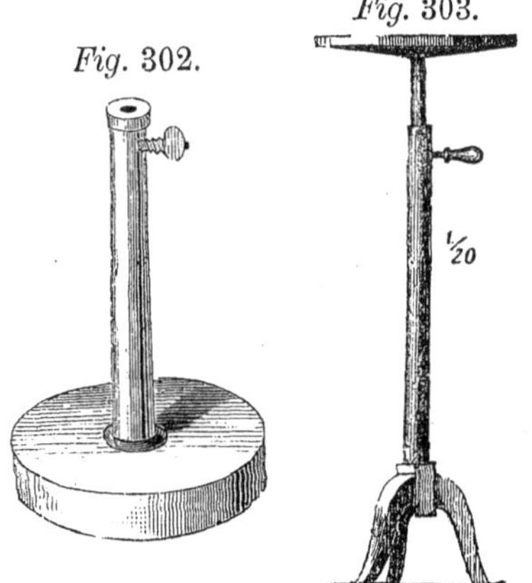


In order to compare the intensity of two lights, they must be placed at the same height as the photometer, about 20 or 30 feet apart, and the photometer placed in a straight line between them. Now observe the illumination of the two sides of the prism, through the hole in the tube *D*, and move the photometer until both sides are equally illuminated. It is well to avoid looking directly at the lights during the process, as this lessens the sensibility of the eye to slight differences in the brilliancy of the illumination. With this precaution, the correct position of the photometer may be attained within 4 or 5 inches. The comparison is most difficult when the two lights are differently colored, *e.g.* when a lamp with a white flame is compared with a wax candle, which always burns with a yellowish flame.

When there are several illuminating bodies to be compared, it expedites the process to have a lamp which burns with a steady flame fixed at an invariable distance as a standard of comparison, and change the position of the other lights until both sides of the prism are equally illuminated. Mechanical lamps are very well adapted to this purpose.

To compare the power of two luminous bodies by moving them until the shadows cast by an interposed body are of equal depth, nothing more is needed than a stick half an inch thick and a white wall. This process is less accurate than the other, because the glare of the white wall makes the eye less sensible to slight differences of light. In all such experiments a trial should be made to ascertain what degree of accuracy may be expected. It may be done very easily in this case by having an assistant move the lights without the knowledge of the experimenter.

In this method of Rumford's the dissimilar color of the flame causes more difficulty than with Ritchie's photometer, but it is better suited for class illustration because more persons can observe it at once. The lights should be so placed that the shadows shall fall close beside each other.

[156] **Bunsen's photometer.**—The flame of a good lamp is surrounded by a tin box with a sliding tube *d*, as seen in fig. 304. The

*Fig.* 304.

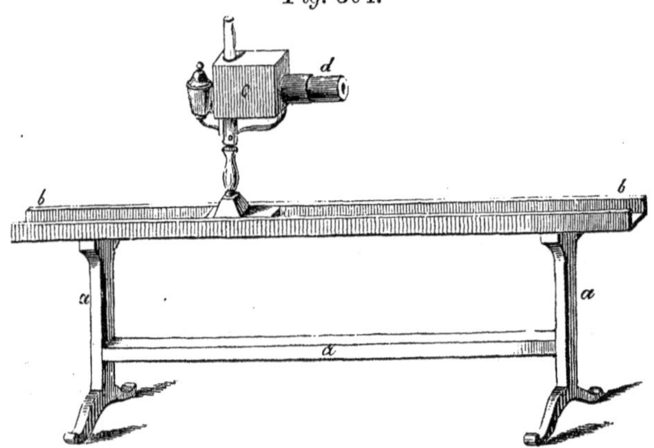


end of this tube is covered with letter-paper, which is rendered transparent by stearine, leaving a ring $\frac{1}{4}$ or $\frac{1}{2}$ an inch wide and $\frac{1}{2}$ an inch internal diameter untouched. The paper should be warmed, and then rubbed with a warm stearine candle. The luminous body to be compared is placed in front of this tube, and moved until the opaque ring entirely disappears; and the same process is repeated with every light compared. The relative intensity of the lights can be ascertained by measuring the distances of the two from the paper screen. This process is less dependent on the color of the lights than the former ones.

[157] **Shadows and half-shadows.**—The simplest way of showing the difference between these two is by hanging up a ball, $\frac{1}{2}$ to 1 inch in diameter, in the sunshine, and allowing the shadow to fall on white paper. The same is very prettily shown by a lamp with a round waste-cup.

Any sharply defined flame may be used for this purpose, by casting a shadow with a body less in diameter than the flame. Hollow flames are not suitable for the illustration, because they cast two shadows, the sides of the flame giving more light than the middle.

The phenomena of the *camera clara* belong here. It needs no special apparatus for its illustration; it is only necessary to attach to the board intended for the heliostat a slide with an aperture half an inch in diameter; on darkening the room inverted images of external objects will be seen on the opposite wall. The smaller the aperture, the sharper will be the outline of the image, and corresponding care must be taken to exclude the least ray of extraneous light. If the room cannot be made dark enough to show the image, a camera clara may easily be made of two tubes of pasteboard *A B*, fig. 305. *A* is closed at one end and pierced with a hole about 1 line in diameter. The end of *B*, which is inserted in *A*, is closed with a plate of ground glass or rice-paper. Hold the opening of *B* close to the eye, and direct the tube toward a strongly illuminated object.

*Fig.* 305.

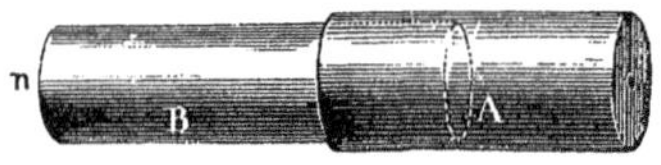

## (*b*.) EXPERIMENTS ON THE REFLECTION OF LIGHT.

[158] **Plane mirrors.**—Besides the ordinary plane mirror, there should be a piece of plate glass thickly coated on the back with indian-ink, a mirror made of two pieces set at an angle with each other, and two parallel mirrors. The latter should be set in black wooden frames. The jointed mirror generally has an inclination of 60°, fig. 306; but it is better to connect the two parts by a hinge, so that they can be set at any angle. To these may be added a kaleidoscope, which can be bought cheaply in any toy shop. This may be easily made by setting two slips of ordinary

looking-glass at an angle of 60° in a tube of pasteboard about 4 inches long, as seen in section in fig. 307. The tube is closed near one end by a disk of clear glass *a*, fig. 308, and at the other by a pasteboard cover

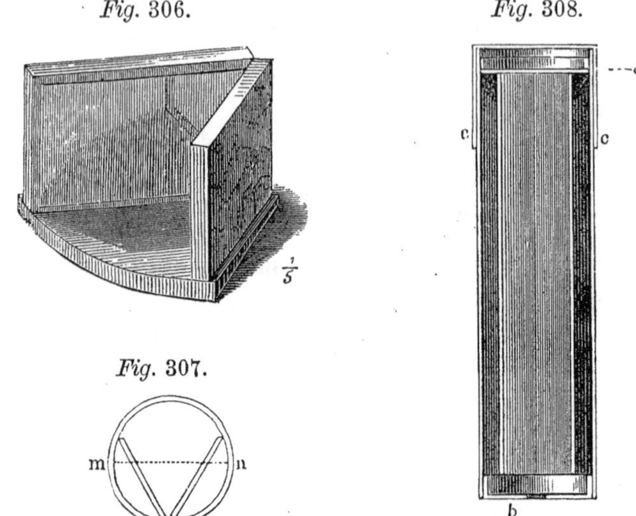

*Fig.* 306. *Fig.* 308. *Fig.* 307.

with a small aperture. Some pieces of colored glass, shreds of colored paper, etc. are laid in the space before the glass at *a*, and a short tube closed with a disk of ground glass is slipped over that end.

[159] The following simple apparatus has been contrived by Müller to illustrate the law of reflection. A strip of brass is tacked around a

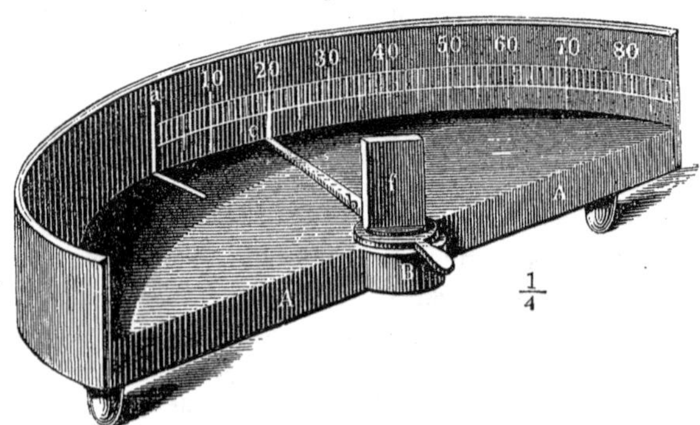

*Fig.* 309.

semicircular board *A*, fig. 309, and one-half of it graduated into degrees. A conical pin, which can be fastened by a screw below, works in the center of the semicircle, and carries the index *b c*, which is bent upright at the extremity. On this index a small plane mirror *f* is accurately centered. The ray of light is allowed to fall on the mirror through the slit *a*. The index shows the angle of incidence, and the point to which the ray is reflected double the same. The accuracy of the adjustment must be tested before the mirror is permanently fixed. The apparatus shows also that when the mirror is turned a certain number of degrees the reflected ray advances twice as many. Upon this principle Wheatstone constructed his apparatus for measuring the velocity of the transmission of electricity.

[160] **The heliostat.**—The heliostat is one of the most indispensable instruments, both for class illustration and for physical research; it should, therefore, be obtained as soon as a permanent place for it can be had. In its simplest form it costs very little. It has already been mentioned that the room in which optics is to be taught must have a proper exposure to the sun. The apparatus room should be similarly situated, so that the experiments can be properly prepared and the teacher have opportunity for research.

The simplest form of the heliostat consists of a square board *A B*, fig. 310, of hard wood with a broad rabbet fitting in a hole in the shutter, and kept in place by two screws like fig. 311. In the middle of this board is a conical hole mounted on the inside with a brass rim, on which a thread is cut for the purpose of screwing on various attachments. Instead of the brass plate a short tube of wood or pasteboard may be used. On the outside is a long, narrow mirror in a wooden frame, hinged to the wooden pin *D E*. This is shown on a larger scale in fig. 312. The pin has a head at *F*, and can be tightened

*Fig.* 310.

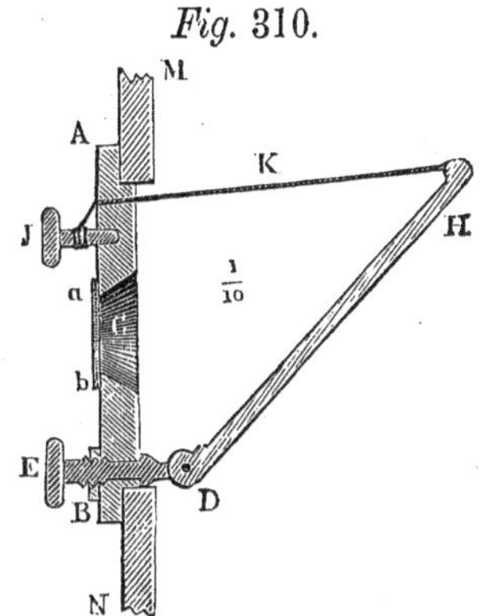


*Fig.* 312.

*Fig.* 311.

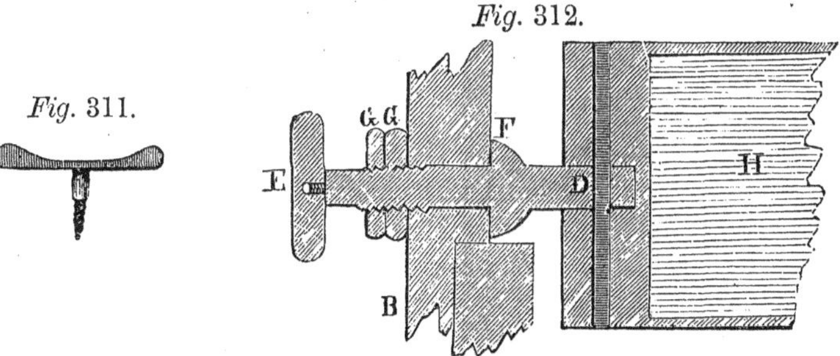


by the nuts *G G*, so as not to turn by the weight of the mirror. When the pin is fixed in its place, the button *E* is glued on. The string *K*, passing through a smooth hole in the board, and winding on the pin *J*, allows the mirror to be set at any angle.

The mirror must be made of very thin, clear glass, so as not to form more than two images, and these as close as possible to each other. It need not be much wider than the aperture *C*, but on its length will depend its power of reflecting the rays of the sun horizontally when the latter is near the horizon.

*Fig.* 313.

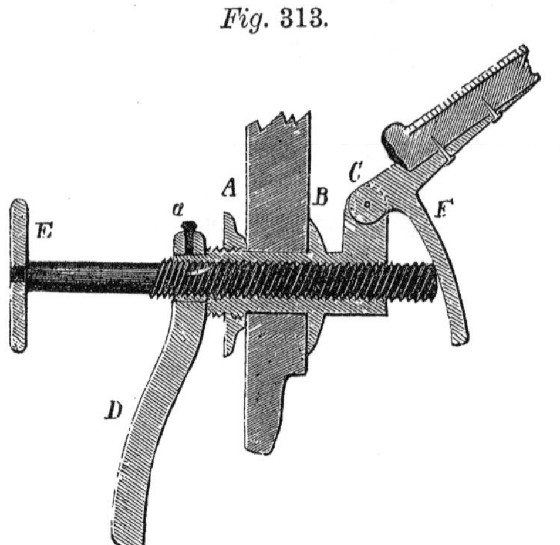


In the simple arrangement just described, the position of the mirror in respect to the pin *J* is varied by turning the pin *E*, and the string *K* becomes oblique to the mirror, which may in time affect the firmness of the apparatus. This difficulty is obviated in the arrangement shown in fig. 313, of half the natural size. Here the support of the mirror is made of brass. A handle *D* is fixed to the square head of this, which serves to turn the mirror. A screw *E*, passing through this brass axis and pressing against the arm *F*, serves to elevate or depress the mirror.

By making the support of the mirror longer, as seen in fig. 314, and fastening it to the frame by a slide and screw, a mirror of blackened glass may be substituted for the ordinary one. This gives a clear image of the sun, and light enough for most experiments.

Neither of the arrangements described costs much. The position of the mirror must be constantly shifted in order to keep the beam of light in the same direction, and some practice is required to do this properly. Heliostats with clock-work and metallic mirrors are very expensive, although they have been of late much simplified.

[161] **Experiments with concave mirrors.**—The wooden stand seen in fig. 315 is very convenient for these and other experiments in this section. *A B* is a straight, slender bar of pine, 12 to 15 feet in length, supported on the tripods $M\ M_1$ which are stuck in holes in the beam so that they can be easily separated. This bar is divided into

inches. The slides *P P P*, fig. 316, slip over it and can be fastened by screws *S*. These slides are a whole number of inches in breadth, and

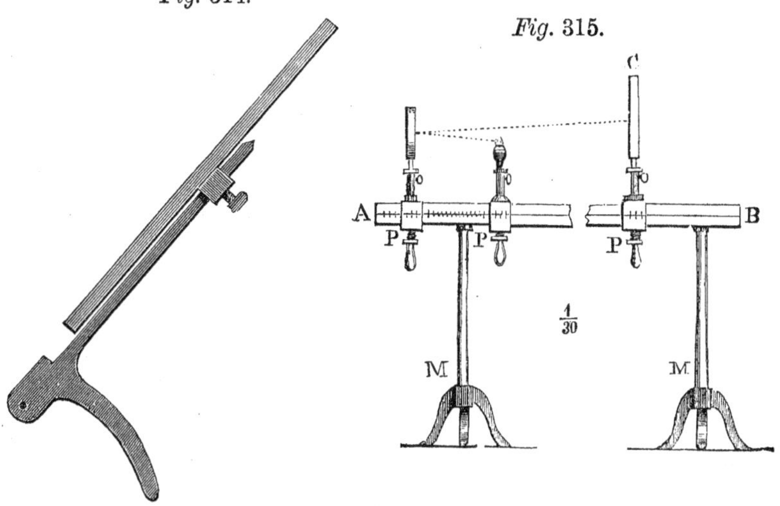


*Fig.* 314. *Fig.* 315.

likewise divided into fractions of an inch. Various pieces of apparatus can be fixed in the sockets *R* attached to these slides. The sockets might be fixed to a board sliding in a groove in the top of the bar, but it would be difficult to make them firm in this way and at the same time easily movable. A small lamp, fig. 317, is fixed on one of the slides. The

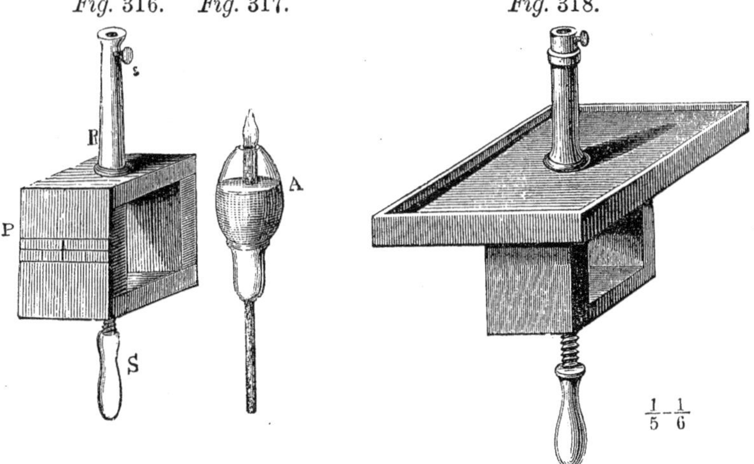


*Fig.* 316. *Fig.* 317. *Fig.* 318.

wooden frame *C*, on which the white paper screen is stretched, must be made large enough to receive the whole image which the mirror reflects from the opposite end of the bar.

Gaslight renders the experimenter in optics in some cases independent of sunshine. This may be arranged by fixing a board with a rim to one of the slides, as in fig. 318; in the socket of this slide the stem of the argand gas-burner, fig. 319, may be inserted, and connected by a flexible

*Fig.* 320.

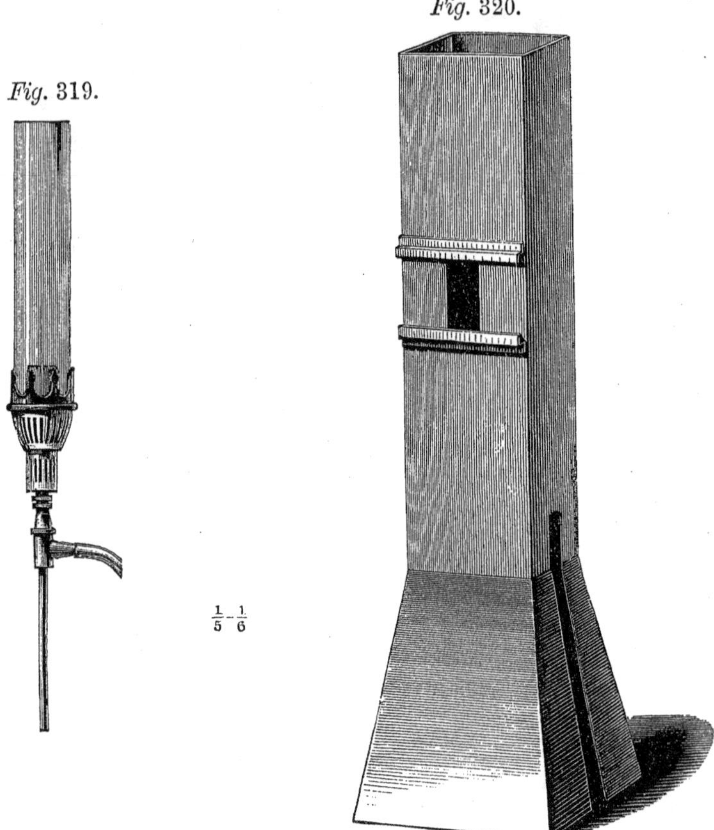

*Fig.* 319.

hose with any gas-burner in the room. Over this burner is placed the sheet-iron chimney, fig. 320, which has a slit on one side for the hose, and an opening at the proper height to place slides before the light. This arrangement prevents the general diffusion of the light in the room.

When the image falls between the object and the mirror, a slip of paper, half an inch wide and an inch long, is fastened to a wire which is bent at right angles at about half the breadth of the mirror from the side of the bar. The mirror must then be turned to correspond, so that the image shall fall outside of the bar.

The screen in these and a number of other experiments may be covered with rice-paper instead of letter-paper; the image, though not so bright, is then visible from both sides. The frames should be about the size of a sheet of letter-paper, and the paper stretched just as on a drawing-board. Any ordinary concave mirror will answer for these experiments. Even watch-glasses coated on one side with black sealing-wax, by heating them until the wax melts on the surface, do very well. But to make the image visible without any screen, the mirror must be wider and have a shorter focus, in order to render it visible to several persons at once. Reflectors with a focal distance of 1 to 2 feet, and at least as great diameter, made of brass and polished by the brazier, are better for this purpose than more highly polished mirrors of less concavity. The experiment succeeds with a mirror 3 or 4 inches in diameter, but the image is visible only to one person. It may, however, be shown to many in rapid succession, by holding the mirror in one hand and the candle at a proper distance in the other, and throwing the reflected light upon the faces of the audience in turn.

The flame of a candle does very well as an object in these experiments, but a bouquet of artificial flowers of bright colors is usually chosen, and so placed as to be highly illuminated and yet invisible to the audience. A glass of water may then be placed where the image will fall, and the flowers will seem to float in it. It is very important, with reference to the future consideration of more complicated instruments, that it should be distinctly understood that this aerial image exists entirely independently of the white wall. For this purpose concave mirrors are preferable to lenses.

[162] **Convex mirrors** need no special illustrations. A clean wine bottle of dark-green glass answers admirably to illustrate the effect of cylindrical and conical mirrors.

### (*c.*) EXPERIMENTS ON REFRACTION.

[163] To exhibit the mere fact of the refraction of light, a rod immersed in water, or a ray of light rendered visible in a dark room by particles of dust, and allowed to fall upon turbid water, is quite sufficient. To prove the law of refraction by actual measurement requires some apparatus. A very simple and instructive experiment is to draw upon a piece of tin painted white, or varnished pasteboard, two black lines at the angle corresponding to the refraction of light passing from air into water, and

immerse the sheet with the drawing, fig. 321, in water up to *a b*; the line *c d e* will then appear straight.

The apparatus shown in fig. 322 enables us to measure the angle of

*Fig.* 321. *Fig.* 322.

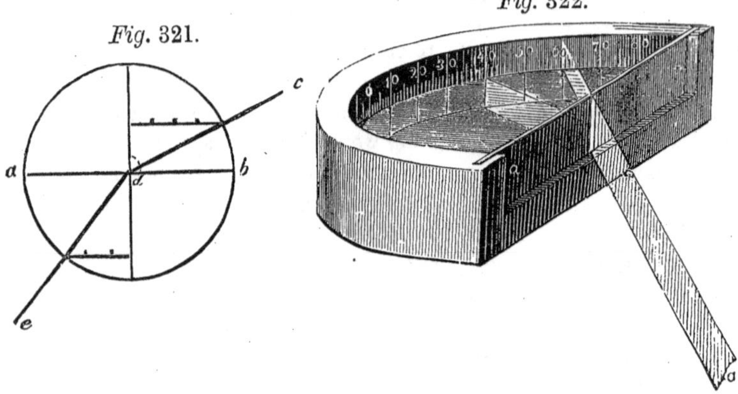


refraction for different fluids. The variety of fluids which can be used depends on the materials of which it is constructed. If the body of the apparatus be made of wood covered with copal varnish, it can be used only for water. The straight side *a b* is made of a piece of plate glass and covered with tin-foil. A very smooth cut may be made in the middle of the foil, leaving the rest opaque. The glass is cemented with putty into grooves on three sides. The apparatus may be placed on a table for use, half filled with water, and the light of a candle, or a ray of light from the heliostat, directed through the slit in the foil. The part of the beam above the water indicates on the graduation the angle of incidence, and the part passing through the water the angle of refraction. Two or three experiments will prove that the sines of these angles preserve a constant ratio in the same liquid.

*Fig.* 323.

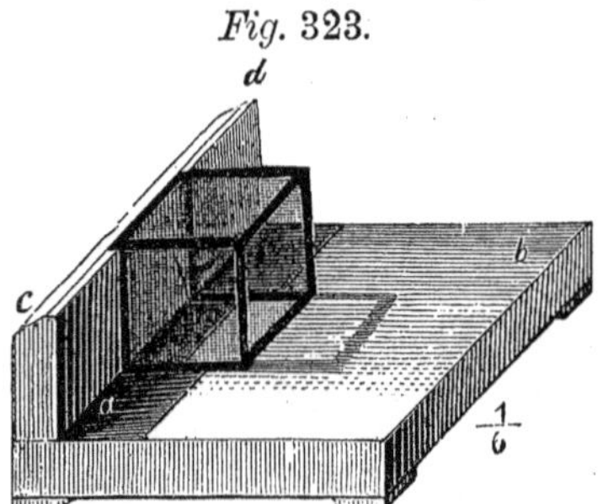


The apparatus shown in fig. 323 is also convenient. It consists of a box made of squares of plate glass, open at top, and let into a board at the bottom and on one side, to a depth equal to the thickness of the glass. When half filled with water and the sun allowed to fall on the side *d c*, the shadow within the water will be narrower than without.

The phenomenon of refraction may also be exhibited by transmitting a ray of light in a dark room, through a wide glass filled with water, into which powdered chalk has been stirred and allowed to stand some hours. The path of the beam through the air will be marked by illuminated particles of dust, and through the water by the particles of chalk.

[164] **Total reflection.**—The same turbid chalk-water may be used to exhibit total reflection. Direct the light first on the plane mirror *a*, fig. 324, so as to give the reflected ray the proper direction for total

*Fig.* 324.

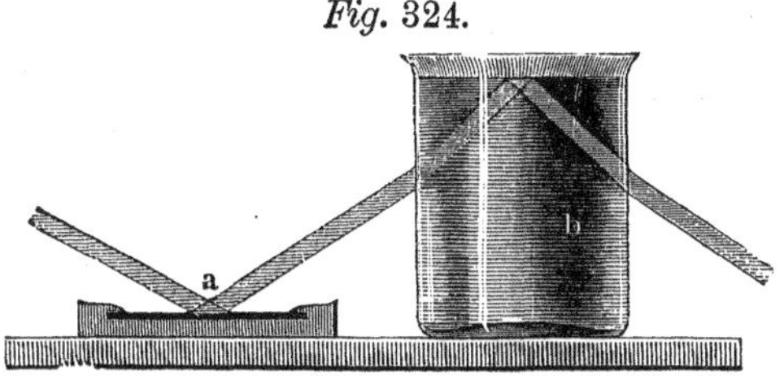


reflection from the surface of the water. The direction of the ray through the water will be visible, and, if the proper angle is chosen, no light will be seen passing through the surface of the water. The vessel should be 4 to 8 inches in diameter, and the aperture in the heliostat not over ½ an inch. A vessel made of plate glass answers the purpose best.

Among the phenomena which depend on total reflection, there are many which may be used as further illustrations. Newton's blue bow on the under surface of a glass prism is one of the most frequently cited, but it is not adapted for class instruction. The three following experiments are easily made:—

(1) An object outside of a glass filled with water cannot be seen through the liquid from above, unless the direction of vision is very oblique to the surface and the vessel very wide.

(2) Fill a glass about ⅔ full of a sirup made of white sugar, which may be tolerably thin. A stratum of water may be poured upon this from a spoon held near the surface, without causing much mixture of the two liquids. Paste a few letters drawn on white paper on the glass just below the contact surface of the two liquids. On look- at this surface from below the letters will be seen double, once direct and once by total reflection.

*Fig.* 325.

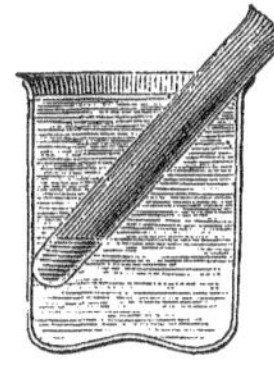

(3) A test tube of thin glass, about ½ an inch in diameter, plunged into water, as seen in fig. 325, appears, when seen from above, to be filled with mercury. When partly filled with water, this metallic luster disappears as far as the water extends.

It is evident from this that much more light is reflected by total reflection than from a metallic mirror, for when the tube is partly filled with mercury that part appears quite gray in contrast with the other.

[165] **Lenses.**—Experiments with convex lenses are made on the stand, fig. 315, with the screens belonging to it. The lens best adapted to the purpose has a diameter of several inches and a focal distance of about 2 feet. The lens is set in a frame with a stem fitting in one of the sockets of the stand. The aerial image should also be shown in this case without the white screen, and the difference between it and the image on the screen pointed out. In this case also, as with the mirror, the image appears behind the glass and enlarged before the rays become divergent, which is apt to cause confusion.

We often wish to measure the focal distance of lenses. For this purpose the glass is held opposite the white wall of a room so as to form a distinct image of a window 15 to 20 feet distant, and the space between the image and the lens measured. For more accurate measurements, fasten a plate of ground glass to a graduated rod and move the lens along this rod until the image is distinct. If the lens be fixed to a slide the image may be inspected with another lens, so as to obtain the greatest distinctness.

[166] **Prisms.**—A highly polished prism of flint glass with an angle of 50° to 60° should be chosen. They are very expensive, but such a prism, even 1 centimeter in length and breadth, is more useful than a great piece of common glass, which will not cost much less. It should be bought on condition that it can be used to exhibit Frauenhofer's lines. It is easily mounted, as shown in fig. 326. The prism is mounted on a piece of wood or brass, terminating in a screw. The nut *c* serves to fasten it to the broad end of the arm *a b*, which is hinged by a screw joint to the rod *a c*. This rod fits in one of the stands, fig. 302. This arrangement admits of its adjustment in any position.

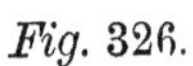

*Fig.* 326.

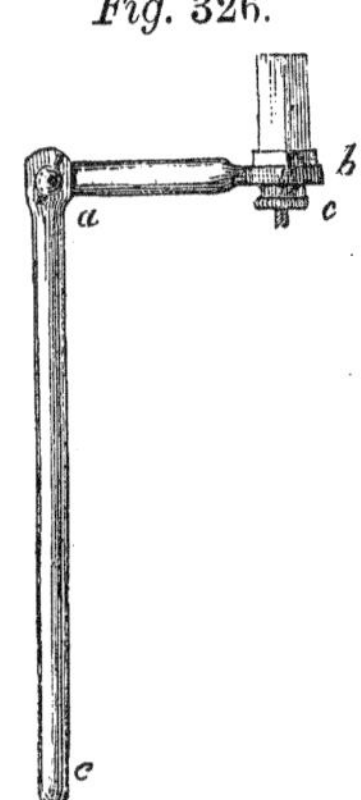


With a little mechanical skill, one can make hollow prisms for himself. Prisms of brass, like fig. 327, cannot be used for acids, and even sulphide of carbon decomposes slowly in them. It is better to take a square bottle with ground stopper, grind off two of the corners, and replace the sides by pieces of plate glass fastened with brass bands and screws. Fig. 328 shows the prism without the plates. If the prism is designed for sulphide of carbon only, the plates may be fastened with isinglass cement. The sulphide of carbon may be left permanently in the prism and covered with a stratum of water,

*Fig.* 327.

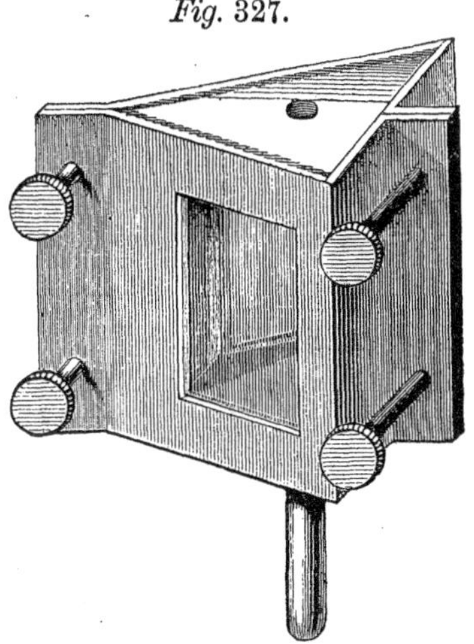

*Fig.* 328.

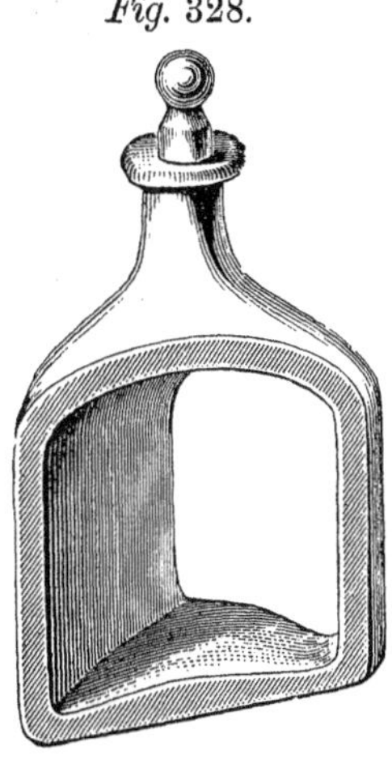

or poured in at an elevated temperature and closed by a stopper cemented in with isinglass. Sufficient space must in any case be left for the expansion of the liquid by heat. The grinding and adjusting such a glass is not so troublesome as one may imagine. It is done first on an iron plate with sand, and finished with fine emery on plate glass.

In filling and emptying the prism, a pipette or glass tube with a fine point should be used. The point of the pipette must reach to the bottom of the glass.

[167] Experiments with prisms can only be made in a dark room, into which light is introduced through a round aperture. To have the colors distinct, this aperture must not exceed 1 or 2 lines in diameter. For most experiments a narrow slit, not longer than the prism, is better. The edge of the refracting angle and the slit should both be vertical.

*Fig.* 329. *Fig.* 330. *Fig.* 331.

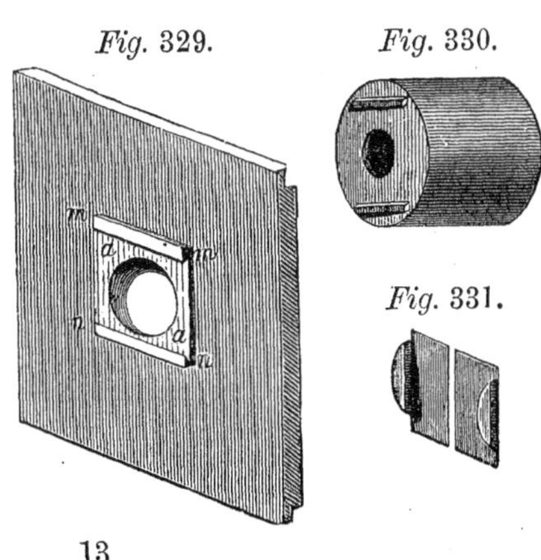


Figs. 329 and 331 show an arrangement of slides for regulating the width of the slit. Fig. 330 is a tube designed to slip over the permanent tube in the aperture, in case there is one. Fig. 332 represents a contrivance by which the edges of the slit are kept always parallel. The little bars *c c* turn on the screws *a a*, and act like parallel rulers.

*Fig.* 332.

The lenses, prisms, screws, etc., may be fixed on stands like figs. 302 and 303, p. 181.

The principal experiments are the following:—

(1) The spectrum itself. A round aperture may be used at first to show the rounding of the upper and lower ends. It is thrown on a screen of white paper.

(2) The combination of all colors makes white. For this experiment the slit is used, the prism placed at a distance of 5 to 10 feet from it, and the spectrum thrown on a convex lens wide enough to receive it. The distance of the lens from the prism must be such that it will give a magnified image of the refracting face of the prism. A white screen is placed where this image falls. If the refracting edge of the prism is parallel to the slit, the lens perpendicular to the direction of the beam, and the screen slightly inclined so that the side where the blue rays fall is a little nearer to the lens, a clear white image of the prism will be obtained, with only a very narrow blue margin on one side and a red one on the other. An achromatic lens acts better, but achromatic lenses of considerable dimensions are very dear. By obstructing parts of the spectrum by strips of black paper, the complementary colors are shown very beautifully.

The experiment may also be made by looking at the spectrum through a second prism. The slit in the heliostat will then be seen in its natural size, and white; but as this can only be seen by one person at a time it is not suitable for class instruction.

Another contrivance for showing the combination of all colors, is the oscillating prism, fig. 333. The vertical axis *a* is made to revolve by a spring and cogged wheel. The bar *ff*, fig. 334, is fastened to the mounting of the prism, and the pin *h* on the disk *c* works in the slit in this bar. By this means the prism is made to oscillate rapidly. The extent of this oscillation must be sufficient to advance the prism its whole length. This must be ascertained by a previous experiment, and the size of the disk *c* and the length of *f f* arranged accordingly. The figure is about half the real size.

*Fig.* 333.

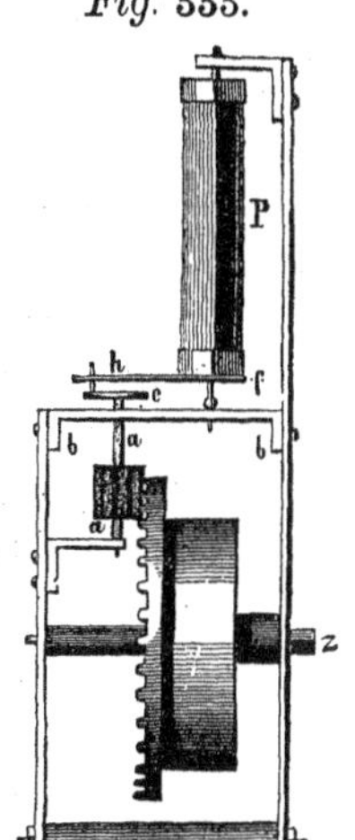


(3) To show that orange, green, and violet are not compound colors, cut in a sheet of paper a slit of less width than the color to be examined, and let it pass through the slit to a second prism.

(4) When the spectrum is thrown on a second prism, whose axis is at right angles to the first, an oblique spectrum is obtained. To make this well-known experiment handsome, the second prism should be large and held near the other, so as to receive almost the whole of the spectrum. The screen should be only a few feet distant from the second prism.

(5) To illustrate the theory of the colors of bodies, paste slips of bright-colored paper on pasteboard and hold them successively in the different colors of the spectrum. Some papers, especially the blue and red title paper of the bookbinders, will appear black in any color but their own.

*Fig.* 334.

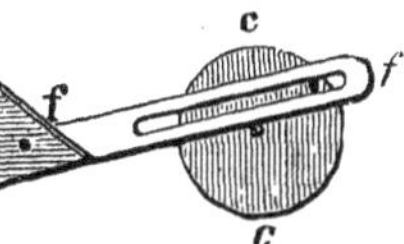


(6) The index of refraction of various substances may be compared approximately, by means of a large rightangled triangle of wood, fig. 335. The handle fits in one of the stands. One of the slides is raised about 2 inches, and divided into centimeters. A hole is made to receive the handle of a prism at a distance of 1 meter from the zero of this graduation. A strip

*Fig.* 335.

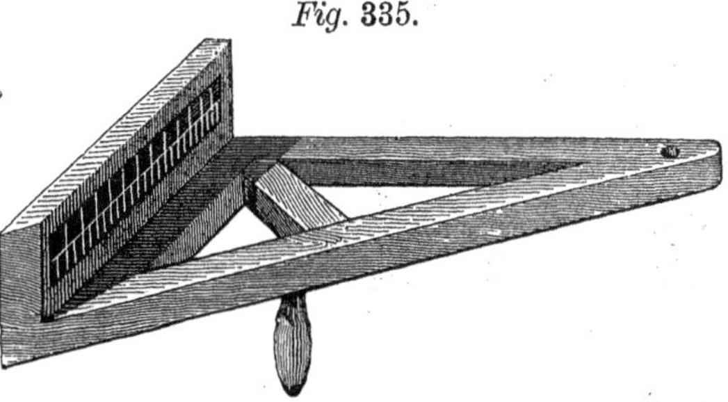

of metal, not seen in the figure, with a long, narrow slit, is placed between the prism and the light.

A ray of light is thrown in such a direction that the part which passes above the prism shall fall on 0 of the scale; the prism is then turned so as to obtain the least refraction, and the point on which the middle of the red falls noticed. This, together with the distance of the prism from the scale, gives the tangent of the angle of refraction; and the index of refraction is calculated from this and the refracting angle of the prism.

(7) **Frauenhofer's lines.**—Any prism whose faces are plane and its material homogeneous, may be used to show these lines. The greater the dispersive power of the substance, the greater will be the number of the lines. A prism filled with sulphide of carbon shows them most beautifully. It is an advantage to place the prism 8 to 12 feet from the slit.

The refracting edge of the prism must be placed parallel to the slit, and turned so as to obtain the minimum of refraction. This is easily obtained by turning the prism slowly on its axis; the position will soon be found in which the spectrum falls nearest to *M*, fig. 336.

*Fig.* 336.

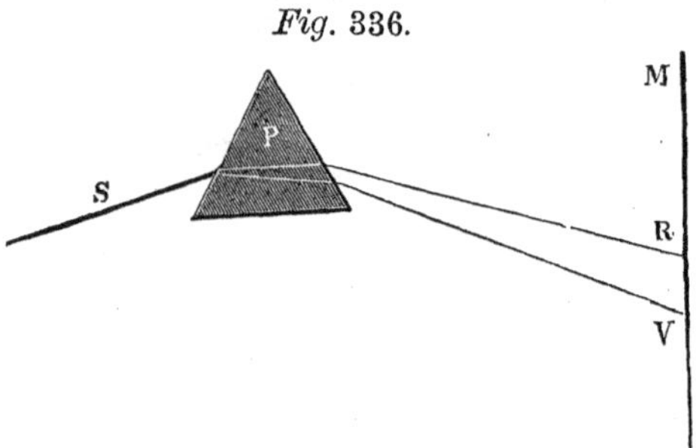


The slit must not be over $\frac{1}{4}$ of a line wide. It gives light enough when it is only $\frac{1}{10}$ of a line, and the lines are more distinct. Its length should be 10 to 15 lines. The spectrum is usually transmitted through the object glass of a telescope which is adjusted as for distant objects, and its object glass placed close to the prism. It is better to fix a wooden ring on the object end of the telescope, and fasten the prism to this in such a way that it can be turned on its axis before the middle of the object glass, as in fig. 337. The prism is thus secured from being overset or displaced, and its position in respect to the telescope can be more accurately adjusted. An achromatic telescope with a magnifying power of 10 to 20 times is quite sufficient. In order to make measurements, the telescope of a goniometer must be used.

*Fig.* 337.

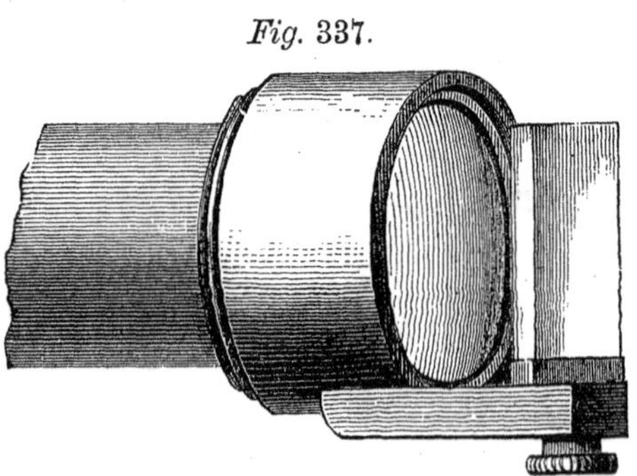

Although the lines appear more numerous and more distinct in this way, it is not adapted to class illustration, because it is visible to but one person at a time.

By placing close in front of the prism a small board, with a large, round aperture contracted to a narrow slit, the spectrum can be thrown on a straw-paper screen, and the broader lines become visible to the naked eye, even with a prism of crown glass. This apparatus is also needed in experiments on diffraction. The slit is arranged just as the first one. The lines may be seen on a screen still better, by placing a convex lens, of from 2 to 4 feet focal distance, behind the prism. The screen must be in such a position that the lens will produce a distinct image of the slit upon it. It increases the distinctness of the lines to move the screen a little back and forward; the lines are the more readily distinguished from accidental shadows on the paper. A second slit is not necessary.

A prism of sulphide of carbon produces most beautiful results, especially with an achromatic lens. The colors are brighter when the image is thrown on a screen of white paper instead of straw-paper, but it injures the eye to look at it long.

A few of the dark lines are perceptible even with the naked eye, especially the two broad ones on the limit between violet and blue, when a very fine slit is looked at through a prism held at the distance of distinct vision. But the experiment is trying to the eye and not suited for instruction.

[168] **The rainbow.**—To illustrate the theory of the rainbow we use glass balls, from 2 to 5 inches in diameter, filled with water. Such balls are easily obtained, as they are used by many artisans to concentrate light on their work. The glass should be of uniform thickness, and the form as accurately spherical as possible. The smaller balls fulfill these conditions better than the larger ones, but they do not show the path of the light through the ball so distinctly. Even a thermometer bulb will show the principle. The balls may either be laid in a wooden socket or hung up. Fill the ball with slightly turbid water, and allow a ray of light to fall upon it through an aperture about two lines in diameter, in a screen placed just before it. The aperture in the heliostat may be from 4 to 6 lines in diameter. The path of the light through the ball will be visible, and the spectrum obtained by two refractions and one reflection may be thrown on the screen. It is best to throw the light in such a direction that the reflected ray shall be horizontal.

To explain the arched form of the rainbow, the apparatus contrived by Reusch, seen in fig. 338, is very convenient. The spherical head *a* turns with considerable friction in the stand *A*, and carries the glass rod *m n* about $\frac{1}{4}$ inch in thickness. The stout cork *b*, which slips over this rod, occupies the place of the eye, and supports the two slender glass rods *r r*,

*Fig.* 338.

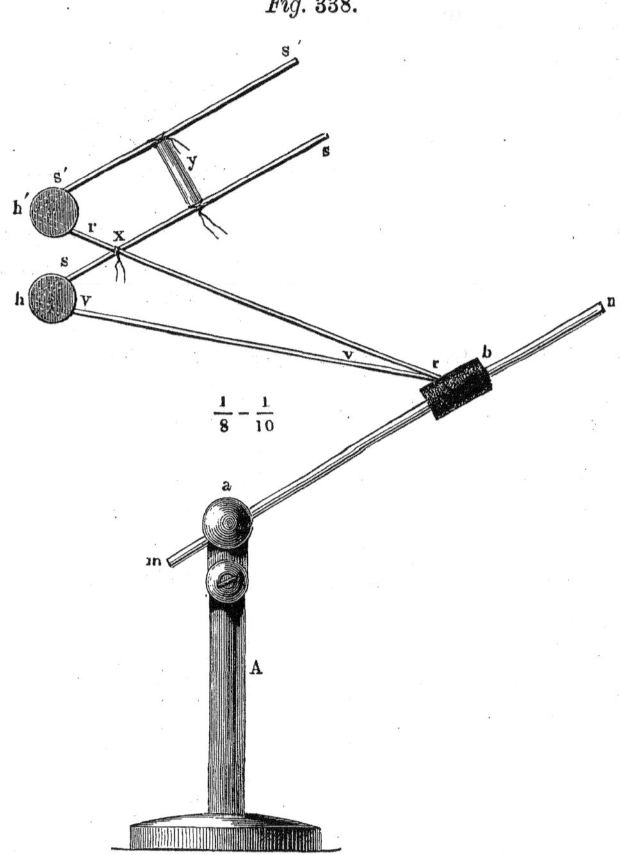


*v v*, which are cemented into it. Each of these rods carries a little disk of wood covered with paper, and each of the disks a slender rod *s s*, *s' s'*. The rods *s s*, *s' s'* are painted white, the rod *r r* red, and *v v* blue. They are set at about the angle required by theory, and the path of the light indicated on the disks by dotted lines, so that *r r* is the prolongation of the red, and *v v* of the violet ray. The rods should be connected together at *x* and *y* to render them firmer. If the cork *b* be turned on the rod *m n*, the rods *v v* and *r r* will describe the cones in whose surfaces the refracting drops lie.

[169] To show the **chromatic aberration** of lenses, take a large

lens, of 4 feet or more focal distance, and throw the image of a window on the white wall opposite. The image will have a blue or red border according as the lens is moved into the focal distance of the red or the blue rays.

This may also be shown very well as follows. Cut in a tin plate two semicircular incisions like fig. 339, leaving the circle within just large enough to receive the image made by the convex lens at the point where it is most distinct and smallest. Set the lens in the hole in the shutter, at right angles to the sun's rays, and let the image of the sun fall upon the plate. A sheet of paper held some distance behind the plate will now receive no light which has passed through the lens. But if both the plate and the paper be brought nearer to the lens the circle will not be large enough to receive the whole image, and the light passing through the aperture will be reddish. If the screen and paper be removed a little farther off, the light on the paper will be bluish. The experiment succeeds better when the lens is set in the heliostat, but this requires a metallic mirror.

*Fig.* 339.

The experiment can be made more conveniently by gaslight, by cutting an arrow in the slide fitting the chimney in fig. 320, and fixing to the slide two guides, in which a plate of green and one of red glass, fitting very closely together at the edge, can be slid over the aperture. Fig. 340 shows such an arrangement half the natural size. The positions in which the upper or lower half of the arrow is distinct are far enough apart with a focal distance of two feet. The experiment can be repeated with an achromatic lens.

*Fig.* 340.

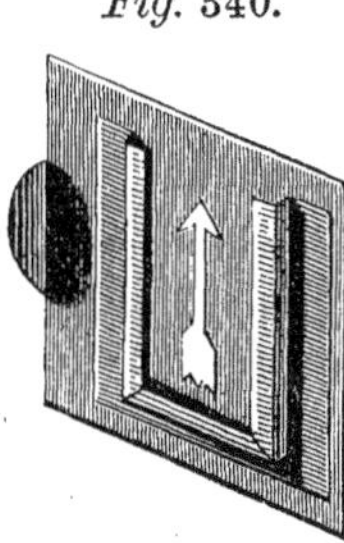

*Fig.* 341.

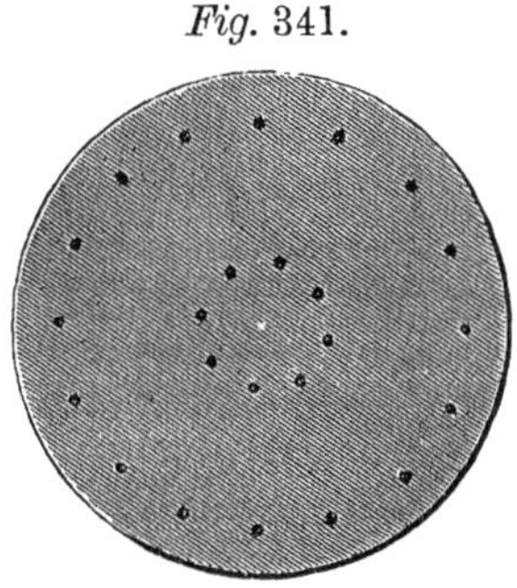

[170] **Spherical aberration** may be very well shown with a lens, a foot or more in diameter and short focal distance, such as were formerly common. Make a screen pierced with two rows of holes, as seen in fig. 341. Place it before the lens and let sunlight fall upon it. If the light

which passes through the screen be caught upon a screen of white paper, which is gradually removed to a greater distance, the rays from the outer holes will be seen to pass through the inner rays, and unite to form an image, the rays from the inner holes forming a ring around it. As the paper is removed still farther, the inner rays will converge to a focus and the outer ones form a ring around it.

[171] **Fluorescence.**—The phenomena of fluorescence or *internal dispersion* will soon be too important to be omitted in physical lectures; and we therefore mention here a few simple experiments which illustrate the important points of the subject.

The following are the substances which are adapted to these experiments: (*a*) Aqueous extract of the hull of horse-chestnuts. A few pieces of the hull digested for $\frac{1}{4}$ of an hour in cold water are enough. (*b*) Chlorophyl, alcoholic extract of nettles. (*c*) Tincture of nightshade [stech apfel.] (*d*) Tincture of curcuma. (*e*) Aqueous solution of sulphate of quinine. (*f*) A piece of glass colored yellowish green by uranium. (*g*) Strips of paper, one-half moistened with the liquids mentioned. When such a liquid is placed in the sun, light of another color seems to penetrate a little way into the interior: *e.g.* a deep-green light into tincture of nightshade, which is wine yellow. This is still more apparent when a beam of light is concentrated upon it by a lens of about 1 foot focal distance. The lens should be fixed on a support. Fig. 342 shows

*Fig.* 342. *Fig.* 343.

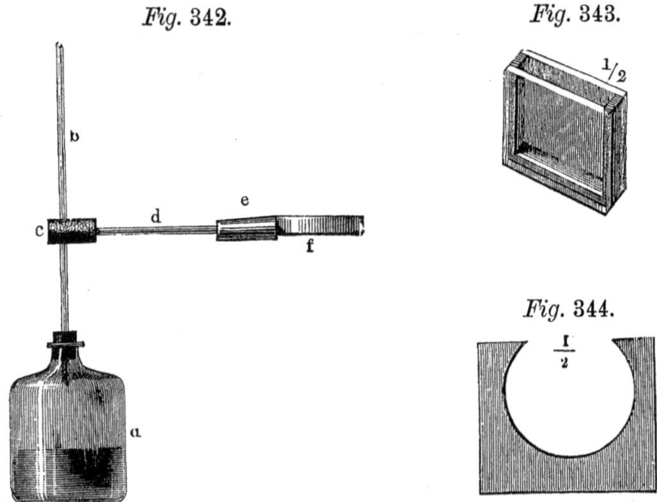


*Fig.* 344.

a simple stand which is available for a great many purposes. The bottle *a* is partially filled with sand; *c* is a cork sliding on *b* and carrying the rod *d*, which supports a ring of pasteboard, in which the lens is fastened in the usual way by two wire rings.

Now take a vessel made of parallel glass plates, kept apart by brass cemented between them, as in fig. 343, fill it with a solution of chloride of copper, and look through it at the green light penetrating into the tincture of nightshade. This light is quite visible through the solution, and yet it is not the green light of the spectrum, for it disappears when the solution of copper is placed between the light and the liquid, although none but green then falls on it. On the other hand, this light remains when ammoniacal solution of copper is placed between the light and the liquid; but it is not visible on looking through the ammoniacal solution. Green light is thus produced by a blue medium. Similar experiments may be made with alcoholic as well as aqueous solutions; and some of the vessels should, therefore, be cemented with isinglass cement and others with shellac. Many liquids attack brass; it will, therefore, be well to substitute strips of plate glass for the brass in some of the vessels.

In a spectrum made with a flint glass prism and a lens, as in § 167, No. 7, the double line *H* in the violet ray will scarcely be visible on white paper. But if the spectrum be thrown on paper colored with any of the liquids above named, light will be visible, with countless lines, far beyond *H*. If such a spectrum be now observed through a second prism, as in § 167, No. 2, the observer will see, besides the oblique spectrum, a number of colors which contain the same lines as the original spectrum, and running in the same direction as the original lines, so as to traverse all the colors. If one of the liquids be poured into a glass vessel with parallel sides, and put in the place of the paper, it will show the same phenomena; only the original spectrum disappears almost entirely, and the new image penetrates into the fluid to a depth inversely proportional to its refrangibility. The dark lines appear like partitions. But if a solution of quinine be introduced between the light and the prism, the original spectrum only remains; because the liquid absorbs the rays beyond the violet. Sulphide of carbon produces the same effect. A prism of this substance cannot therefore be used in these experiments.

## (*d*.) EXPERIMENTS ON SIGHT, AND SOME COMPOUND OPTICAL INSTRUMENTS.

[172] **The eye.**—To show the inverted image on the retina, and illustrate the structure of the eye, take one from a recently slaughtered ox, and free it from the attached fat and muscles by means of a sharp knife or scalpel with convex edge. The optic nerve should be cleaned and left

attached. Then cut out of the sclerotic coat or white of the eye a strip about one to one and a half lines broad and two lines high, in the direction of the axis of the eye. To do this, cut the sclerotic about half through on the four sides of the slip, and then with the point of the knife carefully separate it at one edge, into which introduce the point of a pair of sharp scissors, and follow round the entire outline. Lay the eye on a wooden stand hollowed out to receive it, fig. 345, and remove the choroid coat. This may be done without injuring the retina, by seizing the choroid with a delicate forceps, and cutting through the part thus held, finishing with the scissors. After cutting out in this way a piece of the choroid the rest is thrown back. It will be sufficient to remove the outer black layer of the choroid, but this is more difficult; the image is then more distinct, for the retina is apt to protrude somewhat through the aperture, especially when the stand is not hollowed out to receive it.

*Fig.* 345.

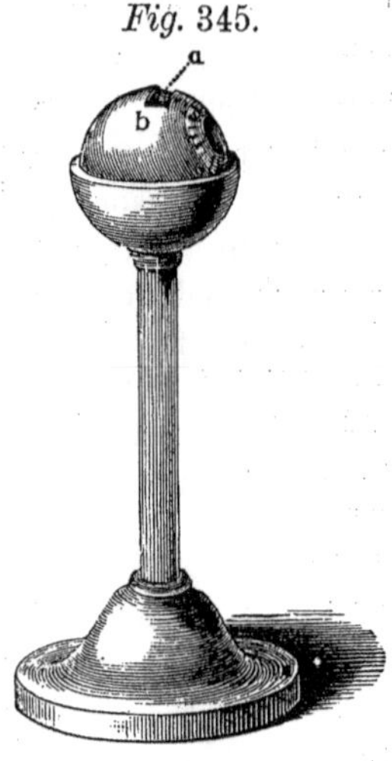


If, after this preparation, the flame of a candle be held before the eye, a very distinct and sharply defined inverted image of it will be seen through the portion thus rendered transparent.

During the preparation of the eye it should be allowed to float on water, that it may undergo no change of form.

After this, lay the eye on the cornea and extend the section of the sclerotic coat to within a line of the cornea, and continue it at this distance circularly around the eye. If the scissors have not a knob on the end, the point must be directed against the sclerotica at each insertion, so as not to injure the choroid. Now take up the sclerotic with the forceps and loosen it from its slight attachment to the choroid by means of the scalpel, all around to the optic nerve, and finally cut through this. The choroid is now completely laid bare. Next make a couple of incisions in the choroid toward the cornea, holding up the coat all the while with the forceps, and laying back the edges. In this way the retina is completely exposed. Finally remove this, by seizing it with the forceps and drawing it aside by piecemeal. It has no tenacity except when the eye is quite fresh, and has only to be pushed aside to show the vitreous humor in perfect clearness. The crystalline lens will also be exhibited and the puckered ring in which the choroid ends anteriorly, and in which the capsule of the lens is fastened. The latter is of no importance in the physical consideration of the eye. On seizing the choroid and sclerotic coats, and holding up the eye by them, the vitreous humor and the crystalline lens become loosened by their own weight with a little assistance, and separate from the iris. Catch them in a dish, the lens uppermost; it is generally encircled by the attached pigmentum nigrum. On spliting the capsule at the side, the crystalline lens will fall out; it may then be stuck edgewise on a pin and used as a lens. To show its foliated texture it must be hardened in alcohol.

Open another eye in front by introducing the sharp point of the scalpel under the edge of the cornea and thrusting it into the anterior chamber of the eye, before the iris, and with the blade parallel to the iris. On withdrawing the knife the aqueous humor will follow, and the whole cornea may be cut away with the scissors to show the iris. When the edges of this are separated from the choroid coat, the crystalline lens and its capsule will again be visible.

This demonstration of the structure of the eye will be very much aided by an artificial eye on a large scale.

[173] **Vision.**—To exhibit the experiment of Father Scheiner, on which Stampfer's optometer is based, take the bar, fig. 315, set a white screen at one end of it, the small lamp, fig. 317, about one third of the length from the other end, and a convex lens between the two, so as to project a distinct image of the lamp on the screen. This image will remain distinct, though a screen with two rectangular apertures, 1 inch high, $\frac{1}{3}$ inch wide, and about $\frac{1}{2}$ inch apart, be introduced close before the lens. The illuminating body is now at the normal distance for distinct vision for an eye whose retina is represented by the white screen, and its crystalline lens by the glass lens. If the distance of the light be now varied, two images of it will appear on the screen, and the distance between them will increase with the change of position of the flame.

[174] **Stampfer's optometer,** in its simplest form, is easily constructed. Have two square wooden tubes made, about 15 inches long, so that one will slide easily in the other and still fit closely. The exterior one should be about an inch square, and the inner one divided into inches and lines. A piece of glass is set in the end *A*, fig. 346, and upon this a

*Fig.* 346.

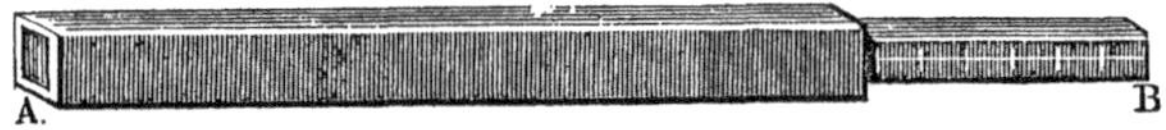


piece of tin-foil pasted in which two slits are cut $\frac{1}{2}$ an inch long, $\frac{1}{5}$ line wide, and about the same distance apart. The inserted end of the inner tube is also closed with a piece of glass covered with tin-foil with a slit in it. The end *B* is closed with a piece of smooth ground glass or straw-paper. When the inner tube is pushed in all the way, the zero of its graduation must project beyond the end of the outer tube as far as the distance between the slit on its glass and the two slits on *A*. If the tubes be held toward the sunlight with *A* close to the eye, and the inner tube drawn out until its slit is seen single through the two slits in *A*, the graduation on *B* will indicate the limit of distinct vision for that eye. The mean of the two positions must be taken in which the slit ceases to appear single when it is moved backward and forward. It is still better to let the eye rest, and to seek the position in which the slit appears single to the unaccustomed eye. The instrument is tolerably long when constructed in this way, for it contains no convex lens like Stampfer's, but it is still quite serviceable. The lens necessary to produce the normal effect, with the visual distance thus measured, can be found by calculation, or a scale attached to the other side which will indicate the focal distance of the required lens. The numbers of the spectacle glasses usually run by Vienna inches.

The following table contains the visual distances in inches corresponding to the running numbers of the glasses used in the trade, the normal visual distance being assumed at 8 inches. In selecting spectacles by this table, the feeblest glass should be used with whose number the slit appears single.

| Number. Concave. | Visual Distance. | Number. Concave. | Visual Distance. | Number. Convex. | Visual Distance. | Number. Convex. | Visual Distance. |
|---|---|---|---|---|---|---|---|
| 1 | 0·88 | 16 | 5·33 | 9 | 72·00 | 18 | 14·40 |
| 2 | 1·66 | 17 | 5·44 | $9\frac{1}{4}$ | 59·20 | 19 | 13·81 |
| 3 | 2·18 | 18 | 5·54 | $9\frac{1}{2}$ | 50·66 | 20 | 13·33 |
| 4 | 2·66 | 19 | 5·63 | $9\frac{3}{4}$ | 44·57 | 22 | 12·57 |
| 5 | 3·07 | 20 | 5·71 | 10 | 40·00 | 24 | 12·00 |
| 6 | 3·43 | 22 | 5·87 | $10\frac{1}{2}$ | 33·60 | 26 | 11·55 |
| 7 | 3·73 | 24 | 6·00 | 11 | 29·33 | 28 | 11·10 |
| 8 | 4·00 | 26 | 6·12 | $11\frac{1}{2}$ | 26·28 | 30 | 10·90 |
| 9 | 4·23 | 28 | 6·22 | 12 | 24·00 | 32 | 10·66 |
| 10 | 4·44 | 30 | 6·31 | 13 | 20·80 | 34 | 10·46 |
| 11 | 4·63 | 32 | 6·40 | 14 | 18·66 | 36 | 10·28 |
| 12 | 4·80 | 34 | 6·48 | 15 | 17·14 | 38 | 10·13 |
| 13 | 4·95 | 36 | 6·54 | 16 | 16·00 | 40 | 10·00 |
| 14 | 5·09 | 38 | 6·61 | 17 | 15·11 | ... | ........ |
| 15 | 5·22 | 40 | 6·66 | ... | ........ | ... | ........ |

[175] **Duration of the impression of light on the eye.**—For this experiment we use the rotary machine with the attachment, fig. 347, upon which the pasteboard disks, painted with the proper colors, are screwed. The disks are ¾ to 1 foot in diameter, and covered with white paper, on which the colors are laid. The best colors are indigo, Prussian blue, gamboge, and carmine; these are mixed to form violet, green, and orange. To obtain white, the disk must be divided in proportion to the breadth of the several bands of the spectrum, and painted with the corresponding colors, laid on thin. The result will probably not be white, *i.e.* gray; but the experimenter can judge from the tint obtained which of the colors must be deepened to produce gray. Indigo and carmine produce a fine violet when carmine covers only 90°. Gamboge and indigo in the same proportions give green. It is advisable to paint the margin of the disk, and a circle of about 3 inches around the center, black. Instead of painting the disks, they may be covered with the bright-colored papers used by the binders for the titles of books.

*Fig.* 347.

[176] **The thaumatrope.**—This toy is too well known to need

*Fig.* 348. *Fig.* 349. *Fig.* 350.

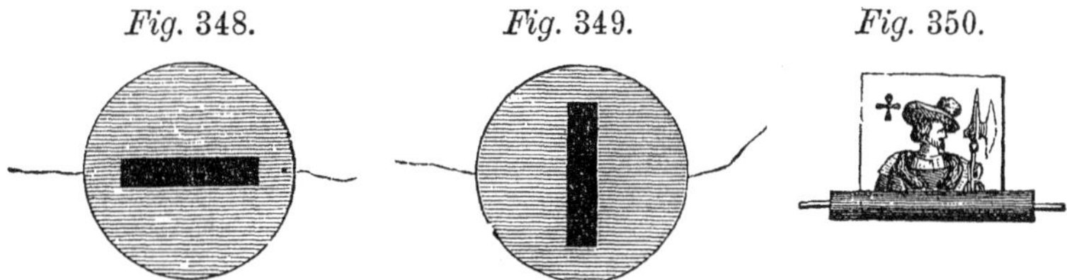

description. Fig. 350 shows a modification of it made by cutting a playing card through the middle, pasting the backs together, and fastening the end in a groove in a round stick. When the stick is twirled rapidly between the fingers by two brass pegs driven in the ends, the whole figure will be visible.

[177] **The phantoscope.**—The principle on which the drawing is made is very simple. The time necessary for a complete revolution of the figure is divided into as many parts as there are holes in the disk, fig. 351, and the figure drawn under the hole in the position in which it is to appear at each of these periods. If several drawings are made, they must be drawn on disks whose diameter is less by about the width of the apertures, and these fastened on the first large disk. Figures may be drawn on both sides of the small disks. The simplest way of setting the disk in revolution is represented in fig. 352: *a b c* is a brass wire, about 1½ lines thick, flattened slightly at top and bent twice at right angles.

The smooth axis *d e* turns easily in the holes *m n*; the button *g* is screwed on this axis, and the disk fastened between it and the nut *x*, fig. 351; the

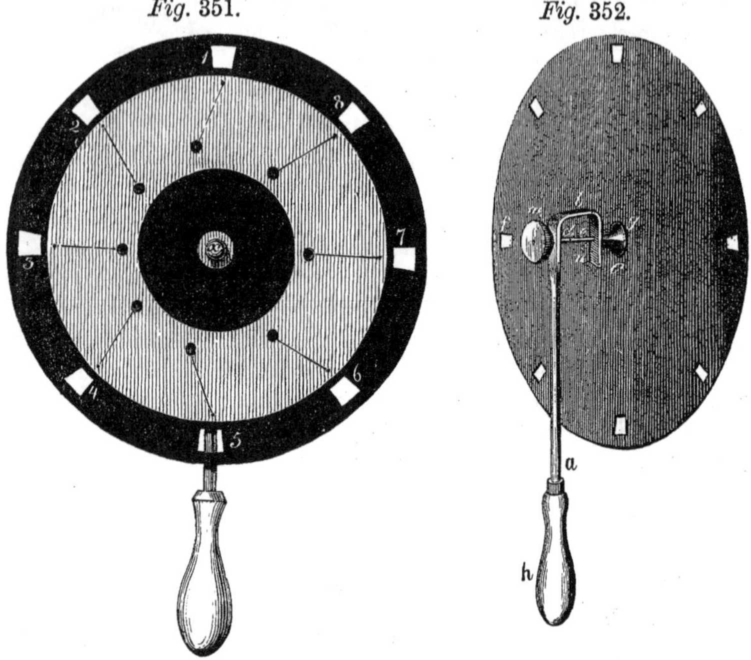

*Fig.* 351. *Fig.* 352.

latter must be made of brass. The disk is turned by the button *f* while held at a distance of 2 or 3 feet from a mirror, with the eye close to the apertures.

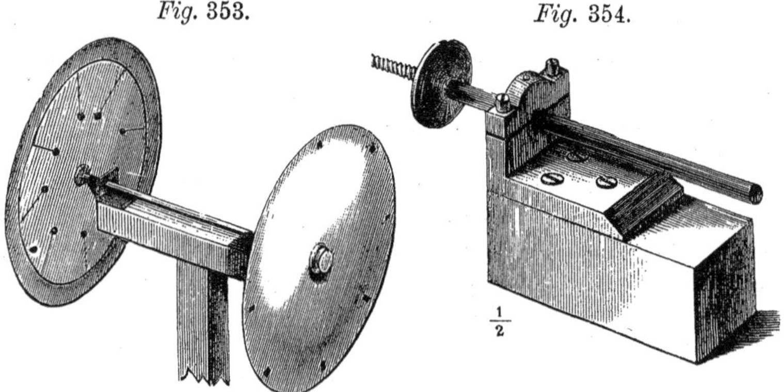

*Fig.* 353. *Fig.* 354.

The mirror may be dispensed with and the experiment made visible to several persons at once, by fastening the disk with the figures opposite that with the holes, on the same axis, as in fig. 353, and on a larger scale in fig. 354.

[178] **Subjective colors.**—The bright-colored paper used by the bookbinders is best for these experiments. Cut simple figures out of such paper and lay them on a brightly illuminated white ground; after regarding them steadily for a moment jerk them away suddenly by a string. An easier mode of trying the experiment is to paste a strip of white paper, 1 millimeter wide and 1 to 2 millimeters long, on a piece of bright-colored glass: the paper will have the complementary color to that of the glass. This appears still more distinct when a piece of very white thin paper, of the same size as the glass, is held behind it; but the shade of color is usually altered. Everything appears to the eye colored after looking for some time through a colored glass. Subjective colors may also be seen very well by looking for a time at the images of the two apertures described in the following paragraph, and then at another part of the white wall.

Nörremberg has contrived an exceedingly convenient and simple instrument for showing the phenomena of subjective colors to a whole audience at once; fig. 355 shows a front view of it, fig. 356 a side view, and fig. 357 a section through the dotted line in fig. 355. It consists of a wooden frame, the lower half of which is double; each part has a groove about 1 line in breadth and depth. The front groove passes through the bars *a a*, *c c*, so that the slide, which is made of cardboard, covered with white paper, can fall to the bottom.

The slide is of such size that when it is held up by a pin passing through the hole *m* no part of it will project below the bar *a a*. The upper part of the frame may, if desired, be covered with paper, so as to conceal the slide entirely. The half frame is closed at the back by a board *b b*. Squares of bright-colored paper, cut like fig. 359, slide in the back groove, so that the square section will occupy the middle of the field: *n* is a black ball, 4 to 5 lines in diameter, fastened by a wire to the bar *a a*, so as to lie in the center of the lower field. The colors of the papers must be as intense as possible, and smaller pieces of the same are provided to slip behind and cover the section of the others.

The experiment is made by placing two papers of different colors in the back groove, so as to have, *e.g.* a green square on a red field, and hanging the apparatus on a well lighted wall. Draw the slide up and fasten it by a wire passed through *m*, with a string attached; let the audience look steadily at the black spot in the center for about half a minute, and then pull out the wire and let the slide fall; it will appear in colors the complement of those of the paper.

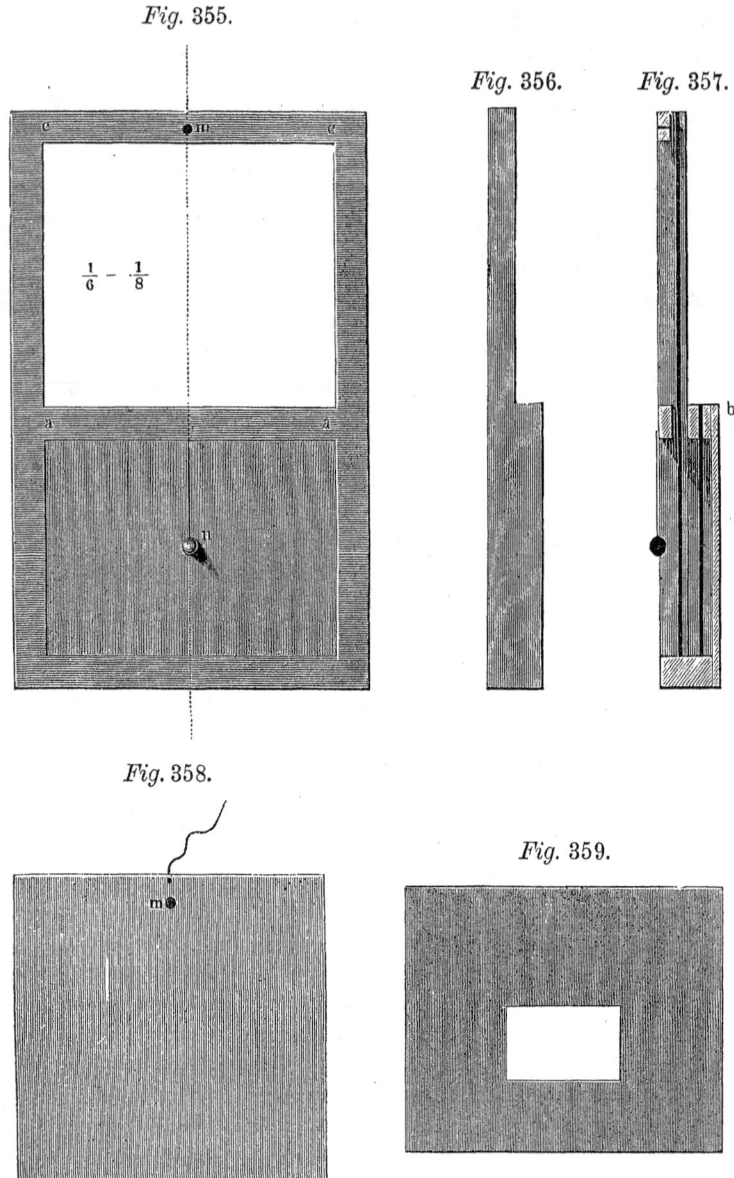


*Fig.* 355. *Fig.* 356. *Fig.* 357. *Fig.* 358. *Fig.* 359.

[179] **Colored shadows.**—This phenomenon may be produced in a striking manner as follows: Make a slide for the aperture of the

heliostat, with two holes, about ½ an inch in diameter and 2 to 4 lines apart; cover one with red, the other with blue glass. Let the images of these holes fall on a white wall 10 or 20 feet distant; they will overlap each other in a great measure, and the parts which overlap will have a more or less white color, according to the shade of color of the glasses. If the images do not overlap sufficiently, the distance may be doubled by reflecting them from a mirror. Hold a slender wooden rod, a lead-pencil, in the rays where they cross each other: it will cast a red and a blue shadow. If the pencil be held near the wall, the shadows will partly cover each other and form a black shadow; a black shadow results from obscuring either one of the apertures.

[180] **The stereoscope.**—To explain the manner of vision with both eyes, we need a cube about one inch on a side, and a Wheatstone's or Brewster's stereoscope. The latter are now very common and cheap. It is well to have also a picture, the two parts of which are painted with complementary colors; it will then appear white when seen through the stereoscope. The pictures designed for the stereoscope also appear solid when the two eyes are separated by a partition, so that each eye sees but one picture. The pictures do not at once unite, but they do so after a moment. Fig. 360 represents an arrangement for looking at the pictures in this way. The pictures are laid on the board $a\ b$; the height of $c\ d$ is regulated by the visual distance.

*Fig.* 360.

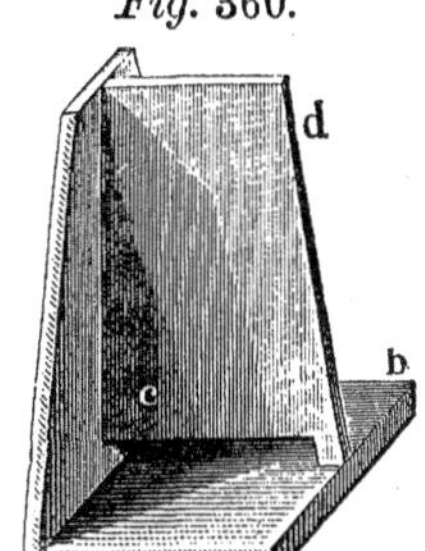


[181] **Camera obscura.**—A surprisingly beautiful result is produced by inserting a convex lens, of 4 to 6 feet focal distance, in a shutter of a dark room and receiving the image on a large white screen. The screen must be adjusted with reference to the distance of the objects and the size of the image. The best effect is produced when the view is of a place about two or three hundred yards distant. In this form the *camera obscura* costs really nothing, for both the pieces can be used for many other purposes. The images are inverted, which is corrected in the ordinary arrangement with the mirror. Erect images may also be obtained with the camera obscura, by letting the light pass through a rectangular glass prism placed close behind the lens, as shown in fig. 361.

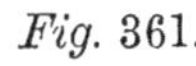
*Fig.* 361.

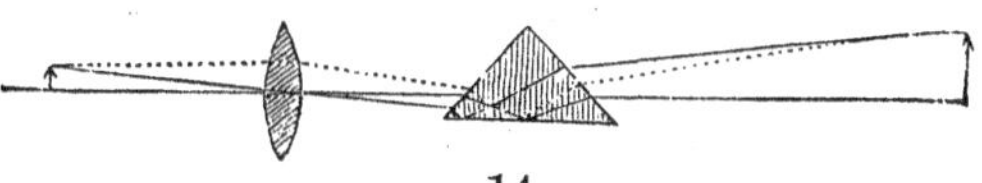

An ordinary camera obscura, fig. 362, is easily made by taking a wooden box whose length corresponds to the focal distance of the lens, and

*Fig.* 362.

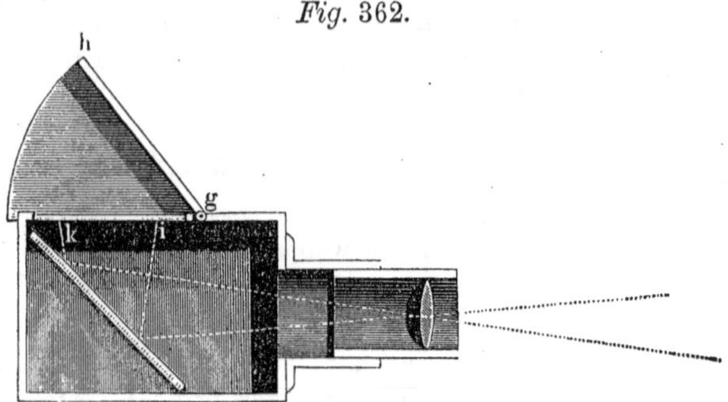


inserting in the front of it a tube of pasteboard, inside of which a second tube carrying the lens can be slipped in and out. A good plate glass mirror is selected and placed at an angle of 45° with the axis of the camera. The image is received on a plate of glass prepared by grinding with emery on another plate.

[182] **The solar microscope.**—This instrument is not adapted to actual microscopic investigations, and it is therefore not best to spend much money on it. But it is entertaining and well adapted to illustrate the action of lenses, and can be easily attached to the indispensable heliostat. The aperture of the latter is furnished either with a screen or a short tube. In the first case, a convex lens, of 4 to 6 inches focal distance, is mounted in a wooden frame *a b*, fitting in a pastebord tube *m m*, *m m*, fig. 363, which is about an inch less in length than the focal distance of the glass. If the heliostat is furnished with a short tube, the lens may be fixed directly in the tube *m m* by pasting a ring on the inside near the end, laying the lens over it and keeping it in place by a wire

*Fig.* 363. *Fig.* 364.

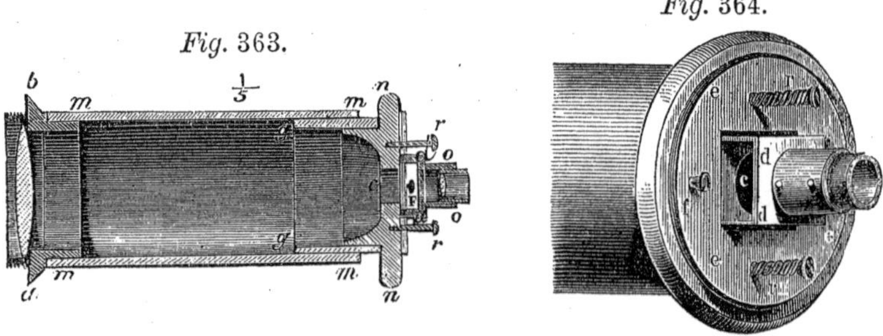


ring. This tube must fit the tube of the heliostat. A second tube, *g g*, (easily obtained from an old telescope,) must fit inside of *m m*, and inside of this again the wooden cap *n n*, shown on a larger scale in fig. 364. The object glass is set in the small tube *o o*, which rests on the plate *d d*. If a compound microscope be at hand, the achromatic lenses from this may be taken and fixed in pasteboard tubes. These small tubes must fit in the tube *o o*, and go to the bottom of it; it is well to have the hole in the plate *d d* as wide as the tube *o o*. The interior is blackened throughout with indian-ink. The plate *e e*, which has a square aperture, is slipped over *d d*. The screws *r r* pass loosely through the plate into the wooden cap; spiral springs, made of hard brass wire, are coiled around these screws to press the plate against the cap; *f f* are knobs by which it may be raised. The slides are introduced between this plate and the cap.

The room must be quite dark and the images be received either on a white wall or a large screen. Dry substances are placed in the ordinary object slides; fluids in drops on a piece of plate glass, or in a little trough formed of two parallel plates of glass, about 3 millimeters apart. For dry and dead objects the tube *g g* must be so adjusted that they will be in the focus, but living objects and fluids cannot endure the heat thus produced.

[183] **The compound microscope.**—In purchasing an instrument of this kind a good one should be selected at once.* A tolerable instrument will answer for the purpose of illustration, but the instructor needs a superior one for his own researches. It is best to buy of a maker of established reputation; the instruments made by Oberhauser in Paris can be recommended as the cheapest in proportion to their power. Before buying a microscope which is casually offered for sale, it is advisable to compare its performance with that of a good instrument whose cost is known. The scales of Bombyx Mori, or, still better, of Lepisma saccharinum, form very good test objects; and, for very good instruments, scales of Hipparchia Janira fem. Care must be taken not to be deceived by preparations presented by the seller, for if they are skillfully made they may give rise to gross deceptions in respect to the power of the instrument. A micrometer scale furnishes a more accurate test of its powers. Such micrometer scales are very costly, and not to be used with a magnifying power below 100, the lines of the first group being only one thousandth of a Paris line apart. In judging a microscope the size of the field of view is to be considered as well as the magnifying power and clearness and distinctness of the image.

Although the optical effect is the chief consideration, the price also

---

* The microscopes made by Schiek in Berlin are very highly commended.

varies with the perfection of the mechanism. In powerful instruments either the body of the instrument or the objective must have a steady motion by a micrometer screw. The object should be illuminated from above as well as by the mirror from below, which should be effected by a convex lens attached to the side, which is far better than Lieberkuhn's mirror. Another consideration is, whether the instrument is provided with an ocular micrometer, or an objective micrometer on glass, or a screw micrometer. For common use an ocular micrometer is most convenient, but it is necessary then to know the magnifying power of the objectives. Objective micrometers are more convenient for determining magnifying powers, but the fine lines soon become dirty and opaque by being used with various objects. Screw micrometers are the most accurate, but cannot be used with all objects. Finally, the instrument must have a firm basis and a suitable case. If it admit of use in a horizontal position, so much the better. All these particulars, and especially the many small, and for the most part quite useless, articles which are sold with microscopes, are of secondary importance, but they materially affect the cost of the instrument.

The power of a microscope is considerably increased by making the aperture in the stage but little larger than the field of view. But as this varies with the different powers used, and the stage must have a larger aperture for some purposes, it must be adapted to receive screens with apertures of various sizes, enlarging below. These are easily made; the edge of the opening must be very sharp and bored smooth. The aperture and the whole upper surface must be blackened. The screens with larger apertures may be made of ebony. Fig. 365 shows a screen of this kind in place.

*Fig.* 365.

Small squares of glass with rounded edges are necessary as supports for transparent objects. The preparations may be laid between two such plates and paper pasted over all but the spot occupied by the object, fig. 366. The name is written on the paper.

*Fig.* 366.

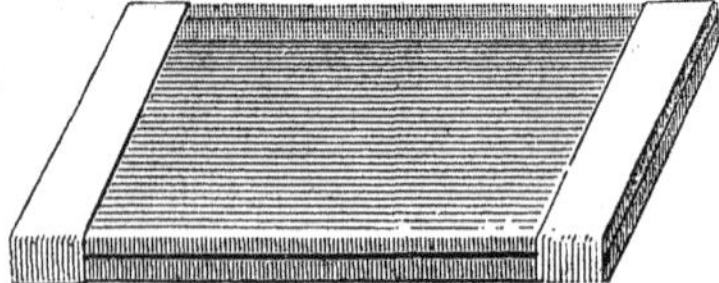

One of the plates must be very thin to enable the objective to be brought close to the object, especially with compound lenses. The opticians sell extremely thin glass prepared expressly for this purpose; it is rather expensive. This glass is not strong enough to bear the pressure of a glazier's diamond; it must be marked with a scratching diamond, and then broken. Many small objects may be made beautifully transparent by being mounted in Canada balsam. Lay a drop of balsam on a slip of glass, warm the glass slightly and press the object into the balsam; add a second drop on the top of this, and press it flat with another slip of glass. The objects to be mounted thus must be previously dried and the balsam quite liquid, otherwise the moisture will adhere to them in minute drops. It is very difficult to avoid air bubbles, but they may be gradually worked out by lateral pressure; the balsam itself must first of all be free from bubbles. The plates must lie long enough for the balsam to harden before they are cleaned off externally. This mode of mounting is not suited to all objects.

Insects of the size of a flea are prepared by squeezing out the intestines through an incision made with a lancet in the abdomen, and carefully removing the extruded parts. The empty shells thus obtained produce surprising effects under moderate magnifying powers, especially with the solar microscope.

The reflected light of white clouds or white houses produces the best illumination: direct sun light cannot be employed for the purpose. The light of a lamp with an Argand burner is excellent, especially for illuminating the object from above.

In order to choose the proper combinations, it is necessary to know the magnifying powers of the several glasses. The statements of the dealer are not to be implicity trusted in this matter. The only mode of determining this accurately is by direct comparison. Lay an object, whose size has been accurately measured with a micrometer, under the instrument, and look at it with one eye while the other eye is fixed on a scale marked with heavy lines on white paper and placed at the distance of distinct vision. With a little practice the observer will be able to see both images distinctly, so that the one will cover the other, and he can then read off the number of graduations the object covers on the scale. The reading is rendered more difficult by the tendency of the two images to move over each other, but this difficulty decreases with the size of the object. The mean of several measurements must be taken, and the powers obtained reduced to the normal distance of distinct vision, unless the operator can observe the scale at the distance of 10 inches, this being the distance usually assumed by the opticians in calculating the magnifying power.

The unpleasant shifting of the images in respect to each other may be entirely obviated by a small steel mirror (Sommering's mirror) over the ocular in such a position that the object can be seen in a horizontal direction; the scale may then be fixed in the same direction, as shown in fig. 367, so that both shall be seen with one eye. A camera lucida, used in the same way, answers the same purpose.

*Fig.* 367.

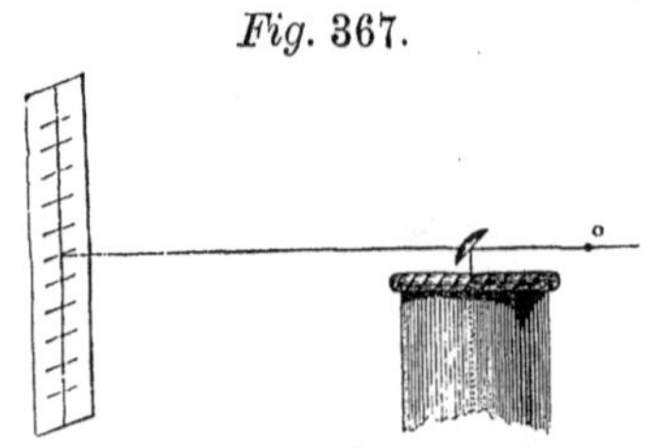

Having measured all the objectives with the feeblest ocular, measure in the same way all the oculars with the feeblest objective, and calculate the coefficients by which the magnifying power of the feeblest ocular must be multiplied to obtain the power of the others. Enter these data in a table similar to the following:—

| Objective, the nearest to the eye. | Ocular, No. 1. | Ocular, No. 2. Coeffic. | Ocular, No. 3. Coeffic. |
|---|---|---|---|
| 1<br>2·1 | | | |

When the instrument has an ocular micrometer, it is only necessary to determine the power of the objectives: this is done by simply noting the number of divisions which an object of known size covers on the ocular micrometer. The position of the ocular micrometer in the instrument varies with the visual distance of the observer, and must be found by experiment. As a rule, the more powerful oculars are only used when the magnifying power is to be increased without moving the object.

The instrument, when not in its case, must be carefully protected from dust. Any dust which settles on it should be removed with a fine hair pencil. When the glasses appear clouded they should be wiped with soft chamois leather, after carefully brushing off the dust, which might otherwise scratch the glasses. It is still better to use a piece of old linen, washed in pure water without soap, and afterwards dipped in water into which prepared chalk has been stirred 10 or 20 minutes before, and then dried. This linen and the leather must both be kept folded in paper and used for no other purpose. Elder pith, cut with a clean knife and freed from the hard rind, is particularly suitable for cleaning out the corners and angles. If the achromatic combinations are not cemented together, dust will sometimes find its way between them; should it be necessary to take them apart, it must be done with great care, and their relative positions first marked with a scratching diamond.

Any compound microscope may be used as a solar microscope by reflecting a ray of light from its mirror in a dark room. The image will be received on the ceiling.

[184] **The telescope.**—The theory of the telescope may be illustrated by lenses fixed in the sockets on the frame fig. 315. The image may be viewed directly by removing the ocular of a telescope and substituting for it a short pasteboard tube, blackened internally. If a screen of ground glass or straw paper be placed at the proper point in the tube, the image will be thrown on it. The theory of the compound microscope may be illustrated in the same way.

In purchasing a telescope, a good achromatic terrestrial instrument, magnifying 10 to 20 times, should be selected, which will be available for experiments on diffraction.

An astronomical telescope must, at least, be sufficiently powerful to show the rings of Saturn distinctly.

Fig. 368 represents a simple method of mounting a telescope equatorially: *a* is a heavy block of wood covered on the bottom with cloth, the angle *a* being equal to the polar elevation of the place. Fig. 369 is an enlarged view of the joint *s*. The telescope is fixed in the ring *p p*. The block is set on a level plotting table, with the axis *a b* in the meridian.

*Fig.* 368. *Fig.* 369.

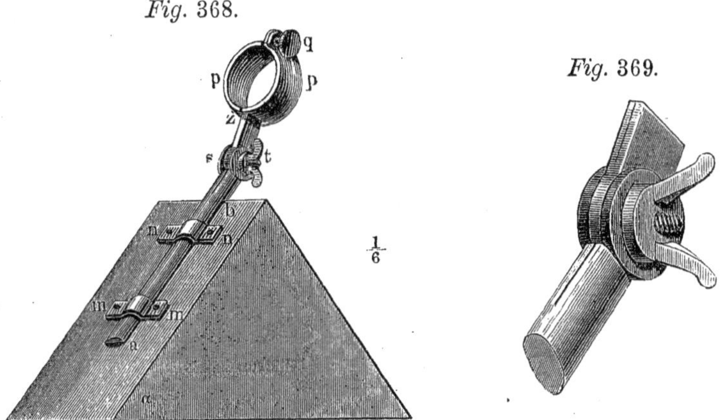


A star being in the field of the telescope, the joint *s* is tightened, and the instrument turned on the axis *a b*, so as to follow the star. When the telescope is mounted on a tripod with vertical and horizontal motion, the tripod may be detached and the upright post fastened immovably to *a* instead of the axis *a b*. It might also be screwed to a block with one face parallel to the equator, as in fig. 370.

The value of a telescope depends less on its magnifying power than on its clearness and definition. The best way of comparing two instruments is to measure the distance at which the same print can be read with each. A clear sharp type on white paper should be chosen for the purpose.

*Fig.* 370.

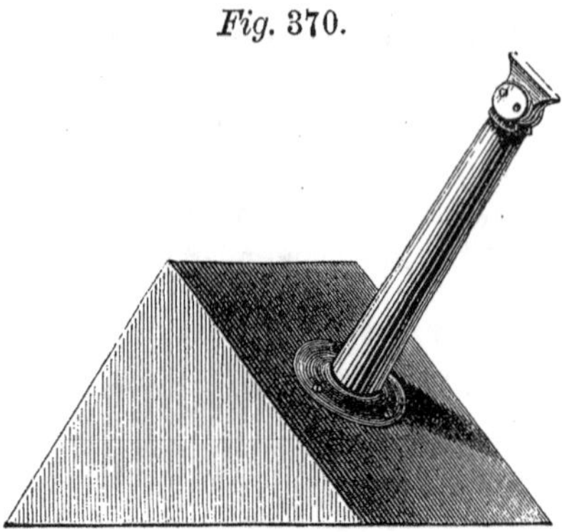

[185] **Reflecting telescopes.**—An instrument of this kind, though no longer made for use, is convenient for illustration and for many experiments. A casual opportunity to purchase an old one should, therefore, be improved. They are easily refitted if the reflector and the lenses are good.

*Fig.* 371.

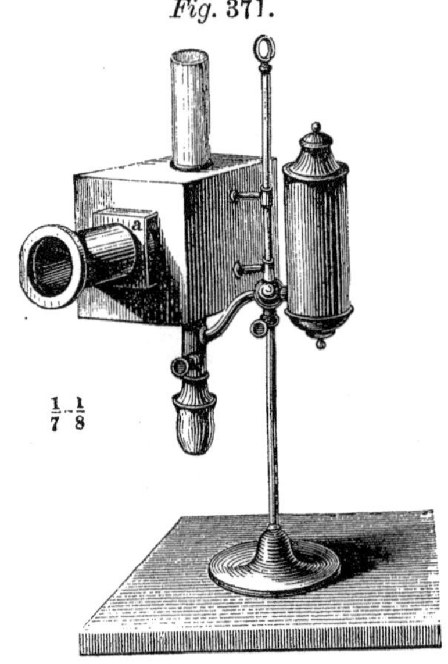


[186] **The magic lantern.** — A powerful instrument may be constructed without much cost, by making use of the lenses and mounting of a camera obscura designed for photographic purposes. Have a tin box made to fit over an Argand lamp, as shown in fig. 371, with a door in the side, a small tin reflector on the back, and a tin tube set on the slide rest *a* to receive the lenses and their mounting. The length of this tube must be adjusted according to the focus of the instrument.

## (*e.*) EXPERIMENTS ON THE INTERFERENCE AND DIFFRACTION OF LIGHT.

[187] **The mirror experiment.**—This experiment is made with two pieces of the same mirror. Take a piece of plate glass which is not very thick, 2 to 3 inches long and 1 to $1\frac{2}{3}$ inches wide, and cut in halves by a clean stroke of a diamond: if it does not break with a straight,

smooth edge, try another piece. Round the edges on the other three sides with a file or sand.

Clean carefully the side on which the cut was made with alcohol and blotting-paper, and apply 4 to 6 thick coats of indian-ink, so as to form an opaque coating. Each successive coat must be applied by gentle parallel strokes of the brush, for the least pressure will loosen the previous coats. The strokes must always cross those of the last coat. Black sealing-wax, or lamp black mixed with oil of turpentine and a little varnish, may be used instead of indian-ink.

Take a block of wood, fig. 372, with a stem fitting in one of the stands, color it a dull black, and smear a little shoemaker's wax in the middle and along the edges of one side. Lay the two pieces of glass on the wax with the cut edges together, and press down the middle so as to give the plates a slight inclination to each other, just sufficient to show a double image of a window bar 10 to 20 steps distant. The edges must fit evenly together.

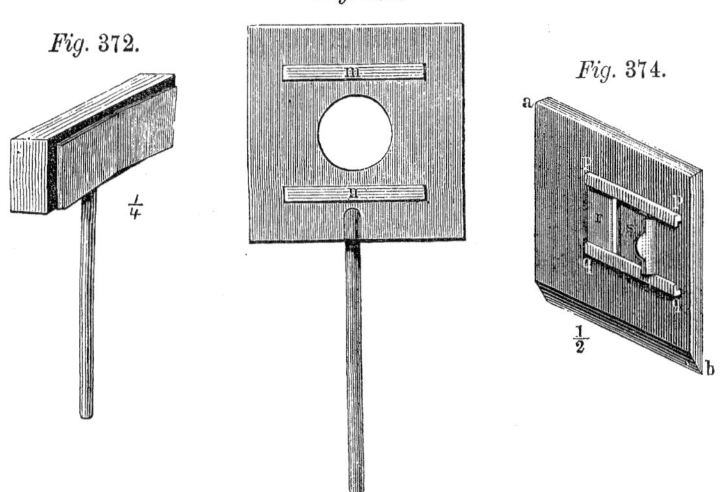


*Fig.* 372. *Fig.* 373. *Fig.* 374.

The experiment may be made with a candle, or a lamp with a narrow flame, without a screen; but the lines are more distinct when a screen with a slit not less than one line wide is interposed. The arrangement with a gas flame in § 161 is very convenient. Fig. 373 shows a screen, between whose bars *m* and *n* either two slides may be inserted or the single slide *a b*, fig. 374. All the parts of these slides must be carefully fitted. The arrangement of the slit shown in fig. 332 is very convenient. The room need not be dark.

Place the two blackened mirrors on a level with the light, at a distance of 10 to 15 paces; at a less distance the lines, though still visible, will be much less distinct. The line of junction of the two mirrors must be parallel to the flame or the slit, and the mirrors set an exceedingly small angle to the rays of light. The light should be observed through a lens with a magnifying power of 4 to 6 times, fixed on a black frame, as seen in fig. 375, and placed 1 to 2 feet distant from the mirrors, nearly in a straight line with the slit and the middle of the mirror. The light, screen, and mirror may be fixed on the stand, fig. 315, and the lens set on a separate stand.

*Fig.* 375.

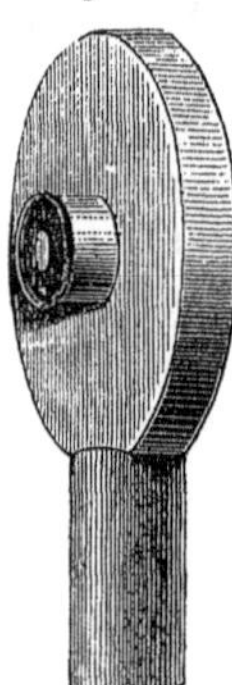

If the experiment be made with sunlight, the slide, fig. 374, is attached to the heliostat and the lines shown on a screen of straw-paper. Objective demonstrations save much time, and the lecturer is sure that the audience really see the phenomenon, which is not always the case otherwise. An apparatus is often displaced, and the fact escapes notice until many persons have looked and seen nothing.

To make the experiment with homogeneous light, sunlight is almost indispensable. A piece of glass of the proper color, or a bottle with parallel sides filled with a colored liquid, is placed before the opening in the heliostat. Ammonio-sulphate of copper yields an admirable blue solution. It is obtained by adding ammonia to a solution of sulphate of copper until the greenish blue precipitate at first formed is redissolved. The solution can be kept unchanged for years in a closely-corked phial. Table salt sprinkled on the wick of a spirit-lamp imparts to the flame a tolerably homogeneous yellow color, which may be rendered completely so by transmitting it through a brownish-yellow glass.

[188] **Experiments with the interference prism.**—It is tolerably difficult to obtain a biprisma both parts of which give an equally small refracting angle. Almost any piece of thick plate glass may be made to answer the purpose, for the surfaces of plate glass are seldom parallel, and it is only necessary to ascertain the direction in which, if prolonged, the faces would intersect each other. This may be done, according to Ohm, by fixing a dark screen with a fine slit before a window, and placing the glass about a foot from it in such a position that the slit can be seen in it. Two images will be seen, one reflected from the nearer and the other from the farther side of the glass, from which the coating has been scraped off. But the two images will only lie in the same straight line when the plane of reflection is perpendicular to the edge of the refracting angle which the two surfaces of the mirror make with each

other. This direction is easily found by turning the mirror, and the direction of the plane of reflection marked upon it. Cut a strip about an inch wide out of the mirror at right angles to this direction, and take two pieces an inch long from this. It is easy to find with the callipers on which side they are thickest; this is ground perfectly plane. To avoid splintering the edge during the grinding, cut a slight bevel on the edge first, and grind it away finally with very fine emery. The other three sides need not be ground so carefully. Fasten a strip of brass *m m* to a board *a b* with a handle, fig. 376, and lay the two prisms against this with their bases together. They may be kept in place by little brass springs *c c c*, or cemented with Canada balsam. Prisms of this kind can be used with a much finer slit than mirrors, and are particularly serviceable with homogeneous light. The whole apparatus can be arranged on the beam, fig. 315, at the same distances as with mirrors.

*Fig.* 376.

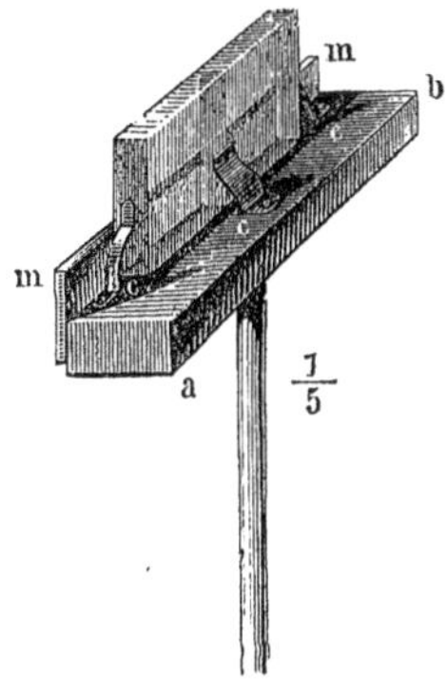

[189] **Newton's rings.**—The rings may be shown very beautifully with a convex lens of about 4 feet focal distance, and a piece of plate glass; their breadth and beauty increase, however, with the focal distance of the lens. For exhibition the lens and plate must be set in a frame, so that the pressure of the two glasses upon each other can be increased at pleasure. Two stout plates of brass with apertures in the center, and held together by four screws, as in fig. 377, is all that is needed. To make the rings more distinct by reflected light, lay the apparatus upon a black ground and view it obliquely. As the rings show but faintly by transmitted light, it is better to project them on a white wall by means of a solar microscope with a low power. The distinctness of the image is increased by covering the lower brass plate with black paper, and leaving an aperture only as great as the diameter of the rings.

*Fig.* 377.

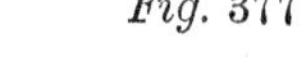

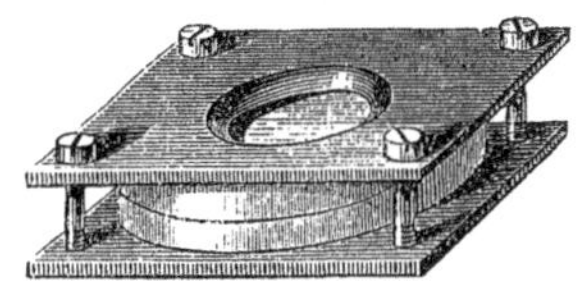

To insert the apparatus in the solar microscope, a tube of pasteboard must be made to slide in or over the tube *m m*, fig. 363, just large enough to receive the glasses, and provided with a smaller tube *a*, as in fig. 378, for the adjustment of the lenses.

Crystals with perfect cleavage, such as gypsum, mica, and calcareous spar, may be used to illustrate the colors more fully. They usually contain air inclosed between the laminæ, or it may be easily admitted by attempting to cleave them with a knife.

*Fig.* 378.

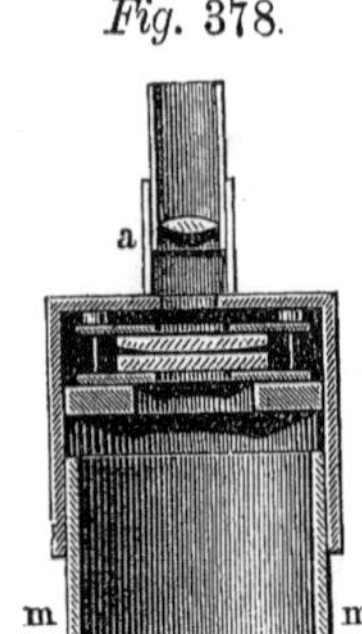


Newton's rings may be shown very beautifully by laying a strip of gold-leaf, about half an inch wide, around the margin of a disk of thin glass, 3 to 6 inches in diameter, laying the gilded side of this disk on a similar one blackened on the back, and pressing them together in the center. The pressure should be applied by a blunt-pointed stick of wood. Slips of glass gilded at the ends may be substituted for the disks; but they do not exhibit the rings so well. The gilding may be applied by breathing on the clean glass and laying it on a sheet of gold-leaf as far as required, and then cutting the gold-leaf by drawing a sharp knife with a convex edge around the edge of the glass without moving it. The two pieces of glass may be united by pasting paper around the edges.

To make the experiment with homogeneous light, the wick of a spirit-lamp may be strewed with salt. The rings are always more numerous in homogeneous light, and two glass disks separated by gold-leaf appear quite covered with them. By placing the apparatus successively in the different colors of a spectrum produced by a prism of flint glass, the rings appear very numerous, and the differences in their breadth in different colors may be perceived without measurement.

To show Newton's rings with soap-bubbles, make a solution of good, dry, home-made soap (the finer kinds are not nearly so good) in pure soft water, in the proportion of 1 to 180 or 200. Filter the solution while hot. It is better to allow it to stand 24 hours, and to avoid stirring up the sediment which subsides. When the solution is cold it may be used as follows:—

(*a*) Place the solution in a saucer over black paper by an open window, and with a small glass tube blow a hemispherical bubble about an inch high. Withdraw the tube before the colors appear, and cover the saucer with a bell-glass. The colors make their appearance on the summit of the bubble and spread in every direction. Often all the rings with the dark central spot form at once, and grow broader. As many bubbles break before the rings are all developed, and the development proceeds slowly, this process is not well adapted for class instruction.

(*b*) Pass the glass tube through a cork and blow the bubble inside of a bottle. Cease blowing before the colors appear, and close the tube with a plug of wax. The rings appear around the tube. The color which occupies the space nearest the ring is often 1 to 2 lines broad. But the rest of the colors are generally narrow, and the successive rings crowd the preceding ones aside. The bubbles remain for a considerable time.

(*c*) The bands may be produced very beautifully and certainly by dipping the mouth of a tumbler into the suds and holding the film thus formed obliquely; the higher parts becoming thinner, the colors appear in bands. This experiment is very easily made, the bands are developed very rapidly and attain a considerable breadth; the experiment is, therefore, well adapted for illustration, although the film which covers the mouth of the glass soon breaks.

(*d*) Boil a more dilute solution of soap ($\frac{1}{400}$) for some time in a white phial, in order to expel all the atmospheric air, cork it quickly, cut the cork even and seal it. If the bottle be shaken after cooling, bubbles will be formed which show beautiful colors and stand a long time. But these colors are at first seldom developed in their regular order. When such a phial has been used for some time, the colors appear with increasing regularity and beauty over the film which covers the whole side when it is held obliquely; and as the phials are always ready for use, they are very convenient. According to Eisenlohr, the colors are exhibited with remarkable beauty when a phial thus prepared is fixed to the center of the wheel of the whirling machine, and a rapid revolution imparted to it, after forming by agitation a film extending across the phial.

Very thin films of glass may be formed by heating a glass bulb and blowing it out suddenly until it bursts; the fragments will glow with the most beautiful colors. Thin films of liquid poured upon a darker liquid exhibit very beautiful colors: especially the ethereal oils poured upon water.

The best means of viewing these rings in homogeneous light is through colored glasses. With colored glass, rings may be detected in the experiments *a* and *b* long before they can be seen with the naked eye. The difference in breadth is very perceptible when blue and red glasses are interchanged. But the difference is most distinct when one-half of the ring is covered with blue and the other with red glass.

[190] **Grimaldi's experiment.**—Make two holes of the diameter of a medium-sized needle, $\frac{1}{3}$ to $\frac{1}{2}$ an inch apart, in a slide of thin metal fitting into the heliostat. If a sunbeam be thrown horizontally on these holes and the image received on a screen of straw-paper, in a dark room, at such a distance that the two images will cover each other to one-third or at most one-half of their extent, a dark border will be perceived on that edge of each image which is next the other, while the middle of the space where the two images coincide is lighter than that which

is illumined by a single hole. If one of the holes be covered, the dark borders will disappear. The holes may be made in tin-foil pasted over a larger aperture in a board. As glass mirrors always give more than one image, several images of the aperture will be seen on the screen; but the main fact will, under favorable circumstances, still be distinctly visible.

[191] **Experiments on diffraction.**—The fundamental experiments on this subject are the objective exhibition of diffraction by means of a slit, by a circular aperture, and a thin opaque body. These must be shown objectively in the simplest and most direct manner, though there are other means of showing the phenomena on a greater scale.

For these fundamental experiments sunlight is necessary. Arrange the slides, fig. 331 or 332, in the heliostat so as to form a vertical slit, about $\frac{1}{2}$ a millimeter wide, through which the ray of light is directed horizontally toward a white wall, 15 or 20 feet distant, in a dark room. About 5 feet from this fix a vertical slit by means of the slides, fig. 373 or 374, or the wooden ring on the telescope, to be described presently. When the second slit is narrow, the spectra will be distinct and broad enough to be seen by several at once. The effect of a change in the width of the diffracting slit may also be shown.

The second experiment is the diffraction caused by an opaque body. Fasten a needle on a stand so that it can be placed in the middle of the beam of light coming through the slit in the heliostat. The shadow of the needle will have colored margins, and similar bands will be perceptible within the shadow. The breadth of the colored margins will increase with the distance from the wall.

A third experiment is made with a round hole. For this purpose make a hole 2 millimeters in diameter in the heliostat, and in the screen, fig. 373, a similar one of half the size. Concentric bands will be visible around the image on the wall, but they will not be so distinct and clear as with the slit: the second or third ring will hardly be distinguishable. A screen of straw-paper may be advantageously substituted for the white wall.

These phenomena, like those of interference, can be better seen through a lens.

[192] To observe the phenomena of diffraction with simple light, colored glasses and a lens or telescope must be used. The author has never succeeded in obtaining lines by throwing simple light upon the second slit by a prism placed between the two. The telescope is the best means of observing these phenomena; it requires a magnifying power of only 8 to 12 times. The telescope of the theodolite is the best, and, for measurements, the only instrument.

Fit to the objective end of the telescope a wooden ring $B$, fig. 379, into which a conical wooden collar $C$, shown in section in fig. 380, fits tightly by a washer of leather. This collar has a flange around the top to hold the disks in which the diffracting apertures are made.

*Fig.* 379.

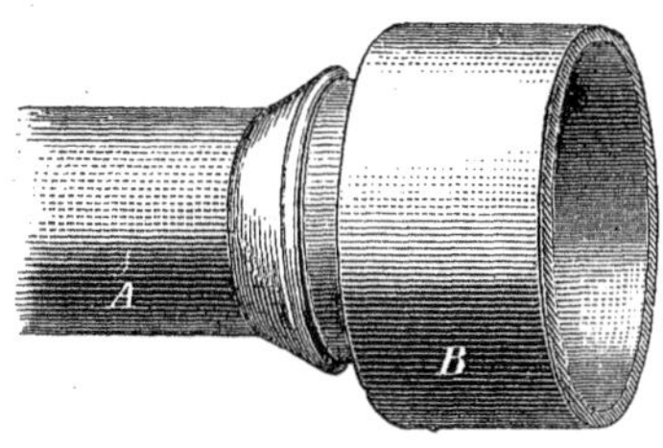


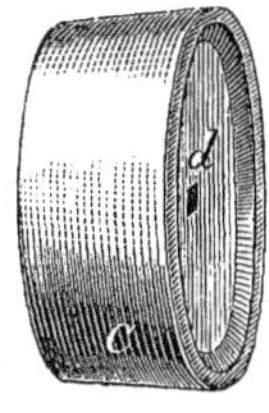


*Fig.* 380.

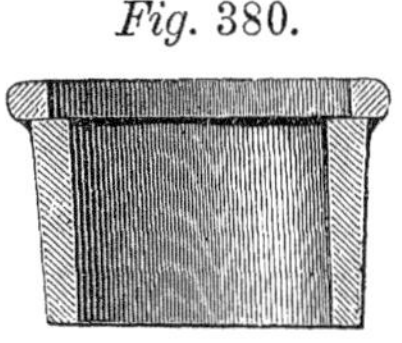

The following arrangements are convenient for instruction :—

(1) A double slit, as shown in fig. 381. A slender wire is fixed at right angles to the bars $m$ $n$, and the slides $b$ and $c$ arranged so as to fit closely against it, leaving a slit on one or both sides of it at pleasure. The slides are about half an inch high, and the square aperture in the disk a little less in size.

*Fig.* 381. *Fig.* 382.

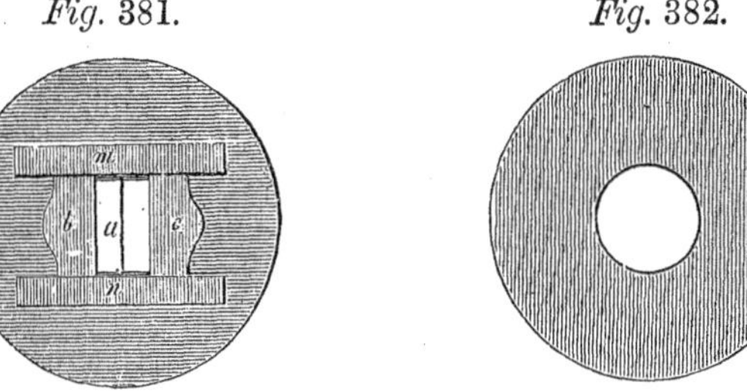


2) A grating made by fastening several pieces of knitting-needle over a square hole, with spaces equal to their own diameter between them. There may be two of these, differing in the thickness of the wires.

(3) A piece of wire gauze fastened in a wooden collar by means of a wire ring.

(4) Several round apertures, varying in size from 2 millimeters downward. The smallest ones may be made with a needle in a piece of tin-foil pasted to a circular disk, fig. 382.

(5) A slip of tin-foil with two round holes at the distance of their diameter from each other, and another with three apertures similarly disposed so as to form an equilateral triangle.

(6) A rhombic aperture, from 1 to 2 millimeters on a side, and two

similar ones with their axes parallel, and separated by their breadth from each other.

(7) Pieces of bobinet, silk, etc., stretched in larger rings.

The experiments may be made either with sunlight or lamplight, but in the latter case all the apertures must be larger. For sunlight the apertures must be set very fine. Only sunlight can be used for experiments with simple red or blue light.

[193] In making the experiments, arrange the telescope, to which the wooden ring *B*, fig. 379, has been attached, so as to furnish a distinct image of the aperture through which the light is entering at a distance of 5 to 10 paces. Insert any of the diffracting arrangements described above, taking care to place the slits and grating parallel to the illuminating aperture. The phenomena of diffraction may now be seen by looking through the telescope without darkening the room. If sunlight be employed, the light which passes through the telescope may be caught on a white screen in a dark room, and the phenomena exhibited to a whole class at once. The double slit and the gratings are well adapted to this purpose, because they permit the passage of sufficient light to make the image very clear. The telescope must be drawn out so as to make a distinct image of the aperture in the blind on the white screen.

It is well to place over the wooden ring already mentioned a screen of pasteboard, from 3 to 6 inches in width, to prevent the interference of the rays of light which pass beside the telescope. These rays are caused by a multiplicity of images from the glass mirror of the heliostat.

By mounting the wooden ring *B*, fig. 379, on a stand, its appendages may be used in the objective experiments first described (§ 191.)

If lamp or gas light be employed, place the screen, fig. 373, before the lamp, or the chimney, fig. 320, over the gas flame, with a movable slide or a round hole at least 2 millimeters in diameter.

Instead of admitting the light through a round hole, it may be reflected from a thermometer bulb, a watch-glass blackened on the inside, or a metal button; and instead of a slit, the reflection from a glass tube, blackened on the inside with sealing-wax, may be used. These may be viewed either through the telescope or with the naked eye, through the diffracting apertures. In the latter case, the source of light must be at the distance of distinct vision and the apertures very small. The simplest way of making regular diffracting apertures is certainly to cut them in tin-foil. Round apertures are made with a needle, and slits by a single cut with a knife over a smooth support.

To observe the simple phenomena of diffraction, one need only look toward the sky through the slit between the fingers; two or three lines will be visible, which are even slightly colored. The same effect is produced by looking with half-closed eyes through the eyelashes, or through a piece of silk, at a distant light: *e.g.* through a silk umbrella at a distant street lamp, in which case the spectra are tolerably distinct.

## (*f.*) EXPERIMENTS ON THE POLARIZATION OF LIGHT.

[194] **Undulations of polarized light.**—The character of the undulations of polarized light is very clearly illustrated by the apparatus fig. 383. It consists of a twisted wooden cylinder, as shown in the

*Fig.* 383.

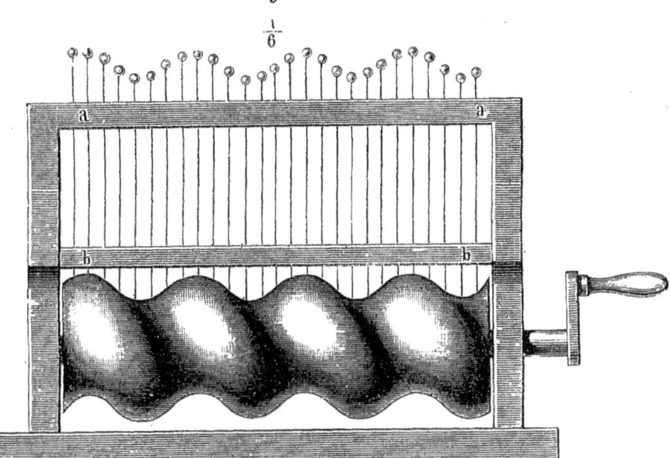


figure, which is turned by a crank. It must be made of hard and sound wood, and carefully polished. A number of straight wires terminating in knobs are passed through holes in the frame, so as to move up and down when the crank is turned, and by their motion to represent the waves of polarized light.

[195] **Polarizing apparatus.**—The simplest apparatus of this kind consists in a tube *A B C*, fig. 384, which may be made of pasteboard, cut away to the middle at *B C*. It is closed at *B* and *E* by diaphragms with a central aperture about half an inch in diameter; the end *C* closed entirely. The diaphragm at *E* is placed a little way from the end, as seen in the figure. The whole tube is blackened internally. A small block, fig. 385, the lower side rounded to fit the tube, and making with the upper an angle of 35° 25′ is glued into the part *B C*. The blackened mirror *D*, fig. 384, is cemented to the upper side with cobbler's

15

wax. Two shorter tubes $A$ and $E$ are slipped over the upper end, the inner one glued fast, and the outer made to support a blackened mirror $H$ between two bars $G\ G'$, $F\ F'$. This mirror may either be fixed in grooves at the polarizing angle for glass, or fastened to a board made to turn on an axis. This outer tube and mirror may be replaced at pleasure by a different analyzing apparatus, such as a plate of tourmalin or Iceland spar, etc. The whole apparatus may be easily mounted on a stand, as shown in fig. 386.

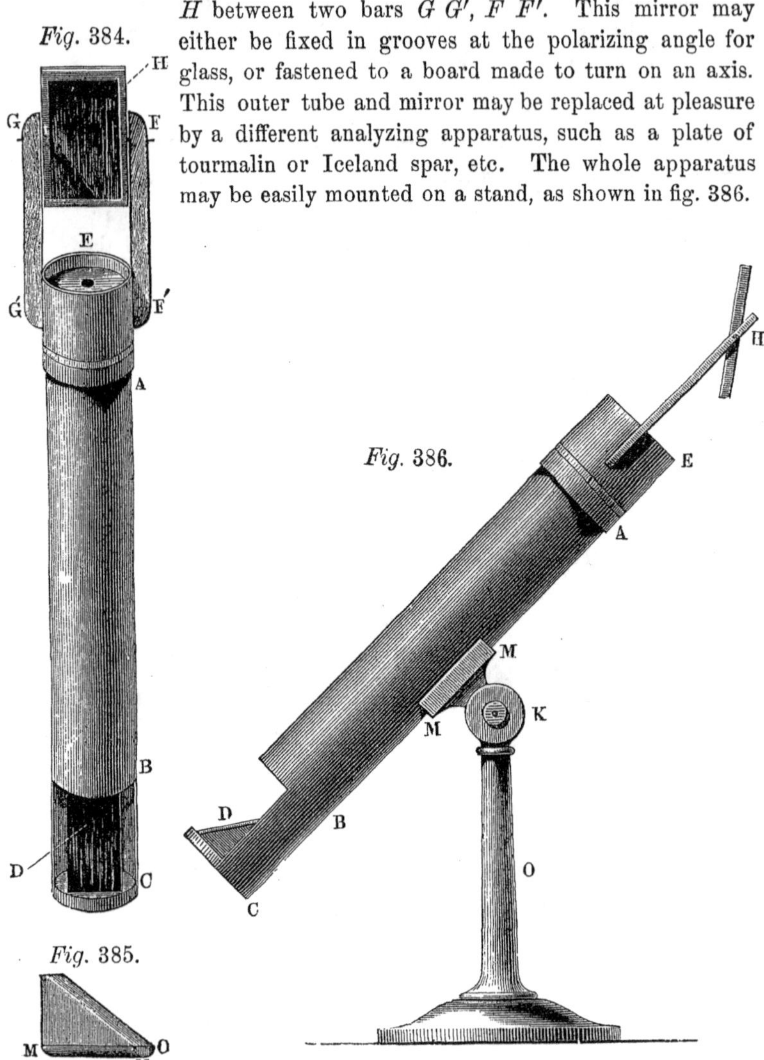

*Fig.* 384.

*Fig.* 385.

*Fig.* 386.

The upper mirror of this apparatus cannot be seen conveniently in every position, and when standing vertically it must be illuminated by a lamp placed on a lower stand. This is not advantageous even for the fundamental experiment of polarization, and entirely inapplicable to the experiments with plates of crystal; when the apparatus is inclined, however, the plates will not lie in their proper position on the top of the tube. The most convenient apparatus is that described by Nörremberg, which the experimenter can construct for himself well enough if not able to purchase a complete instrument; for some of the instruments sold under this name cannot be used for all experiments. The apparatus is shown in figs. 387 and 388, $\frac{1}{4}$ of the real size.

*Fig.* 387.

*Fig.* 388.

Two cylindrical rods, made either of old, thoroughly seasoned wood, or of brass, are fastened into a block of wood. A round or square mirror *C* is laid over paper on this block and fastened with brass bars, or strips of paper pasted around it. Two cubes of cork sliding over the rods serve to support the mirror *A B*, which turns somewhat stiffly on its axis. More simply still, the mirror may be fixed, without a frame, in grooves cut in the corks at an angle of 35° 34′ with the perpendicular. Before fixing these corks on the rods, a cork supporting a wire bent into a ring at the end, fig. 389, is slid over one of the rods. This wire is designed to hold a lens, the center of which must be in the axis of the instrument. It is, however, not needed in most experiments.

*Fig.* 389.

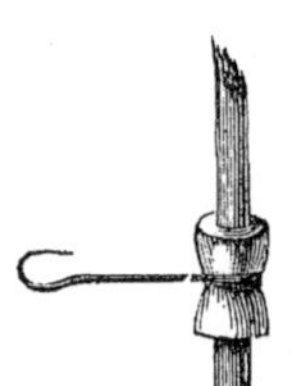

Above the mirror is a disk of brass, wood, or pasteboard, supported in a similar manner on two rings of cork. Binder's board may be strengthened for the purpose by pasting paper over it, which should be white above and black beneath. This ring is perforated in the middle to receive a second ring, which is shown in fig. 390. It consists of two pieces pasted together, the lower one smaller than the upper and perforated with a smaller aperture. The edge of the upper ring is beveled off, which can be done with a file if the pasteboard is saturated with glue. A circular piece of plate glass is fitted loosely into the ring *F F*. A disk of tin or pasteboard of the same size, with a circular aperture in the center about 2 lines in diameter, is also provided, and both it and the ring blackened.

*Fig.* 390.

Before setting on the upper ring *G G*, a wire like fig. 389 is fixed on one of the rods in the same way. A shoulder is filed on the ends of the rods to receive the ring *G G*, which is kept in place by rings of corks, or glued fast after the adjustment of the apparatus is completed. The ring *G G* is covered with white paper above and black below, and supports another ring made exactly as *F F*, except that the aperture has the same dimensions throughout; into this aperture is fixed a short tube *H H*.

The rings *E E* and *G G* are divided into eighths, with the zero in the plane of reflection of the mirror. A mark made on the beveled edge of the ring *F F* serves as an index. The analyzing apparatus is set in the tube *H H*, and made to revolve with it around the ray of light.

[196] The principal analyzing apparatus is the second mirror, which may be made just as was directed for the other polarizing apparatus. The tube which contains it may be made to fit over *H H*. A mark is made on the tube which coincides with the mark on *H H*, when the two mirrors are parallel.

Another tube is made to fit in *H H*, closed at top by a disk with an aperture large enough to hold one of the tourmalin plates from the apparatus to be described presently. A mark is made to indicate the position the plate must have when its axis lies in the plane of polarization, and the index on *I I* coincides with zero of the division. In polished crystals this position is readily recognized by its preventing the passage of light polarized by the mirror *A B*.

Another tube contains 8 or 10 plates of very thin glass, set at an angle of 35° 34′ with the axis of the tube. This is most easily accomplished by fitting into the tube another which is cut off at this angle, laying the plates on this and fastening them below by a ring cut at the same angle. Bits of very thin glass may sometimes be obtained from the dealers in glassware, but they are not often white enough to allow sufficient light to pass through 8 or 10 plates. Glass suited to the purpose may be obtained from an optician. Plates only half an inch in diameter will answer if larger ones cannot be had. The tube is previously blackened on the inside and closed at the top, leaving an aperture a quarter of an inch in diameter. A mark is made on the outside which coincides with the zero when the plates are parallel to the polarizing mirror. A tube of this kind is represented in fig. 391.

*Fig.* 391.

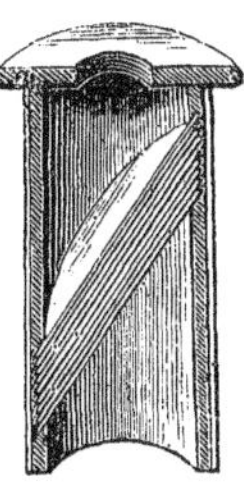

A fourth tube fitting on *H H* contains a prism of Iceland spar rendered achromatic by means of glass. It is better to buy this prism from an optician, for the very facility with which it yields to the means employed to grind it renders it very difficult to do this with exactness.

Clear rhombohedrons, 15 to 20 millimeters long and 10 to 15 millimeters broad, are used for the purpose. The section is made through the obtuse angles of the rhombohedron, and the part cut off replaced by a piece of glass of the same size and shape. As Iceland spar, which is clear enough for optical purposes, is rather expensive, the crystals should be sawed in two with a fine saw, so that both pieces can be used. The surface is ground with the finest emery on a new piece of plate glass, and polished with levigated crocus on fine felt or canton-flannel. The glass must be cemented to the spar with colorless Canada balsam, and the four sides through which the section is made covered with black paper. It is well also to cover the other surface of the crystal with a plate of very thin glass, in order to protect it from scratches. Canada balsam hardens by exposure to the air; if it is already thick, the pieces to be cemented should first be warmed. If the balsam is too thick, it must be diluted with rectified spirits of turpentine. It requires several weeks for the balsam to harden sufficiently to operate further upon the crystals. The time required for hardening may be very much lessened by dissolving solid Canada balsam in twice its weight of sulphuric ether; the pieces may then be used in a few days; but this process does not answer well for microscopic objects.

The prism is fitted accurately into a cork of the thickness *a b*, fig. 392, and this inserted in the polarizing apparatus, so that the normal line to the crystallographic axis will be nearly vertical. The position of the principal section of the crystal must be indicated on the outside, and this mark made to correspond with the mark on the polarizing apparatus.

*Fig.* 392.

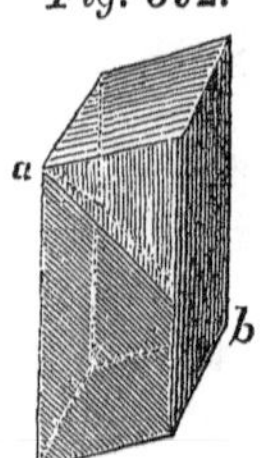


The fundamental experiment in polarization must be made by throwing the polarized light on the upper mirror of the apparatus, whether it be Nörremberg's or any other. For this purpose a candle is placed in a suitable position, and the pupils allowed to watch singly the course of the flame in the mirror while the latter is slowly revolved around the axis of the apparatus. The phenomena cannot well be exhibited objectively with the mirror, on account of the constantly varying position of the image on the walls of the room, which is of itself sufficient to cause the intensity of the light to vary; besides that, when sunlight is used, too much unpolarized light reaches the second mirror and produces an illumination in any position. With any other analyzer the experiment may be shown objectively. The achromatic prism of calc spar is best for the purpose. A screen, with an aperture 2 lines in diameter, must be laid on the middle stage of the apparatus.

[197] **The tourmalin apparatus.**—Fig. 393 shows the whole apparatus on a reduced scale, and fig. 394 a single plate with its mounting of the natural size. The plate, cut parallel to the optical axis of the crystal, is set in cork and mounted with a rim of brass corresponding exactly to its thickness. The aperture *a b* is made so large that it just embraces the plate. The cork and the anterior surface of the brass frame must be blackened. The thickness indicated in the figure is sufficient, even for the light-green varieties. Transparent crystals, although more costly, are preferable, because they need not be sliced so thin; but the dark varieties polarize more perfectly. In order to make the dark crystals transparent enough, they must often be sliced as thin as paper. The plates are secured from breaking during the grinding by cementing

*Fig.* 393.

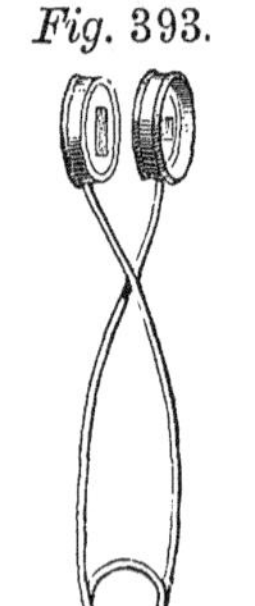

*Fig.* 394.

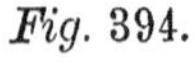


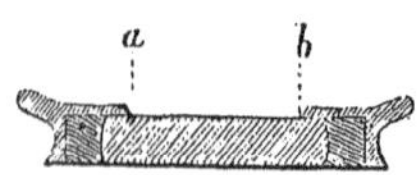

them to plates of glass. The plates may be connected by a wire, bent as in fig. 393, and provided at each end with a ring fitting the mounting. The elasticity of hard-drawn brass wire will be sufficient to hold the plates fast and allow them to be turned round with a certain degree of friction. This form of the apparatus is liable to the objection that the tourmalins cannot be brought into close contact with a thick crystal; and, moreover, that being only supported below, the crystal is apt to fall out and be injured. Besides this, the tongs are opened by pressure, which is awkward, and liable in unskillful hands to occasion damage to the crystal by falling out. Both of these objections may be obviated, by setting the plates between pointed screws in the ends of an elastic brass plate bent into tongs. Fig. 395 shows one end of this arrangement, and fig. 396 a section of the mounting, through the screws. The tourmalins are set at right angles to each other. Figs. 397, 398, and 399 show an arrangement by which one of the plates may be made to revolve for the sake of instruction. The ring, fig. 398, turns on the points of the screws, but does not revolve in its own plane; the ring, fig. 397, is made to turn in fig. 398, and into it is screwed fig. 399, in which the tourmalin is mounted.

*Fig.* 395.

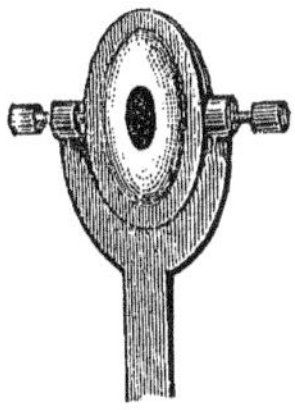

*Fig.* 396.

*Fig.* 397.

*Fig.* 398.

*Fig.* 399.

[198] **Cutting crystals.**—The hardness of tourmalin makes the cutting plates of it a difficult operation. But as it is often necessary, in studying the phenomena of polarization, to cut plates of hard stones, it is worth while to incur the expense of the necessary arrangements.

Take a circular plate of copper, about a line in thickness and 2 inches in diameter, and through the center of it drive a tapering square iron axis, as in fig. 400. This is either fastened into a chuck on the lathe, or worked with a roller and bowstring. The disk must be ground off to a thin edge.* A tin trough, like fig. 401, must then be made, large enough to admit the copper plate and embrace about $\frac{5}{6}$ of its circumference. It should be fixed on a wooden block with the opening toward the workman, the copper plate reaching nearly to the bottom of it. In this trough is placed a thin paste of emery and oil. With a plate of

* In turning and boring copper it should be constantly wetted with soapsuds.

this kind, tourmalins and other hard minerals may be cut with great rapidity.

For grinding the plates, cover a block of wood, about 3 inches in

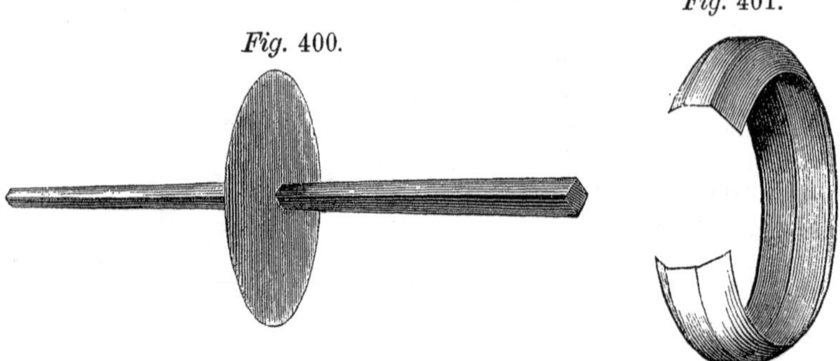

*Fig.* 400. *Fig.* 401.

diameter, with sheet copper, made like the top of a box, and screw this on the spindle of the lathe. A better and more desirable arrangement is to have a disk of the size named forged out of copper, and a screw soldered into it which fits the spindle of the lathe; or, more simply still, to clean the surface of the copper, heat it until sealing-wax melts on it, and cement it to a wooden chuck.

This copper disk is now to be turned to a perfectly plane surface with a slightly convex edge. Emery is supplied to this disk by a ring like fig. 401, and the crystalline plates are ground true on its cylindrical edge, making their surface always a little concave rather than convex, which the shape of the wheel facilitates. The plates are cemented with sealing-wax to corks in order to handle them more easily.

When one surface has the proper shape and direction, remove the ring, fig. 401, and finish on the front of the wheel. The wheel must be turned very slowly and emery applied with a stick; the crystal should be moved in epicycloids over the surface of the wheel, and its surface constantly tested by a steel ruler. For polishing, the wheel must be carefully cleaned and finer emery applied, and the final polish is given with the finest emery on plate glass. As the value of the plate depends entirely on the accuracy of the grinding, one must not spare labor nor pains. Inequalities cannot be removed in polishing, unless the substance is quite soft. Tourmalin is too hard to be ground smooth on hones of any sort, and even quartz is only slightly affected by them; with glass and other softer substances they facilitate the work very much. Such substances are polished on the polishing wheel described under the article "pendulum," with the aid of water and colcothar applied at last nearly dry.

Tourmalin usually occurs in prisms, and the optic axis is, as in quartz, parallel to the length. The experiment with crossed and parallel axes may be made with any thin transparent prisms without cutting them.

All other hard minerals are cut in the same way. Softer substances are ground on glass with emery, or else on a hone. A fine yellow oil-stone (novaculite) is excellent for fine grinding. Calc spar is apt to crumble when rubbed on a stone or on glass without some cutting powder. Substances which are soluble in water may be roughly ground with water, but the fine grinding must be done with oil. The polishing should in all cases be done at once. All substances will bear the application of a more or less moist polishing powder, except sugar, which must be treated quite dry or slightly moistened with oil. It is best always to polish on the lathe, although soft substances may be polished with rouge and a soft linen rag.

[199] **Mounting the crystals.**—All these crystals are to be set in smoothly pared or filed corks, but little thicker than the crystals themselves, and large enough to project a little way over the edges of the tourmalin tongs. It is a good plan to insert the corks in rings cut from brass tubing.

Plates which are very thin and fragile, such as mica and gypsum, or deliquescent or liable to decomposition, such as saltpeter, sugar, carbonate of lead, etc., should be cemented with Canada balsam between two plates of thin glass. The same treatment is advantageous for all crystals, as it increases their transparency and compensates for inequalities in the polish. A drop of the balsam is laid upon the glass, the crystal set edgewise on it and gradually pressed down; another drop is then laid on the upper surface of the crystal and the second glass applied in the same way. When the plate is thin, the balsam fills up the space between the glasses completely and cements them together; but with thicker plates a strip of paper should be pasted around the glasses to hold them, at least until the balsam hardens. One side of the glass should be covered with white and the other with black paper, with apertures of the same size as the mineral. The black side is always turned toward the observer. The name of the substance and the position of the optic axis should be written on the white paper, or the cork setting.

## (*g*.) EXPERIMENTS ON DOUBLE REFRACTION.

[200] Before explaining the laws of double refraction, the crystalline form of calc spar, and the meaning of the terms principal section and crystallographic axis, must be explained. For this purpose a double six-sided pyramid of wood, about 4 to 6 inches in length, is needed, and

a similar one, with the basal edges truncated, and a rhombohedron about 4 inches square, with the angles 74° 55′ and 105° 5′. Mark on this rhombohedron the principal section, and two others parallel to it, and likewise two others at right angles to the crystallographical axis, and cut through them with a fine saw. Fig. 402 represents such a rhombohedron. Paste thick paper on the cut faces to fill up the cuts, and drive in little pins to hold the parts together. Draw upon the principal section and those parallel to it the direction of the axis, and the direction of vibration of the ordinary and extraordinary rays in a double reflected ray of light, as in fig. 403.

*Fig.* 402.

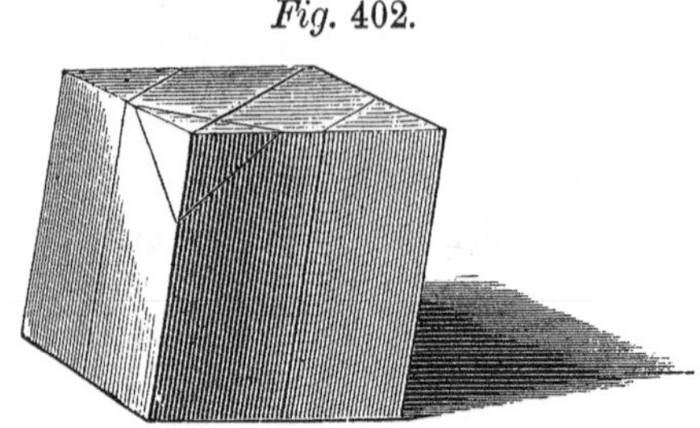

*Fig.* 403.

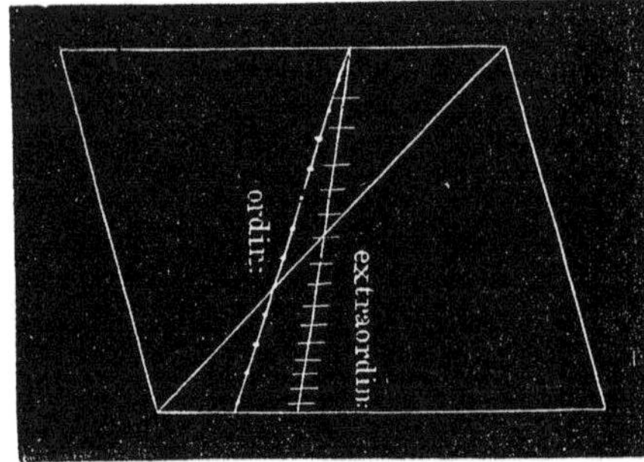


For experiments with calc spar it is only necessary to have two opposite faces clear and smooth. Double refraction is much more difficult to observe in quartz, because of the unfavorable position of the faces with regard to the axis, and the peculiar grooving of the mineral, which obscures the effect. It is best to grind and polish all the faces of the crystal, or at least two which are not adjacent nor parallel. A crystal polished in this way may be used as a prism, and two spectra, lying very close together, produced in a dark room. A prism made of a fragment of Iceland spar shows the double refraction very well in a dark room, the spectra being widely separated.

[201] To show the relative positions of the two images, lay a crystal of Iceland spar on white paper over a black line not so long as the distance between the images, which will depend on the thickness of the crystal. The line must be parallel to the principal section of the crystal. Several persons may see the phenomena at the same time. Turn the crystal with the paper slowly around, so that the principal section

will be toward each in turn; he will then see the two lines as one continuous line, while in every other position one line will gradually revolve around the other.

To show the polarization of the two rays, cover the polarizing mirror of Nörremberg's apparatus with white paper, and lay upon the support in the middle a screen with an aperture only 1 millimeter in diameter, upon which place the calc spar. The aperture in the screen may be larger for thicker crystals, but the two images must not even partially cover each other. Using the second mirror as an analyzer, when the plane of reflection of the mirror is parallel or at right angles to the principal section, only one image of the aperture in the screen will be visible; but in every other position two, which will be equally bright when the principal section makes with the plane of reflection an angle of 45°.

The same result is obtained by passing polarized light through the screen, and viewing it through a crystal of Iceland spar. Upon this is founded the use of calc spar as an analyzer, as already described. When unpolarized light is thus viewed, two images are seen in every position.

To these experiments may be added that of Huygens, with two rhombohedrons of calc spar laid one upon the other. One of these rests upon a sheet of white paper with black spots, while the other is turned round, so that the principal sections make successively every angle from 0 to 360.

[202] Nichol's prism furnishes an excellent means of observing the phenomena of polarized light. Two of these prisms mounted on cork in brass tubes, as described for the achromatic prisms of calc spar, act like the tourmalin apparatus in allowing light to pass only when their principal sections are parallel. They possess, moreover, the great advantage of transmitting more light, because the material is colorless. The cost of these prisms varies with the size of the field of view; they should not be less than 1 centimeter on a side. A pair of Nichol's prisms may be mounted as seen in fig. 404. The prisms with their settings are pushed into the elastic collars *m m*, until they touch the mounting of the crystal which rests in the support *b b*.

*Fig.* 404.

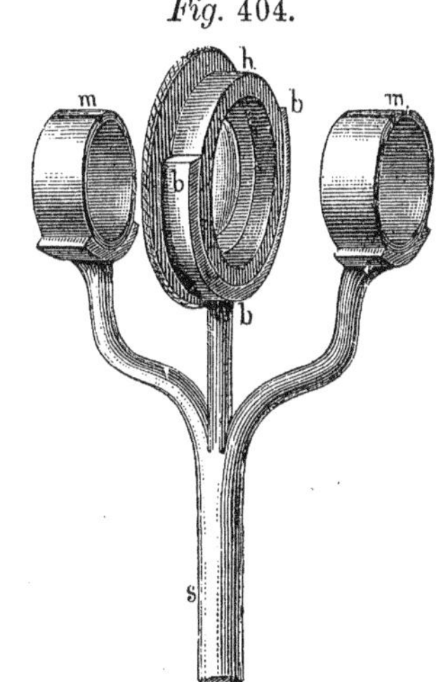


The price of a pair of Nichol's prisms is double that of the tourmalin tongs.

## (*h.*) EXPERIMENTS ON THE COLORS OF DOUBLY REFRACTING CRYSTALS IN POLARIZED LIGHT.

[203] **Colors of films.**—Films thin enough for this purpose are difficult to obtain from any minerals but mica and selenite; the selenite from Montmartre answers best. It is necessary to have *one* film of considerable size, so as by cutting it in two to obtain two films of the same thickness. They should be left in the rhombic shape into which they naturally cleave, in order the more readily to determine the position of the axis.

The films are cemented with Canada balsam between plates of clear glass, not over an inch in diameter, and inscribed with the color of the film. It is difficult to grind other minerals parallel to the axis thin enough to show the colors, for even in quartz the film must be less than $\frac{1}{4}$ of a millimeter in thickness.

It is easy to grind films of selenite into a wedge shape by taking a film, $\frac{1}{2}$ to $\frac{2}{3}$ of a millimeter in thickness, laying it flat upon the polishing-wheel and pressing strongly on one edge until it is ground off and the cut surface is about 1 centimeter broad. These films are also covered with glass. To obtain broader films, cement a film of selenite, about 2 centimeters square, to a plate of thin glass, and allow the balsam to harden, which will require some weeks. Then rub some emery quite fine with one piece of glass upon another, and on this grind the selenite and polish on the wheel.

[204] The experiments may be made with Nörremberg's apparatus, using the mirror as an analyzer. The films are laid on the middle stage in such a position as to show the colors best, and the upper mirror is revolved around the polarized ray. The colors seen in gypsum, when the mirrors are at right angles to each other, are complementary to those seen with the mirrors parallel. The middle stage must be movable, to accommodate the position of the films to short-sighted persons.

Films of variable thickness exhibit mingled colors: when they are wedge shaped, they show Newton's rings, and in homogeneous light, alternate light and dark bands. This is the best means of studying the succession of colors in Newton's rings. If the film of selenite be not acute enough to make the colored bands tolerably wide, fix a lens in the ring between the upper and lower stage, at the right distance from the film for distinct vision. This aids very much also in determining, by means of a uniform film laid over a wedge-shaped one with the axis crossed, which colors of the latter are absorbed by the former, in order to ascertain to what order the color of the former belongs. This investigation is also facilitated by laying the uniform film over the other in such a way that a portion of the latter shall project over the former. A stripe of the color of the uniform film will then correspond on the uncovered portion to the spot which appears black in the covered space. The succession of the colors is, according to Newton, as follows:—black: (1) blue, white, yellow, red; (2) violet, blue, green, yellow, red; (3) purple, blue, green, yellow, red; (4) green, red; (5) blue, red; (6) blue, red; (7) blue, red.

[205] When the films are laid upon each other so that the corresponding planes of oscillation coincide, the color is that of a film equal in thickness to the sum of both; when the planes of oscillation cross each other, the color is that of a film equal to the difference of both. This furnishes the means of distinguishing the different planes of oscillation already determined by the total disappearance of color. The corresponding planes in all the films should be marked on the mounting. When the achromatic prism of Iceland spar is used as an analyzer, thin films appear double, the colors of the two being complementary. As the prism is turned, the colors change, and when the two images cover each other, they appear white. This is a good way of ascertaining the right position of the prism in the tube, the spot not appearing white unless both images are equally bright. When this is not the case, the prism must be moved.

The effect of a Nichol's prism is the same as that of the mirror.

The experiments may even be made by laying the films on a table near the window, and looking at them obliquely through an analyzer.

[206] **Colored rings in crystalline plates,** cut perpendicular to the axis.—Calc spar stands before all other substances. A very small rhombohedron is sufficient, and it facilitates the grinding very much to cleave it into a regular shape, leaving the three edges which form the obtuse angle equally long. It must be ground by very gentle pressure with emery on a glass plate, as the corners are apt to break off. The phenomena can be observed very well when the triangular face is not over 5 millimeters on a side; but the thickness must not exceed 1½ lines, otherwise the rings have too small a diameter.

The colors are usually observed through the tourmalin tongs, by laying the crystal between the plates. The crystal should be mounted, for protection, in cork a little thicker than itself. The apparatus must be held close to the eye, toward the clear sky. The experiments with homogeneous light may be made either with colored glasses, or by the flame of a spirit-lamp colored with common salt.

Sulphate of nickel has a cleavage perpendicular to the axis, which makes it easy to obtain plates suitable for this purpose. If the cleavage faces are not quite even, they may easily be made so on the polishing-wheel, which must not be very moist.

The rings may be seen very well in plates of ice an inch thick, held in the tourmalin tongs.

[207] The phenomena may be shown objectively very beautifully by placing the crystals between the tourmalin tongs, or two of Nichol's prisms in the focus of the illuminating glass of the solar microscope, without any magnifying glass, and catching the image on a white screen. To avoid holding the tourmalins, a wooden block may be fitted into the tube *m m*, fig. 363, with a slit in the side to receive the tongs and a spring to hold them. Fig. 405 exhibits this arrangement. To see the rings in the ordinary polarizing apparatus, the crystals must be placed in immediate juxtaposition with the analyzer. This can be done with the prisms but not with the mirror. In the latter case the rings may be observed by laying a lens of 1 to $1\frac{1}{2}$ inches focal distance upon the middle stage of the polarizing apparatus, fixing the crystal in a ring above this at the focal distance or a little less, and a second lens in the tube *H H*, fig. 391, at exactly its focal distance from the crystal. The rings can be seen, but less distinctly, with the latter lens alone.

*Fig.* 405.

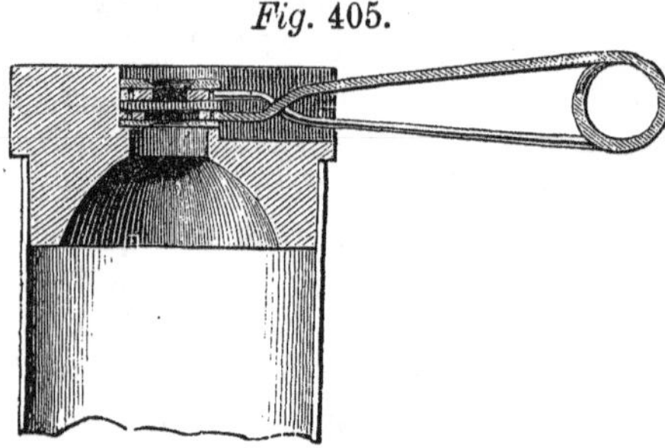

[208] Among the **biaxial crystals,** whose axes make so small an angle that one can look parallel to both axes in a plate cut perpendicular to the meridian line, saltpeter is the only one generally accessible. It is not difficult to find glassy crystals of sufficient thickness adhering to the sides of larger crystals, which are usually porous in the center. They may be ground on a hone or on ground glass, polished as usual, and then cemented between glass. A thickness of $1\frac{1}{2}$ to 2 lines is necessary in order to exhibit 4 or 5 closed curves around each axis.

*Fig.* 406.

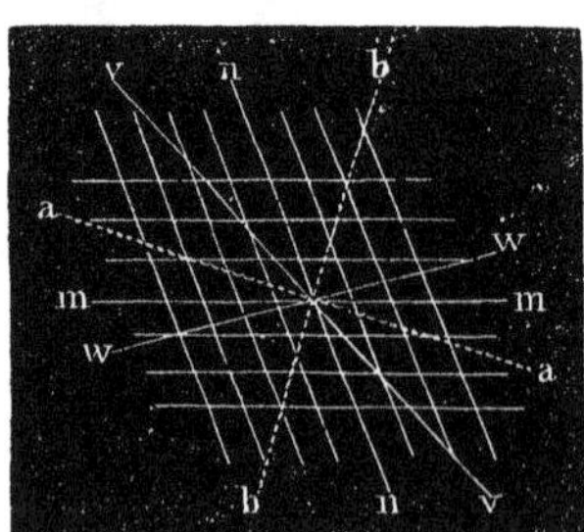


Among the crystals whose axes make a considerable angle with each other, selenite, sugar and chromate of potash are most easily had. Selenite occurs in prisms or plates of tolerable thickness and purity. Besides the principal cleavage parallel to the optic axis, there are two others, *m m* and *n n*, fig. 406, which give the plates a rhombic shape. If one of these plates be placed between the tourmalins, with their axes crossed, and turned to one of the two opposite positions in which no light is transmitted, the axes of the tourmalins will correspond to the two lines *a a*, *b b* on the selenite, and one of these will make with one of the directions of cleavage an angle of 16°. This line bisects the angle of 60° which the optic axes of the selenite *v v* and *w w* make with each other, rendering it easy to mark their position on the plate. Saw out a plate of selenite, 2 or 3 lines square, perpendicular to one of the axes and to the plane of both, wrap it around with strong waxed thread to prevent it from splitting, and grind and polish it. The thickest plates need not exceed 1 line in thickness. Splitting may also be prevented by cementing the plate with Canada balsam to glass.

Sugar crystallizes in prisms terminated by two sloping faces. The cleavage bisects the angle formed by these faces. The cleavage faces are not smooth enough without being polished. The rock-candy of the shops furnishes crystals of sufficient purity. They must be protected by glass from the action of the air.

Chromate of potash cleaves readily parallel to the broad face of the prism, and one of the optic axes is nearly perpendicular to this plane. If successful in splitting a smooth plate, $\frac{1}{4}$ to $\frac{1}{2}$ line in thickness, it may be used directly, if it is only a few lines square. If a suitable piece cannot be obtained by splitting, it must be cut with a knife and polished.*

[209] The bands in crystalline plates cut parallel to the axis can be best seen by the flame of a spirit-lamp colored with salt, in a dark room. Plates of calc spar, 1 line thick, show the hyperbolas very distinctly, although the lines are fine; quartz crystals may be much thicker. Crystals as thick as one's thumb show tolerably distinct lines, if two opposite faces be polished. All the lines are apt to be irregular, on account of inequalities in the structure of the crystal. Plates of selenite, 2 to 3 lines thick, exhibit the lines most beautifully.

[210] **Circular polarization.**—We need for this, besides plates of quartz cut perpendicular to the axis, and 1 to 3 millimeters and upwards in thickness, and the monochromatic plates of gypsum already mentioned, very thin films of mica, which will scarcely lose their white color by the rotation of the mirror of the polariscope. Fresnel's parallelopipeds must be purchased, if desired, from an optician; but they are not necessary to show the principal facts.

---

* To effect the cleavage of a crystal, lay a knife edge on the crystal, parallel to the direction of cleavage, and strike a smart blow with a small steel hammer. If the crystal is very hard, it should be laid on a plate of steel.

Whether a crystal of quartz polarizes to the right or left, may be determined by the secondary trapeziform faces, which replace the corners of the prism and the pyramid. Holding the vertex upward, if the crystal rotates to the right, these faces are directed downward and from left to right, and vice versa. The spirals seen by viewing together a right and a left rotating plate of the same thickness through the tourmalins, can be exhibited objectively in the manner decribed in § 207.

[211] **Circular polarization in liquids.**—Take a glass tube, about half an inch in diameter and as long as the polariscope will admit, grind one end smooth and cement to it with sealing-wax a round piece of plate glass, fitting into the lower stage of the instrument. Solutions of gum and sugar serve best for illustration: the former rotates right, the latter left. The tube and the projecting edge of the glass plate must be blackened. The depth of the liquid in the tube may be measured by a graduated rod.

*Fig.* 407.

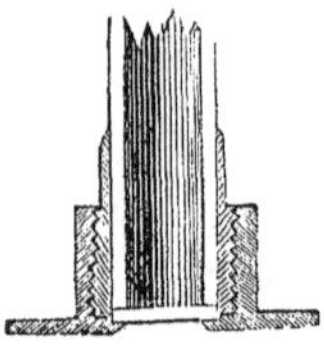

To adapt the tube to any kind of liquid, the glass plate must be ground more accurately and kept in place by a brass collar and cap, as seen in fig. 407.

[212] **Polarization in unannealed and pressed glass.**—Any fragment of unannealed glass will answer the purpose; bits of Bologna flasks ground and polished do well. Or take a piece of plate glass, about an inch square, fix it in a sling of wire and heat it gradually red hot over a coal fire, and leave it to cool in the air. To show the phenomena in pressed glass, make a stout brass frame, such as is shown in fig. 408 in natural size. A moderate pressure with the screw is sufficient to impart the power of polarization to a plate of glass. All these phenomena are seen to most advantage when the pieces of glass are laid on the lower horizontal mirror of Nörremberg's apparatus.

*Fig.* 408.

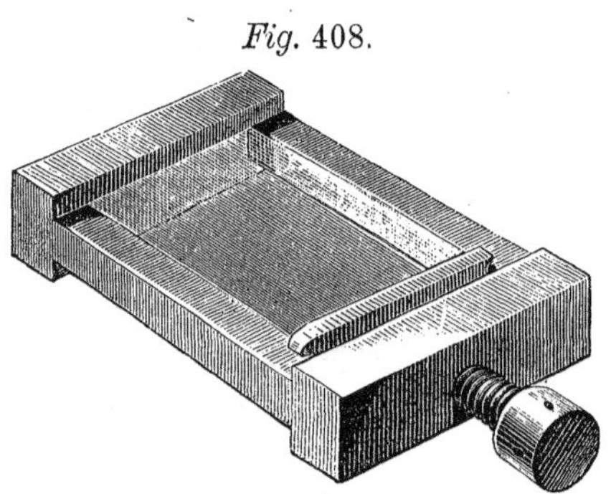

# CHAPTER V.

## EXPERIMENTS ON MAGNETISM.

[213] **Treatment of iron and steel for magnetic experiments.**—Take good charcoal iron, forge it into the proper shape, and, after coating it with clay, heat it to redness. It becomes still softer if left to cool among the coals. Artificial magnets are made of the best English or German cast-steel, which may be had in almost any desired shape. It must not be raised above a dull-red heat, nor kept long even at this temperature. After forging it into the desired shape, finish it with a file. Great care is necessary in forging cast-steel; it must never be heated above a dark cherry red, nor must it be hammered cold.

In tempering the bar it must be held upright, and, if it be in the shape of a horseshoe, both poles must be plunged in together. This is the only way to prevent it from warping. The scale of oxide usually falls off in tempering, but the steel may be hard although this does not occur. It should be so hard as not to be attacked by the best files.

*Fig.* 409.

In this condition the steel is so brittle as to break to pieces if it should fall on the floor. In order to prevent this, it must be polished and slowly tempered to a straw-yellow, over a bright charcoal fire. The fire may be fanned, but not blown with a bellows, which makes an irregular heat. When the steel attains its straw-yellow color, it must be quenched in cold water.

If the steel has warped in cooling, so as to be useless, it must be tempered blue. It loses thereby a part of its coercive force, but can be used for many purposes, and may be hammered into shape with a sharp-edged hammer, like fig. 409. The bar must be laid on the anvil and the blows struck close together on the concave side. With a little care, this may be done even when it is tempered yellow.

The north pole should be marked by a stroke with a file, or a letter N stamped upon it, before the steel is tempered. It may, however, be etched in with acid afterwards.

[214] **Form of the magnets.**—Bar magnets should be only about one-third or one-fourth as thick as they are broad, and not over a foot long. Horseshoe magnets are made somewhat longer, each limb being often a foot in length, and $1\frac{1}{2}$ to 2 inches in breadth. The poles of

these magnets, whether they have the shape of a horseshoe, as in fig. 410, or straighter, as in fig. 411, should not be more than the breadth of the

*Fig.* 410.

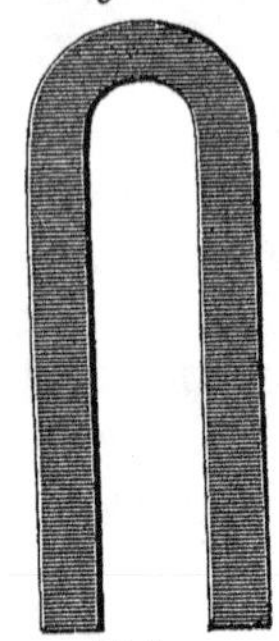
*Fig.* 411.

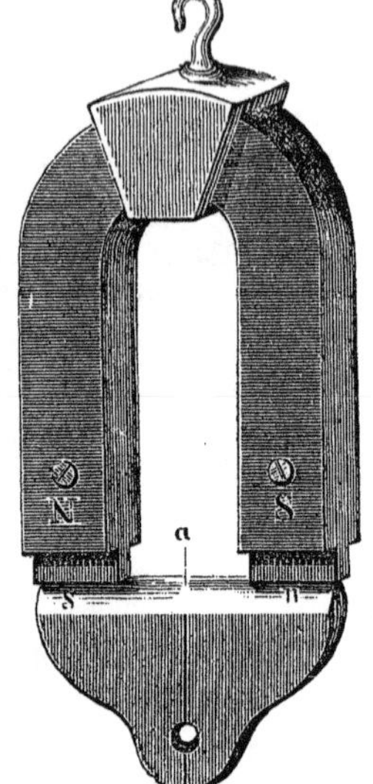

*Fig.* 412.

bar from each other. Bars which are much curved are difficult to magnetize, and very apt to warp in tempering. It is best to have large steel bars tempered by a file-cutter.

[215] **Magnetic magazines.**—Powerful magnets are made by combining several bars in one. The middle bar is sometimes made longer than the rest, fig. 412, but it is better to have them all of the same length. Each bar is magnetized separately, and the whole bound together by brass bands or screws. It is of no advantage to separate them by strips of brass, as is sometimes done. If the bars are warped and cannot well be straightened without softening them, it is well to separate them, under the bands or screws, by a thin plate, in order to compensate for the irregularity.

Straight bars may be combined either by simple brass bands, or by caps of soft iron screwed over their ends. A simple plan is to bind them up tightly with brass wire, and drive little wedges of brass between them at proper places. Fig. 413 represents a cap for bars of equal, and fig. 414 one for bars of unequal length. The binding screws are made of brass, because screws of strongly magnetized iron are unpleasant to drive. These caps are called the armature of the magnet. They may be attached also to horseshoe magnets, but they must be very carefully fitted on.

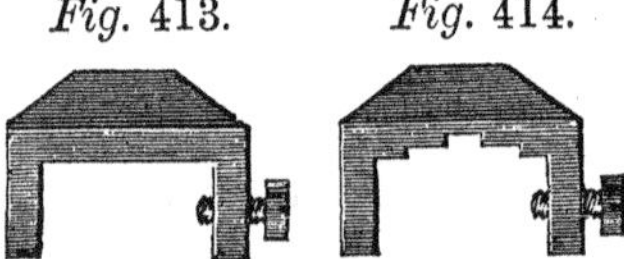
*Fig.* 413. *Fig.* 414.

[216] **Preservation of magnets.**—The poles of both horseshoe and bar magnets must be connected by soft iron, in order to preserve their power. Straight bars are armed by laying them parallel, at a little

distance apart, with the opposite poles next each other, and joining the two ends, as in fig. 415. For horseshoe magnets, this piece of iron is called the anchor, and is usually furnished with a hook, to which weights may be attached. Fig. 412 represents a horseshoe magnet with its anchor. The anchor is sometimes made with a slightly convex face, but an even surface seems to be better. The anchor must be comparatively thick, so that when attached, its lower side will have very little attractive power. It is not necessary to keep the anchor of the magnet always loaded with weights in order to preserve it in good condition. Rust does not diminish the power of a magnet, except by preventing the close contact of the anchor. To keep them from rusting, they may be rubbed with a greasy rag after being used. Magnets should be guarded from blows or concussions, and not rubbed with any hard substance, especially with iron, nor laid upon iron. Heat is still more injurious to the power of a magnet; even a temperature of 40° C. has a perceptible effect.

*Fig.* 415.

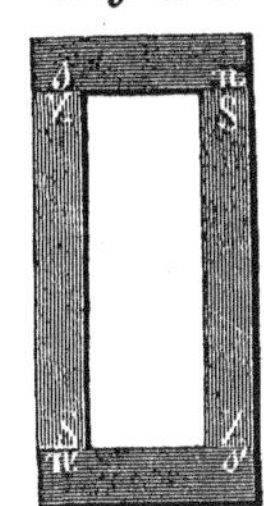

[217] **Armature of natural magnets.**—The position of the poles is ascertained by rolling the loadstone in iron filings, and two parallel faces cut perpendicular to the axis. The other sides of the magnet are also cut, so as to give it somewhat the shape of a parallelopipedon. Over the poles are laid two iron plates, figs. 416 and 417, the size of the face, bent over at the top, and terminating below in projections *p*. The under surfaces of these projections must lie in the same

*Fig.* 416. *Fig.* 417. *Fig.* 418. *Fig.* 419.

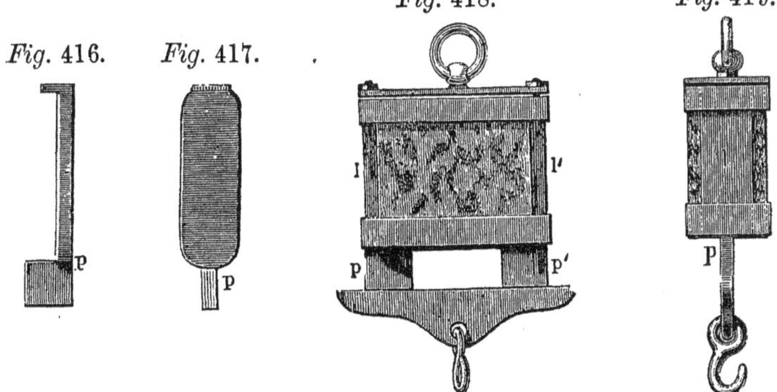


straight line. The magnet is supported by shoulders on these projections. These two plates are connected by brass bands, and an anchor laid on the two ends. The magnet may be hung up by a ring attached to a cross-piece of iron, as shown in figs. 418 and 419.

The power of natural magnets may be increased by laying them between the poles of a powerful electro-magnet.

[218] **Magnetic needles.**—There should be several of these, of various lengths and different degrees of sensitiveness. The only difficulty in making them is in suspending them on pivots. It is very easy to suspend the needle by a cocoon-fibre, and cover it with a glass case, but this arrangement does not answer for most purposes. Delicate needles should be supported on pivots turning in caps of agate; and as the manufacture of these is difficult, it is better to buy the needles ready made, or at least the agate caps, which are not very expensive.

To set an agate cap in a needle, fix a piece of stout brass wire to a chuck on the lathe, bore in it a hole, a little smaller than the hole in the needle, and deeper than the thickness of the cap and needle together. Enlarge this hole to the depth of a millimeter, just sufficiently to admit of the cap being forced into it, and turn the brass off thin to the same depth on the outside. Set in the cap, and press the brass tightly down upon it with a burnisher. Fig. 420 represents an enlarged section of the wire and cap. Next, turn off the wire to just the size of the hole in the needle. If the needle is of the shape represented in fig. 427, the apex of the conical cavity need not be much above the needle; but if it be thin, like fig. 425, the cavity must project 1 or 2 millimeters. In the latter case it is well to make a little shoulder on the brass ring, as shown in fig. 421.

*Fig.* 420.

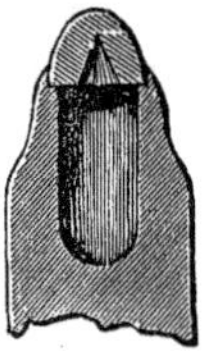

*Fig.* 421.

*Fig.* 422.

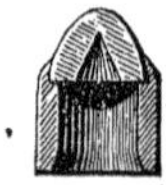

*Fig.* 423.

After cutting off the wire, fix the cap in a block and widen the hole a little below, as seen in fig. 422. If the cap be too large to pass through the hole in the needle with its brass collar, the wire must be turned off, as represented in fig. 423, inserted from above, and fastened by driving a conical steel punch into the hole below. Glass caps answer tolerably well. A suitable piece of glass may be cut out of a thick plate, by means of a copper ring (§ 15,) and the conical cavity drilled with a three-cornered file and spirits of turpentine. It must afterwards be ground smooth with fine emery on wood, and polished with rouge. The cavity may be bored after the glass is set in brass.

Where great delicacy is not required, which is most frequently the case, a very simple contrivance answers the purpose. Bend a strip of brass, about a line in width, fig. 424, over the hole in the needle, and make a little depression in it with a steel punch. This is both simpler and better than to drill out a cap of brass or steel. A similar depression made in the needle itself, while lying on a plate of lead, will insure sufficient stability and mobility.

*Fig.* 424.

The needles may be made of various shapes, as represented in figs. 425, 426, and 427. The first two are made of watch-spring, the last of bar

*Fig.* 425.

*Fig.* 426.

*Fig.* 427.

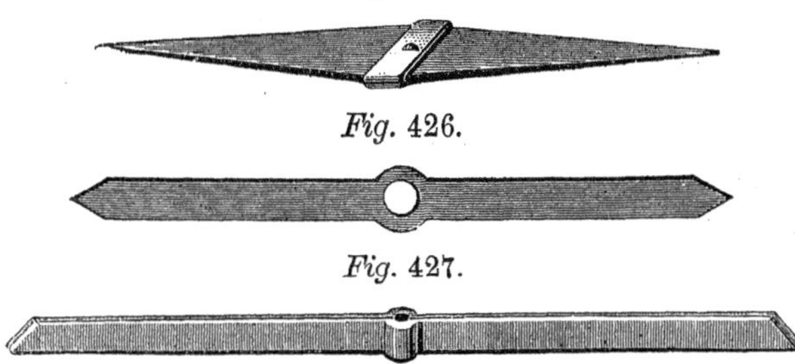

steel. The hole in the center is made to receive the cap. They must be cleaned and polished before tempering. In tempering, the needles should be laid on a sheet of iron, or fastened to a wire, instead of being held in the tongs, which prevents their being uniformly heated. The south end of the needle may either be made heavier than the north end, or loaded with a strip of brass, like fig. 424, made to slide upon it.

Fine sewing needles answer very well in most cases. They may be fastened into a base of wood or brass. For very delicate magnetic needles, the support must be made of steel. File a piece of steel sharp, and screw or rivet the blunt end into a brass foot; fasten this on a lathe, and sharpen the point with a fine file moistened with oil, turning meanwhile very rapidly, and moving the file back and forth. The point of the needle is next hardened and afterwards polished with a fine whetstone on the lathe. The point must appear sharp when examined with a lens. The support need not, generally, be very high. Fig. 428 represents one of natural size.

*Fig.* 428.

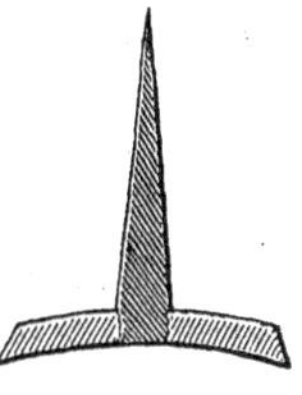

[219] **The compass.**—For many experiments in magnetism and electricity, a compass, that is, a delicate magnetic needle moving over a graduated circle, and protected by a glass case, is indispensable. For Weber's experiments, the needle must not be over 1 or 2 inches in length. With so small a diameter the divisions of the arc cannot be made less than 2 degrees, and even with the greatest care, the errors must be very great. It is better, therefore, in this case, to fasten a

thin thread of black glass to each end of the needle, parallel to the axis. The form represented in fig. 427 is well adapted to this purpose by filing a groove in the back, as represented in enlarged section in fig. 429. The circle may now be made 4 inches in diameter, and a greater degree of accuracy be attained, even by hand-division, than is possible with a smaller circle on the graduating machine. If the circle be made on paper, it should first be pasted on a board, about ½ an inch thick, and then graduated. The pivot upon which the needle turns must not project much above the board, and even then a large error of parallax may occur in reading off. This is best avoided by excavating the inner surface of the circle, and inserting in it a mirror: the point where the glass thread and its image coincide, will give the true reading. A narrow ring of pasteboard, pasted around a circular plate of glass, may serve as a case for the instrument. The needle should be taken off its support when not in use. If the board is in the shape of a parallelogram, and the median line of the graduation parallel to one of the sides, the compass will serve for the approximate determination of the declination or of the geographical meridian.

*Fig.* 429.

To preserve the pivot of a compass from injury, the needle must be raised from it whenever it is not in use. An arrangement is usually made to effect this without lifting the glass cover. In fig. 430 *b a c* represents

*Fig.* 430.

a bent brass lever, terminating at both ends in a perforated disk. It rests at *a* in a little cavity in the base of the compass. By pressing the screw *c*, the end *b* raises the needle against the glass case cover of the compass; when the screw is loosened, the lever sinks by its own weight.

To test the delicacy of a magnetic needle, set it in vibration by bringing near it a piece of iron. Its position on the graduated circle may be accurately noted, and a piece of iron brought near, so as to draw it a little to one side; when the iron is slowly withdrawn, the needle must return exactly to its former position.

Instead of supporting the needle on a pivot, it may, for many purposes, be suspended by a cocoon fiber, as represented in fig. 431. The needle is thus made very sensitive, but cannot be used to show the effect of a second magnet upon it, because the attraction of the magnet draws it out of the center of the graduation.*

* It may be remarked, by the way, that the magnetic needle must not be observed through steel spectacles.

[220] **Magnetizing.**—At the present day large steel magnets are seldom or never used; in place of them we have electro-magnets, which can be made of any degree of power required. These electro-magnets enable us to magnetize to saturation the largest bars of steel, by a very simple process. The many methods which have been devised for making strong magnets from weak ones are therefore of little importance to us, the so-called single and double touch and Hoffer's method answering every purpose.

*Fig.* 431.

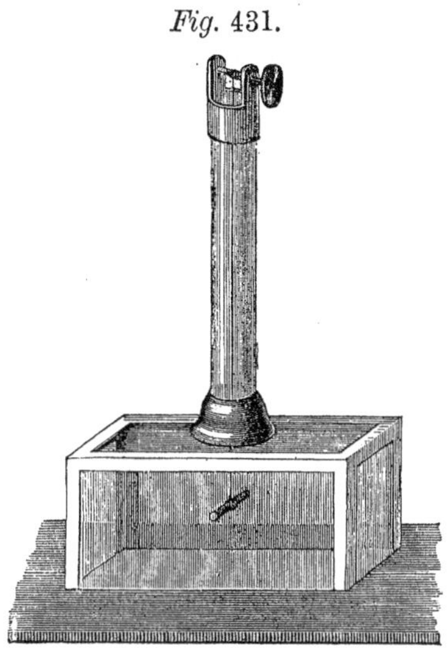

[221] **The single touch.**—The simplest method of performing this, is to stroke the end intended for the north pole on each side about 10 times with the south pole of a magnet, beginning in the middle and bringing the magnet back in a curve through the air. The same process is repeated with the north pole, on the other end. From 10 to 20 strokes will give all the power which the magnet is capable of imparting. The new magnet may be fastened to the table, by a piece of iron in the center, which will have a tendency to determine the neutral point to this position. The magnetism is stronger and more uniformly distributed, by setting the opposite ends of two equally strong magnets upon the center of the bar, so as to make an angle of 80° with each other, and drawing them slowly asunder, preserving the same inclination. In this case also the magnets must be returned to the starting-point through the air. The end which is touched with the north pole becomes the south pole. This method is better than any other for making magnetic needles or bars to be used in measurements. A horseshoe bar is treated in the same way, only the anchor is placed on it when it is touched with two magnets at once, or with both limbs of another horseshoe magnet. The anchor is held fast, and the pole of the touching magnet carried over it. When a straight bar is touched with a horseshoe magnet, the surface of the latter must be held perpendicular to the bar.

*Fig.* 432

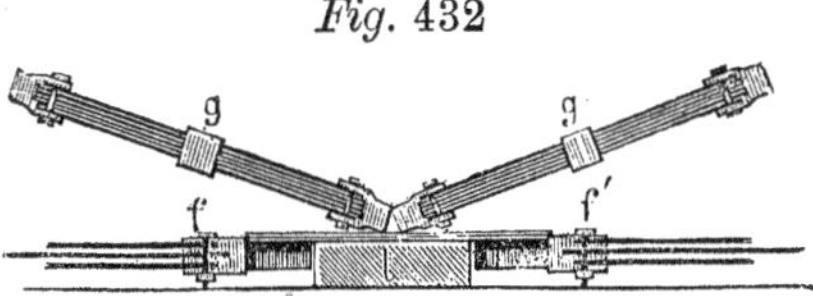

A bar treated in this way becomes still stronger by placing it between the opposite poles ($ff'$, fig. 432,) of two powerful magnets. The bar must be supported by a wooden block *l*.

[222] **The double touch.**—This method consists in laying the two poles of a horseshoe magnet, or two equally strong bars, together upon the middle of the bar to be magnetized, and moving them slowly back and forward 10 or 20 times over its whole length; this is repeated on both sides of the bar. The motion must cease in the middle of the bar, so that each end shall receive an equal number of strokes. If a horseshoe magnet be used for this purpose, its two poles must be very close together, otherwise it almost inevitably produces succession points. If two bar magnets be used, their poles may be brought very near together, but should be kept from actual contact by a bit of wood (*l*, fig. 433,) placed between them. The magnets are held at an angle of 15 to 20° with a bar.

*Fig.* 433.

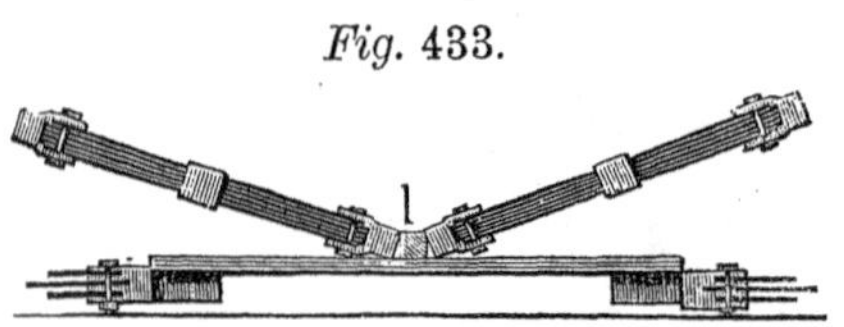


In this case, as in the other, the effect is considerably increased by laying the ends of the bar on two strong magnets. The touching poles must be the same as those of the supporting magnets next them respectively; the poles of the new magnet will be the opposite of those upon which they rest. In magnetizing a horseshoe bar, it is an advantage to attach the anchor, the magnet being stronger even after this is removed.

[223] **Hoffer's method** of the double touch consists in placing an anchor before the horseshoe bar to be magnetized, and passing over it the two poles of a horseshoe magnet, either from the end toward the curve or in the contrary direction, as represented in figs. 434 and 435. In the

*Fig.* 434.

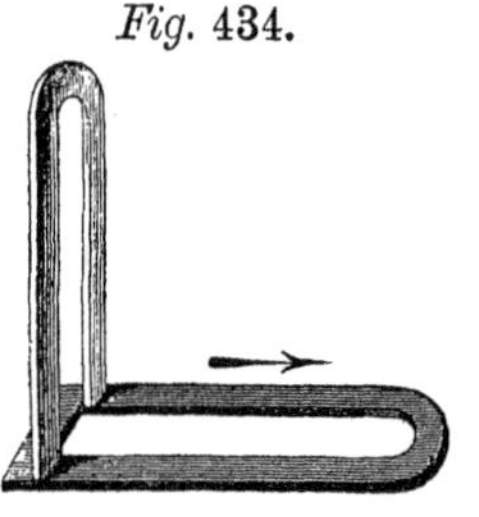

*Fig.* 435.

first case, the poles of the new magnet will be the same as the inducing magnet; in the latter, they will be the contrary. The inducing magnet must, of course, be of the same width as the other. By this method a maximum effect can be produced in ten strokes, and it is one of the best known. Bars can be magnetized in the same way by laying them parallel, with an armature at each end. If the inducing magnet be heavy, to avoid the labor of moving it, it may be clamped to a table with its ends projecting, as seen in fig. 436, and the bar carried over it. With electro-magnets the conductors render this arrangement necessary.

*Fig.* 436.

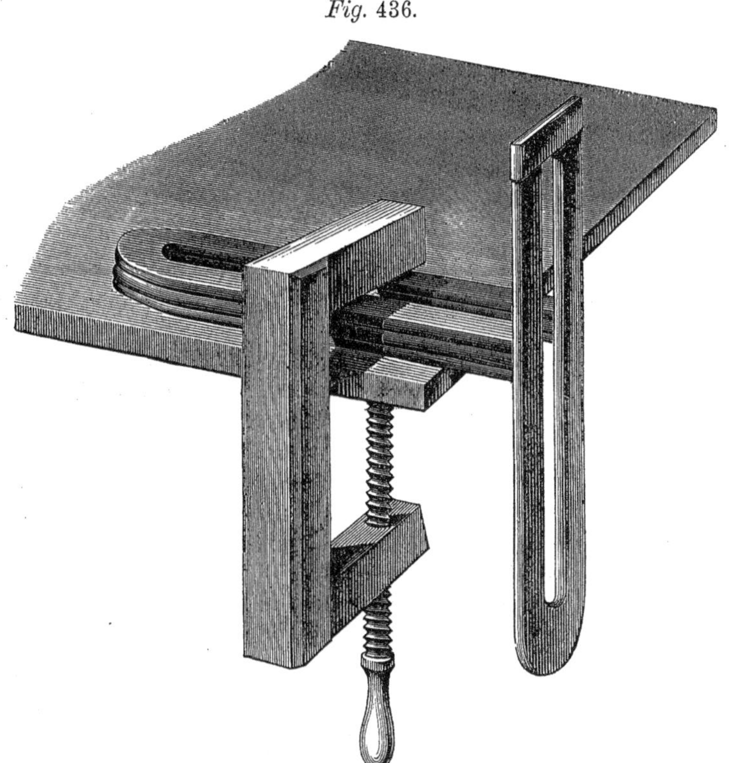

[224] Large magnets have always less power, in proportion to their weight. According to Häcker, if $M$ be the weight supported, and $p$ the weight of the magnet, $M = a \sqrt[3]{p^2}$, in which $a$ is a constant factor, varying with the quality and temper of the steel. Häcker found this factor in some cases equal to $39\frac{8}{10}$, which may be considered very satisfactory, though it has been exceeded in a few instances. This gives us an easy method of testing and comparing the strength of magnets. The experiment may be made, by laying successively small weights (coarse shot for instance) in a pan attached to the armature of the magnet. The point of attachment must be so adjusted that the armature will part at the same time on both

sides. Of course, the contact surface of the armature must be horizontal. This method, however, does not give exact results. The only good method of testing the quality of magnetic needles and bars is by oscillation. The bars may be suspended for the purpose in a brass stirrup, by a thread of unwrought silk. According to Häcker, if $T$ be the time of one oscillation in seconds, $p$ the weight, and $a$ the length of the magnet, $T = c \sqrt[3]{p} \sqrt[6]{a}$, in which $c$ equals $2 \frac{35}{100}$, $p$ being expressed in Bavarian loths, and $a$ in Paris inches. To show the attraction and repulsion of magnets, suspend the bar magnet, about 6 or 8 inches in length, in a stirrup of paper or brass, by several threads of unwrought silk, as shown in fig. 437. To make the same experiment with a natural magnet, fasten a ribbon around its middle and suspend it in the same way.

*Fig.* 437.

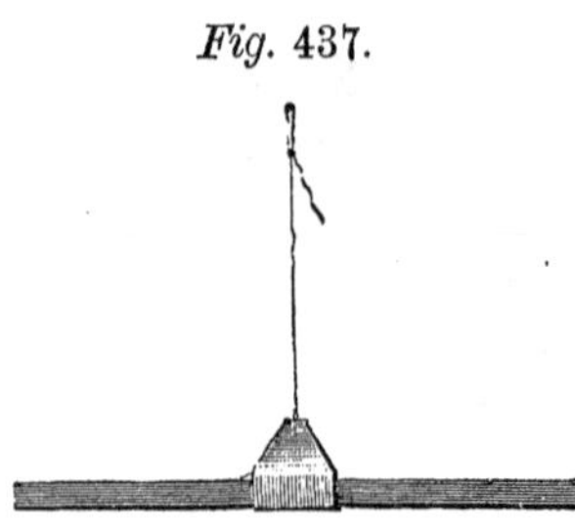

[225] **Distribution of magnetism.**—For these experiments we need clean, fine iron filings, and a number of small bars of soft iron, from ½ to 2 inches in length and of various thicknesses. These may be made of good iron wire, their ends filed round and well annealed.

(*a*) *Magnetic curves.* Sift fine iron filings upon a plate of glass or a sheet of stiff paper, under which a magnetic bar is held; a few gentle taps of the finger will cause the iron filings to arrange themselves in curves. The same process serves to show the succession points in a magnetic bar.

Magnetic bars with succession points are produced with most certainty by means of a powerful horseshoe magnet and the double touch.

(*b*) A magnet brought near two needles, suspended near each other by silk threads, causes a mutual repulsion.

(*c*) Dip the similar poles of two magnets in iron filings and bring them close together, the projecting filaments of iron filings will repel each other; but if the opposite poles be taken, the filaments will interlock, like the arms of a polyp around its prey.

[226] Iron exposed to the inductive influence of a magnet while red hot, and suddenly cooled, remains magnetic. This is most easily shown with iron filings and a horseshoe magnet. A thick bunch of the iron filings may be taken up between the poles of the magnet; reduce this to about the breadth of a finger and half that thickness. Heat this with a spirit-lamp and a blow-pipe, and cool it suddenly. The middle portion will cohere, and will be found to have the properties of a magnet. A powerful electro-magnet answers best for this purpose.

[227] **Declination of the needle.**—In order to give a distinct idea of the direction of a magnetic needle, it is well to designate the meridian in the lecture room by means of plummets. For this purpose, fasten to the ceiling three firm iron hooks, upon which the plummets may be suspended, with a sharp incision in each, to designate the position of the thread. Balls of lead may be used for the plummets. Place a long sensitive needle on a rectangular board, and divide the circle by hand, so that the diameter from which the graduation begins shall be parallel to the side of the rectangle. This diameter can be easily fixed in the line of the astronomical meridian, and the declination determined approximately in degrees. The simplest mode of determining the meridian depends upon the situation of the room, and therefore no general directions can be given for it.

[228] **Dip of the needle.**—The following experiment is instructive as preparatory to the explanation of the dip of the needle: Lay a bar magnet, about a foot long, on the table, and balance a bit of magnetized knitting-needle on a fine thread. If the needle be carried back and forth over the bar, it will stand horizontal over the middle, while toward either end the opposite pole will dip downward. The best mode of suspending the needle is by passing it through a cork, and balancing it on an axis supported on wire hooks, as shown in fig. 438. Neither the declination nor the dip need be determined accurately for purposes of illustration; for measuring the latter approximately, we may use a magnetic needle, 5 or 6 inches in length, with an axis pointed, and furnished with screw threads on both sides, fig. 440. The axis passes loosely through the hole in the needle, and is fastened by two nuts, *a* and *b*. The axis must be adjusted accurately in the center of gravity of the needle, after screwing the nuts moderately tight. The needle is fixed on a support like fig. 439, which consists of a brass fork fastened in a wooden base. Fig. 441 shows the upper end of natural size. The inner superior edge of each side of the fork is filed out cylindrically, and polished with emery, so that only the points of the steel axis will rest

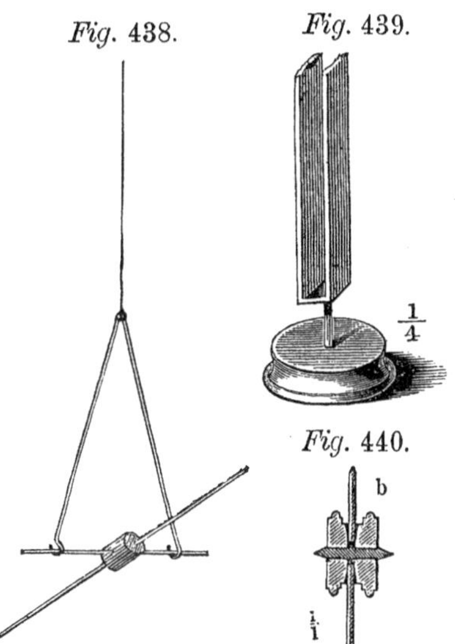

*Fig.* 438. *Fig.* 439. *Fig.* 440.

upon these surfaces. The plane of rotation of the needle must be made to coincide with the magnetic meridian. As it can be no longer shown, after the needle is once magnetized, that it was previously in equilibrium, the poles must be reversed, to show that the north pole will always dip toward the earth.

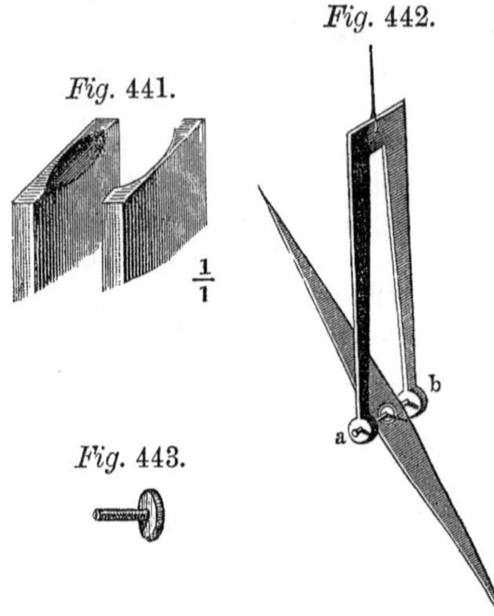

*Fig.* 441. *Fig.* 442. *Fig.* 443.

The needle may also be suspended in the manner seen in fig. 442. The axis is supported by two steel screws, like fig. 443, each of which has a little conical cavity in the end. This apparatus assumes of itself the direction of the magnetic meridian, but must be very carefully constructed.

The simplest apparatus is the one in fig. 438, already described. Thrust a bit of unmagnetic knitting-needle through the cork so that it will rest in a horizontal position, then cut away the cork below, or stick wax on it above, so as to throw the center of gravity in the axis. The needle may then be magnetized and its poles reversed at pleasure.

[229] **Influence of terrestrial magnetism on iron.**—This may be shown by means of a delicate magnetic needle, and a bar of soft iron, 3 or 4 feet in length, which must be well heated, and used for no other purpose, because any mechanical action upon iron gives it some retentive power for magnetism. Hold the bar in the direction of the dipping needle, and bring a small sensitive needle alternately near the upper and the lower end. The poles are reversed by inverting the bar. Hammering it while under the inductive influence of the earth will impart to it a degree of permanent magnetism.

# CHAPTER VI.

## EXPERIMENTS ON ELECTRICITY.

### (*a*.) GENERAL REMARKS AND MANAGEMENT OF ELECTROMETERS AND THE ELECTRICAL MACHINE.

[230] **Elder pith.**—This is obtained in the winter from the year-old shoots of elder, by carefully splitting off the wood with a good knife. The pith may also be forced out with a stick, but it remains somewhat compressed. It is cut into little balls with a sharp knife, and afterwards rolled between the hands, carefully avoiding any pressure. The pith of the sunflower is still lighter, and may be managed in the same way.

[231] **Silk.**—Silk varies in value for electrical purposes, especially those sorts which are dyed with metallic colors, such as Prussian blue. It is easy to ascertain whether or not the silk insulates, by attaching two pith balls to the end of a silk thread, 8 or 10 inches long, hung on a wire, and imparting the same kind of electricity to both; they ought, in a heated room in the winter, to approach each other very slowly. In using silk cord, care must be taken that it has not a core of cotton, which is almost always the case.

[232] **Glass.**—The conducting power of glass varies very much. Common green glass (not white glass colored green by copper or chromium) generally insulates best. Some sorts of white glass also, the Bohemian among others, are good insulators. Its insulating properties may easily be tested, by attaching to it pith balls by fine linen threads, and imparting to them electricity; or by touching a well-insulated electrometer with a rod of the glass.

All glass is more or less hygroscopic; its surface attracts moisture strongly. For this reason, solid columns of glass are better than tubes. This difficulty may be temporarily obviated by warming the glass and rubbing it with warm cloths; but it can only be permanently overcome by coating it with a varnish of shellac or sealing-wax. If shellac is used, the coat must be thin, and the glass previously warmed, just as in varnishing metals. Solution of sealing-wax must be applied until it forms a uniform opaque coating; very little must be laid on at a time. Shellac is not a perfect remedy, and sealing-wax disfigures the apparatus very much; the red uniform of electrical apparatus has gone out of fashion. All experiments in frictional electricity succeed best in winter in a heated room: this subject should therefore be reserved for the winter term.

[233] **Gutta-percha** is also a useful substance for electrical purposes; it may be had in sheets as thin as paper, in threads, or in almost any form. When recently kneaded in warm water and dried, it insulates admirably, and becomes negatively electrified by friction with wool. Unfortunately it loses this property after awhile, and even becomes brittle, so that its value is really not so great as was first supposed.

[234] **Amalgam.**—Kienmayer's amalgam consists of 1 part tin, 1 zinc, and 2 mercury; while others recommend 2 tin, 3 zinc, and 4 mercury. The tin must first be melted in a Hessian crucible, and the zinc added in successive portions. The mercury must be heated previously, and slowly poured into the crucible after it has been taken from the fire, the mixture being constantly stirred. The whole is then slowly poured into water and kept in constant agitation. The amalgam is thus obtained in a granular condition, and may be reduced to powder by folding it in paper and striking it with a hammer. It must be well dried, and kept in air-tight bottles. It always acts best when recently pulverized, as it oxydizes superficially after a time, and the oxyd injures its effect. The following amalgam is highly recommended: 1 part of zinc is melted and slowly added to 4 parts of mercury, previously heated in a mortar; the mixture is rubbed with the pestle until it is cold, when it has the consistency of butter. The author has made no experiments on the subject, but it seems that the effect of the amalgam must vary with the quality of the glass, otherwise the directions on the subject could not differ so much.

The old amalgam should always be removed with a dull knife before fresh is applied; the leather should be thinly smeared with grease, and the amalgam spread on it with a knife as evenly as possible. It is still better to sprinkle the finely pulverized amalgam on the leather, and rub it with paper. In this way no more amalgam adheres than is necessary. The end of a cylindrical woolen swab, 1 or 2 inches in diameter, does very well for applying the amalgam. The amalgam need not acquire a metallic luster.

[235] The use of chains as conductors is to be avoided as much as possible, since their countless angles and points cause too great a loss of electricity; pliable brass wires, about 1 millimeter in thickness, are preferable, and may be wound around at the proper points. When the electricity acquires a higher tension, brass wires, 2 or 3 lines in thickness, must be employed, the ends of which should be carefully rounded and bent

into large hooks. The whole surface should then be filed smooth, polished with pumice and emery, and thickly varnished with shellac.

[236] **Fundamental experiments.**—Attraction and repulsion, the difference between conductors and non-conductors, and between positive and negative electricity, are illustrated by pith balls fastened to silk or linen thread. The thread should be drawn through with a needle, and fastened with a knot, which can be drawn into the ball after being cut off close. The threads must be fine enough to be drawn straight by the weight of the balls. Very fine wire may be substituted for linen. The experiments will be visible at a greater distance if hollow cylinders of gilt paper be used instead of pith. Two little balloons of collodium, suspended by silk threads, serve well to show the mutual repulsion of similarly electrified bodies. They become electrified by being merely drawn a few times through the hand. A simple wire fastened into a foot, bent over at the top, like fig. 444, serves for a support. The wire may be insulated by cementing it with sealing-wax into a glass tube, as in fig. 445. The

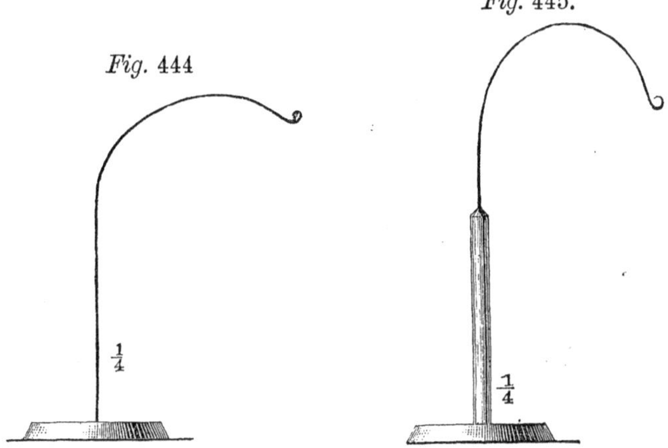

*Fig.* 444

*Fig.* 445.

simplest support of all is made by half filling a tolerably wide bottle with sand or iron filings, and thrusting a wire, bent at right angles, through the cork. In making experiments with the linen threads, a conducting wire must be attached to the support, fig. 445. For other purposes insulating stands must be used. The threads must be at least 6 inches long. To give greater firmness to the wooden base, a circular groove should be cut on the under side and filled with lead.

Sealing-wax becomes highly electrical by rapid gentle friction with a woolen cloth. Glass requires more friction, especially if it be not warm beforehand. A piece of amalgamated leather is the best rubber.

The experiment may be varied by suspending a short, thick glass rod by a thread. When this is rubbed it will be repelled by the other glass rod and attracted by the sealing-wax. The thread must be tied in a notch, cut with a file around the middle of the rod, so that it will not slip when the rod is rubbed. A stick of sealing-wax may be used in the same way. If both these rods be rubbed they will attract each other.

A good way of supporting the rods, is to fold a slip of card-paper, an inch long, into a stirrup, and suspend it by a fine wire or several strands of raw silk. A very thick or very highly excited glass tube brought near the small glass rod, will almost always attract instead of repelling it.

To show that any friction develops both kinds of electricity, cement a stick of sealing-wax to a disk of wood, and glue a piece of leather to the other side. Amalgamate the leather as usual, and then rub it upon a plate or rod of glass. The wood and the glass become oppositely electrified, as may be proved by the pith balls, or any other electroscope.

The influence which the condition of the surface exercises upon the kind of electricity developed by friction may be readily shown by grinding one-half of a glass rod on a stone. Friction with a woolen cloth will excite the ground surface positively, and the smooth surface negatively; each part must, however, be rubbed separately: if the whole rod be rubbed at once there is generally no electricity developed. The success of this experiment seems to depend upon many accidental conditions, and is never sure; even glass rods which have been formerly used with success for this experiment often yield after a time the same electricity throughout. A glass rod is more certain to be negatively electrified by drawing it once lightly through a woolen cloth. It should be first passed through the flame of a spirit-lamp to remove all traces of electricity, and its indifference tested by an electrometer.

[237] **The electrical needle.**—This can be made in the following simple method: *a a*, fig. 446, is a wooden stand, about $1\frac{1}{2}$ to 2 inches in diameter, into which a well insulating rod, *b*, is fastened. A fine sewing-needle is cemented to the upper end of this as a pivot for the needle. This may be connected with the ground, when desirable, by a fine wire. The electrical needle may be made of a piece of brass wire, about $\frac{1}{2}$ a line in thickness and 2 inches long, in the center of which a conical cavity is bored almost through, and polished with a conical stick and emery. After balancing the needle the ends must be carefully rounded off. The needle will be lighter and more sensitive if the cap alone be made of brass and two tolerably stout

*Fig.* 446.

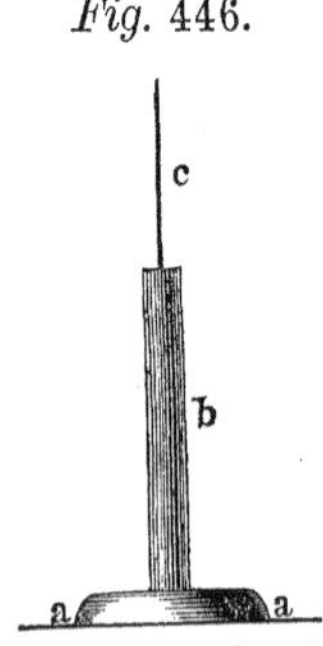

knitting-needles soldered to the opposite sides of it. The cap can in this case be made on a lathe and more carefully polished.

A very sensitive needle is made by bending a slip of brass, about a line in breadth and 0·2 line thick, into the form of fig. 447, and making a little dent in the middle of it with a punch. Two needles are soldered to the ends of this.

*Fig.* 447.

*Fig.* 448

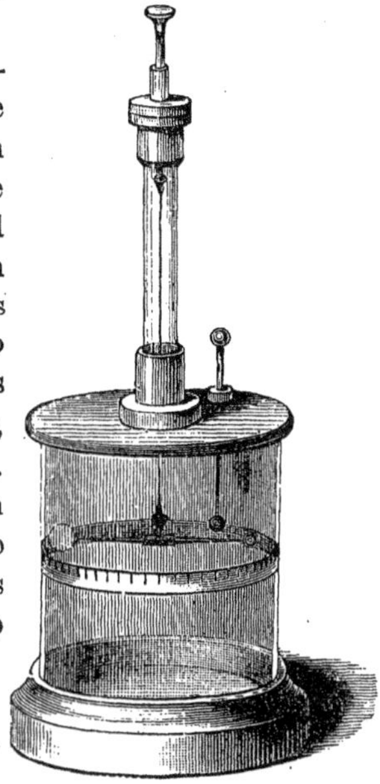

[238] **Coulomb's electroscope.**—This is made most simply out of a large white tumbler, the insulating properties of which have been ascertained by setting upon it the electrical needle or an insulated stand, and observing how long it remains electrical in dry weather. A wooden cover, fig. 448, is fitted somewhat tightly to the glass. Into an opening in the middle of this cover, a glass tube, 2 to 4 inches long and $\frac{1}{4}$ to 1 inch wide, is cemented and ground even at the top. This is also covered by a movable wooden cap, which has an opening in the center to receive a small cork, through which a brass pin carrying a cocoon fiber can be moved up and down.

Cocoons may be obtained almost everywhere, and the silk fibers can be unwound from them upon a roll of paper after they have been soaked in warm water. As these simple fibers are indispenable for many purposes, this trouble must be submitted to. It is not easy to purchase suitable fibers, since those designed for this purpose are at least tripled in reeling off. For many purposes stronger threads are needed, and for such the latter are especially adapted. They may easily be had from the silk factories. It is hardly possible to obtain suitable fibers by simply untwisting wrought silk.

A fine thread of shellac, made by drawing out the material when softened by heat, is fastened horizontally to this cocoon fiber. A little disk of tinsel, about $1\frac{1}{2}$ or 2 lines in diameter, is gummed in a vertical position to the shellac needle, at a distance from the point of suspension equal to the radius of the glass. A fragment of shellac may be attached to the under side of the needle opposite the thread, in order to lower the center of gravity a little. The two arms of the shellac lever may be brought into equilibrium by leaving the vacant one at first a little longer than the other, and afterwards melting the excess into a bead by a flame held near it; if this be too heavy, it is only necessary to melt it back a little toward the thread. An opening, about ½ an inch wide, must be left in the cap *a a*, through which the body to be tested can be introduced by means of a slender rod of shellac. This rod is attached above to a little disk, by which it is supported in the aperture at such a height as just to reach the tinsel. The amount of repulsion is measured by a strip of paper, graduated into 360°, pasted on the inside of the glass, the zero point coinciding with the point where the body tested is in contact with the tinsel disk. Generally, however, the body itself is not introduced, but a wire inclosed in a well-insulating tube, and ending above and below in a knob, is introduced through the opening, in precisely the same manner as the rod of shellac, and the electricity communicated to this; this requires, of course, stronger electricity. The disk of tinsel, when not electrified, may always be made to coincide with the zero point of the division by turning round the upper cap. In most experiments it is well to introduce a capsule, containing chloride of calcium, into the glass, in order to dry the air thoroughly, though this is not absolutely necessary, and in really damp weather is not sufficient. In such weather the fundamental experiments in electricity never succeed well.

Besides these three instruments for showing the presence of electricity, viz.: the elder pith electroscope, the electrical needle, and Coulomb's electroscope, there are the following, called electrometers, some of which are used only for special purposes, while others admit of some measurements and comparisons of the force of electricity, and are simpler in their use than Coulomb's.

*Fig.* 449.

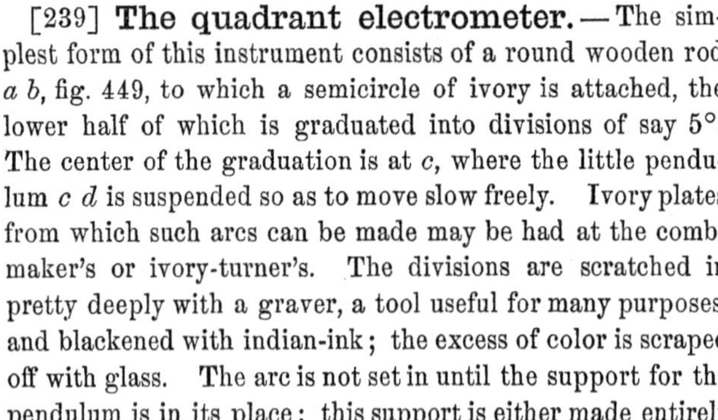

### [239] The quadrant electrometer.

— The simplest form of this instrument consists of a round wooden rod *a b*, fig. 449, to which a semicircle of ivory is attached, the lower half of which is graduated into divisions of say 5°. The center of the graduation is at *c*, where the little pendulum *c d* is suspended so as to move slow freely. Ivory plates from which such arcs can be made may be had at the combmaker's or ivory-turner's. The divisions are scratched in pretty deeply with a graver, a tool useful for many purposes, and blackened with indian-ink; the excess of color is scraped off with glass. The arc is not set in until the support for the pendulum is in its place; this support is either made entirely of brass, as is shown in natural size in fig. 450, or a piece of sheet brass

*Fig.* 450.

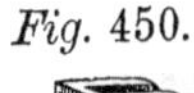

is bent up for the front part and fastened by an iron screw to the rod. The pendulum consists of a thin splinter of whalebone or wood, which is left a little broader at the end where the axis (the end of a knitting-needle) passes through it, or of fine wire. The ball is made, according to the power of the electrical machine, of elder pith, cork, or metal. A metallic pin is driven into the lower end of the rod *a b*, and reaches to where the rod is cut away, so as to allow the ball to come in contact with the pin when the pendulum is at rest. By means of the peg *e*, the instrument is fixed into a suitable opening in the conductor, or if necessary, screwed in. It is evident that the force of the electricity is not proportional to the degrees of the graduation.

[240] **Straw and gold-leaf electrometers.** —Both are easily made out of narrow-necked bottles, fig. 451. The glass tube which contains the conducting wire must insulate well, and is fitted through a cork. It is a good plan to paste strips of tin-foil on the sides *a b*, reaching up through the neck to the outside, in order to discharge the leaves when they strike against them. Strips of rush pith may be used instead of gold-leaf or straw; they are less sensitive than gold-leaf, and more so than straw. They are, like the straw, suspended by fine wires, for which purpose the end of the conducting wire must be pierced with two holes. When the glass has become electrified by drawing out the cork, it is hardly possible to introduce slips of gold-leaf, if gilder's leaf be used, as they are always attracted by the glass; it is necessary to wait then until the electricity is dissipated. The slips of gold-leaf may be cut and pasted on the wedge-shaped end of the conducting wire by the bookbinder. They are made $1\frac{1}{2}$ to 2 inches long, and 1 to 2 lines broad. In most cases it will be advisable to use somewhat thicker leaves, and for this purpose mixed foil, made of alternating gold and silver leaf, is good. When the straw electrometer is to be used for comparative experiments, a large, clean glass tumbler is taken and fitted with a cap of wood or metal, through which the glass tube is inserted. Upon the bottom of the glass is placed a graduated arc of ivory fixed in a piece of wood, fig. 452. Slips of straw can, however, be employed only when the instrument is

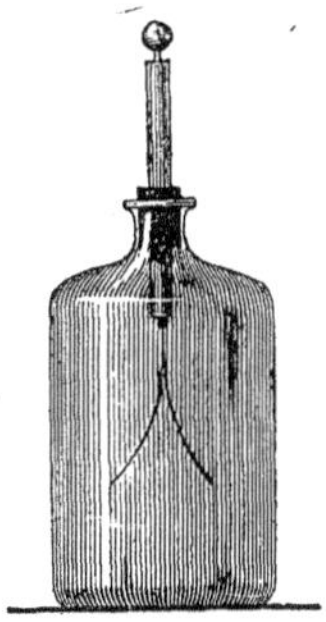

*Fig.* 451.

*Fig.* 452.

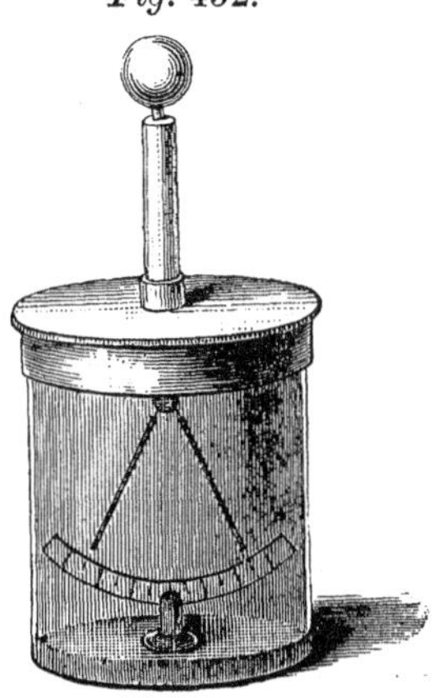

designed to indicate very strong electricity; generally the upper, thin ends of slender grass blades are used. When the blades of grass are very delicate, the end of a fine silver wire, bent into a hook, is thrust directly into the hollow of the blade; if the blade be coarser, the end of the hook may be pushed obliquely through the blade and bent over.

The upper end of the conducting wire must in every case terminate in a screw, in order to attach at pleasure a little ball of $\frac{1}{2}$ inch diameter, or the plates of a condenser. Provided the divergence does not exceed 30°, the electrical tension may, without perceptible error, be considered proportional to the angle.

[241] *Andriessen's very sensitive gold-leaf electrometer* is shown in fig. 453. It differs from the ordinary one only in having a brass wire *a b c d*, 1 line thick, bent three times at right angles, inserted through a hole 2 lines in width, bored through the side of the glass by means of a copper rod and emery. This wire is secured in its place by being cemented at *a* into a piece of turned wood, about 1 inch thick, fitting into the hole by a suitable plug, which is cemented into the hole and to the sides of the glass. The apparatus would have a neater appearance if a brass tube were soldered to a round plate, through this a glass tube inserted and the wire cemented into it, as is shown in fig. 453. As it is easy to communicate the contrary electricity to the knob which terminates this wire, the gold leaves diverge the more readily when the knob *e* is also electrified.

*Fig.* 453.

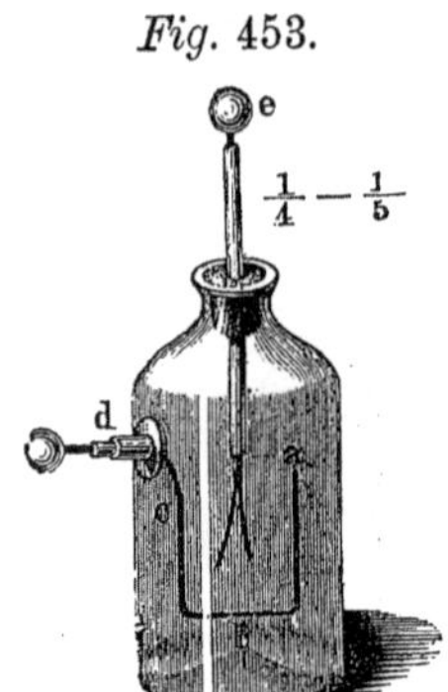


When gold-leaf is used for the electrometer, the slips must not be long enough to reach the sides, because it is difficult to detach them. Even carefully drawing out the wire to which they are attached will often tear them. At all events, such delicate electrometers as those described in this paragraph, where the slips of gold-foil are only 1 to $1\frac{1}{2}$ inch long and 1 line broad, and of the finest gold, must never be exposed to strong charges of electricity, and must always be handled with the greatest care. One can easily make several electrometers with heavier gold-foil, mixed-foil, or tin-foil; all, of course, without the separate bent wire, in order to have an instrument suitable for every purpose.

In using this electrometer for very feeble electricity, when the bent wire is therefore charged beforehand by induction, it is not advisable to screw the condenser on the instrument. The lower plate had better be fixed on a separate support, and after the separation of the plates, the electricity of one is communicated to the knob *e*, as in fig. 454, where a separate tube for chloride of calcium is also introduced.

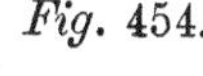

*Fig.* 454.

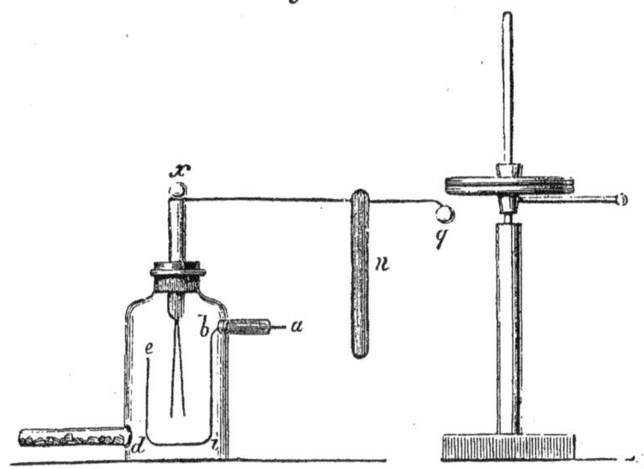


[242] A very good electrometer for many purposes is represented in

*Fig.* 455.

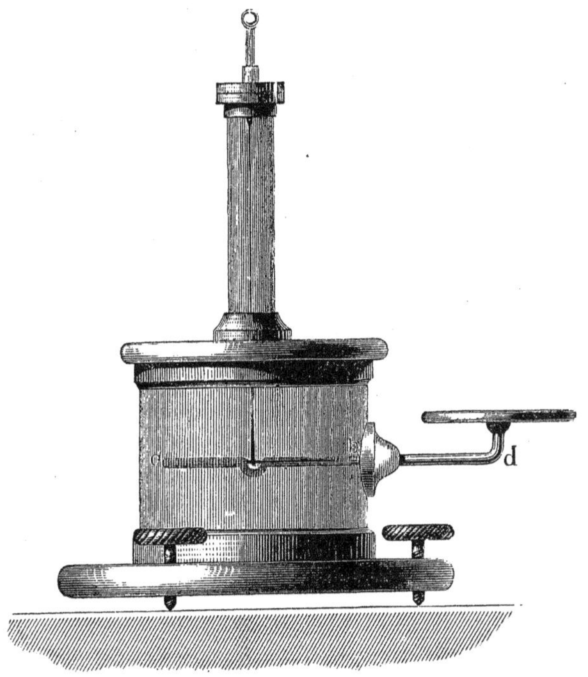


simplified form in fig. 455. It was originally proposed by Dellmann. A well-insulating glass is provided with a wooden base and a movable cap, which has an aperture in the center, in which a glass tube *a* is cemented; this tube is also closed by a movable cap, through which the brass pin *b* passes. To this pin a single or double fiber of silk (according to the use to be made of it) is glued, carrying below a needle of the finest wire, an inch in length. A hole is bored in the side of the glass about an inch from the bottom, and a wooden or brass cap *c* cemented over it, through which a glass tube is passed. In this tube the conductor *d e* is fastened by corks; this consists externally of a stout brass wire, which is bent upward and ends in a screw, in order to attach at pleasure a knob or the plate of a condenser; within the glass a thin strip of brass, 1 line in breadth, carefully rounded and smoothed in every part, is soldered vertically to the wire with tin, and reaches to within $\frac{1}{4}$ of an inch of the opposite side of the glass. In the middle of the glass it is bent, as is shown in fig. 456. The instrument is placed on a board with adjusting screws, and so adjusted that the thread reaches just to the middle of the bend in the strip of brass, and the needle assumes the position indicated in fig. 457. The pin *b* is then turned until the elasticity of the thread just keeps the needle in contact with the brass. This is accomplished by raising the pin *b* so that the needle can turn freely, turning the pin until the needle, when at rest, fulfills the required condition, and then depressing the needle.

*Fig.* 456.

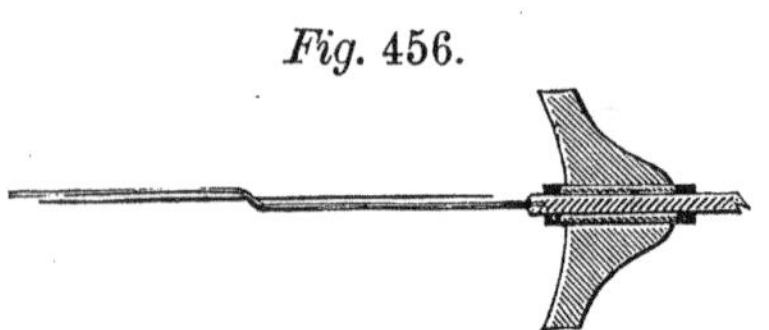

If electricity be now imparted to the conductor, it is communicated to the needle, and the latter repelled. This instrument is exceedingly sensitive and very reliable; it does better service in Volta's fundamental experiment than any other; but it is not adapted to class illustration, because the fine needle can only be seen near at hand.

[243] **Bohnenberger's electrometer.**—The experimenter can easily make this apparatus for himself. A large tumbler may be taken, and two Zamboni's piles attached to the cover; but in this case the gold-leaf hangs between these and sticks fast too easily, and in shaking it loose is very apt to be torn. It is better to take a lamp chimney, about 2 inches wide, as shown in fig. 457, for which a wooden cap and base are turned; in the foot, fig. 458, two grooves are made to receive the tubes of the Zamboni's piles; the grooves are connected by a slit, so that both may be lined with an unbroken strip of the foil. The tubes are left open below, and first filled with the disks of paper and then cemented in their places; as they are connected by the strip of tin-foil so as to form one pile, the poles have always equal strength. The insulating qualities of

the glass tubes must previously be carefully tested, and they must be of such a size that the pieces of paper will not be very tightly compressed. Disks 2 to 4 lines in diameter are large enough; they are best cut with a punch of corresponding size, out of imitation silver and gilt paper, pasted together with starch. If the sheets are pasted together by a bookbinder, he must be directed not to use the paste mixed with glue which they use for other purposes. The upper end of the tubes may be closed by a cork pared off smoothly, and a wire thrust through this, which is bent into a ring on both sides and presses on the paper. Of course the two projecting ends must be of the same length. If it be desired to furnish the upper end with a brass mounting, having a screw on the outside, in order to attach to it a brass cover with a rounded knob, the space above the papers may be filled with a spiral coil of brass wire. When the wires are a little longer than the vacant space, they press on the papers. The pole which has copper uppermost is positive, because the last copper and the last tin are not coupled.

*Fig.* 457.

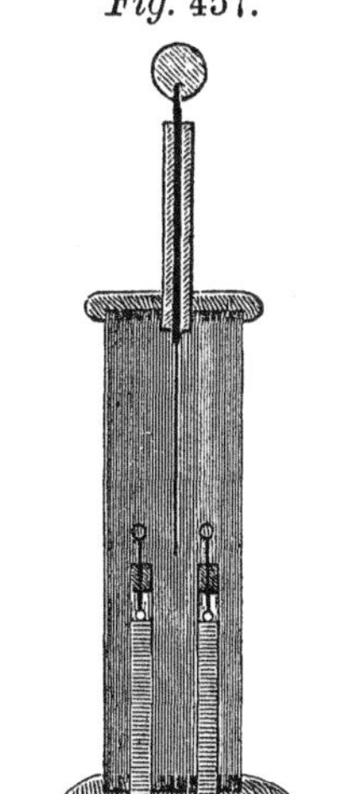

*Fig.* 458.

The conducting wire is in this, as in other electrometers, either wrapped with silk, or coated with sealing-wax and cemented in with the same. The screw on the end of all of them should have the same thread, in order that the same pair of condensing plates may be used with all.

The large glass tube is not cemented in its base, but a strip of velvet or linen, a little narrower than the depth of the cavity in the base, is pasted around the glass in order to stick it in fast. The disks of paper of the Zamboni's pile lie loosely upon each other, and more or fewer of them may be introduced into the tube, according as the wires which form the poles press upon them more or less tightly. By this means the strength of the poles, and therefore the delicacy of the instrument, can be regulated at pleasure. With the dimensions given, the poles can easily be made so strong, that a strip of tin-foil suspended between them will oscillate for a considerable time, which, of course, must not occur when the instrument is used as an electrometer.

This limit should, however, be approached as nearly as possible even with gold-leaf, when a very sensitive instrument is required.

This instrument has also been made with an unbroken horizontal pile, instead of a broken vertical one, and the two poles brought toward the middle. This construction has the difficulty, that one must either have a glass of proper shape made, or construct one by putting together plates of glass; figs. 459 and 460 show such an instrument. This form has

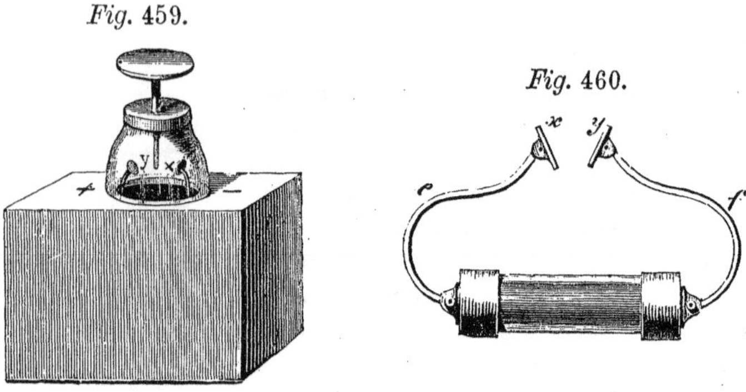

*Fig.* 459. *Fig.* 460.

certainly the advantage that the pole wires *e f* are longer. If they are made elastic and acted on by screws attached externally terminating in glass rods, the distance between the plates, and consequently the delicacy of the instrument, may be varied at pleasure.

[244] A Coulomb's torsion balance, though not a very perfect one, may be made by using for the electroscope fig. 448, a glass, 6 inches in diameter, graduating the upper mounting of the glass tube, and attaching an index to the brass pin. The shellac needle is suspended by a very fine silver wire, instead of a cocoon fiber, and the torsion is effected by turning the pin. The zero points of the upper and lower graduations must coincide, and the disk of tinsel must stand opposite this when the mark on the cap is at zero. It is a good plan to counterpoise the disk of foil by one of paper on the other end of the needle, which acts as a vane and brings the needle to rest sooner. The cap of the glass vessel must be loose, so as to admit of introducing a capsule with some fused chloride of calcium. It is indispensable, before using the instrument, to wipe the cylinder and the tube with a woolen cloth; the same precaution is necessary with all electrometers. Experiments with the torsion balance before a large audience do not often succeed, for the air in a crowded room soon becomes too moist to admit of uniform results.

[245] Charging a gold-leaf, straw, or Dellmann's electrometer, in order to judge of the effect of a body held near it, is best effected by induction. The knob of the electrometer is touched with a conductor, while an excited rod of glass or sealing-wax is held near it, and the contact broken

before the inducing body is removed. The attempt to charge it directly seldom succeeds well. If the body is highly electrical, it often acts upon the electrometer too strongly at a distance; and if it be too feebly charged, it communicates scarcely any electricity to the knob, on account of its low conducting power.

[246] **The proof plane.**—When it is required to test the electricity of a body which is too highly charged to be brought near the electrometer, or to investigate the diffusion of electricity over the surface of a conductor, it is done by means of the proof plane. This is a disk of foil, tinsel, or gilt paper, $\frac{1}{2}$ an inch in diameter, fig. 461, cemented to a thin glass rod which insulates well, and is coated with shellac, or better still, a rod of shellac, 4 to 6 inches long. The body to be tested is touched with this disk, and the charge communicated to the electrometer, which has been discharged with the finger. If the electricity be too feeble, the communication may be repeated a number of times.

*Fig.* 461.

[247] **The electrical machine.**—The electrical machine is one of the most indispensable pieces of apparatus; it admits at the same time of the greatest variety in construction and size. The size of the machine depends, other things being equal, upon the length of the sparks desired; and it cannot be denied that all phenomena of theoretical importance can be exhibited as well with a machine which yields sparks $\frac{1}{4}$ to $\frac{1}{2}$ inch long as with one which gives 8 to 12 inch sparks, for it is immaterial whether a single sheet of paper be penetrated or a whole pack of cards. In the school the matter cannot be always regarded so abstractly, as it is precisely the variety of phenomena which often rivets the attention of the pupil, and serves to illustrate the theory. But the philosophic amateur is as little satisfied with a minimum of effect as the scholar. On the other hand, all unnecessary expenditure must be avoided, and in view of this we may assert that an electrical machine, the conductor of which affords sparks 1 to 2 inches in length, is sufficient for all experiments, except such as Von Marum instituted.

The length of the sparks depends also partly upon the construction of the machine, and it may be assumed in general that a plate machine which has a plate 15 to 18 inches in diameter, and does not give sparks fully 2 inches in length, is to be considered defective either in material or construction. If the machine be intended for instruction, care must be taken that it furnish negative electricity as well as positive, and if possible in equal strength.

It is immaterial whether it be a plate or a cylinder machine. Cylinder machines can be furnished somewhat cheaper than plate machines of equal power, provided the sparks required do not exceed 1 to 2 inches.

The question whether it be better to undertake to construct a machine or to buy one ready made, is a different one to the teacher and to the amateur. The former must either provide an entirely new apparatus, and in this case to make an electrical machine, which is the very first piece of apparatus needed, is out of the question; or he finds a machine suitable, except that it has not as much power as might be expected from its size, and he has to improve it. The amateur will take pleasure in constructing his own machine. The principles by which both should be guided shall be developed in the following pages. In ordering a machine, it should be stipulated that it shall be capable of being used both for positive and negative electricity, and in a heated room in winter give sparks of a certain length.

[248] **The plate or cylinder.**—Plates for this purpose are usually made of mirror glass, and may be had, ready rounded and pierced, from the manufactories of mirrors; they may be obtained through a dealer in mirrors. In general, greenish glass is to be preferred. Plates may also be had from plate glass factories, but these are seldom of sufficiently even thickness, and the rubber must be made very yielding; such plates are, at all events, not to be recommended. The axis is usually made of iron. A thread is cut in the middle of it, and the plate fastened by two stout plates screwed on the axis, and separated from the plate by leather washers. In small machines the length of the axis must be nearly equal to the diameter of the plate, and it should be covered with a stout wooden cover varnished with sealing-wax, as shown in section in fig. 462. Since the hole in the plate is always somewhat larger than the axis, it is easy to adjust the latter exactly in the center of the plate, before screwing up the metallic plates very tightly. Glass axes are preferable to iron, but they require very careful workmanship, and it is better to leave them to the instrument maker. A simple axis of this kind is described in § 252.

*Fig.* 462.

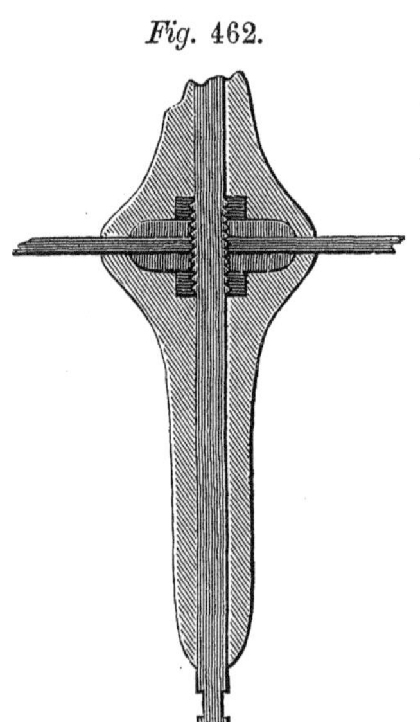

The bearings of the axis are made like all other bearings, of two plates of gun-metal, which can be approximated to each other by screws, and are best supported on glass columns with well-turned wooden heads. Bearings of beech-wood cut across the grain are, however, quite good enough. The crank has also an arm of glass. It is not absolutely necessary to insulate the bearings, but it increases the power of the machine considerably, for the electricity is apt to pass off from the collecting arm of the conductor, or from the rubber to a non-insulated axis, of which one may easily satisfy himself in the dark.

The axis of the cylinder machine is passed through the cylinder; screws are cut upon each end, and instead of plates, wooden caps are used, which inclose the necks of the cylinder. The caps must generally be made somewhat large and the necks cemented into them, in order to fit the axis exactly in the center of the cylinder, which is even then seldom perfectly attained.

[249] **The rubber.**—To plate machines, either one or two pairs of rubbers may be attached. Experience seems to prove that two pairs of rubbers increase the quantity of electricity liberated; but that, except where the plates are more than 24 to 30 inches in diameter, the length of the spark is diminished thereby.

As the chief object of an electrical machine is to produce the greatest possible tension, all machines of less dimensions than this should have but one pair of rubbers. The Leyden jars may be more rapidly charged when two pairs are used, provided again that high tension be not required; this inconvenience may be remedied by a few more turns of the crank, but nothing will increase the length of the spark. The rubbers need not be broad in the direction of rotation; 1 to $1\frac{1}{2}$ inches is quite enough, but they should be $\frac{2}{3}$ the radius of the plate in length. They are best made of $\frac{1}{2}$-inch boards, the corners well rounded off and the surfaces planed smooth. The boards are covered with thick, soft-dressed calfskin, with the rough side outward to receive the amalgam; the leather is glued only at the sides, and the edges pared off; a double fold of flannel is stuffing enough. The backs of the boards are covered with tin-foil, and two knobs are screwed into each by which they can be let into a slit in a broad brass spring; this is fastened to the conductor, and may be compressed by two spherical screw-nuts, in the manner

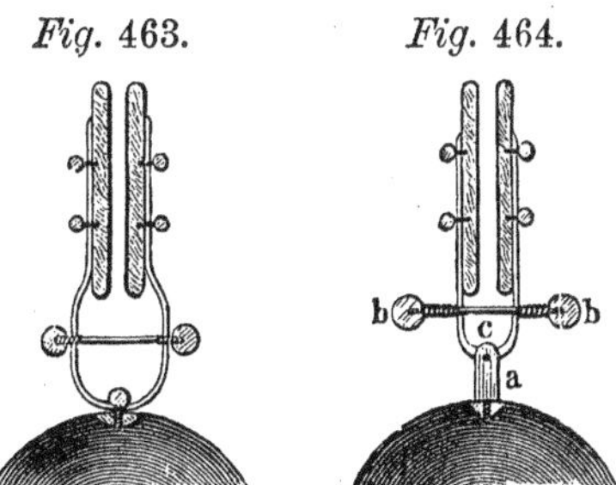

*Fig.* 463. *Fig.* 464.

shown in fig. 463. Another mode of fastening them is seen in fig. 464, where the rubbers are attached to two strong brass rods connected with the conductor by a joint at *a*. The pressure is here maintained by two spiral springs, coiled around the wire *c*, between the brass knobs *b* and the rods. The wire *c* is thickest in the middle, and filed square toward the ends, so as not to turn in the holes in the rods, which are also square. However the rubbers may be fastened, each pair must have a common spring, so that they may the more easily yield to the inequalities of the glass plate.

The rubbers are fastened to an insulated ball of brass. If the size of the ball, which depends upon taste, makes brass too expensive, a wooden ball may be substituted, and the rubbers fastened to a brass rod which passes through the ball and terminates in a brass knob, 1 to 2 inches in diameter.

The manner of attaching the rubbers to a machine is not easily changed without entirely reconstructing it; but badly constructed rubbers can easily be exchanged for better ones. The excited part of the plate must be covered with oiled silk from the rubbers to the collecting points of the conductor. This flap is usually fastened immediately to the rubber, and supported by silken cords fastened to the pillars of the machine. This flap of oiled silk cannot entirely prevent the union of a part of the positive electricity of the glass with the negative electricity of the rubber, even when the latter is not insulated; this takes place where the glass leaves the rubber. The quantity of electricity lost in this way is greater when parts of the rubber are amalgamated which are not in contact with the glass. It is necessary, therefore, to avoid amalgamating more of the rubber than is actually in contact with the plate. This loss is especially sensible when the rubbers are made of thin plates of metal covered with leather, and for this reason such rubbers should, notwithstanding their elegant appearance, be avoided. This loss may be most effectually counteracted by pasting a strip of thick silk to the side of the rubber, and allowing it to pass over this and extend a few inches under the flap. The amalgam is spread upon this silk, and connected by strips of tin-foil with the back of the rubber. The strip of silk which is placed under the flap requires no further fastening; it will be immediately attracted by the electrified plate. This improvement is easily made on almost any machine, and increases the effect surprisingly. When the silk becomes soiled by particles of the amalgam, which are carried forward by the plate, it must be rubbed off with a woolen cloth, or exchanged for a clean piece. In spite of all precautions, the electricity from the rubber of many machines will be found stronger than that from the conductor, owing to the loss in transmission. Instead of the oiled silk, which is not always of the same quality, gutta-percha paper may be used with the same effect, and is much cheaper; it lasts, however, only about two years, and then gradually becomes so tender that it falls to pieces. A better substitute still is silk, varnished with shellac while stretched on a frame.

The mode of applying the amalgam has already been given; its renewal is generally necessary when the full power of the machine is required to be exerted after long disuse. Rubbing off the cushions with blotting-paper, or simply rubbing them against each other, is often sufficient to restore their effect.

Cylinder machines are furnished with but one rubber, the length of which is about $\frac{3}{4}$ that of the cylinder; the breadth should be about 2 inches, or in small machines not more than $\frac{1}{8}$ of the circumference of the cylinder. It is made of wood hollowed out to fit the convexity of the cylinder, and stuffed with horse-hair, in order to yield more readily to the inequalities of the glass. A flap of oiled silk must cover the cylinder from the cushion to the conductor, and in this case also the strip of silk mentioned above is a decided advantage.

If it be required to insulate the cushion, the springs must be set between it and a small carefully rounded board attached to the glass column; but it is difficult to produce a variable pressure in such a case. In smaller machines the glass column itself can be supported upon a movable wooden base against which the springs act, in order to press the cushion against the cylinder, as is seen in fig. 465. Large machines of this sort do not

*Fig.* 465.

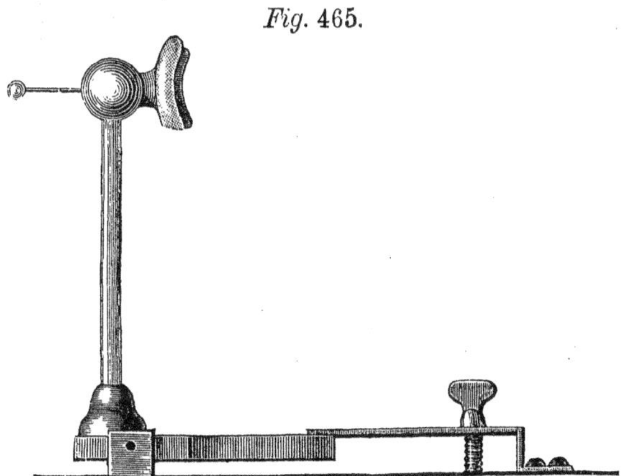

usually have the cushion insulated, and then it is easy to attach the necessary springs to the support.

[250] **The conductor.**—The conductor usually consists, in machines where it has no symmetrical position with the rubber, of a short metallic cylinder, from 2 to 5 inches thick and 10 to 20 inches long, ending in a hemisphere, or of a sphere, 2 to 5 inches in diameter. The latter shape is always preferred when the conductor stands opposite the support of the rubber, as it does in the plate machine. It generally has on the side opposite the rubber a stout brass wire, 2 to 3 inches long, terminating in a knob, 1 inch in diameter. When the support of the cushion is a wooden ball, the conductor may be of wood too, and in this case the support of the collecting arm passes through the ball and terminates in a small brass knob. When the conductor consists simply of a metallic or wooden ball, it is very desirable to have a cylindrical conductor on a separate stand, which can be placed in contact with the cushion or the prime conductor at pleasure.

This conductor should have at each end a projecting brass wire, 2 to 3 inches long, terminating in a knob. The effect of the simple spark from the conductor is greatly increased by such an arrangement. The prime conductor carries on the side next the glass an arm which is forked for plate, and T-shaped for cylinder machines, with sharp points directed toward the glass. The points should be as sharp as possible. The rods

*Fig.* 466.

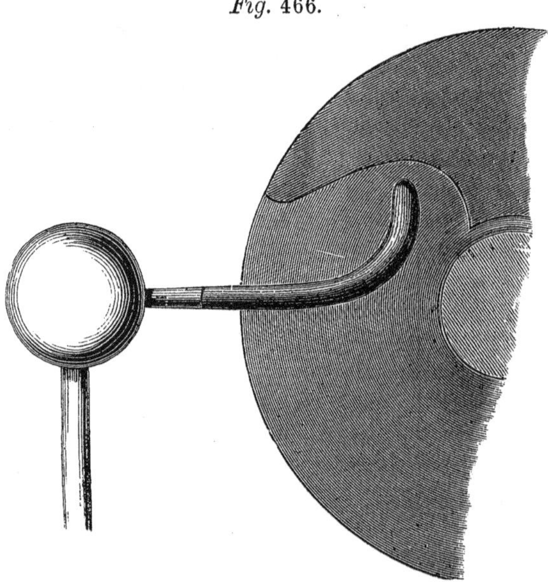

to which these points are attached should be, at least toward the end, not only varnished, but thickly coated with a layer of shellac or sealing-wax, in order to prevent the escape of electricity toward the axis. Instead of setting the collecting points in a straight metallic fork, it is still better to employ a bent wooden one, as in fig. 468, and to make the points of pins set in a slit lined with tin-foil, and filled up again with sealing-wax. Where this wooden arm joins the metal one, it is connected with it by strips of tin-foil, and the whole arm varnished with shellac.

[251] **The insulation.**—The best mode of insulating the conductor, the rubbers, and the bearings of the axis, is by means of columns of green glass. These columns are, however, very expensive, since with a plate 15 or 20 inches in diameter they must be nearly that high, and both strength and symmetry require that they should not be too slender. It will, therefore, often be necessary to use instead of solid columns, stout tubes; these can be more readily procured straight and tolerably uniform in thickness. Tubes answer the purpose quite as well as solid columns when varnished internally with shellac and cemented in, during cold, dry weather, or a piece of fused chloride of calcium inclosed.

The conductors have frequently a neck to receive the head of the columns, as in fig. 467, which is very objectionable, causing a considerable escape of electricity along the glass, which may be readily observed in the dark. The form of the insulated body should rather, as Von Marum directs, be depressed at this point, as is shown in section in fig. 468. The defect may be partially remedied, by cementing around the lower edge a smooth roll of shellac or sealing-wax, 6 to 8 lines thick, rolled out between boards while warm.

*Fig.* 467. *Fig.* 468.

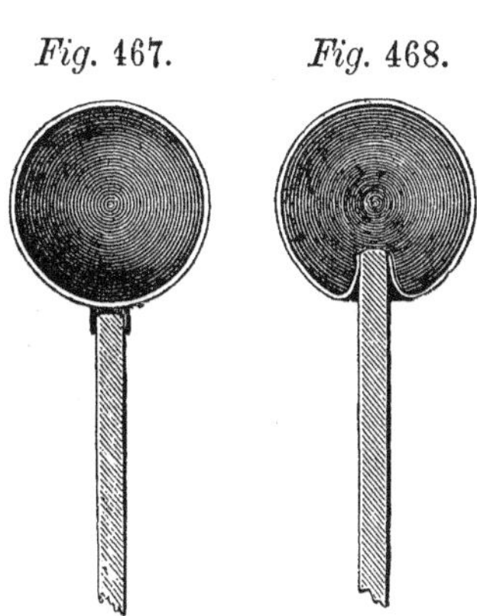

Defects of insulation influence chiefly the length of the spark, and have very little effect on the charging of the jars, unless the tension be great. Many machines with which a battery may be charged very rapidly give only short sparks. The perfection of the insulation may be judged by the brushes of light which stream from the conductor in the dark, especially toward the axis and along the columns, which ought not to be visible if the insulation is good.

[252] **Winter's electrical machines.**—The electrical machines constructed by Winter, in Vienna, have attracted much attention, on account of the length of the spark yielded by them in proportion to their

dimensions. The author has one of Winter's machines, with a plate 47 centimeters ($18\frac{1}{2}$ inches) in diameter, which, with proper management, will give sparks 9 to 10 inches long; and as its construction can be imitated in many respects in other machines without much expense, a full description of it is given. Fig. 469 is a perspective view of it, $\frac{1}{9}$ to $\frac{1}{10}$

*Fig.* 469.

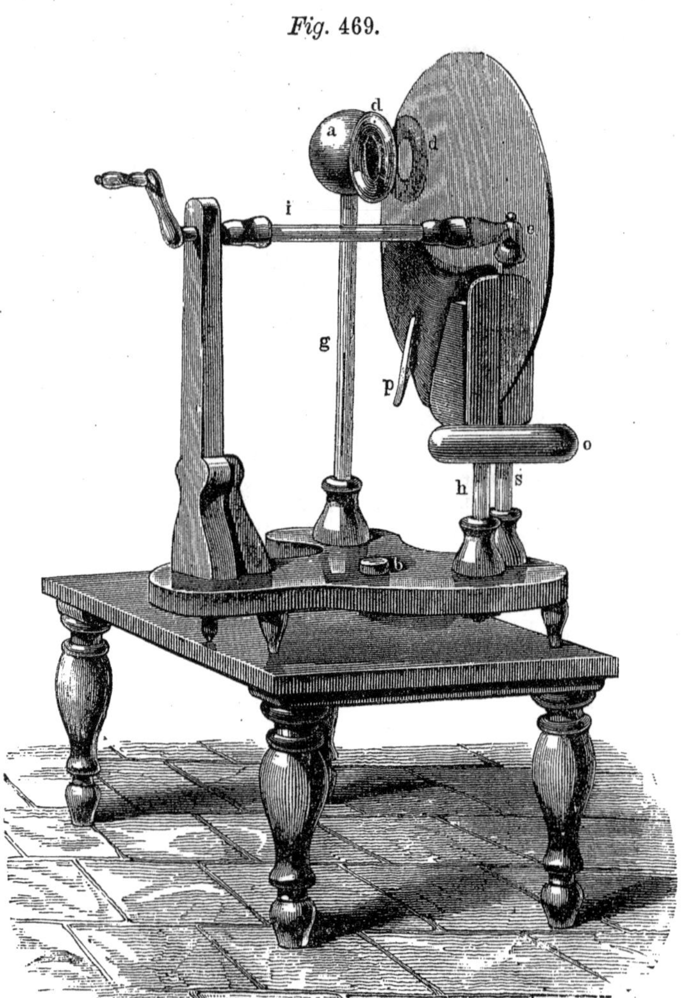


of the natural size. The plate, which is 8 millimeters thick, is secured by screws between wooden blocks, one of which is connected with the winch by an axis of green glass *i*, 24 centimeters long; the other ends in a gudgeon turning in a wooden knob *e*, which is supported on a glass column *s*, 40 centimeters high. The gudgeons are all of wood and turn in wooden bearings. The rubbers are made of wood covered with leather, and stuffed with a little cotton, and are broader toward the edge of the plate.

Fig. 470 shows a side view of one of the rubbers. The board *a*, upon which two springs are fastened, fits into the grooves of the socket, fig. 471.

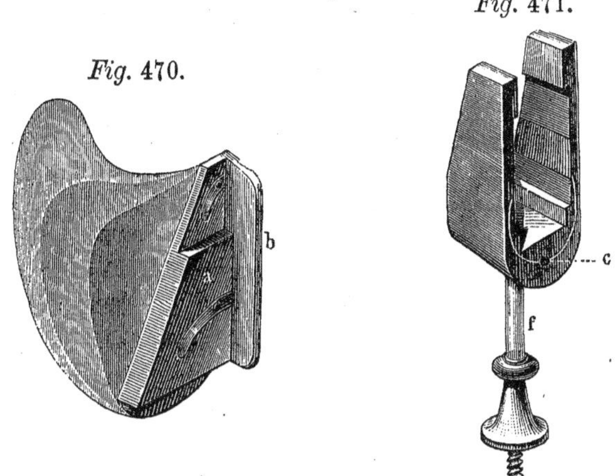


*Fig.* 470.

*Fig.* 471.

This socket is lined with metal, and connected by wires with the stalk of the negative conductor *o*, fig. 471, which is inserted in the hole *c*. This conductor is supported on a short glass column *f*. A flap of oiled silk is attached to the rubber which extends to the collecting arm of the conductor, and is double throughout, three or four fold at first. (Oiled silk can be cemented together by solution of shellac.) As these flaps fall off from the plate and curl up when the machine is not in use, little pins of split cane are used to keep them in constant contact with the plate *p*, fig. 471. The conductor is supported on a glass column *g*, 47 centimeters high, and consists of a hollow globe of brass *a*, 10 centimeters in diameter. The globe is depressed below, as is shown in fig. 472. It has a cylindrical tube inserted in it above to receive the handle of the wooden ring, and opposite the collecting arm a hemisphere only 5 millimeters in diameter, which slides in and out in a narrow tube, 1 inch long. The collecting apparatus consists of two thick, polished wooden rings, from 2 to 5 centimeters thick, with an external diameter of 13 centimeters. They are fastened to a T-shaped arm of brass, the handle of which, *z*, fig. 473, slides into the ball of the conductor. A groove is cut on the inside of each ring where the excited surface first approaches it, as is seen in the ring, fig. 474; this groove is lined with tin-foil which extends to the

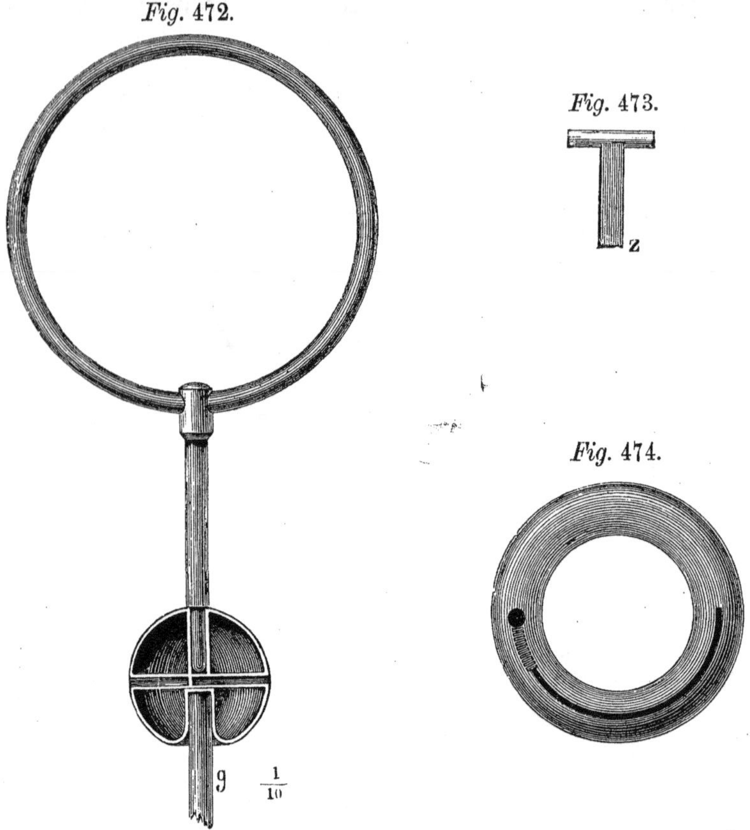


*Fig.* 472.

*Fig.* 473.

*Fig.* 474.

cross-piece of the arm, and set with a thick row of fine pin points which reach only to the surface of the ring. The support of the conductor is movable on the base, being, like all the other glass columns, secured to the base board by a screw from beneath; the rings can, therefore, be made to extend more or less over the glass. The longest sparks are obtained when they do not extend with their whole diameter beyond the edge of the glass.

A polished wooden ring, 3 to 5 centimeters thick and 68 centimeters in diameter, can be inserted into the conductor by a handle 50 centimeters long, ending in a brass ferrule, fig. 474. The ring is made of several pieces of wood glued together, and has concealed in the inside a wire which reaches down to the ferrule. Care must be taken that this ring, when in use, does not reach within 2 or 3 feet of the ceiling of the room.

*Fig.* 475.

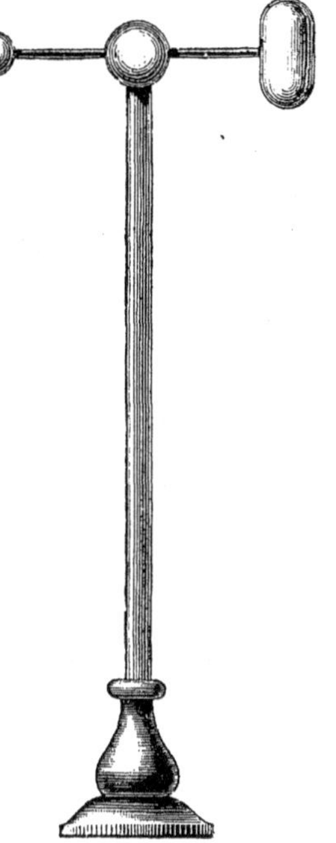

A separate *spark-drawer*, fig. 475, is attached. It consists of two brass balls connected by a metal rod and supported on a wooden stand. The larger ball is placed opposite the little knob of the conductor and connected with the rubber; it is an oblate spheroid, and represents on the side next the conductor a segment of a sphere of great diameter.

The whole machine is fastened by a wooden screw *b*, fig. 469, to a low table.

Although the machine deserves full credit for the excellence of its construction, the chief cause of its powerful action must lie in the quality of the glass. Being entirely free from superfluous ornament, these machines can be sold cheaper than others. A price list of Winter's machines is given below. Each machine is furnished with a *spark-drawer*.

| Diameter of plate. | Length of spark. | Price. | Diameter of plate. | Length of spark. | Price. |
|---|---|---|---|---|---|
| 40 inches. | 22 to 24 inches. | $150 | 15 inches. | 7 to 9 inches. | $25 |
| 36 " | 20 " 22 " | 100 | 12 " | 5 " 7 " | 20 |
| 30 " | 16 " 18 " | 80 | 10 " | 4 " 5 " | 15 |
| 24 " | 12 " 14 " | 00 | 8 " | 3 " 4 " | 10 |
| 18 " | 9 " 10 " | 30 | 6 " | 2 " 3 " | 6 |

Less elegantly finished machines, but equally effective, cost $\frac{1}{6}$ less. The table is not included.

[253] **Management of the electrical machine.**—In order to obtain the greatest effect from an electrical machine, it must be carefully freed from dust and from particles of amalgam adhering to the glass, and the insulating columns rubbed off with warm woolen cloths. The bearings must be cleaned and oiled from time to time. It is very advantageous to warm the whole machine, and for this reason it should be placed in winter near the stove. The best effects are obtained when the building is warmed by heated air, and the machine is placed near the flue. This will be readily understood when one reflects that the air in houses

which are warmed by furnaces is so dry as to be a fertile source of disease to the occupants. The machine may be warmed in summer by fitting to a common charcoal furnace, fig. 476, a jacket of sheet iron, fig. 477, terminating in a pipe and pierced with numerous holes below. The furnace is filled with well charred coals, and the pipe so directed that the air which is heated between the furnace and the jacket shall flow out against the machine.

*Fig.* 476.

*Fig.* 477.

The flaps of oiled silk must also be removed from time to time, and freed from adherent particles of amalgam; it is also well to lift them up before beginning to turn the plate, as they sometimes stick fast and are easily torn.

It need hardly be mentioned that the rubber and the conductor must not both be insulated at the same time. The communication must be sufficiently established with one side, and for tolerably powerful machines it is not enough to allow a chain to rest upon the dry floor; the chain must either be spread out upon the floor, or put in communication with other larger masses of metal. The velocity of rotation depends upon the size of the plate, as the question is not so much how many revolutions are made, as how rapidly the rubber passes over the surface of the glass. It does not appear that the limits are very narrow; the velocity may be tolerably great, but it should not properly exceed 10 feet in a second, and an average of 5 feet is enough. When the rotation is too slow, the electricity of the plate is dissipated under way.

In experiments in which the maximum effect of the machine is required, it is useful to ascertain previously, in the manner described under Lanne's jar, in what condition the machine is; for which purpose a card is attached to the jar, noting the number of revolutions necessary under favorable circumstances to produce a spontaneous discharge when the balls are at a certain distance apart.

Where no separate lecture-room can be had, there is no resource, when the whole power of the machine is to be brought into play, but to clear the room one or two hours before the lecture on electricity, and let it be properly aired and heated. Experiments with high tension, which require good insulation, never succeed well in a room occupied for several hours by a crowd of persons.

[254] **Preservation of the electrical machine.** — The

machine is generally so high that it is inconvenient to work when placed on an ordinary table, and is too low to be placed on the floor. It is better to have a special table of such a height that the winch can be easily turned; the machine is fastened to this by a pair of screw clamps, remains upon it, and is covered with a large cotton cloth.

[255] **The steam electrical machine.** — Any small steam boiler will serve to show the action of this on a small scale. A simple arrangement of such a machine in which the heat is applied externally, will be given in the section on heat. The steam is allowed to escape through a lead tube, about 3 feet long, closed by a plug of wood with an aperture, about $\frac{1}{2}$ a line wide, and impinges against an uninsulated network of fine brass wire placed about a foot distant.

Even with very small dimensions (which will be given hereafter) and a pressure of only 3 to 4 atmospheres, such a boiler gives an abundance of small sparks. The effect is increased by attaching two brass plates *a b*, fig. 478, to the inner side of the plug in such a manner that the steam in escaping between them must take the direction of the arrow. If the boiler be not insulated, the steam may be directed against an insulated wire gauze, which gives the opposite electricity from that given by the boiler.

*Fig.* 478.

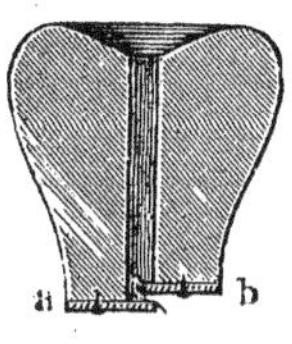


[256] **The insulating stool.** — This is made of a board, 1 inch thick and $1\frac{1}{2}$ to 2 feet square, well rounded at the corners and edges, and strengthened at the corners by four additional pieces. The glass legs must be 1 foot in height and not less than 1 inch thick. Such supports are somewhat expensive and may be very cheaply replaced by 4 champagne bottles, which are set in at the corners, as shown in fig. 479, and cemented fast with ordinary sealing-wax. Besides this large insulating stool, a smaller one, 4 to 5 inches high and 24 to 30 inches square, is needed for many experiments. Both should be well varnished.

*Fig.* 479.

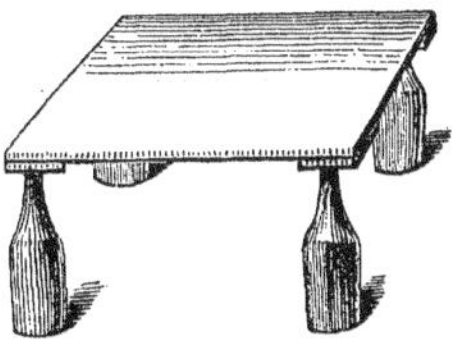

[257] **Experiments with the electrical machine: attraction and repulsion.**—(1) *The electrical spider.* A ball of cork, with a few appendages like spider's legs, is suspended by a silk thread near the conductor; the palm of the hand or a metallic plate is held on the other side of it. The ball will be first attracted by the conductor, then repelled against the hand, then attracted again, etc.

(2) *The electrical chime.* This is easily made out of two clock bells,

into the screw holes of which plugs of wood are driven; a wire is thrust through the plug and bent into a ring close to the metal above and below, as is seen in fig. 480. Two such bells are suspended, as in fig. 481, from a strong bent wire, the one by a silk string, the other by a fine wire. Between the two is suspended, by a silk thread, a little metallic clapper, which can be made of a heavy button. The bell which is suspended by silk is put in communication with the floor, and

*Fig.* 480.

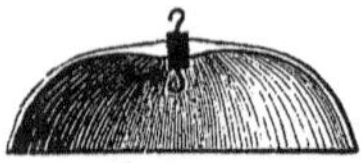

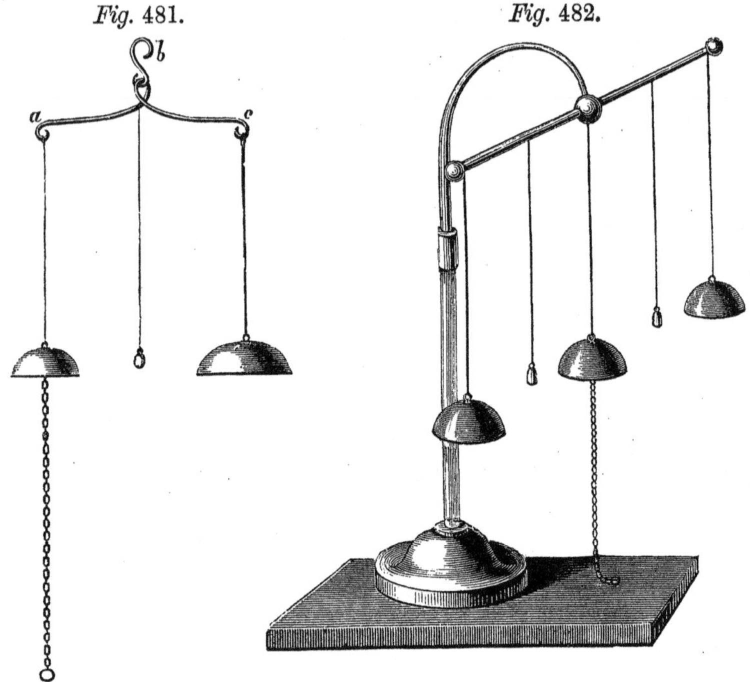


*Fig.* 481. *Fig.* 482.

the whole apparatus is hung upon the conductor. With a little more expenditure of time and money, the bells may be hung upon an insulated stand, as seen in fig. 482, and polished and varnished. Of course, when it is possible, bells which accord should be selected.

*Fig.* 483.

(3) *The electrical fly.* For this experiment, the little insulated stand of the electrical needle, fig. 446, is used. The fly consists of an S-shaped brass wire, fig. 483, sharpened at both extremities and supported on the pivot by a little conical depression bored in the center; this needle is then brought into communication with the conductor by a chain. A

simpler plan is to set the fly on a pivot fixed in one of the holes of the conductor *E*.

(4) *The golden fish.* A figure of a fish, fig. 484, is cut out of tinsel, and fastened with gum to a silk thread. If the head be turned to the conductor it will be attracted, but if the tail be presented to the same, it will be repelled.

*Fig.* 484.

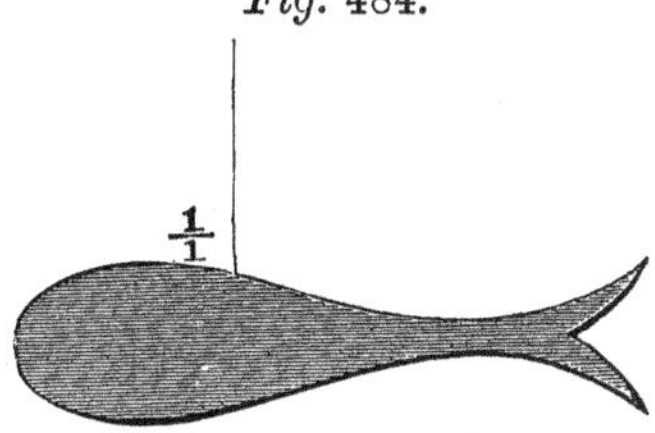


(5) The expansion of a stream of water by electrical repulsion is readily shown by means of a tin funnel suspended from the conductor; the outlet is closed by a cork, through which a capillary tube passes. As soon as the machine is set in motion, the water, which before scarcely issued in drops, begins to flow out rapidly.

(6) *The electrical dance.* A number of little balls of cork or elder-pith are laid on a metallic plate, which is placed at a short distance below a similar plate suspended from the conductor of the machine. It is well to have the upper plate tolerably thick; it may be made of a circular board, well rounded off at the edges, and covered with tin-foil pasted on smoothly, as in fig. 485. Figures cut out of paper or pith may be substituted for the balls, and the distance between the plates must then only slightly exceed the length of the figures. The pith balls are apt to fly off from the plates; it is, therefore, better to inclose them in a cylinder

*Fig.* 486.

*Fig.* 485.

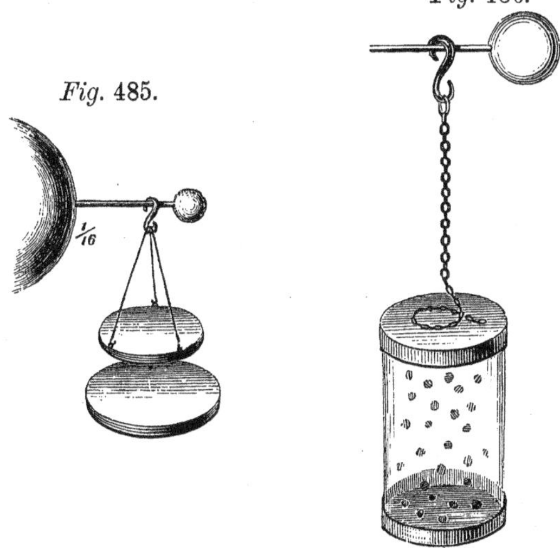


of glass closed above and below by a cover of pasteboard coated with tin-foil, upon which a chain from the conductor is laid, fig. 486. Bits of paper or sand may be used in the same manner, and the experiment is then called the electrical hail-storm.

(7) The repulsion of similar electricity is exhibited by fixing in the conductor or an insulated stand a wire, about 10 inches long, carrying at the upper end a small disk of wood coated with tin-foil. To this disk are pasted alternate red and white strips of tissue-paper, 2 or 3 lines broad and somewhat shorter than the wire, fig. 487. When the machine is set in motion the strips will spread out like an umbrella.

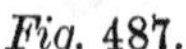

*Fig.* 487.

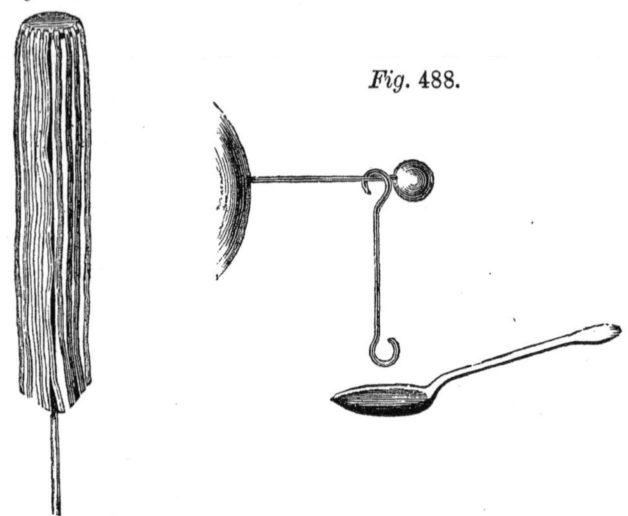

*Fig.* 488.

(8) The electrical spark inflames spirits of wine, ether, or a mixture of oxygen and hydrogen. For feeble machines, which give a spark of 2 inches or less, spirits of wine must be previously warmed, or lighted, and, after being allowed to burn a moment, blown out again. Both spirits of wine and ether should be held in a capsule or tablespoon near a knob of the conductor which is directed downward, or a stout wire bent into a ring and suspended from it, fig. 488. Ether need not be previously warmed. It generally requires several sparks to cause inflammation, probably because some of them go to the side of the spoon. Explosive gases are ignited in a metallic vessel like fig. 489 or 490, or else in a small cannon, fig. 491; the latter is least liable to burst, but even for the one represented in fig. 490, thick tin plate is strong enough, provided the capacity be not above 20 or 30 cubic inches. A short tube *t*, about 2 to 3 lines wide, is soldered to the outside of the apparatus, and into this a glass tube *t t* is cemented;

the wire which passes through the tube is bent at both ends to a ring, and placed about 1 millimeter from the bottom of the chamber.

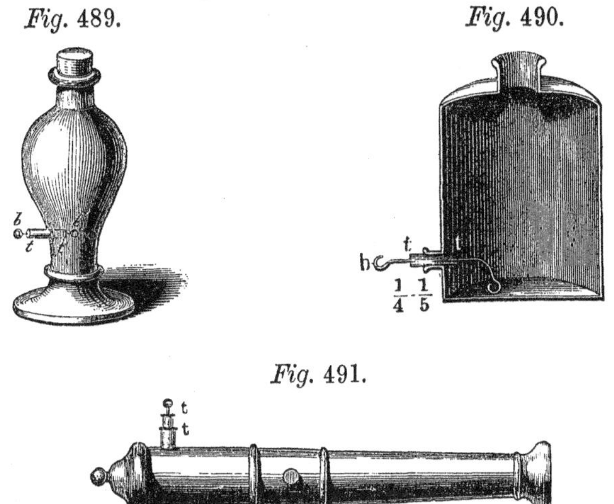

*Fig.* 489. *Fig.* 490. *Fig.* 491.

Before cementing the wire in its place, it should be ascertained by inspection whether sparks really pass from the wire to the vessel when the external ring is connected with the conductor.

The pistol may be loaded either by mixing hydrogen gas directly with the atmospheric air, and then pressing in the cork pretty tightly, or by filling an ordinary bottle with the explosive gas and the pistol with peas, shot, or dry sand to displace the air; the mouth of the pistol is then placed over the neck of the bottle and the sand shaken into the latter, holding the hand tightly around the necks of the two to prevent as much as possible the escape of gas. In this way a much louder detonation is produced. The smallest electrical spark is sufficient to ignite oxy-hydrogen gas. In charging directly with hydrogen, care must be taken to mix it in proper proportion with the air, for a mixture of little hydrogen with much air will not explode. It must never be omitted to test the strength of the pistol beforehand with pure oxy-hydrogen gas. This is done by placing it behind a door, connecting chains with the pistol and the wire and *b*, and discharging it by a Leyden jar.

(9) One of the most striking experiments with the electrical machine is made by placing a person on the insulating stool and connecting him by a stout brass wire with the conductor. The sparks are peculiarly pungent, both to the person insulated and to one who receives them when they are drawn out through the clothing. If a second person hold a spoon with

ether, the person insulated can inflame it by touching it with his finger. If a second person hold a metallic plate or simply the hand over the head of the insulated one, his hair will stand on end; and the same thing will take place of itself if the hair be dry and the electricity strong.

(10) *Penetrating glass.* A three or four ounce phial is filled with oil and closed with a cork, through which a pointed wire is passed and bent, as in fig. 492, so that the point presses slightly against the glass. The phial is then suspended by the outer ring from the conductor. If a metallic knob be now suddenly held near the point of the wire, a spark will pass through the glass, even when the striking distance of the machine is only an inch. The subsequent sparks pass through the aperture to the usual striking distance. The same phial may be used repeatedly, for the aperture is exceedingly fine, so that it can only be seen with a lens; and after standing for days, only a drop of oil exudes through it. Under the microscope, the aperture presents the same appearance as those made through the Leyden jar, which will be mentioned hereafter. Unfortunately, this experiment which can be performed with so simple means, is not adapted for illustration before a class, because of the fineness of the hole. The wire should not press too hard against the sides of the phial, otherwise there is danger of breaking and spilling the oil.

*Fig.* 492.

(11) *The back stroke.* A frog, which has just been stunned by a blow on the head, is suspended by a wire from a conductor in the neighborhood of the prime conductor of the machine; as often as the conductor is suddenly discharged, a twitching of the legs of the animal will be perceived. To make this experiment suitable for class illustration, the machine must be very powerful.

[258] **The distribution of electricity over the surface** of conductors, and the action of points, may be illustrated by the proof plane applied to various parts of the surface of an insulated globe. The globe may be of wood coated with tin-foil, and suspended by a silken cord; the tin-foil must be smoothed with special care. A long cylinder may be employed for the same purpose. The electrometer used may be a gold-leaf or straw electrometer. The experiments can be easily shown to a class, when accuracy is not required.

[259] **The action of points** may also be demonstrated by fasten-

ing a fine needle to the conductor with wax. A candle flame is held against this to show the effect of the electrical wind, and attention may be called at the same time to the difference in the length of the spark. The latter may also be done by holding a needle in the hand near the conductor of the machine. Fig. 493 shows a very convenient apparatus for illustrating the action of points. Into the board *m m* a bent glass rod is fixed, carrying on the upper end the block *b*, through which the sharp-pointed wire *d* is passed. The wire fits tightly enough to maintain any position. The wire terminates in the metallic ball *e*, 1 to 2 inches in diameter; but a tin ball cast in a bullet mould will answer. Opposite this ball another ball *f* is fastened to the board with a conductor. If the point of the wire be directed toward the charged conductor, it will silently draw off the electricity which passes to the second ball, in sparks of greater or less length, according to the distance.

*Fig.* 493.

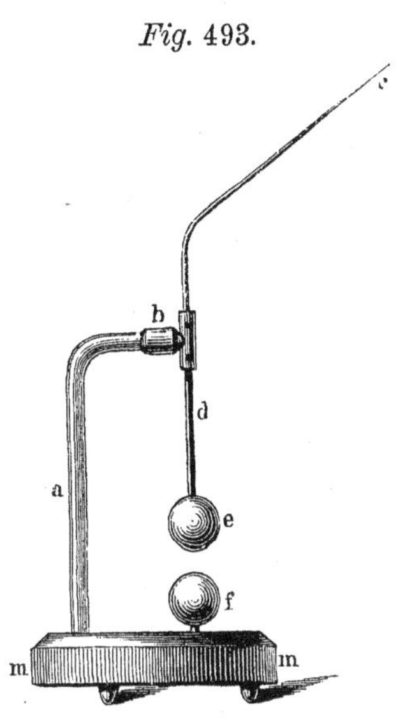


[260] Two contrivances may be employed to prove that electricity resides only on the surface. One consists of a conducting globe, furnished with two hemispherical caps, also conductors, with insulating handles. The internal diameter of the caps is at least $\frac{1}{2}$ an inch greater than that of the globe; and the latter is insulated on a glass rod or silk cord, which passes through incisions in the hemisphere, fig. 494. The globe is electrified either before or after covering it with the caps; in either case the latter are suddenly removed by drawing them quickly in opposite directions, after first moving them so that they are not in contact with the globe. Only the hemispheres will be found to be electrical. The experiment is never a certain one, because one is very apt in drawing off the covers to touch them against the ball. The following apparatus is surer. A strip of genuine gilt paper, the length of a sheet, and 2 or 3 inches wide, is pasted to a metallic cylinder *m m*, fig. 495, in both ends of which grooves are turned. The other end of the paper is fastened around a piece of thermometer tube *a b*, to which the ends of a silken cord, about 2 feet

*Fig.* 494.

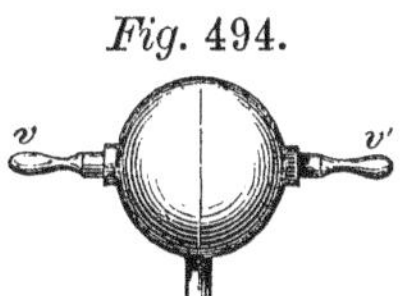

long, are fastened, and which supports also a pair of pith balls on linen thread. The ends of another longer silk cord are passed through the

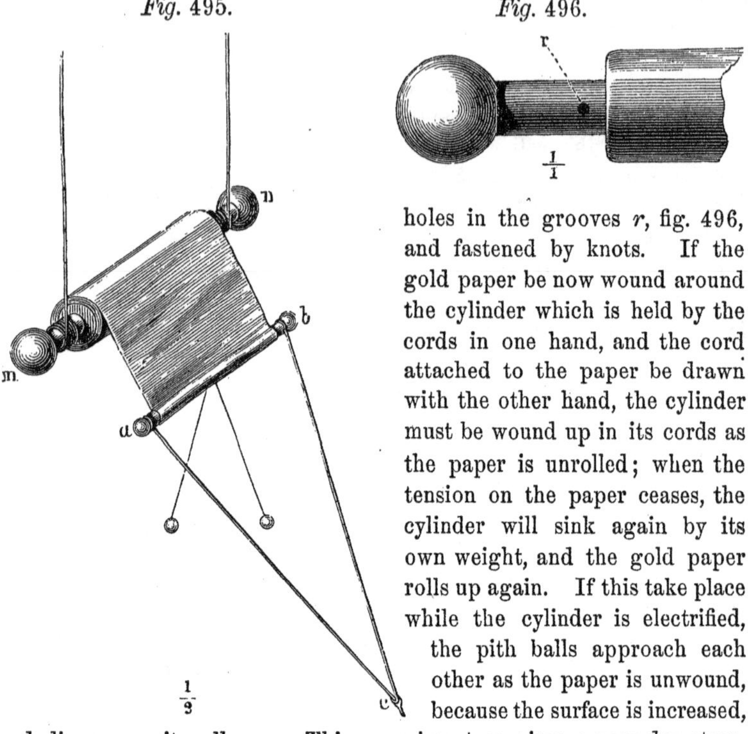

*Fig.* 495. *Fig.* 496.

holes in the grooves *r*, fig. 496, and fastened by knots. If the gold paper be now wound around the cylinder which is held by the cords in one hand, and the cord attached to the paper be drawn with the other hand, the cylinder must be wound up in its cords as the paper is unrolled; when the tension on the paper ceases, the cylinder will sink again by its own weight, and the gold paper rolls up again. If this take place while the cylinder is electrified, the pith balls approach each other as the paper is unwound, because the surface is increased, and diverge as it rolls up. This experiment requires a very dry atmosphere.

## (*b.*) EXPERIMENTS ON ELECTRICAL INDUCTION.

[261] Take either two brass wires, the thickness of a finger and 10 to 12 inches long, rounded off at their extremities and polished all over with pumice and fine emery, or else two larger conductors of brass smoothly soldered. Cement these conductors to green glass rods, 10 to 12 inches long, well coated with shellac varnish, and set in a wooden foot, as in fig. 497. Four elder-pith electrometers with linen threads are tied by silken strings to each conductor. These experiments must be made only in very favorable weather, and it is better in many cases to withdraw the electricity from the conductor of the machine with the finger, rather than to remove the other conductor from it so as to withdraw it from the

electrical atmosphere. A cake of rosin furnishes a better source of electricity for such experiments than the conductor of the machine; even a rod of sealing-wax or glass is preferable to this, because less actual transfer of electricity takes place from them. To prove that the conductor which is electrified by induction has different kinds of electricity at its two ends, an electrified pith ball is brought near the conductor at different points. If the conductor be touched with the finger while thus electrified, no matter at what point, it will show throughout its whole length, but stronger in the neighborhood of the inducing body, the contrary electricity; if the finger be taken off and the inducing body then removed, the electricity which was before confined will distribute itself over the conductor according to the usual law, *i.e.* it will accumulate on both ends. The quality of the electricity on the induction conductor may be most certainly ascertained by means of the proof plane, fig. 461, by communicating electricity with this to a delicate Bohnenberger's or gold-leaf electrometer, to which an excited glass rod is approximated.

*Fig.* 497.

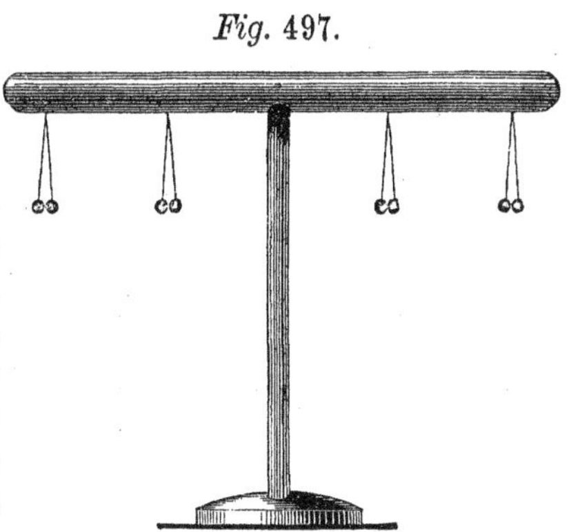

If both induction conductors be placed in contact end to end, and separated suddenly while in a state of induction, each will possess but one kind of electricity, which is not the case when they are only placed near each other in a straight line.

*Fig.* 498.

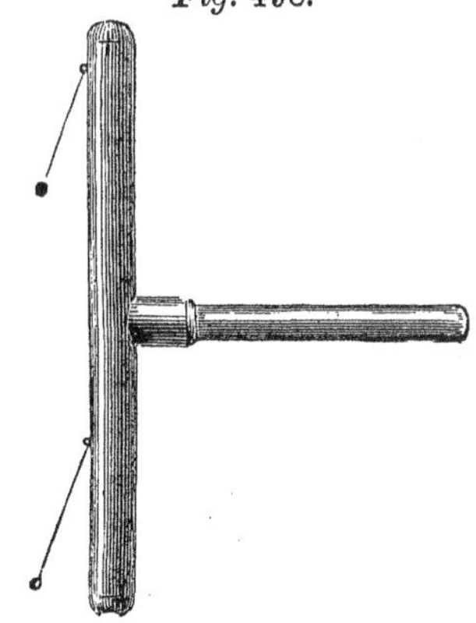

As the pith balls which are nearest the electrified body are strongly attracted by it, it is best to bring the excited glass tube, or whatever the electrified body may be, toward the induction conductor from above, or place the conductor in a vertical position, and employ only two pairs of pith balls, as is seen in fig. 498, where the induction is produced by a cake of rosin. The conductor must always be so arranged that the hands may be free to examine the quality of the electricity. The induction should, moreover, never be too strong, otherwise an error may be caused by a partial transfer of electricity.

Any delicate electrometer may be used for experiments on induction. As soon as an electrified body is brought near it, it will show electricity of the same kind as the excited body, because this is driven down into the lower end of the electrometer. With the removal of the electrified body the evidences of electricity disappear.

If the knob of the electrometer be touched with a conductor while in a state of induction, all signs of electricity disappear; but if the communication be broken before removing the exciting body, the electrometer will show the contrary electricty, after this is taken away. This is at once a simple and instructive experiment for the doctrine of disguised electricity.

Connect the knobs of two similar straw electrometers by an insulated conductor, (wire attached to sealing-wax,) then bring an electrified body near one of them and remove the wire while the induction continues, both electrometers will show, after removing the inducing body, contrary electricities.

*Fig.* 499.

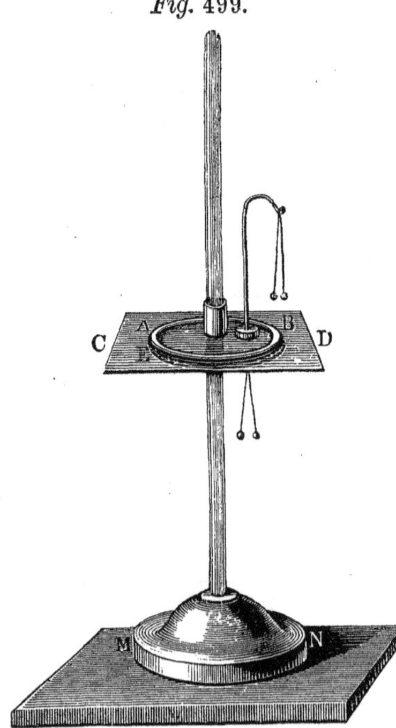


Many experiments on the distribution of electricity over surfaces, the mutual neutralization of two kinds of electricity, etc., may be made with two similar straw electrometers, by electrifying one and connecting it by an insulated wire with the other, which is either not at all, or less strongly, or oppositely electrified.

In illustrating the doctrine of electrical induction the fundamental experiments with the pith balls may be tried again, and attention called to the greater force with which an insulated ball is attracted, than one not insulated.

[262] **Disguised electricity.**—The theory of the Leyden jar is best illustrated by the apparatus fig. 499. *A B* and *E F* are two metallic disks, with thick, rounded rim, and well insulated glass handles, one of which is fixed in a wooden foot. The glass plate *C D* is

laid between them, extending at least an inch beyond the edges of the disks on every side; its edges need not be rounded. Polished wooden disks, covered smoothly with tin-foil, may be substituted for the metallic plates. Each plate is provided with an elder-pith electroscope, which can be easily removed; for this purpose the upper one is suspended from a wire on a metallic foot, and the lower one from a hook. Take *A B* by its glass handle and draw off a spark with it, and bring it near the glass plate *C D* while its balls are divergent, fig. 500, its balls will sink nearly together, and those on *E F* will diverge; conduct off the electricity from *E F*, and the balls on *A* will sink to within an almost imperceptible distance. Take *A B* off, and both pairs of balls will diverge again. Conduct electricity to *A B*, by means of an insulated wire while resting on the glass plate, and both pairs of balls will diverge once more; continue the charging so long as *A B* will receive any electricity; then touch *E F* with a conductor, the lower balls will fall together, the upper ones nearly so, and *A B* will receive more sparks, etc. If the two plates be then connected by a wire, a lengthened spark will be produced; but if the two be separated from each other and the glass and discharged, and the apparatus put together again, it will be found still to contain a residual charge. A thin, uniform plate of glass must be selected, and the charge not urged too far for fear of a spontaneous discharge.

*Fig.* 500. *Fig.* 501. *Fig.* 502.

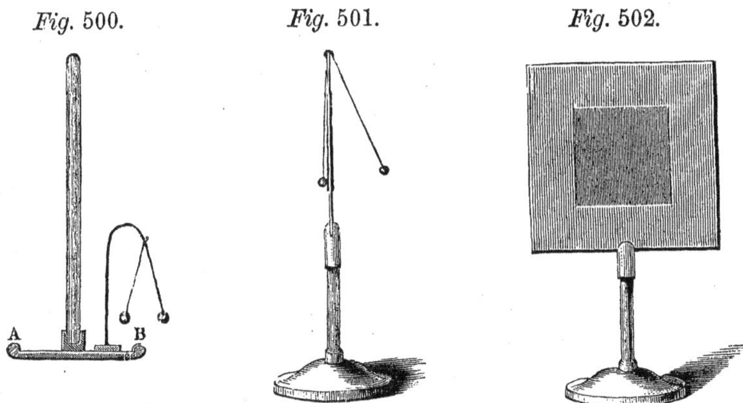


A plate of glass, coated on both sides with tin-foil to within 2 inches of the edge, serves also to illustrate the theory of the Leyden jar. It may be set upright on a wooden stand, figs. 501 and 502, and a pith ball with linen thread attached with wax to the tin-foil. It is usually charged by holding one coating against the conductor, while the opposite coating is touched with the finger.

Remove the plate from the conductor and touch each coating alternately, the ball will fall to the coating each time on the side which is touched, and the other will fly off. The margin of the glass should have several coats of sealing-wax varnish. This should be made with 90 p. cent. alcohol and an excess of sealing-wax. When the coating has become so thick as not to show the glass anywhere, which should be effected by repeated rather than thick coatings, one or two coats of pure shellac are applied to give it a firm, glossy surface. Even the best sealing-wax is apt to remain a little sticky on the surface.

[263] **The Leyden jar: its construction.**—Wide-mouthed specie jars are the best for the purpose. They should be of various sizes; a small one, 4 to 5 inches high and 2 to 2½ inches wide, a larger one of about 1 quart capacity, which may be used as a measuring jar, and several others of different sizes, besides one of 2 to 4 quarts capacity made of thick glass, for experiments with high tension.

A greater effect is produced by one jar with a large coated surface than by several smaller ones whose aggregate surface is equal to it, combined into a battery. For large jars, those which are high and narrow are preferable to wider and shallower ones, because they occupy less space on the table. Economy of space should always be considered. Glasses of uniform thickness and free from bubbles should be selected, as they are easily broken by a spontaneous discharge at thin spots or where bubbles occur — which will happen sometimes at all events, and makes it the more necessary to be able to construct jars for one's self. Jars with thin glass are, indeed, preferable, but should only be used when quantity of electricity is needed, which is obtained by increasing the number of jars, and not for high tension. The jars designated for experiments with electricity of high tension should have walls at least 1 line in thickness, and a broad uncoated margin.

A greater quantity of electricity may, it is true, be also obtained by charging the jars to a high tension, *i.e.* till they yield sparks nearly as long as those of the electrical machine; but on account of the imperfect insulation, much electricity, and consequently time in charging, is lost, and with a small machine this consumes time enough. One may easily satisfy himself how far a jar may be charged without loss of time, by insulating it well and connecting the outer coating with Lanne's self-discharging jar, the knobs of which are placed very close together. When the number of revolutions necessary to cause a discharge of Lanne's jar increases rapidly, the quantity of the charge will not be increased in proportion to the labor expended, and it will therefore be better, when the success of an experiment requires it, to increase the number of jars. Four square feet of coated surface are sufficient for all illustrations of the effect of the spark, if the machine give sparks at least 2 inches long. The uncoated margin must be 3 to 4 inches high, to admit of considerable tension in certain cases.

The next requisite after the glass is tin-foil, which is, unfortunately, not to be had everywhere. Imitation silver paper is a bad substitute for tin-foil, for the tin on it is broken in many places and the paper is a bad conductor.

The tin-foil is pasted on with thin starch. The paste should be applied to the tin-foil very thinly and the foil immediately laid on the glass, covered with a sheet of paper and pressed down with a cloth rolled up into a ball, so as to lie smooth. Any bubbles still remaining are caused by inclosed air or starch; they should be cut open with a sharp knife, and the foil pressed out smooth. As the bottom of the jars is usually convex within, the beginning is made with it: cut a round piece of tin-foil large enough to reach up a little way around the sides, and make incisions around the edge. The sides are coated with upright strips of tin-foil, not over 2 to 3 inches wide. When the inside is completely coated, begin on the outside. As the bottom is concave on this side, the tin-foil stretches evenly across it; when the edges are pressed down, endeavor, by gentle friction with a cloth, to force the foil gradually into the cavity, which generally succeeds unless the cavity be very deep. Any rents which occur must be patched with foil. When the coating is completed, place the jar on the table and draw a line all around it at a uniform height. Cut off the foil evenly by this line, first outside and then inside. The small jars and the measuring-jar must have an uncoated space around the top, about 2 inches wide, which should be covered with shellac or sealing-wax varnish. The smaller ones may be previously warmed. The uncoated margin of the larger jars must be 3 to 4 inches wide, and for high tension the margin must be 6 to 8 inches wide. The conducting wire should also be surrounded by a glass tube extending from the knob into the interior of the jar. A coating of sealing-wax is very useful.

After the jars are coated, make for each of those which are to remain single a case of pasteboard, reaching to about $\frac{1}{3}$ to $\frac{1}{2}$ the height of the coating, and cover it inside and out with silver paper. This serves as a protection to both coating and glass. A ring of thin brass wire may also be laid around the jar just above this case, to serve for attaching chains, etc.

Finally, procure from the turner a tight-fitting cover of hard wood, through which the conducting wire is inserted, reaching nearly to the bottom of the jar. For this, brass wire 1 line in thickness is employed, or for very large jars, wire 2 to 4 lines thick. The outer end of the wire terminates in a knob, $\frac{1}{2}$ to 2 inches in diameter, screwed or soldered to the wire. The knob should be hollow, and may be made by any brass-worker.

In case of need a ball of tin may be cast on the wire in a bullet mould. A piece of gold lace, or a couple of little chains, are attached to the lower end of the wire to connect it with the inner coating. The tops of the small jars and of Lanne's measuring-jar are cemented on. This is effected simply by coating the edge of the cover thickly with sealing-wax in solution, and allowing the same to run in between the wood and the glass after applying the cover. The covers should all be well varnished to protect them from moisture.

To combine several jars in a battery, place them in a box which will just hold them all when placed in order, and coat it internally with tin-foil, with which a brass ring communicates on the outside. Then connect the inner coatings by brass wires extending from knob to knob; the hooked ends of these wires must be mounted with balls as large as a bullet.

These batteries are very convenient, because the inner surfaces of any number of them may be connected by wires. They are not convenient for charges of high tension, for in this case one of the jars is frequently broken by a spontaneous discharge, and one cannot always be found to take the place of the broken one. But batteries are, in fact, unnecessary, for any number of jars may be quickly placed together and united by a chain embracing them all. The inner coatings must be connected by thick brass wires bent into a hook at each end, and furnished with knobs. Care must be taken to make the connection perfect.

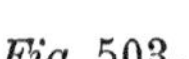

*Fig.* 503.

Very small jars are made of medicine phials, coated with tin-foil only on the outside. Instead of the inner coating they are filled with iron or brass filings, or else a thick solution of gum is poured into them and spread over the surface to the required height, and then filings poured in and shaken up; what does not adhere is shaken out after drying. The conducting wire is stuck through a cork. The wire of such bottles may be bent as in fig. 503. By holding in the last three fingers a piece of amalgamated leather, with which a glass tube held in the other hand can be rubbed while the bottle is held between the thumb and fore finger, the tube passing through the ring in contact with it, such a small bottle may be charged sufficiently to ignite oxy-hydrogen gas, and give a slight shock.

[264] **The discharger.**—The most convenient form of the discharger is that shown in fig. 504, in which the two wires *b c* and *b′ c* are connected by a hinge at *c*, and each has its own insulating handle, by means of which the distance between the knobs can be altered at pleasure. The arms must each be 8 to 10 inches long, to admit of their being used for the largest jars. The handles are made of green glass rods, 4 to 5 inches long and $\frac{1}{2}$ to 1 inch thick, cemented into brass collars, which are soldered to the wire. When the discharger has only one handle attached to the hinge *c*, it is very inconvenient, more so than the simplest

*Fig.* 504. *Fig.* 505.

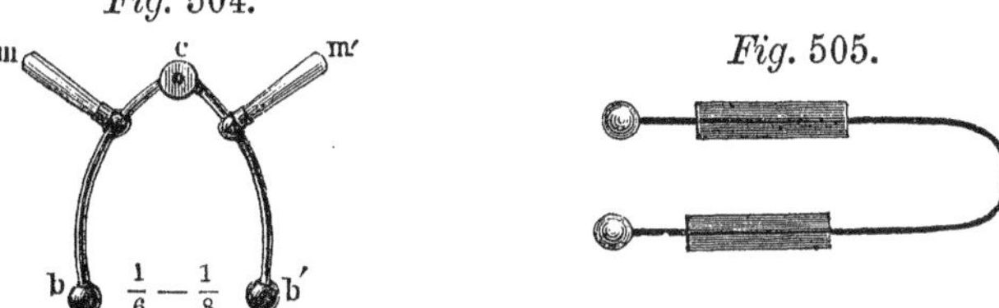


form of the discharger. The simplest form is a small brass chain attached to a short wire cast in a ball of lead or tin, and inclosed in a short glass tube. The chain is hung around the outer coating, and the ball applied to the knob of the jar. A more convenient arrangement is a wire, 2 to 3 millimeters thick, furnished with two knobs and glass tubes, as seen in fig. 505. The wire should be hardened so as to have some elasticity.

*Fig.* 506.

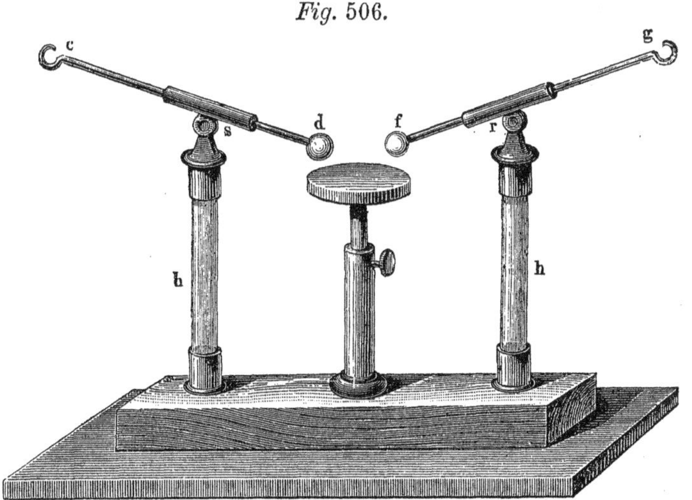


[265] *Henley's universal discharger* is a very convenient arrangement; it is shown in fig. 506. It consists of two insulating columns set in a

base-board, and supporting the conducting wires, between which a movable stand is placed. The wires are so arranged as to slide back and forward in the caps of the glass columns, and a joint permits them to be inclined at any angle. They must also be capable of being fastened in any position, which may be done very easily in the manner shown in figs. 507 and 508, in half the natural size. The piece *a a*, in which the wires slide, turns around the screw *c* between the heads *b b*. The whole may

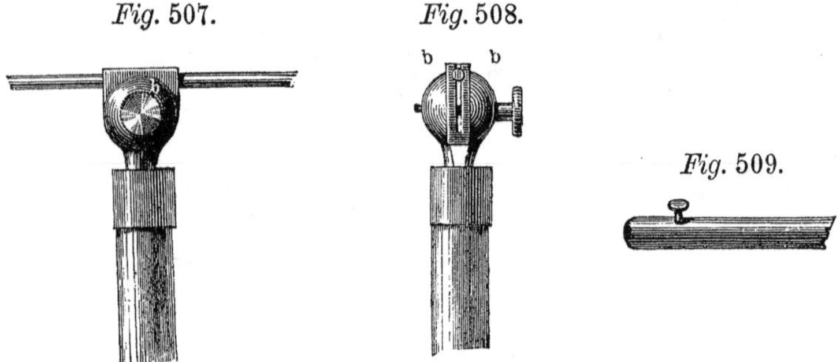

*Fig.* 507. *Fig.* 508. *Fig.* 509.

be drawn tightly together by the screw *c*. The wires need not necessarily terminate in knobs; they may simply be rounded off. It is very convenient when they have a small opening, as seen in fig. 509, in which fine wires may be fastened by a binding screw; the rings at the other end will always serve instead of the balls, by inverting the wires. The little stand need not be insulated, but it is convenient in some experiments to have a glass plate large enough to cover the top. It is also convenient to have the stem made of brass, extending down through the base.

[266] **The discharging electrometer, or Lanne's measuring-jar.**—To make this, select a jar holding 1 to 2 quarts, of clear, thin glass, and coat it to within an inch of the edge, as it will never be used for charges of high tension, and it must be so arranged that a discharge will take place over the edge sooner than through the glass. It is very advantageous to use always the same jar—the same measure of the charge—for this purpose. The cover must be fastened on, because the knobs must retain their relative positions unchanged. The jar, with its case of pasteboard, is placed on a board excavated $\frac{1}{4}$ of an inch to receive it, fig. 510; and into the same base the wooden column *a*, fig. 511, is fastened, which ends in a wooden clamp with a binding screw, of which a front view, half the natural size, is given in fig. 512. This clamp is arranged to hold a brass wire *c* a line in thickness, and terminating in a ball at one end, at any given distance from the knob of the jar. These distances are marked in tenths of an inch on the wire itself. As the wire

may be bent, or the column *a* warped, the distances between the balls may also be measured with a pair of dividers, and the wire adjusted by

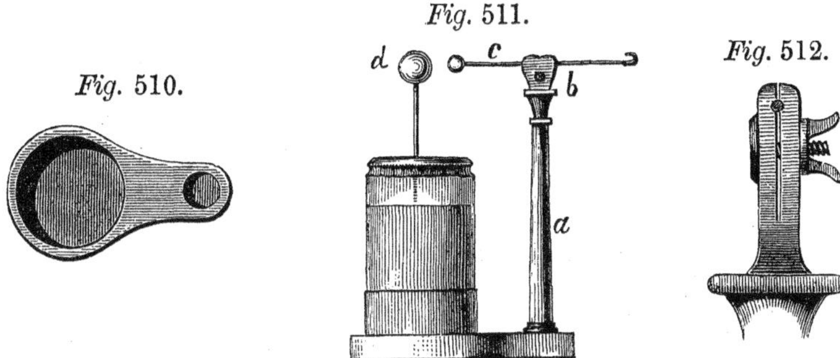

*Fig.* 510. *Fig.* 511. *Fig.* 512.

a fine thread cut on it. The wire *c* is either connected with the outer coating of the jar by a small brass chain, or, still better, by a strip of tin-foil passing down the column and along the board to the jar.

The charge of a jar or a battery is measured with Lanne's jar, by insulating the jar or battery, and connecting its outer coating with the inner coating of Lanne's jar, while the knob of the wire *c* is placed at a very small distance from the knob *d* of the jar. When the same number of sparks have passed over in two different cases, the charged jars have the same quantity of disguised electricity, let their size and number be what they may. The condition of the electrical machine may also be examined by means of Lanne's jar, by observing how many revolutions are required to effect a discharge at different times with the same distance between the balls. In the same way two machines may be compared with each other, but only in respect to quantity.

[267] **Experiments with the Leyden jar.**—*General observations.* A distinction must be drawn in general between experiments which require electricity of high tension and those which require quantity: to the first class belong all mechanical effects, such as breaking through glass; to the latter, the evolution of heat in good conductors, like the burning of gold-leaf between glass, etc. It is, moreover, a good plan to make every experiment which requires long-continued turning of the machine, with insulated jars, and connect the outer coating with the measuring jar, the balls of which preserve a constant interval. If the experimenter knows then how many discharges of the measuring jar are necessary to charge the larger jar sufficiently for a given purpose, he can proceed much more certainly in subsequent experiments, and loses no time either by a premature discharge or by superfluous labor. The number

of turns of the crank is, on account of the variable condition of the atmosphere, and other disturbing influences, not so safe a guide. Henley's quadrant electrometer, fixed on the conductor, gives much more reliable indications when a known jar is sufficiently charged. It is indifferent whether the jars be charged with positive or negative electricity; if the machine affords both, choose the stronger, which is often the negative, since, as already mentioned, certain sources of loss do not exist on this side which occur on the other, *e.g.* the loss in passing from the rubber to the conductor.

If it be necessary, for any purpose, to charge a jar with negative electricity and the machine affords only positive, one has only to hold the jar by the knob, and let the positive electricity pass into the outer coating while the jar is held up in the hand, or stands on an insulated support. It must, at all events, be set down on an insulated support; if this precaution be neglected, the charge will pass through the arm. The charge produced in this way is somewhat weaker than the positive charge.

The spark always seems to go from the side on which free electricity is accumulated.

The interior of the jar is connected with the prime conductor by brass wire, 2 lines in thickness, rounded off and bent into a hook at each end. Several of these wires of different lengths should be provided, some of which may have the hooks placed at right angles to each other. A considerable loss of electricity is incurred by using thin wire for connecting the inner coating with the conductor, as may easily be proved with Lanne's measuring-jar, by connecting an insulated jar at one time by thick, and another time by thin wires, and counting the number of turns which gradually become necessary to cause a discharge as the tension increases.

Chains must be entirely avoided for this purpose. They are very suitable for connecting the outer coatings, and for connecting the outer coating with the discharger and the various parts through which the shock is to pass. The simple brass chains used for this purpose cost little more than the price of the wire.

It should be remarked, that the effect of the discharge *is much increased by bringing the knob of the discharger suddenly near the knob of the jar*, as if about to strike it; this precaution is necessary when the force of the charge is only barely sufficient for the required purpose.

[268] **Experiments with the Leyden jar.**—(1) *Perforating paper.* Single cards can be perforated with very light charges. The card is laid against the outer coating of the jar and one knob of the discharger, figs. 504 and 505, held against it while the other is suddenly brought near the knob of the jar. To perforate several cards at once it

is more convenient to place them between the balls of Henley's universal discharger.

(2) *Perforating glass.* In this experiment electricity of high tension is required. It can be effected without difficulty with a jar having only 70 or 80 square inches of coated surface, if the margin of the jar be 6 or 8 inches wide, and the electrical machine give sparks 2 inches long; whereas with less tension, *i.e.* with a margin only 3 or 4 inches wide, 4 square feet of surface are not enough to effect it, the jars in both cases being charged as highly as possible. The glass perforated must be a pane of ordinary window glass, at least 6 inches square, otherwise the charge will go around it.

The best plan is to wind a wire around the stem of the stand on Henley's discharger, fig. 507, and allow one end of it to reach to the middle of the plate, while the other, bent into a hook, is connected with the outer coating. If the stem of the stand be of metal, it is only necessary to attach the chain to it. The pane of glass is laid on the stand, and one or two inches of its surface smeared with oil to make it insulate better. A pointed wire screwed into the rod of the discharger is laid on the glass opposite the middle of the stand, fig. 513, so as to press upon the surface with some force. The other end of the rod is connected by a chain with the knob of the discharger, which is held in the hand and laid on the knob of the jar.

*Fig.* 513.

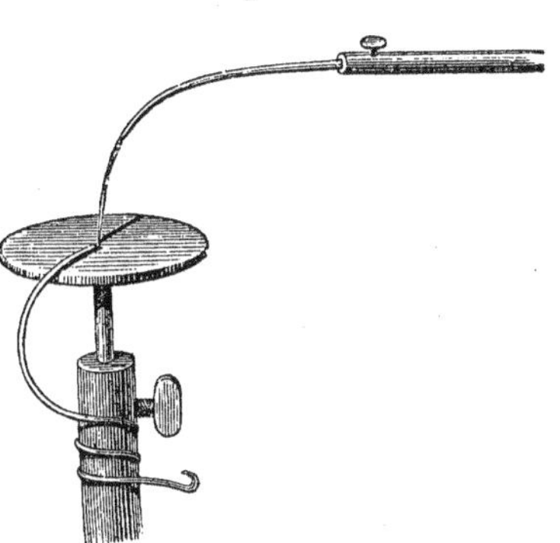

(3) *Wood* may be perforated in the same way. The board should be varnished on both sides.

(4) *To burst glass tubes* requires a greater quantity of electricity, but is easily effected. Brass wires, bent into hooks at the ends, are inserted through two corks fitting tightly into the ends of a green glass tube, which is filled with water and the corks pressed in tightly, fig. 514. The ends of the tube should be contracted over a spirit-lamp, the corks then fit tighter, and the tube is not so apt to burst in driving them in. The ends of the wires are placed about 2 or

*Fig.* 514.

3 lines apart, and the apparatus interpolated in a chain connecting the outer coating and the simple discharger. As the fragments of glass are sometimes thrown to a considerable distance, some arrangement must be made to prevent this. Even open glasses may be burst by an electrical spark, but a tolerably strong charge and 3 or 4 square feet of coating are necessary to effect it. The arrangement is made somewhat as shown in fig. 515. The wires are bent so as to hold on to the glass by their own elasticity, and their ends brought within 2 or 3 lines of each other. The glass is often broken off at the stem, unless the stem taper gradually.

*Fig.* 515.

(5) *Melting iron wire.* This experiment, to succeed without powerful apparatus, must be made with wire much finer than can be purchased. A piece of wire about 3 inches in length may be laid in nitric acid so as to project at both ends, and eaten down to the required fineness; it is then washed off with water, and its thick ends fastened into the rods of the universal discharger. A good charge is still required to fuse the wire, but it may be effected with 4 to 8 square feet of coated surface.

(6) *Igniting gunpowder.* Bore a hole, ½ to 1 inch in diameter and 1 inch deep, in a block of hard wood; introduce into the bottom of this two fine brass wires through holes in the sides, and let the ends of the wires be about 2 lines apart, fig. 516. Pour powder in loosely and press a tolerably tightly fitting cork directly upon it. Tie a piece of ordinary twine, 3 to 5 inches long, and thoroughly moistened, to one of the wires, and attach the chain of the discharger to this; this interruption is not necessary on the other wire. Without this precaution the experiment will not succeed. The charge required is about the same as in No. 4.

*Fig.* 516.

Gun-cotton may be ignited in the same way, with about ¼ the charge necessary for gunpowder. The cotton must be placed between the ends of the wires. This experiment may be made with the weakest machines.

(7) *Melting gold on glass.* Have a strip of gold-leaf laid on a piece of even glass, as in fig. 517. This may be done by laying the glass, previously moistened with the breath, on the edge of a leaf of gold, and

drawing a right sharp, clean, dry knife along the edge of the glass; the excess of gold-leaf is then cut away from the glass. The glass is covered with another similar piece and laid between two strips of felt, one of which is coated at both ends with tin-foil extending to the glass. The

*Fig.* 517 *Fig.* 518.

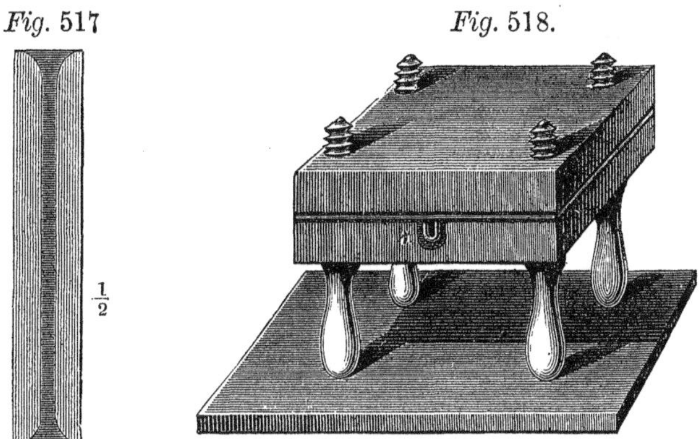

strips are then tied firmly together, or screwed up in the little press, fig. 518. This press consists of two boards and four wooden screws, which serve at the same time as feet; the female screw is cut in the upper board. Strips of tin-foil *a* are pasted on opposite sides of one of the boards extending some distance over the inner surface, and wire hooks are driven in through them. The coated ends of the felt are laid on these strips and the screws drawn moderately tight. The apparatus is intercalated in the chain connected with the outer coat of the jar. The glass is generally shivered and the gold purple is burnt into it. This experiment does not require so strong a charge as to ignite gunpowder.

(8) *Combustion of metallic ribbons.* Wires of other metals than iron and platina can only be fused with powerful machines and large batteries; but a strip of fine tin-foil, about $\frac{1}{4}$ line wide, makes an experiment which is successful even with a feeble apparatus. A strip of this width, and an inch or two long, fastened between the rods of the universal discharger, is burnt by the electric spark, and the oxide forms a light white cloud. The metallic coating on genuine and imitation gold and silver paper may be burnt in the same way; false gold paper requires a rather stronger charge. If the strips are inclosed between strips of white paper in the little press, fig. 518, so as to touch the slips of tin-foil, they will leave colored streaks on the paper. Similar but broader streaks are made when fine wires are melted between the points of Henley's discharger over a sheet of paper.

(9) *Igniting resin.* Pulverize resin finely and mix it intimately through a wad of cotton, the size of a walnut, by rolling the cotton in the

powder, picking it to pieces and rolling it together again repeatedly. Lay a ball in a flat metallic capsule on the stand of the universal discharger connected with the outer coating of the Leyden jar, and place the ball of one of the rods of the discharger so as to stand 1 or 2 lines above the cotton. If the discharger have no ball on the end of the rod, a small ball with a bent wire should be screwed into it, as seen in fig. 518; the chain of the ordinary discharger is then connected with the rod, and the charge passed through the cotton. A much lighter charge is required than to ignite gunpowder.

*Fig.* 519.

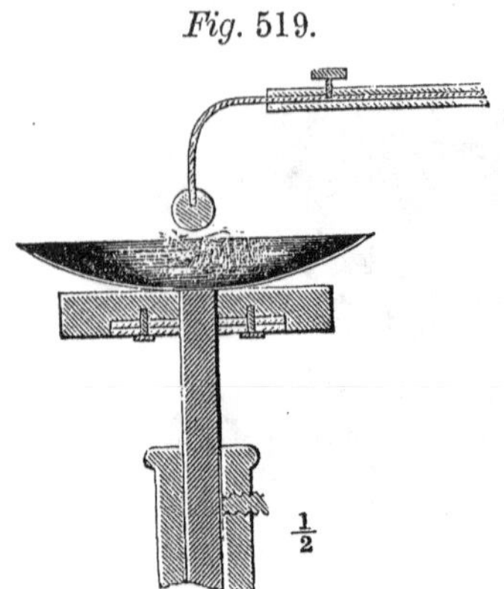


If the cotton be wrapped around the wick of a wax candle placed between the knobs of Henley's discharger, the candle may be lighted by the spark.

(10) A lighted candle placed between the knobs of the discharger will be extinguished by the shock; one just extinguished will be relighted by a very strong charge.

(11) *The thunder house.* Make a little wooden tower *a a*, fig. 520, with a movable pointed roof, through which the wire *c d*, terminating in

*Fig.* 520. *Fig.* 521. *Fig.* 522.

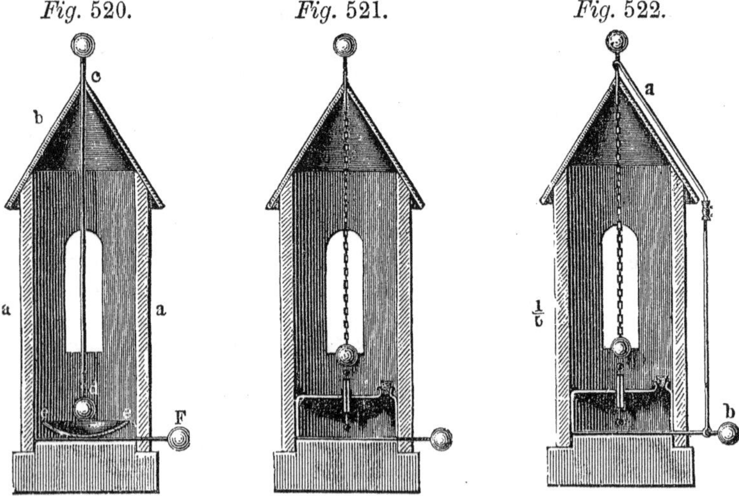


little knobs at both ends, extends downward to within a short distance from the capsule *e e*, filled with alcohol or ether, which is connected by the wire *F* with the outer coating of a jar. A slight charge is sufficient to inflame the contents of the capsule.

An electrical pistol of tin may be placed in the bottom of a similar tower, the roof of which will be blown off by the discharge, fig. 521. If the tower be provided with a conductor well connected at *a* and *b*, fig. 522, the two parts of which fit with slight friction into each other at the bend, as seen in fig. 523, the explosive gas will not be ignited if the conductor is good enough to carry off a slight charge. If, however, too thin a wire be inserted between *a* and *b*, the charge will be divided, and the gas will be fired. The chain of the discharger should be attached to the wire at the apex of the roof. The pistol should not hold much gas. This experiment is especially instructive for the theory of the lightning-rod. It may be performed more simply without the tower, the conductor

*Fig.* 523. *Fig.* 524. *Fig.* 525.

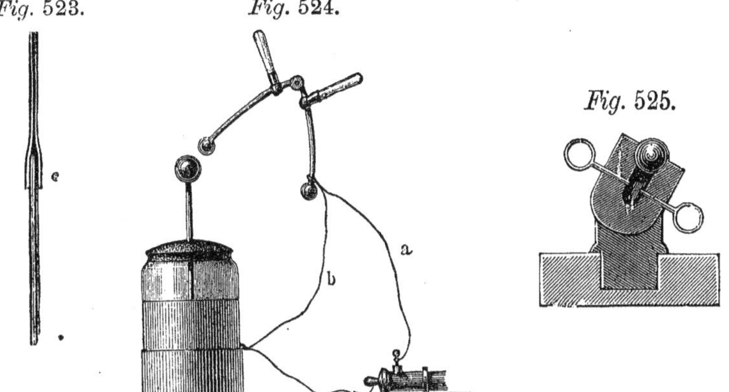


of which is troublesome on account of the roof being blown off, by fastening to one limb of the discharger two wires *a b*, fig. 524, one of which is connected with the knob of the pistol, and the other directly with the outer coating of the jar, the pistol itself being also connected with this. If the wire *b* be thick enough, the charge all passes through it, but if it be too slender, a part of the charge will pass through the other wire and discharge the pistol. This wire should be short and kept stretched during the discharge, to insure a good connection. The striking distance within the pistol must not be too short.

(12) *The electrical mortar.* In a cylinder of boxwood, shown in fig. 525, of $\frac{1}{3}$ the real size, bore a hole, with a hemispherical enlargement above, to receive a ball of ivory, wood, or elder-pith which lies loosely in it so as to close the mouth of the tube entirely. Pass two wires, about

1 line in thickness, into the cavity, about the middle of its length, and insert them in the connection between the inner and the outer coatings. The ball will be thrown out with more or less force according to its weight and the strength of the charge. This very instructive experiment succeeds with a very moderate charge, if the ball be of pith.

(13) *Physiological effects.* To give the shock to a number of persons, let them join hands, and the first one taking hold of the chain connected with the outer coating of the jar, let the last touch the knob of the jar with a wire ending in a knob. The chain must be of considerable length, so that the jar will not be jerked over. The greater the number of persons, the heavier the charge must be. The effect is diminished somewhat toward the middle of the row, because a portion of the charge goes through the floor. For this reason one should beware of approaching too near to the parts connected with the inner coating of a heavily charged jar, when separated from the outer coating only by a small space of the floor. The effect of the shock upon men is exceedingly variable, according to their constitution. The effect of a certain charge upon ourselves should always be known before giving the shock to others; very unpleasant effects might otherwise be produced. One has involuntary opportunities enough to gain such experience, so that the precaution is only necessary with machines and jars whose power is not known.

In order to make the experiment of killing animals by an electrical shock, the extremity of the vertebral column must be connected with the outer coating of the jar, and one knob of the discharger laid on the head of the animal, while the other is suddenly brought near the knob of the jar, so as to pass the charge through the brain and spinal marrow. Cats and other animals of this size can only be killed by a powerful apparatus. Birds are killed more easily, if it be deemed necessary to make such experiments at all.

(14) The following experiment is interesting as illustrating the action of the apparatus for giving shocks by galvanism, and its connection with the physiological effects of the Leyden jar. Set the knobs of Lanne's measuring-jar, about $\frac{1}{3}$ to $\frac{1}{2}$ millimeter apart, and place the person who is to receive the shock, or a whole row of persons, in communication with the outer coating of the jar on one side, and with the wire *c*, fig. 511, on the other side, having first interrupted the connection of the wire with the outer coating. A single discharge of the jar will scarcely be perceived, but if the discharge be turned rapidly, numerous discharges succeed each other at exceedingly minute intervals, and the effect is the same as is produced by an induction apparatus with the lightning wheel.

(15) To show the effect of the electrical current from the Leyden jar

on the magnetic needle, pass the charge through a multiplier of very long, fine wire, (at least 200 turns,) interposing several inches of wet hempen twine in the circuit. It needs only a moderate charge.

(16) *To magnetize steel by an electrical shock*, coil fine, well-wrapped wire closely around a small glass tube, 2 or 3 inches long, and varnish it; solder to each end a piece of thicker wire bent to a hook. Lay a piece of steel knitting-needle, the length of the tube, inside of it, and pass the charge from a jar through the wire. If the charge be not too strong, the needle will have its north pole where it should be according to Ampere's theory. With a stronger charge, variations will occur, which cannot be explained in this place.

(17) *Induction by the Leyden jar.* Have two circular boards made, 10 inches in diameter and 1 to $1\frac{1}{2}$ inches thick; they should be made of two or three thinner pieces glued together so as not to warp, and turned round on the lathe, and planed even on one side. On the even side cut concentric grooves, $\frac{1}{4}$ inch deep and $\frac{1}{2}$ inch apart, and connect them so as to form a spiral, as shown in fig. 526. The grooves must be cut symmetrically in both boards, so as to fit together when laid upon each other.

*Fig.* 526.

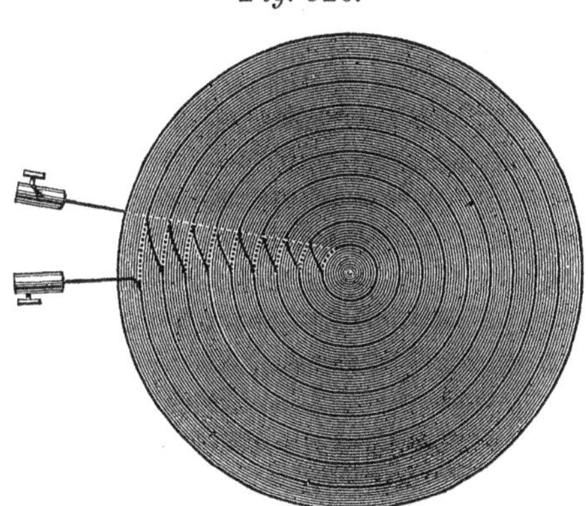

Varnish the grooves and lay a copper wire, about 2 millimeters thick, in them, and fasten it down with little pins driven in slanting. Pass one end of the wire through the board near the middle, and bring it toward the edge on the outside; carry the other end of the spiral over the inner surface so that the two ends will come near together, and fasten both firmly by driving in staples. Then fill the grooves with resin or shellac melted in with a hot iron, so as to project above the surface of the board. Smooth off the surfaces clean and varnish the boards with shellac, which will impart a gloss to the resin also. Furnish one of the boards with three low feet, and the other with a knob to serve as a handle. They need not be insulated, except for experiments where measurements are to be made, and then only the lower one. Even the laying the wires in resin might perhaps be dispensed with; at all events, I have hitherto used such a coating for one disk only. The experiment is made by laying a plate of glass between the boards, bringing the two ends of one spiral near together, and discharging a jar through the spiral in the other disk. At each discharge a spark will be seen to pass between the ends of the other wire. If a person complete the circuit of the second wire, he will receive feeble shocks. It is desirable, though not necessary, to make the connections with binding screws. Fig. 527 shows the apparatus complete.

*Fig.* 527.

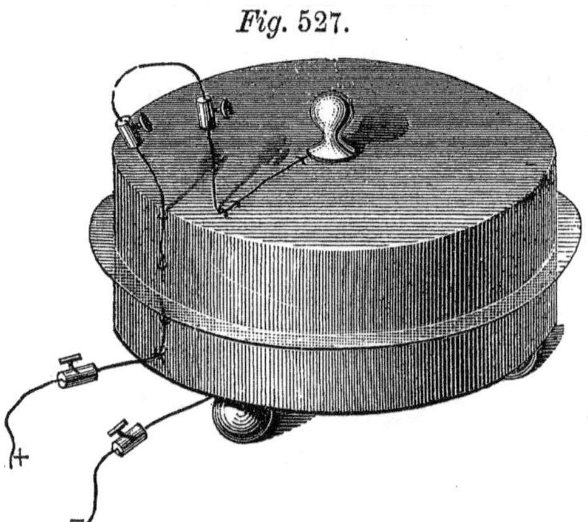


(18) *Slow discharge of a Leyden jar.* This experiment is very well adapted to illustrate the action of points. Attach a chain to the outer coating of the jar, and hold the end with a fine needle between the thumb and fore finger. Present the point of the needle to the knob of the jar: it will be gradually discharged, and the point of the needle will be luminous in the dark.

(19) *Electrical breath images.* Upon a small stand lay a coin, and upon this a piece of common window glass wiped clean, and upon this a second coin exactly covering the first. Place the end of a wire bent into a ring under the lower coin, and bend the other end of the wire over toward the other coin, so that its ring will stand about a line from it. Set the stand under the conductor of the electrical machine and let a chain hang down to the coin. As the machine is turned, discharges of the little apparatus will take place in quick succession. After several hundred such discharges, take the little apparatus apart, and the glass, when

breathed on, will show the impressions of both coins. These images are often visible when the glass is breathed on, after the lapse of many months and after being repeatedly wiped.

[269] **The electrophorus.**—This is generally made at home. The mould may be made either of wood or tin. In the first case, a wooden rim is tacked to a round disk of dry wood, so as to project about $\frac{1}{4}$ inch above the bottom; the whole mould is then covered, inside and out, with false gilt paper or tin-foil. Tin moulds are much more easily bent than wood, and they expand more by heat, causing the cake of resin to crack in every direction. Wood expands most from moisture, especially in the direction perpendicular to the fibers, but this may be prevented in great part by selecting well-seasoned wood, and coating it thickly with hot oil varnish; it may warp somewhat during this process, but should be planed off even and varnished again.

The materials usually prescribed for the resin cake consist of various mixtures of shellac, pitch, wax, rosin, turpentine, etc. It is very doubtful whether the proportions of these are the result of extensive experience, and the ingredients are in themselves incompatible. What good can turpentine, rosin, and pitch do together, since rosin is only pitch freed from its oil of turpentine, and of itself a brittle substance? Shellac is unquestionably the most suitable substance, and needs only an admixture to render it less brittle. This is effected completely by a mixture of 5 parts shellac, with 1 part of wax, and 1 of pitch; 5 shellac with 1 wax, or 5 to 10 shellac and 1 pitch would probably answer as well. The mixture given above certainly furnishes a mass which is not brittle, and very electrical.

The materials should be melted in a shallow earthen pot or a brazen pan, and the more fusible ingredients melted first over a gentle equable heat: the shellac may then be added in successive portions, increasing the heat gradually and stirring constantly, waiting until each portion of shellac has become at least pasty before adding more. By exposing the shellac to the heat too suddenly, it assumes an almost infusible condition.

Before filling the mould it should be slightly warmed, so that the mass may not cool too rapidly, and set level. It must be filled to the brim. It is not easy to prevent the formation of bubbles on the surface, but they occur mostly around the edge, and may be removed by holding a red-hot iron over them and melting them down, so that the projection disappears and they form only shallow cavities with a rounded margin. Such spots do no other injury than diminishing the surface of the electrophorus. Instead of melting down the bubbles, they may be pared off with a sharp tool.

Such an electrophorus soon becomes cracked, on account of the unequal expansion of the form and the resin: those made of wood crack in the direction of the fibers of the wood; the tin ones, in every direction. So long as the cracks are not too numerous they do not render the instrument worthless, though they impair its effect. When its action becomes too feeble, it must be melted over again, which may be done by passing a mass of red-hot iron over it, within an inch of the surface. A plowshare answers the purpose very well; or a piece of sheet iron larger than the form may be held about an inch above it, and coals heaped upon it. Care must be taken not to soil the surface with ashes. Bubbles often form during the fusion, and must be removed as before directed.

These difficulties may be avoided by having the cake of resin free, and without a case; when in use it may be laid on a smooth board coated with tin-foil. The mould for such a cake may be made of paper. Both sides of the cake are afterwards to be ground smooth with sand and water on a stone or plate of glass, using very fine sand at last. It may be polished with a strip of felt tacked to a board, and Tripoli and water; but this is not necessary. The edges may be rounded with a knife and file. The bubbles may be ground away, but this is superfluous labor; turn the cleaner side, which was beneath in casting, above for use. The surface of the resin need be perfectly even and highly polished only when the electrophorus is required to retain its electricity for weeks together even when covered.

Such an electrophorus should be kept lying horizontally on a board, otherwise it may bend by its own weight in the heat of summer.

Gutta-percha has not kept its reputation as a material for the electrophorus long: it becomes crumbly; gutta-percha paper falls to rags, and an electrophorus made of it furnishes from the outset less and less electricity continually, and may under certain circumstances even yield positive electricity.

In respect to the size of the electrophorus one must be moderate. Its effect is always feeble, and especially so in charging a jar, even when its striking distance is considerable. An electrophorus 10 to 15 inches in diameter, will answer any reasonable demands. One with a 2-inch plate will yield sparks ½ an inch in length.

The cover of the electrophorus should be 2 to 4 inches less in diameter than the cake of resin. It is made either of a smooth plate of metal (zinc is the cheapest) with an upturned rim soldered on, or of a disk of well-seasoned wood, ½ an inch thick, well-rounded at the edges, and coated with tin-foil. The simplest insulating handle is three silken cords.

The electrophorus is excited by beating it with a foxtail, or a catskin held by the four paws and made to glide over the cake at every stroke.

In winter the electrophorus must be warmed before any electricity can be obtained from it. When the cake gives little sparks to the knuckle, it is sufficiently electrical. In laying on or removing the cover, it must be held parallel to the cake, and should be touched with the finger after laying it on, upon which it will emit a small spark. By touching the cover and case with the thumb and finger at the same time a shock will be felt, and this takes place also when the case is touched with one finger and the raised cover with the other.

The electrical condition of the cover may be ascertained by placing the electroscope of the upper plate, fig. 499, upon it, and making a trial with a stick of sealing-wax. To test the electricity of the case, it may be insulated. It may be done more readily, however, with the gold-leaf electrometer, fig. 452, by screwing on the plate of the condenser instead of the upper knob, and laying on it a cake of resin a little larger than the plate. If this cake be struck gently with a small piece of catskin, the gold leaves will diverge, and the quality of the electricity may be ascertained by bringing a stick of sealing-wax near the lower knob, which is connected with the bent wire. The second condenser plate may be used as the cover of the electrophorus. If the plate be touched, the gold leaves fall together, but they diverge again when the resin cake is taken off, or the cover laid on. If the cover be laid on without touching the plate, the leaves will fall. The electrical condition of the second plate, used as a cover, may be investigated by imparting electricity from it, by means of the trial plate, fig. 461, to a second electrometer, or connecting it with this by an insulated wire. Bohnenberger's electrometer is the best one for this purpose.

Although an electrophorus may easily be made to give sparks an inch long from its cover, the quantity is always small, and the charging of even small jars proceeds very slowly with it. The fundamental phenomena of electricity may, nevertheless, be all shown with this cheap apparatus. The electrical chime may be set in motion by holding the knob of the charged jar against it. To charge a jar with negative electricity, it must be held by the knob, as already explained in § 267.

[270] **Lichtenberg's figures.**—These are produced in the following simple manner: Describe any figure upon a cake of resin, with the knob of a positively, or moderately negatively charged jar, held by the outer coating; dust semen lycopodii over it from a gauze bag, and blow away with a fan, or the breath, what does not adhere. The figures are very permanent; they show themselves again after the surface has been freed from all dust, and dusted over again. Flowers of sulphur may be used instead of lycopodium; both attach themselves equally to positive

and negative figures. If red-lead be strewed over the surface, a positive figure only becomes visible by wiping the dust off it; it then shows black on a red ground. If positive and negative figures be drawn on the same surface, and red-lead and flowers of sulphur strewed over it at the same time, the latter sticks chiefly to the positive figures, and the former to the negative.

*Fig.* 528.

This experiment does not succeed well on an old, long-used surface; it must be melted over, at least superficially. To make small figures, it is best to melt the resin on a sheet of tin, upon which it can quickly be remelted after each experiment. The mass need not be over $\frac{1}{2}$ a line thick, and impure resin answers best, being less brittle, but the experiment will succeed with any resin.

[271] **The condenser.**—This indispensable instrument is attached immediately to an electrometer. For the sake of convenience, the lower plate should be capable of being screwed on any of the various electrometers, instead of the knob, fig. 528. Take for this purpose brass plates, about a line in thickness and 2 inches in diameter. It is very unadvisable to make the plates only an inch in diameter, as is commonly the case in those purchased in the shops, since the effect increases in much greater ratio than the surface. In galvanic experiments, smaller plates will also be needed. On one plate solder a brass socket to receive a glass rod, 3 to 4 inches long; on the other solder a small plate, in which a female screw is cut to fit the thread of all the electrometers. Fig. 529 shows this plate and the stem of the electrometer. Turn off the plates smooth and grind them on a piece of plate glass with emery;

*Fig.* 529.

polish them finally with fine pumice powder and water on the lathe, to give them a metallic luster and circular streak. The edge must also be rounded on the lathe. When the plates are too thin, they are very hard to grind even, because the outer edge bends up continually; such plates should be supported during the grinding and turning by cementing them to a block of wood of the same size, which will serve also as a handle in grinding.

Especial care must be bestowed on varnishing the plates. A coat of moderate thickness, which is generally sufficient, may be applied on the lathe. When a thicker coating is required, it is better not to warm the plates previously, for it is very difficult to lay a coat of varnish evenly on a heated surface: lay the plates level and brush them over with a tolerably concentrated solution of shellac, which has, in this way, time to spread itself uniformly over the surface of the plate. The varnish becomes glossy when dry. A second coat may be applied in the same way, without disturbing the first. A piece of fine sponge is more suitable for varnishing than a brush. The varnish on the plates of the condenser must never be too thin.

*Fig.* 530.

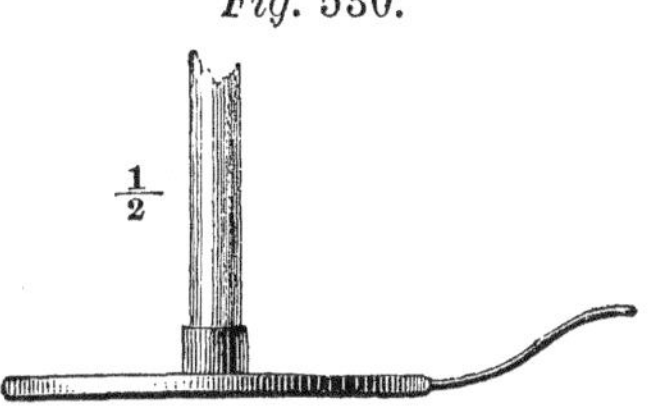


If the edges and upper surface be varnished, it should be done previously, and the varnish applied warm. A conducting wire with a rounded end must, in this case, be screwed into the edge of the plate and left unvarnished, fig. 530. It is convenient to have such wires on both plates; they need not be more than 1 or 2 millimeters thick. They are desirable because their ends can easily be cleaned with a file, which is not so easy with the plates, which are soon soiled by contact with the fingers, often purposely wet, and both their appearance and effect is injured thereby. On the other hand, the wires have the disadvantage that they are in the way and apt to cause a displacement of the plates. A good condenser should retain a weak charge at least 12 hours in good weather.

Condenser plates of copper and zinc are needed in galvanic experiments; it may therefore be as well at once to make one plate of copper and the other of zinc, and to make the rods also of the same material, as it is immaterial in other experiments what kind of metal is used. In this case, however, both plates must have female screws. The handle is made of a glass tube sealed at one end; a wire furnished with a screw is cemented into the other. The whole tube should be filled with bits of sealing-wax and heated over a lamp so as to coat the inside entirely.

The upper plate must always be held parallel to the lower one in lifting it off.

In charging one plate, the other must always be touched with a conductor. Other precautions will be mentioned when describing experiments on galvanism.

When the varnish of a condenser once becomes electrical, it often remains in this condition for days together, and causes annoying errors. The speediest remedy is to pass the plate a few times through the flame of a spirit-lamp. If this does not remedy it, and the condenser is to be used at once, the varnish must be washed off with alcohol, and freshly applied: the electrical condition disappears of itself after awhile. It will be seen whether the condenser has any charge, by noticing whether the leaves diverge when the cover is touched with a conductor and then lifted off. In using the condenser, never fail to test it in this way.

In experimenting with very feeble electricity, it is often necessary to use two condensers, in the following way: the first is charged from the source of electricity, the plates separated, and the upper or under plate of the second electrometer touched with the upper plate of the second, and this process repeated from 6 to 10 times. Any less delicate electrometer may be used as a stand for the first condenser. The old instruments called duplicators are no longer used.

## (*c.*) EXPERIMENTS ON ELECTRICAL LIGHT, AND ELECTRICITY BY HEAT AND PRESSURE.

[272] **Experiments in the dark.**—(1) *The lightning tube.* Take a glass tube, $\frac{1}{2}$ to 1 inch in diameter, fit it at one end with a wooden plug with a hemispherical head covered with tin-foil, and beginning at this paste little diamond-shaped bits of foil, with the points about 1 millimeter apart, in a spiral down the tube, fig. 531. This tube is taken in the hand, and held near enough to the conductor to draw copious sparks, which are repeated at every interruption in the spiral. To prevent the bits of tin-foil from scaling off, the tube may be set inside of another to which the knob and handle are fitted.

(2) *Illuminated figures.* Fix a knob of wood, covered with tin-foil, to the edge of a pane of glass by sticking the glass in an incision in the wood, and from this as a starting-point describe any figure by pasting on bits of tin-foil as already described. The portions not designed to be illuminated are made with strips of foil. A larger strip leads to the opposite edge of the glass where it is held in the hand. As the spark overleaps a considerable space on the plate, the lines of the figure or name must be

drawn at some distance apart, and part of the circuit made on the other side of the glass. The dark-shaded parts in fig. 532 are made on the other side of the glass.

*Fig.* 531.

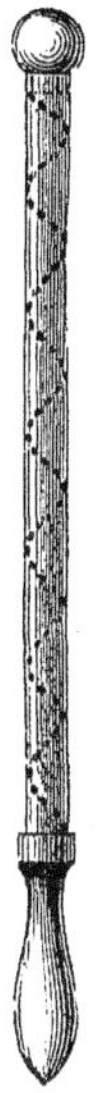

*Fig.* 532.

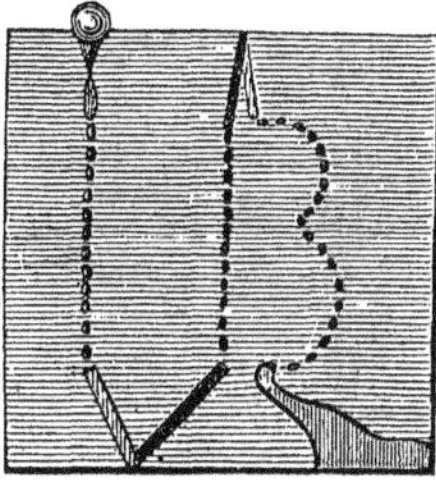

(3) Coat a pane of glass on both sides with tin-foil, to within 2 inches of the edge, and blacken one coating with indian-ink. Cut this coating, when quite dry, into little lozenge-shaped pieces, by cutting out very narrow strips of the foil. Make the cuts with a sharp knife and a ruler, and peel off the strips. In the middle of the glass, on the same side, cement a little disk of tin with a ring in it. Lay the pane on the table, and connect the conductor of the machine with the ring, and the uncut coating with the discharger. While the plate is being charged, zigzag flashes of light will dart over the cut surface, and when the discharger is brought near the conductor and the apparatus discharged, the whole surface will be covered with flashes of light going in a zigzag direction toward the ring.

The pane of glass is usually set in a dull-black wooden frame, and the uncoated margin covered with black paper. A strip of tin-foil pasted to the back coating connects it with a hook driven into the edge of the frame, to which a chain may be fastened.

(4) *Brushes of light.* To observe the brush of light from the positive, or the stars from the negative conductor, it must end in a small ball, 2 or 3 lines in diameter, near which the hand or an even metallic surface is held. Wires with rounded ends, successively less in diameter, may be used instead of the ball. The brush may also be obtained, with some modification, by holding the little ball or the rounded wires in the hand near the conductor. These phenomena are only exhibited by machines of considerable power, and the less the striking distance, the smaller must be the wire used.

(5) *Experiments in vacuo.* The simplest experiment of this kind is made with the tube described for the guinea and feather experiment, § 102 (26.) The experimenter has only to hold it in the hand, and allow sparks to pass to it from the conductor; they fill the whole tube with a bluish light. With feeble machines, the tube must be grasped near to the end next the conductor; the light of course only reaches to the hand.

The experiment may be varied, by laying on the plate of the air-pump (when of glass) a metallic plate connected with the body of the air-pump, and covering it with a bell-glass provided with a stuffing-box. Various devices, such as stars, balls, etc., may be screwed to the rod of the latter, to vary the form of the light which streams through the bell. In making this experiment, the experimenter must be sure that the bell-glass does not conduct electricity; and with this view warm it before use and rub it dry with blotting-paper. A glass vase, such as is shown in fig. 533, makes a very beautiful experiment.

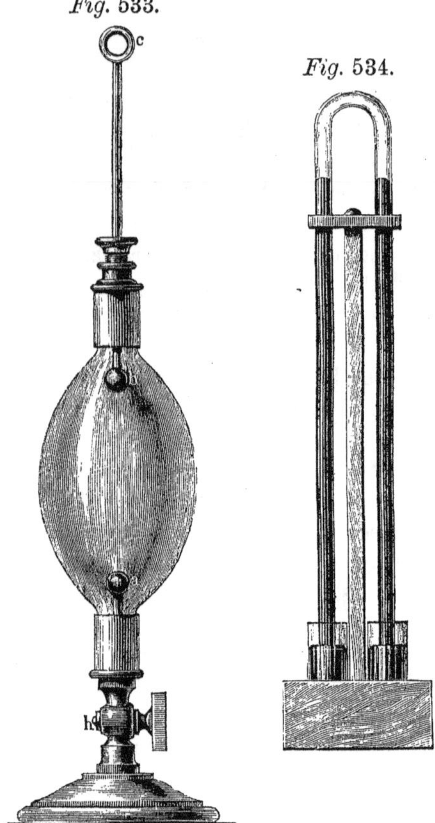

*Fig.* 533. *Fig.* 534.

A double barometer, constructed as in fig. 534, may be made luminous, by connecting one cup with the conductor and the other with the ground.

(6) If a piece of fluor spar be placed between the rods of Henley's discharger, and a vigorous shock passed through it, a feebly luminous streak will be visible on the fluor spar, but it disappears in a few seconds. A piece of white sugar, $\frac{1}{2}$ to 1 inch in diameter, will show a similar appearance. An orange may also be made feebly luminous between the points of the discharger.

[273] **Electricity of tourmalin.**—The electrical properties of tourmalin may be exhibited very easily, by hanging a piece, about an inch long, (which may, however, be very thin,) by a silk string, and then warming it by holding a piece of sheet iron $\frac{1}{4}$ to $\frac{1}{2}$ an inch below it, and heating the iron with a spirit-lamp. The kind of electricity may be tested with a stick of sealing-wax, rubbed on one end. The electrical condition is often less evident during the warming, on account of the ascending current of air; but during the cooling it is always very marked and distinct: the effect of the sealing-wax is

felt at a considerable distance. Many other crystals become electrical by warming, but none of them show the property so easily as tourmalin.

[274] **Electricity of calcareous spar.** — Calc spar becomes positively electrical by being pressed, for a few seconds, between the fingers, and retains its electricity a long time. This is shown most easily by bending a narrow strip of tin like *a b c*, fig. 535, and making a little depression at *b* with a punch. At *a* cement a thread of shellac about the thickness of fine twine; on the end of the shellac cement a fragment of calcareous spar with wax, and set the apparatus on the little pivot, fig. 535. It may be balanced by a little wax at *c*. Press the spar between the fingers before setting it on the needle: it will be briskly repelled by an excited glass rod. Some other crystals exhibit the same phenomenon.

*Fig.* 535.

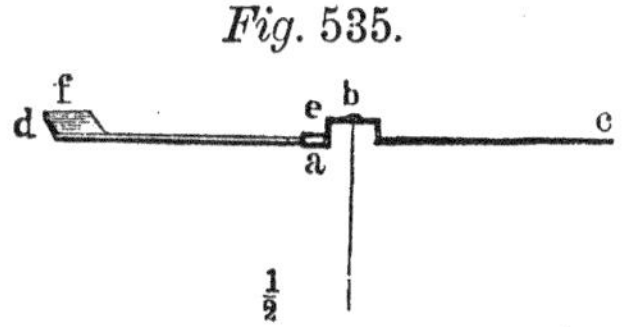


## (*d.*) EXPERIMENTS ON GALVANISM, (CONTACT ELECTRICITY,) AND THE ACTION OF THE GALVANIC PILE.

[275] **The frog experiment.**—Take a living frog, stun him by a blow upon the head, cut him in two in the middle with a pair of shears or a sharp knife, remove the entrails from the hinder portion, and strip off the skin: the nerves which issue from the lower end of the spinal column can then be seen very distinctly, and must be dissected clean from the surrounding cellular tissue with a sharp knife. Thrust a strip of clean brass-foil under the nerves, and lay the preparation on a glass plate. Muscular twitchings ensue whenever the brass plate and the muscles of the leg are touched with a bent iron wire, or a strip of zinc. The preparation may also be hung up by a silk string, and a brass hook thrust under the nerves.

[276] **The fundamental experiments.**—For these experiments, a plate of copper and another of zinc are needed, and one of copper and zinc soldered together. The first pair of plates must be made of metal about 1 line in thickness. Zinc plate of this thickness is not usually sold in the shops; it must be ordered specially; it is so often necessary for the construction of galvanic apparatus, that it will be well to have a few square feet of it on hand. It is convenient to have these plates of the same diameter as the plates of the condenser; they may then be used as top plates for the condenser when the other plate is well varnished. They ought in this case to have a socket soldered with tin to

the back, in which a female screw is cut with the same thread which all the electrometers have, fig. 536. They must have insulating handles; but if they are not to be screwed on the electrometers, simple sticks of sealing-wax melted on to the hot plate will answer the purpose. It would be neater, however, to solder to the back brass sockets, and cement into them glass tubes closed at one end, fig. 537. If they are arranged for screwing on, the tube should have a brass wire cemented into it with a screw to fit the plate, or the glass tube may be cemented into a cap on which the screw is cut, as in fig. 538. The edges of the plates must be well rounded, and one surface made even with a file and straight-edge, and then ground on plate glass. The soldering on the back must be done before the grinding, as it might cause the plates to warp; for the same reason it is well to cement them to a block of wood while grinding. The back is finally cleaned and varnished. Zinc is easily worked on the lathe, but an even surface can be given to copper only with a file moistened with oil, or with a tool fixed in a rest. The ground side and the edge must have a clean metallic surface when used. Without these precautions, the fundamental experiments are very likely to fail, because the contact surfaces are either not even or not clean. The surface may always be cleaned by strewing finely powdered pumice on clean paper resting on an even surface, and rubbing the plates upon this. The adherent powder is removed with a feather. The surfaces become soiled, even when the plates are kept wrapped up in paper, and it is, therefore, advisable to clean them as directed, each time before using them. When in use, they must on no account be rubbed together, because the copper plate immediately becomes contaminated with zinc, which disturbs their action. When the apparatus is brought into a room already occupied for some time by a numerous audience, especially in winter, these experiments, as well as all experiments with condensers, are apt to fail, and attention must be paid to this fact.

*Fig.* 536.

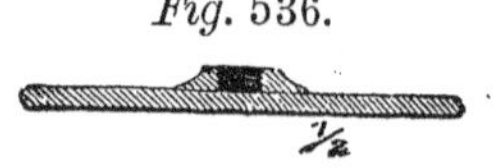


*Fig.* 537. *Fig.* 538.

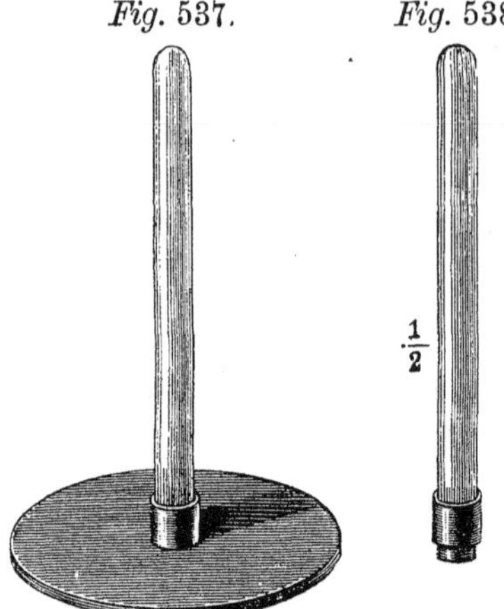


The compound plate of zinc and copper is made in the shape of fig. 539; part of the edge must be cleaned with a file or scraper before using it.

The experiments admit of many modifications, of which only the most important can be described.

*Fig.* 539.

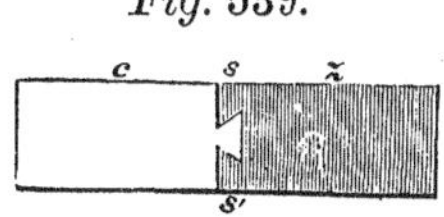


(1) Lay the insulated zinc and copper plates together, and separate them again, keeping them parallel. Bring one of them in contact with the lower plate of the condenser, while the finger is held on the cover, discharge the other plate also by touching it, and lay the two together again, etc. At each separation, an adhesion will be felt if the plates are good. After from 5 to 20 repetitions of this process, remove the cover of the condenser, and the electrometer will show the electricity communicated to it. To prevent deception by the electricity of the human body, a conducting wire of the same metal may be attached to the wire projecting from the cover, fig. 530, and pushed off with a glass rod when the electrometer is charged. If the condenser be touched with the copper plate, the former may be made of brass; but for contact with the zinc plate, the base of the condenser must also be of zinc. As it is more convenient to change the cover than the base, the cover may be made of zinc, and the electricity of the zinc plate communicated to it; but in this case the electrometer indicates the contrary electricity on raising the cover.

With plates of 2 inches diameter and upwards, and a gold-leaf electrometer, no condenser is needed; the gold leaves diverge more and more perceptibly when the contact is repeated a few times. This is especially easy with Andriessen's electrometer, fig. 454, by touching the knob connected with the gold leaves with one plate, and that connected with the curved wire with the other. With a delicate Bohnenberger's electrometer a single contact is sufficient. The same is true with Dellmann's electrometer, fig. 455. This leads to the second mode of making the experiment. When the condenser itself is made of zinc and copper plates, touch the base and the cover each with the like metal, and repeat this several times without touching either the condenser or the exciting plates with the fingers. This experiment gives the most decided results, and succeeds, though less certainly, even when the condenser is of brass.

(2) Screw one of the plates on the condenser and lay the other upon it. Whenever the upper plate is raised by the insulated handle, the electrometer will indicate the electricity of the lower plate. This experiment serves at the same time to show the condition of the electricity during the contact: it requires plates at least 2 to 3 inches in diameter.

This experiment requires a very sensitive electrometer, and will hardly succeed with a simple gold-leaf electrometer; whereas the former experiment is easier, because a double condensation occurs, and gives surer and more decided results. By using the gold-leaf electrometer with the induction wire, fig. 453, and imparting electricity to the latter, so that the gold leaves diverge from induced electricity, and then, after touching the zinc plate screwed to it, laying on the copper plate and removing it, the increased or diminished divergence of the gold leaves will indicate the electricity. With Dellmann's electrometer, electricity is first imparted to the conductor after the two plates are set on, so that the needle is repelled but little. By removing one plate the repulsion will be increased or diminished, according to the kind of electricity imparted and the plate which is screwed on. A delicate electrometer, with Zamboni's pile, is the simplest of all for the purpose; it is only necessary to screw on one plate and make the experiment without further preparation.

(3) Touch the brass part of the condenser with a piece of brass or copper held in the hand, or touch the zinc part of the condenser with a piece of zinc in the same way, while the other plate is touched with the finger. If the condenser was perfectly neutral before, it will show no electricity after this contact. But touch the brass part of the condenser with zinc, or the zinc part with brass or copper, and electricity will be manifested. For this also a very delicate electrometer; is required, and disturbances may be caused by contact with the fingers.

(4) Touch either the base or the top of the condenser with one end of the compound plate, fig. 539, holding the other end in one hand and touching the condenser with the other hand. The end held in the hand must be of different metal from the part of the condenser touched with the other end. The contact need not be continued long, but the surfaces in contact must be clean. As only a single condensation takes place, the effect is less in this experiment than in No. 1, and about like No. 2 and No. 3. To produce a stronger action, two condensers must be used. (See "Condensers.")

(5) The two plates used for the second experiment will also serve for some experiments with fluids, by laying between them a broad, very thin plate of glass, and spreading the fluid over the surface of this, or moistening with it a disk of paper laid on the glass, just the size of the lower plate. If the lower plate is of zinc, connect it with the fluid for a few moments by a strip of zinc bent into proper shape, which must be scraped clean each time, and can be opened and shut like a clamp, by means of two sealing-wax handles; lift the glass by its projecting edge, and the electrometer will show the negative electricity of the zinc, if the liquid was acidulated or pure water. This experiment is surer by using two condensers, and charging the second with the first, as already described. The second condenser may be of brass.

It should be remarked finally, that all these experiments give certain results only in favorable, dry weather; a condition which applies equally to all experiments on electrical tension, especially when dealing with very feeble electricity, as in this case.

[277] To show that electricity is liberated likewise by contact between liquids and metals, the chemical and magnetic effects of the galvanic current may be employed in anticipation of their full illustration. The experiment is made by mixing dilute nitric acid in a tumbler with a little sulphuric acid, and placing in it an amalgamated zinc plate bent over above, with a folded blotting-paper soaked in solution of iodide of potassium laid on the bend. Dip a platina plate, with a platina wire soldered to it, into the liquid, and bend the wire over against the paper, as in fig. 540: a brown spot will immediately appear on the paper in consequence of the separation of iodine from the compound. Pass the current from the platina previously through a galvanometer; the electrical current will be indicated at the instant of contact with the paper by a corresponding deflection of the needle.

*Fig.* 540.

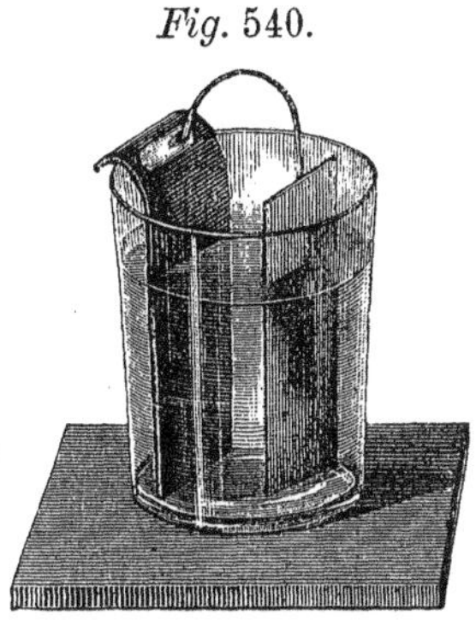

[278] **The galvanic pile.**—For actual use the form of the pile in galvanic apparatus is long ago obsolete; it is, however, a standing article in all text-books, and valuable both from theoretical and historical considerations, and therefore indispensable in all collections of physical apparatus. This is the more true, because other more powerful arrangements have not usually so many elements, so that they exhibit the phenomena of tension in a less marked manner than the pile. For this reason, however, small plates may be chosen in procuring a pile, and the number of pairs not exceed 50 to 100.

The single plates are cut out of sheet copper, about 1 millimeter thick, hammered flat, and rounded accurately with a file; and the same with the zinc plates. The latter must, however, be made of metal about 1 line thick, because they wear out faster than the copper plates, both by the action of the pile and by frequent cleaning. Zinc plate of this thickness cannot, however, be cut easily with the shears. The best plan is, to divide the sheet of zinc into squares, about a line longer on the side than the required diameter of the plates. Scratch in the division lines deeply with a steel tool, (made of a worn-out file of good stuff.) Then lay a drop of mercury on the beginning of one of the lines, and draw a sharp-pointed stick, dipped in muriatic acid, slowly along the line: the mercury will amalgamate the zinc along the track of the stick, so that it will break easily in this line. This process is convenient for cutting thick zinc plate for any purpose. The further rounding of the plates is easily effected with a coarse, deep-cut file, after cutting off the corners with a chisel.

The copper and zinc plates must now be soldered in pairs, which, besides increasing the action, saves nearly half the labor in cleaning them. They must, however, be soldered over the whole surface, and not merely around the edges. For this purpose, clean the plates on one side with the file and scraper, strew them with powdered rosin, or moisten them with sal ammoniac, or the soldering liquid, (§ 42,) then heat them, and when a piece of soft solder melts on them, distribute it over the surface with a soldering iron. From 3 to 6 such plates may be heated at once on a piece of sheet iron, but the copper plates must be heated separately, as they require more heat than the zinc. When all the plates are tinned in this way, strew a little rosin over their surfaces, and hold each pair successively in a small pair of smith's tongs over the fire until the solder melts, which takes place very soon, and press them together, to squeeze out the excess of solder; clean the edges of the plate finally with a file.

Short copper wires should be soldered to several of the couples, or projecting ears left on some of the plates, for the purpose of attaching conducting wires. These couples should be distributed at proper intervals in the pile.

The moist conductors may be disks of woolen cloth or pasteboard, about 2 lines less in diameter than the plates, saturated with a solution of sal ammoniac or salt, or sulphuric acid diluted to $\frac{1}{20}$. Pasteboard has the advantage of not being so compressible as cloth, and can, for this reason, be kept moister, the liquid not being pressed out by the weight of the plates. The liquid must never be allowed to trickle down the pile, and whenever it is seen to do so it should be wiped off with blotting-paper. The disks, after they are saturated with liquid, (which for pasteboard requires about an hour,) must be squeezed out, so that the pressure of the plates above will not press out any more liquid; the upper disks can be made wetter than the lower ones.

Cloth disks may be cut with the scissors by a tin pattern. Pasteboard becomes so tender, after repeated use with acids, as to fall to pieces; it is well, therefore, to have a punch with which to cut the disks. Pasteboard must be selected which is easily penetrated by the liquid.

The pile may be built up in a frame like fig. 541, in which three rods *B B B* are fastened upright in the block *A*, at such a distance from each other as just to inclose the plates. A barometer tube is half sunk in a groove on the inside of each rod, and fastened with sealing-wax, so that the plates will not come in contact with the wood. The rods are connected above by the block *C C*, and fastened with wooden pins. A wooden screw *D* works through this block, by which a moderate pressure can be applied to the pile.

*Fig.* 541.

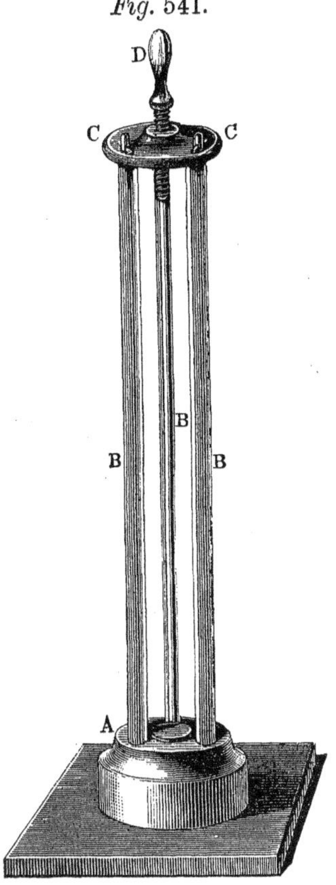


If one is not disposed to incur the expense of such a stand, a simple substitute may be made by cementing three barometer tubes at proper distances, into any block of wood; these tubes may be held together above by a triangle of brass wire, like fig. 542. A half-pound weight may be used to press down the upper plate.

In building up the pile, lay first a couple of glass plates, or a disk of sealing-wax, on the base *A*, so as to make a foundation about ½ inch thick, and on this lay one of the pole-pairs. The cloths must be squeezed out by hand to the requisite degree, which is soon learned by experience, and the plates laid on by some one with dry hands. The series is terminated by another of the pairs with coupling wires, and a disk of glass or sealing-wax laid over all.

The pile should be taken down immediately after use, the pasteboards laid out to dry, and the plates laid in water preparatory to cleaning them. To facilitate the cleansing, drive three pegs into a board in a triangular position, so as to hold the plates fast between them without projecting above. Scrub each plate with a bit of wood and fine sand, and lay it in water until all are ready; then wipe them perfectly dry. Pile up the clean plates in the frame, laying copper on copper and zinc on zinc, which will prevent rusting. Keep the dry pasteboards by themselves.

*Fig.* 542

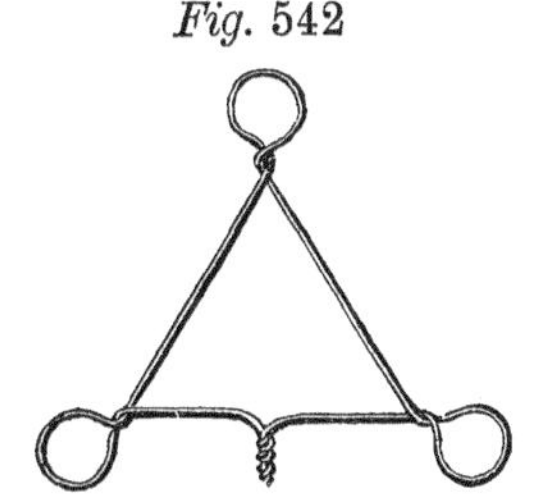

Besides the shocks given by the pile, the decomposition of water and its magnetic effects should be shown.

[279] **Binding screws and mercury cups.**—The electricity of the galvanic battery being of feeble intensity, mere contact is not sufficient to establish a communication between the conductors, as with frictional electricity. A more intimate connection is produced by dipping the ends of the conductors in little cups containing mercury, or pressing them together by binding screws.

Mercury cups may be made by simply boring in a little block of wood, with a center-bit, a hole ½ an inch deep and wide. It may be varnished internally with sealing-wax, but it is not necessary. It is well to have in the bottom of the cup two small holes, in which to insert the ends of the wires, or little hooks in the side to which they can be fastened. Common iron thimbles set in wood answer better, especially when there are several cups in the same board. Each cup can then be taken out separately and emptied, thus avoiding a loss of mercury. This direct loss of mercury and the deterioration caused by the solution of the metals dipped into it, make mercury cups, at the present price of mercury, dearer than binding screws. The latter have the further advantage of allowing the whole to be moved without disconnecting the conductors, and not requiring the ends of the wires to be amalgamated. The amalgamation is effected by dipping the metal into nitric or sulphuric acid and then into mercury, and rubbing it over the surface.

Binding screws are always preferable, except where a sudden interruption and renewal of the contact is required, or where the parts must have a motion independent of each other.

Besides the binding screws which are attached to the various pieces of apparatus, a supply of separate ones must be on hand, whose shape depends on the special use for which they are required, whether to connect plate with plate, plate with wire, or wires with each other. The screws should be thick and well made, with a moderately close, deep thread. Their construction has already been described in § 38.

*Fig.* 543.

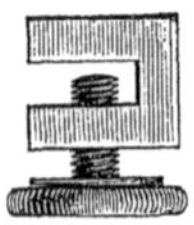

*Fig.* 544.

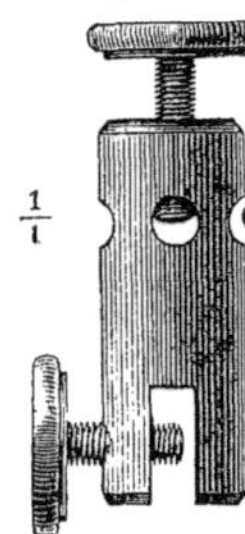


To connect plates or ribbons, simple iron screws, of the form and size represented in fig. 543, are used. To connect plates with wires, they must have the form shown in fig. 544; the screw *a* holds the plate, and *b* the wire which is thrust through one of the holes *c*, bored at right angles to each other. The holes must be large enough to admit stout wires, and the hole for the screw must not be cut deeper than their upper edge. When the hole extends lower, thin wires are forced into it and bent, and finally broken off. Wires may be connected by binding screws like fig. 545; but the simpler form, represented in section and in natural size in fig. 546, is sufficient for all purposes. Nörremberg's wire clamps, figs. 547 and 548, are simpler still. They are made of bright hard-drawn wire, $1\frac{1}{2}$ to 2 millimeters thick; silvered copper wire is best.

*Fig.* 545. *Fig.* 546.

Binding screws designed to remain permanently on an apparatus are often made in the form of fig. 549. The wire from the apparatus is bent

*Fig.* 547. *Fig.* 548.

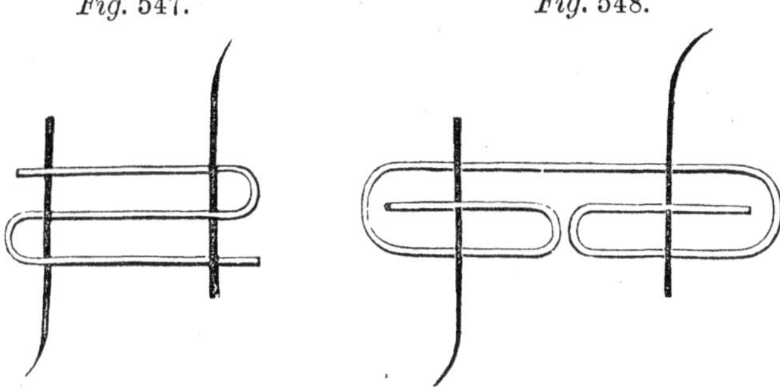

into an ear and hammered flat, and screwed in either at *a b* or *c d*, the wire from the electrometer being secured by the screw *f*. To save room, the wire and the nut *e* may be let into the wood, in which case it is made in the form of fig. 550. The ends of the wires should be cleaned with a file or scraper; the screws must be always turned very tight, especially with brass wires. The effect of the tight pressure becomes evident when the electrometers are inserted. For this reason, the heads of the screws should not be made too small.

*Fig.* 549.

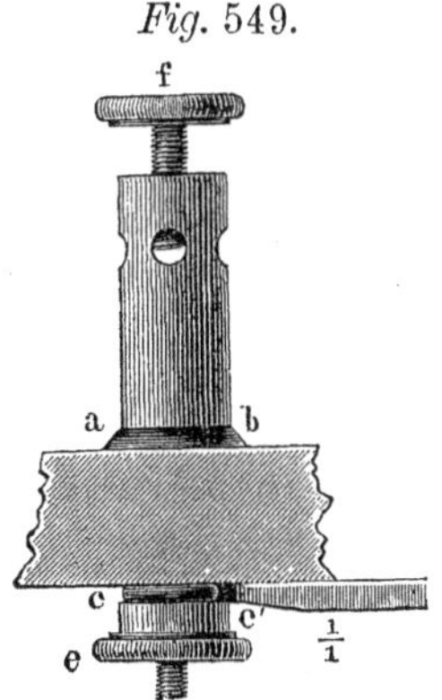


*Fig.* 550.

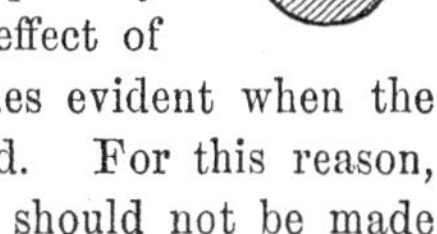

[280] **Batteries.**—Galvanic piles being very difficult to use, other more convenient arrangements have been substituted for them. Those of most common use are modifications of Volta's cups. The

simplest mode of arranging them, is to cast balls of zinc on copper wires, in a bullet mould, then bend the wires as shown in fig. 551, and place them in common tumblers, so that the elasticity of the wire will hold each pair of glasses together. The tumblers are partly filled with the usual

*Fig.* 551.

*Fig.* 552.

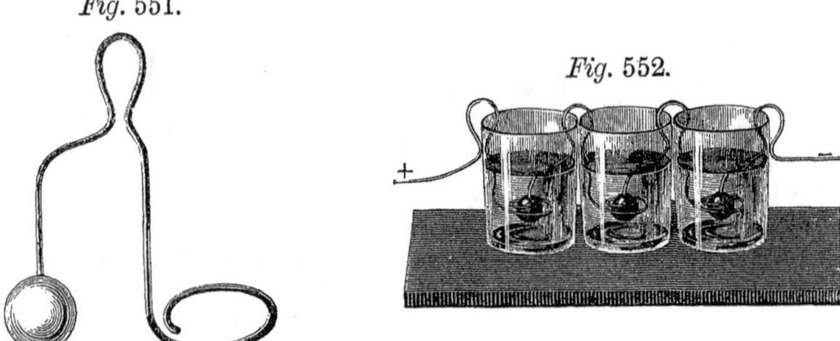


exciting liquid, water acidulated with $\frac{1}{20}$ to $\frac{1}{10}$ part sulphuric, and $\frac{1}{30}$ to $\frac{1}{20}$ of nitric acid. Twenty or thirty pairs are sufficient to give shocks. Fig. 552 shows the arrangement of the apparatus. A simple copper wire forms the positive pole. To determine which is the positive pole in any pile-like arrangement, we have only to consider the complete couples; it is always on the zinc side, as in the pile. Wollaston's modification of this apparatus is much more convenient when greater quantity is required.

To construct Wollaston's apparatus, cut plates of copper, $\frac{1}{2}$ to 1 millimeter thick, like fig. 553, with 12 to 15 square inches surface. Two

*Fig.* 553.

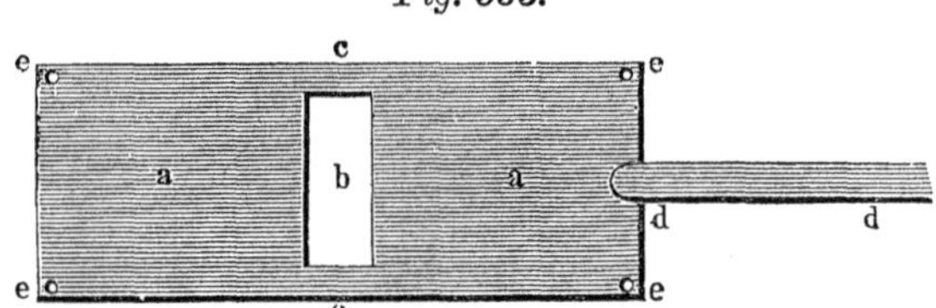


strips *c c* are left to connect each pair of plates. A narrow strip *d d*, of somewhat thicker copper, is soldered with tin to one of these. The zinc poles *a a* are made of the same size but somewhat thicker. The strips *d d* are made all of the same length, and a zinc plate soldered to one end of each; one copper plate is left uncoupled, and likewise a zinc plate. Each pair is then bent as shown in fig. 554, and the zinc plate of the succeeding pair laid between the two.

The zinc is separated from the copper by blocks of wood, represented in natural size, fig. 555, which are pushed under the lower edge. Two similar blocks are laid above, and the copper plates tied over these by wires inserted through the holes *e e e e*. Fig. 556 represents one of these elements.

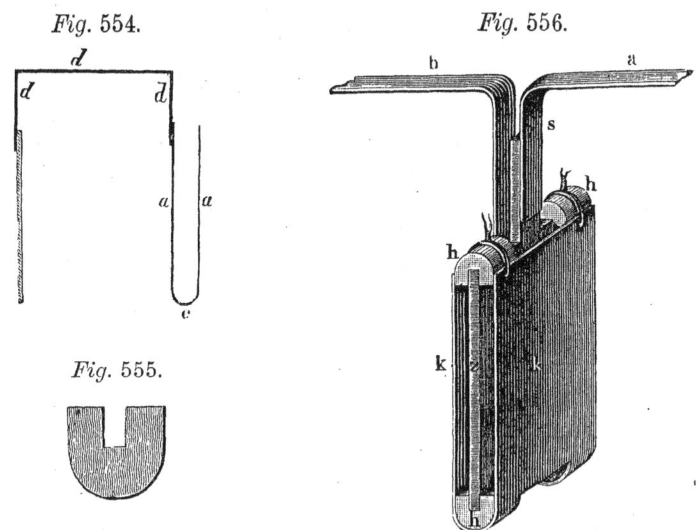


*Fig.* 554. *Fig.* 555. *Fig.* 556.

After thus combining the plates, each pair is fastened with wooden screws through the strip *a b* to a bar of oak, about 1 inch square, as represented in fig. 557. It is evident that the length of the strips *a b* must depend on the size of the glasses. The apparatus would be more compact if the glasses were square, and only about 1½ inches wide. Ten

*Fig.* 557.

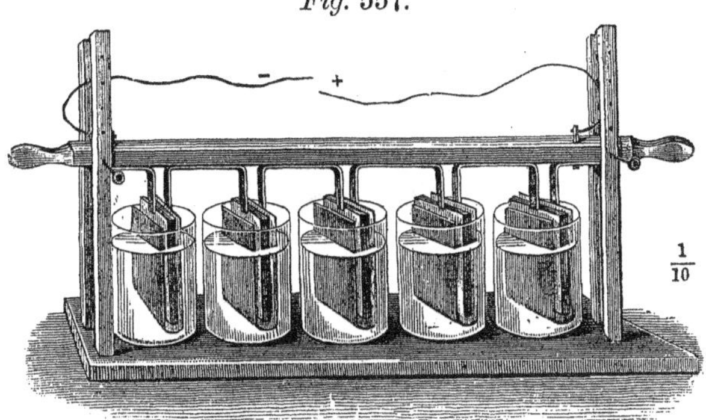


or twelve pairs are fastened to one bar, and the copper strip of the first copper plate and the last zinc plate thrust through the bar, so that conducting wires can be more conveniently attached to them.

The glasses are placed on a wooden stand, so that all the plates may be lifted at the same time and supported above the glasses. Fig. 557 represents an arrangement of five cells.

By connecting the first copper of one such series with the last zinc of another, we obtain a battery of as many pairs as there are cells; by combining the uncoupled copper plates of each series, and likewise the uncoupled zinc plates, we produce the same effect as with a single series of plates of twice the size.

The advantage of this arrangement is, that the plates can be at once removed from the liquid when not in use, and thus avoid the waste of zinc; and as they are all immersed simultaneously, the first action, which is the most powerful, is not lost. The power of such a battery remains tolerably constant after the first violent action is over. The plates should

*Fig.* 558.

be washed off with water after being used; if any further cleansing is necessary, the apparatus must be taken apart.

A trough of well-glazed earthenware, divided into cells, may be substi tuted for the separate glasses. A battery of 6 to 12 pairs is made very compact, and needs no stand; it is always ready for use, is easily cleaned, and will last for many years.

Muench has contrived a very good apparatus for constant use, which resembles in some respects Wollaston's, and in others Faraday's arrangement. It is represented in fig. 558, and consists of 6 copper and 4 zinc plates, set in separate grooves in two wooden frames, made to slide up and down on a rod. The plates dip into two glasses, in which they can be immersed to any depth required. The form of the plates is shown in fig. 559, and each, with the exception of one zinc plate, has a strip of copper soldered to it, to which binding screws may be attached. Fig. 560

*Fig.* 559. *Fig.* 560.

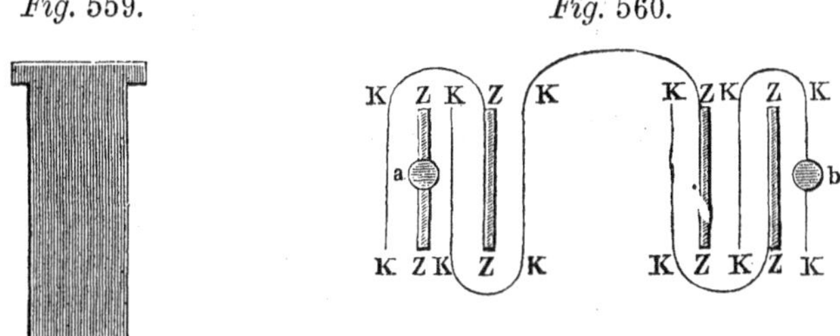


shows the arrangement of the plates, the poles being at *a* and *b*. Each plate may be taken out separately and cleaned or amalgamated.

[281] **Hare's calorimoter.**—This apparatus, even with a surface of only 1 or 2 square feet, gives a very powerful current for a short time. Its comparative cheapness and convenience are great recommendations. It need only be placed in the tub to produce a powerful current, and is easily cleaned by pouring water through it.

A large glass jar is sufficient to contain an apparatus of moderate dimensions, which is constructed as follows: The plates of zinc and copper, which are about the same thickness as for Wollaston's elements, are made as wide as the depth of the jar will admit, and 3 feet long, for a vessel 4 inches in diameter. The copper being on the outside is made a little longer. Any excess can easily be cut off with the shears. Lay between the sheets of zinc and copper, and upon the latter, strips of soft pasteboard, about 1 line thick and as long as the plates, and clamp the pack between two iron bars, about 1 inch wide and 3 to 5 lines thick, fastened by a hand-vice at each end. After heating the metal to a moderate degree on the side thus fastened, beat the plates into a coil around the bars with a wooden mallet, which can be done by a tinner or coppersmith in a few minutes. The inmost turn will not be quite round, but that is of no importance. The pasteboard is then removed.

A wooden cross, fig. 561, is wedged fast to the end of a round bar of proper thickness, in the upper part of which two square holes are cut to receive two other smaller bars at right angles to each other. These bars are let into each other, and one of the holes made twice as high as the other to admit them; they are fastened by a wedge. The distance between these crosses must be 2 or 3 lines less than the width of the coil.

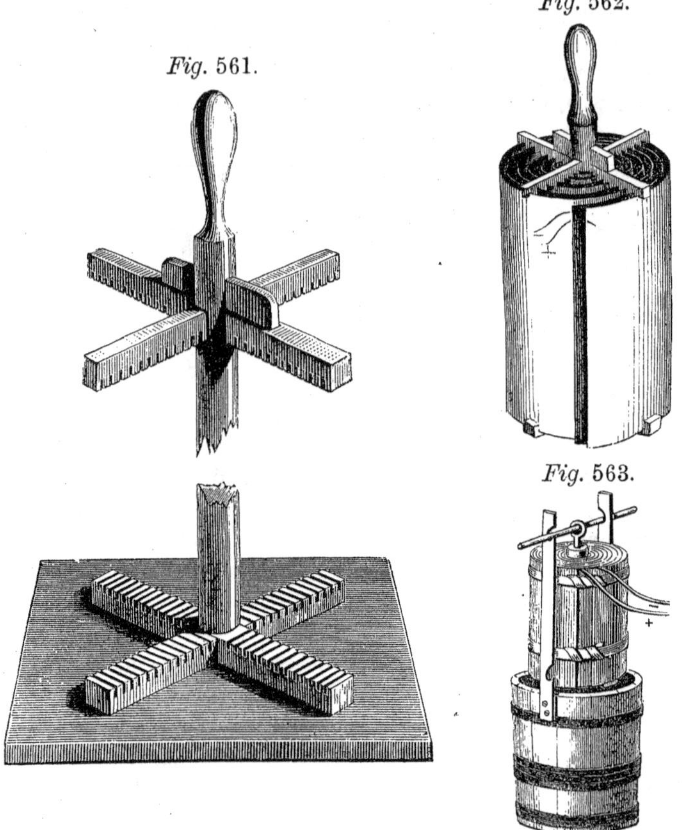

*Fig.* 562.

*Fig.* 561.

*Fig.* 563.

This is placed between them, and the places where the plates touch the wood marked; grooves are afterward cut in these places to keep the plates apart. The plates can at any time be taken apart to amalgamate the zinc.* The poles are formed by stout copper wires soldered to each

* The best mode of amalgamating zinc plates is to cover some mercury with strong sulphuric acid, and pour the two together over the plate with an iron spoon. Frequent rubbing with a swab facilitates the process.—*Trans.*

plate. Fig. 562 represents the apparatus complete. For large coils, wooden tubs are used, and arranged as shown in fig. 563.

A single Wollaston's element is very convenient for similar purposes, when a strong current is not needed. A handle is fastened to the copper plate, and short copper wires soldered on as poles.

*Fig.* 564.

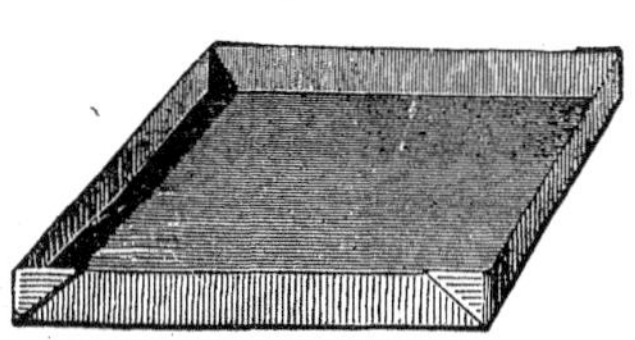

Oersted's trough apparatus is very convenient when a powerful current is required for a short time. Take a sheet of copper and bend up the sides so as to form a rim all around an inch high, as shown in fig. 564. In this lay a zinc plate on a couple of sticks, and the apparatus is complete.

[282] **Amalgamating the zinc.**—The amalgamation of the zinc not only increases the power of any galvanic apparatus, but effects a great saving of zinc. New plates are amalgamated by simply placing some mercury and diluted sulphuric or hydrochloric acid in a dish, and applying both together to the surface of the metal with a small brush or a swab. When the plates have once been amalgamated, it is only necessary to place them in the exciting liquid and pour into it a little mercury; it will soon spread over the surface. After the plates have stood for some time unused, the mercury collects in little drops; but it spreads over the surface again as soon as the plates are immersed in the exciting liquid. When the zinc is in a separate cell, it requires no more labor than simply to place some mercury in the cell. To amalgamate the Wollaston's elements described in § 280, dip them into dilute acid, lay them in a dish so that the longest edge of the zinc will be uppermost, and pour a few drops of mercury on each zinc plate from a pipette; then immerse the elements again immediately.

Mercury which has been used for amalgamating zinc must be kept by itself, as it is not fit for any other use without purification.

[283] **Zamboni's pile.**—The essential points of the construction of this pile have already been described when speaking of Bohnenberger's electrometer. Take a glass tube coated on the inside with sealing-wax,* and fill it with disks, $\frac{1}{2}$ to 1 inch in diameter. The tube must be wide enough to admit disks without crowding. The pile is most effective when

---

* Heat the tube gradually, beginning at one end, and drop bits of sealing-wax into it, it will melt and spread itself very uniformly over the surface. The tube must be turned constantly. This is the surest way to make the glass insulate well, but it makes the apparatus neater to take a tube which insulates well already and varnish it on the inside with bleached shellac.

the disks are made of silver paper, coated on the back with very finely pulverized oxide of manganese rubbed up with thin gum water, and applied very thinly. If the disks rub against the glass in going in, they leave a fine metallic coating, which almost destroys the effect of the pile. The tube needs no mounting; it may be closed with dry corks, through which slightly tapering, smooth wires are thrust, the ends of which press on brass plates laid over the disks of paper. The friction on these wires keeps the disks sufficiently compressed. The corks are afterwards coated with sealing-wax. Such a pile, consisting of upwards of 1000 pairs, causes a divergence in a delicate straw electrometer. A couple of dozen such disks, strung on a thread and tied together, give indications of electricity with a condenser. The dry pile is very convenient for showing the use of the condenser, and the difference in the tension at its poles in an insulated and uninsulated condition may be shown also. A pile of 1000 pairs may be built up on a board between three glass rods and compressed by silk strings tied below, leaving little scraps projecting between every 100 pairs, to show the gradually increasing tension.

*Fig.* 565.

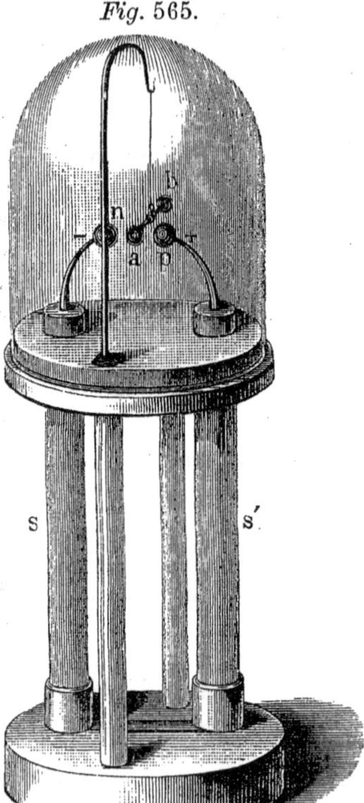


Two piles of 2000 pairs each, set on a board and connected below by a strip of foil, as in Bohnenberger's electrometer, will set in motion an electrical pendulum. The arrangement is represented in fig. 565. The upper poles pass through a round board, which rests on two pillars, and over which the pendulum is suspended from a bent wire. The pendulum is made of a light, hollow brass ball *a*, fastened to a slender glass rod; another ball is used to balance the first. The ball *a* hangs, when at rest, between the poles, and the glass rod must be long enough to allow it to reach both poles by turning on the thread as an axis. A bell glass is placed over the pendulum to keep off currents of air. Once set in motion, this pendulum will continue to oscillate for years, sometimes coming to rest and beginning again of itself.

[284] **Constant batteries.**—In order to obtain a strong galvanic current which shall remain constant for several hours, batteries must be constructed of two metals separated by a porous partition, each acted on by a separate fluid. Porous animal membranes, such as the bladder, may be employed for this purpose; but their use is attended with so many inconveniences, that they are now almost universally supplanted by cells of porous clay. Bladders are seldom used except in the galvanoplastic process, for which they are very convenient; the elements are arranged for this purpose in a peculiar way, which will be described in its proper place.

The porous clay cells are made of porcelain clay, pipe clay, or ordinary clay free from iron, with the addition of a little quartz sand. They may also be made of plaster cast in moulds. The porcelain cells are very good, and stronger than the others. Any potter can make them out of white clay and fine sand, the proper proportions of which he will learn by a few trials. They must be uniform, and as thin as possible. Plaster cells must be handled very carefully when they are wet.

The experimenter can make the cells for himself over a wooden mould on the lathe. The mould is covered with paper, and bored through parallel with the axis, in order to remove the cells more easily. They can be made more uniformly thin in this way than by the common potter, to whom the preparation of the clay and the burning should be intrusted.

Good clay cells should become quite damp on the outside in 1 minute after water is poured into them; those made of pipe clay require only 20 to 30 seconds. They must, however, be free from cracks which allow water to trickle through them; otherwise the nitric acid will penetrate to the zinc and blacken it. Imperfect cells are readily detected in this way.

To preserve them in good order, it is absolutely indispensable to wash them out each time after use, and let them lay 24 hours in a tub of clean water; otherwise they soon become so brittle as to break with the slightest jar.

*Fig.* 566.

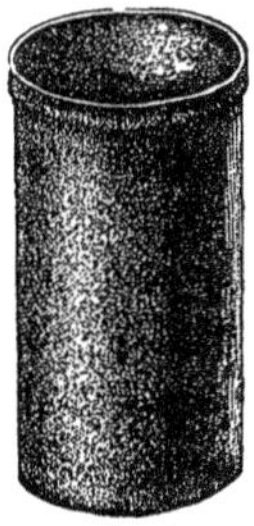

Their shape must be accurately cylindrical, fig. 566, so that the liquid may form a stratum of uniform thickness between them and the metal. Their size depends, of course, on that of the elements; but they should be just high enough to allow their thickened rim to project above the metal. The best cells which the author has seen, in respect to the resistance opposed to the current, were made by Stöhrer, in Leipsic. They far excel the Paris porcelain cells.

[285] **Daniell's constant battery.** — This consists of copper and zinc. The copper is acted on by a solution of

sulphate of copper, and the zinc by dilute sulphuric acid. It is indifferent which metal is placed on the outside, that is, which has the greater surface. If a Bunsen's battery is also at hand, the zincs should be so arranged that they can be used for either.

Large tumblers are used for the cells. The zinc and copper are both bent into a cylindrical shape, so that both will approach as near as possible to the clay cells. The zinc must be heated to between 200° and 300° F. before bending it, otherwise it will break. It is afterward amalgamated. A great saving of zinc is effected by coating the side which is turned away from the copper, with a varnish of asphaltum dissolved in spirits of turpentine. Strips of copper are soldered to both metals, as shown at *a*, fig. 567, to which binding screws may be attached. The

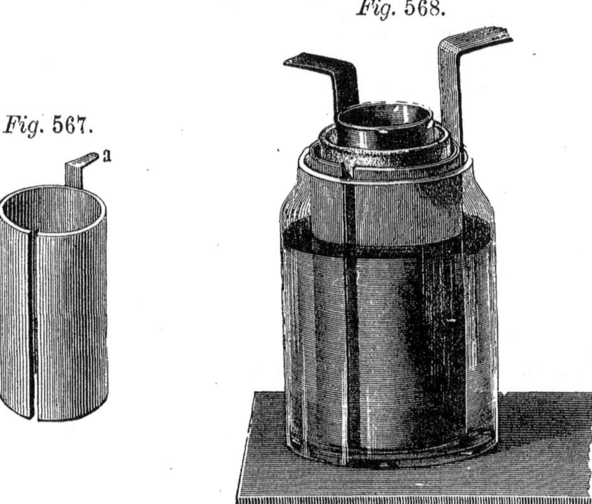


*Fig.* 567. *Fig.* 568.

outer metal is usually made to project a finger's breadth above the glass, and the inner the same distance above the clay cell. Fig. 568 represents a Daniell's element complete.

As the uniform action of this battery depends upon the constant saturation of the solution of copper, it must always contain an excess of crystals of sulphate of copper. If this is simply thrown in the bottom of the cell, only the lower stratum becomes saturated; the vitriol must be hung in a gauze bag in the upper part of the vessel. The action of this battery is very constant; but it is so much weakened by its specific resistance, that it is not nearly so powerful as Bunsen's zinc-carbon battery of the same size. On the other hand, it has the advantages of greater cheapness, more permanent action, and freedom from nitrous acid vapors. For the reason last mentioned, it is especially adapted for use in workshops, etc., where the acid vapors would be injurious. The zinc may be acted on by a solution of common salt instead of sulphuric acid; a bag of the salt must then be hung in the solution, but the substitute is not very good. When the elements are kept a long time, several weeks, in constant action, a part of the old liquid should be withdrawn by a siphon, and replaced by fresh. This applies especially to zinc.

The elements may be so connected, as either to form a battery or a single element of large surface. In the former case, the copper of one element is connected by binding screws to the zinc of the next, and the elements so arranged as to bring the two poles as close together as possible. In the latter case, all the copper plates are cemented with one strip, and all the zincs with another. The cells should be arranged in a tray, fig. 569, just large enough to contain them.

*Fig.* 569.

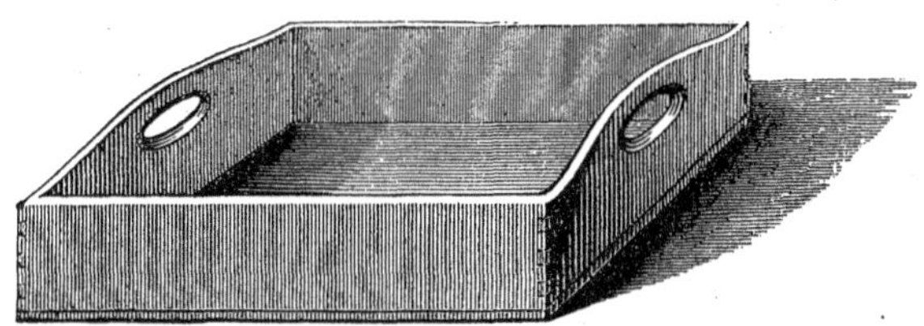

[286] **Grove's battery.**—This has platinum instead of copper, which is acted on by concentrated nitric acid, but it is arranged in other respects like Daniell's battery. The zinc is acted on by sulphuric acid diluted to $\frac{1}{10}$ or $\frac{1}{5}$. To produce very strong currents the sulphuric acid is used stronger, even as high as one-half; more depends upon this than upon the strength of the nitric acid. This is the most powerful combination known, in proportion to the size. Its first cost is considerable, but not so great as might be supposed, when made at home. The platina is used in the form of very thin foil, as it depends only on the surface, and 40 or 50 square inches of such foil need not cost over four or five dollars, $\frac{5}{6}$ of which is the permanent value of the metal. Such very thin foil requires care in handling, which must be observed, in fact, with any apparatus. Varrentrapp advises to bend this thin foil around a piece of lamp-chimney, and tie it above and below with fine platina wire; one corner of the foil is bent over to the inside of the cylinder, and a piece of thicker foil laid on it. This foil is soldered to copper and varnished all over with asphaltum, except where it is in contact with the thinner foil. The two pieces of

foil are pressed together by a cork fitted in air tight, and varnished. The nitric acid consumed costs rather more than the sulphate of copper required to produce a current of equal strength with Daniell's battery. When the foil is thicker, it is better to retain the arrangement described in the preceding paragraph for Daniell's battery, substituting platina for copper. It is still better to fasten the foil to a wooden cover fitting on the clay cell. Leave a piece projecting from the middle of the foil, fig. 570, to which rivet a double slip of copper of the same width. Make a slit through the middle of the cover, and from this cut an S-shaped groove, about $\frac{1}{2}$ a line deep, fig. 571. Pass the strip of copper through the slit, bend the foil into the shape of an S, and insert it in the groove, fig. 572.

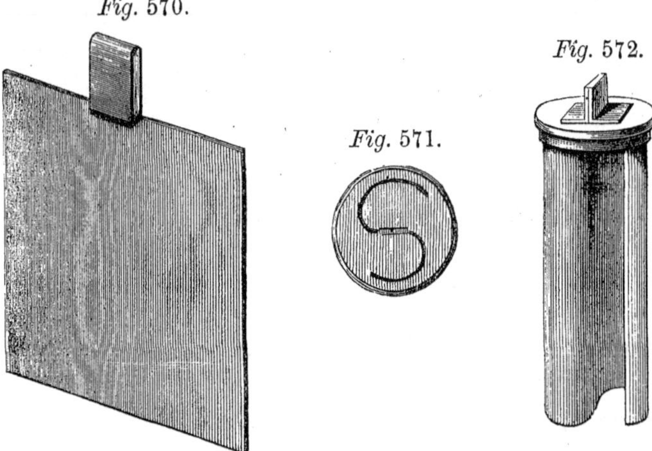
*Fig.* 570. *Fig.* 571. *Fig.* 572.

Finally, fill the lower side of the cover with melted rosin, to keep the foil in place. The cover serves to retain the nitrous acid vapors partly in the cell. The platina foil may have from 6 to 20 square inches surface. Fig. 573 represents a Grove's element complete. Another form of the same battery is represented in fig. 574. In this the cell is a rectangular trough, around which the zinc is bent. The platina foil is straight, and fixed to a cover as before. The elements may be placed in wooden troughs coated with pitch. If the zinc is bent twice at a right angle, the succeeding platina foil may be attached directly to it, at *d*.

The clay cells of these elements must be soaked well after use, and the foil washed with water. Each foil should be kept in its cell.

[287] **Bunsen's zinc-carbon battery.** — In this battery the platina is replaced by carbon. The exciting liquids are the same as in Grove's battery. The cylinders of carbon require a special preparation.

They are made as follows: Take two parts by weight of coke and one part bituminous coal, pulverize and sift them. The powder must then be placed

*Fig.* 573.

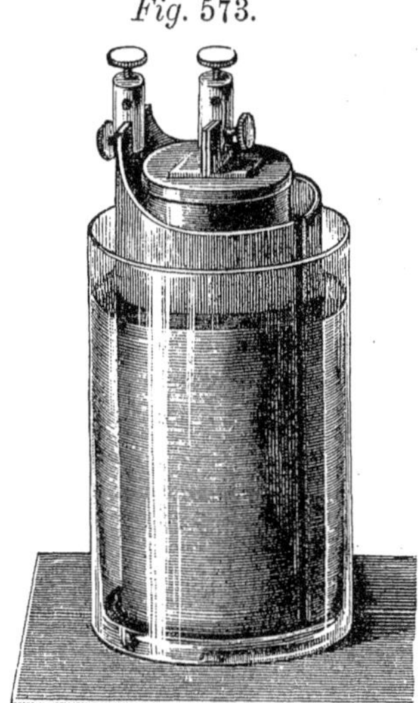

*Fig.* 574.

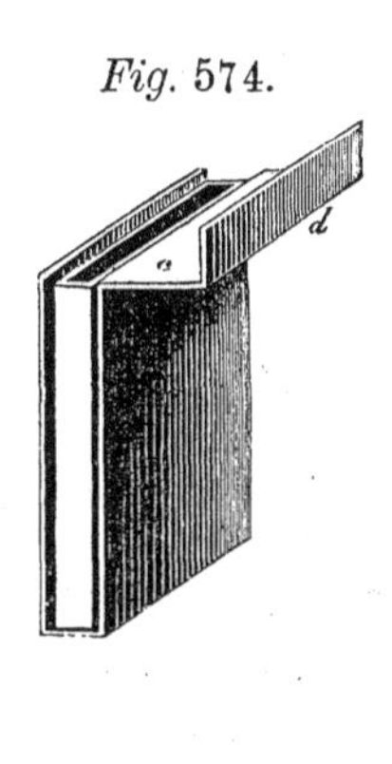


in a mould of sheet iron, which is a little wider and about an inch higher than the glass in which they are to be used. This mould is not riveted together, but simply the edges turned up to the width of $\frac{1}{2}$ an inch and clamped together, fig. 575. The mould is closed above and below by pieces bent up in the same manner and pushed in. After one end is closed, place in the mould a cylinder of pasteboard, so as to leave a space of about 3 centimeters between it and the mould, and fill this space with the coal dust shaken together but not compressed; then close the other end, and plaster the joints with clay. Heat a number of moulds filled in this way to a red heat in a furnace, and keep up the heat until the evolution of carbureted-hydrogen gas ceases, which requires about $\frac{3}{4}$ of an hour. After cooling, the coal will be firm enough to be turned on a lathe. Turn them off to the shape of fig. 576. The length of the narrower part of the cylinder must be a little less than the height of the glass, so that

*Fig.* 575.

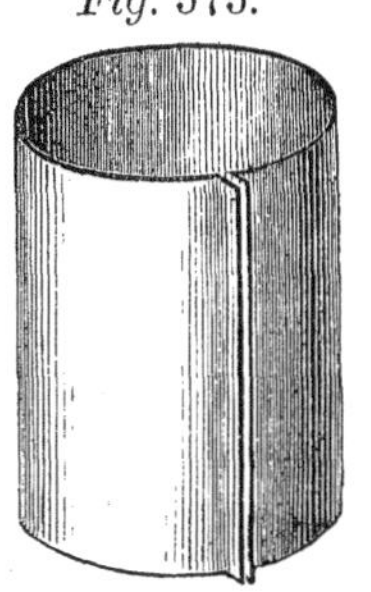

*Fig.* 576.

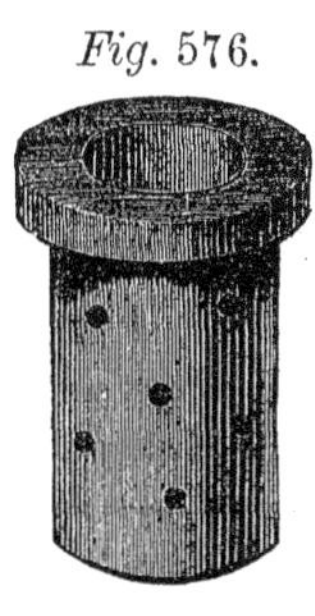

the broad rim will rest on the glass and cover it. The smaller part must slip easily into the glass. The inside must also be turned off until the smaller part is only 5 or 6 millimeters thick. The coal cylinders are fastened on the lathe by a wooden chuck with a wide slit and a sliding iron ring, an arrangement familiar to every turner.

After this, saturate the cylinders with molasses or coal-tar, and dry them again. When dry, place them in a crucible among charcoal powder, lute on a clay cover, and burn them in a potter's furnace.

After the cylinders are burnt, bore a few holes, about 5 millimeters in diameter, obliquely downward through them; then heat the broad rim and saturate it with melted wax, to prevent its absorbing nitric acid. A copper ring, either riveted or hard soldered, with a projecting ear, fig. 577, is then to be driven over the rim of the cylinder with a few gentle blows of a hammer. When the ring fits well, and the cylinder is held in the hand, this can be done without injury. It is better to tighten the ring by a binding screw, fig. 578, so that it may be removed and cleaned after

*Fig.* 577. *Fig.* 578.

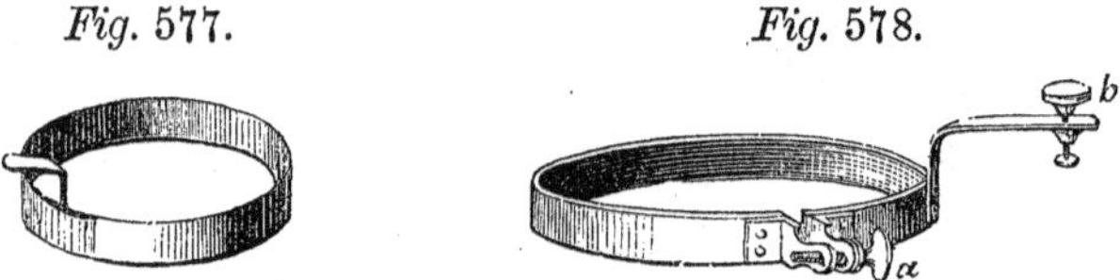


use. It is advisable, when the ring is screwed together, to lay under it a strip of wire gauze of equal width, to increase the points of contact, and then varnish the ring and the upper part of the cylinder with sealing-wax. The same varnish should be applied when the ring is driven on.

The carbon is sometimes placed in the clay cell and the zinc made the outside element. In this case, the carbon is in the form of a straight cylinder, with a copper ring driven on one end. It is not necessary to gild this ring to protect it from the action of nitric acid; saturation with wax is sufficient to prevent the acid being drawn up, without injuring the conducting power of the coal. A coat of varnish will protect the ring from acid accidently dropped on it. As the rings are destroyed by constant use, let them be made as they may, the simple pattern shown in fig. 577 is preferable on account of the ease with which they can be renewed.

It is not advisable to connect the zinc of one element permanently with the carbon of the next. The contact should always be made by binding screws. The form seen in fig. 579 is well adapted for this purpose. The slit in the strip connected with the zinc slips over the screw, the head of which presses upon it. When the carbon is on the outside, the jars should be contracted at the neck, so as to contain a considerable quantity of acid without exposing much surface. To obtain the maximum effect, the clay cells must fit into the carbon cylinders as closely as possible. Jars of the form of fig. 580 are made specially for Bünsen's battery. Fig. 581 represents the whole apparatus, in section.

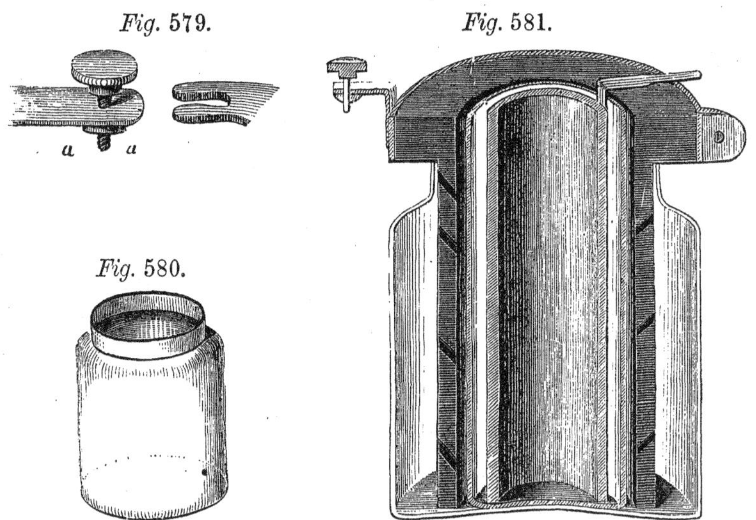


*Fig.* 579.

*Fig.* 581.

*Fig.* 580.

From the directions given for the preparation of the carbon, it will be seen that it is a tedious process. It generally requires some preliminary trials, for the proportion of the coke to the bituminous coal varies according to the quality of the latter; if it is too bituminous, the cylinders crack during the first heating; if it is not bituminous enough, they are too friable. In the former case, more coke must be used; in the latter, less. Carbon of equal density also differs very much in electro-motive power. It is, therefore, best to buy these cylinders, unless a large number of them is required. The best are made by Deleul, in Paris, and Stöhrer, in Leipsic. The former sells the whole element, including glass, very heavy zinc, porcelain cell, a copper ring like fig. 568, and binding screws, for 4 francs. Stöhrer makes more powerful elements for 3 thalers ($2 25.)

The prime cost of this battery is less than that of Grove's of equal strength. The carbon absorbs, however, a considerable quantity of nitric acid, amounting to $\frac{1}{3}$ of the whole, which is lost in the subsequent soaking. The loss by conversion into nitrous acid is not more than in Grove's battery. The inconvenience from nitrous acid vapors is about the same in both. Smoking tobacco, and especially cigars, partially obviates this inconvenience.

When the battery is used frequently at short intervals, it is an advantage to have the carbon on the outside for the clay cylinders may then be removed and the carbon left standing in the acid, covered over with a plate of glass, so that nothing is lost by soaking. Carbon recently soaked and dried yields, it is true, a stronger current for the first instant, but it soon falls to its ordinary force. To obtain a feebler current, which shall continue constant for a long time, Bunsen's battery may be charged with sulphuric acid diluted to $\frac{1}{20}$ or $\frac{1}{15}$ its volume, in both the zinc and the carbon cell; thus charged, the current will continue nearly uniform for four weeks.

[288] **The zinc-iron battery.**—The platina foil in Grove's combination may be replaced by sheet iron, provided very strong nitric acid of specific gravity 1·4 be used. Tin plate may also be substituted for zinc, and a strong current obtained even with iron in both cells. Notwithstanding the ease with which batteries can be constructed in this way, they have not come into such general use as might have been expected. The difficulty in their use is, that one must notice constantly whether the acid is strong enough not to attack the iron; but during the action the acid is constantly growing weaker, until it reaches unexpectedly the point where the iron is violently attacked, and a cloud of nitrous acid vapor is evolved.

Besides the batteries already mentioned, a multitude of others have been recommended, which, not having found their way into general use, may be passed over here.

[289] **Experiments on the tension of the pile.**—A tolerable demonstration of the law that the tension increases as the number of pairs, may be given by the aid of a delicate gold-leaf, or Dellmann's electrometer, with only 8 to 12 pairs. The pile is built up on the hand, with moistened blotting-paper, and the projecting ear of the upper plate brought in contact with that part of the condenser which consists of the same metal. By touching any part of the pile with the fingers of the hand on which it rests, only the plates above the finger come into play By increasing in this way the number of pairs, the increased divergence of the electrometer is very evident. With a very delicate electrometer, not more than 8 to 12 pairs must be used, even when the blotting-paper is moistened with simple saliva. With pasteboard the piles would be too high for this process. With 40 to 50 pairs a delicate straw electrometer may be used. Zamboni's pile of 1000 pairs affects such an electrometer without a condenser.

[290] **Physiological effects of the galvanic current.**—(1) The simplest experiment is to pass the current through any branch

of the fifth pair of nerves, which communicate with the optic nerve. The most convenient apparatus is a couple of plates of zinc and copper, not more than an inch in diameter, fig. 582, with handles. The copper may be silvered for the sake of cleanliness. Hold these plates in the mouth between the jaws and the cheeks, so that when the mouth is closed the handles will be in contact. At every contact a faint spark will be seen, especially when the eyes are closed.

(2) The same plates may be used to illustrate the curious taste produced by placing the tongue between them with the handles in contact.

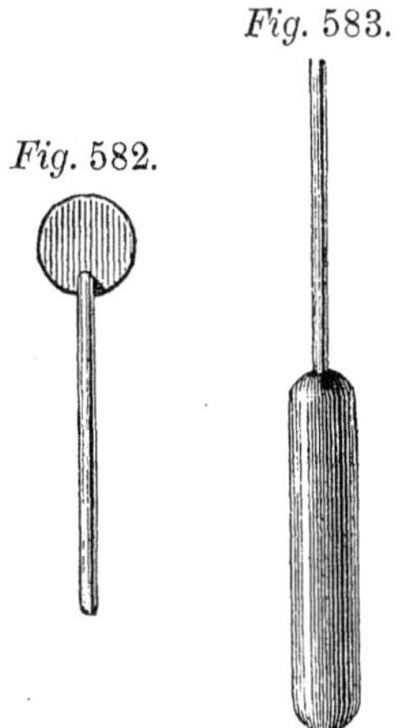
*Fig.* 582. *Fig.* 583.

(3) Lay a copper or silver coin on a plate of zinc, and on this place a leech. It will begin to creep, but always retreat as soon as it touches the zinc.

(4) A single pair is not sufficient to produce shocks in the human system, this effect depending upon the number of pairs; but with 10 or 12 pairs a slight tremor is felt, and with 40 to 50 pairs it becomes very sensible, especially when the poles are touched with metallic handles, fig. 583, held in the hands, moistened with acidulated water. These handles are made of sheet brass, and the wires should pass entirely through and be soldered to both ends, otherwise they soon break off. The cylinders may be left open at both ends, and the wire soldered to the side. With a strong pile of 50 to 60 pairs, the circuit cannot be kept closed by these handles; the twitching which ensues as soon as the circuit is closed, immediately breaks it again. When the shocks have become weaker, so that the circuit can be kept closed, the current produces a tingling sensation in the fingers and wrists. If the wires from the poles be made to terminate in two basins filled with acidulated water, the circuit may be completed by plunging the hands in these. The wires should in this case terminate in plates 4 inches square, to increase the surface of contact. The shock will pass through several persons, if they hold each other with wet hands; but it is then very much enfeebled.

[291] **The spur wheel.**—The effect of quite feeble shocks becomes intolerably great when they follow each other in quick succession. For this purpose the circuit must be rapidly closed and broken. This can be accomplished very readily by connecting a rasp with one pole and drawing

the other terminal wire rapidly across it. Any body inserted in the current will then receive shocks in rapid succession.

The interruptions may be produced with more regularity, and the interval between them regulated at pleasure, by means of the apparatus, fig. 584. It consists of an old clock wheel, the iron axis of which rests

*Fig.* 584.

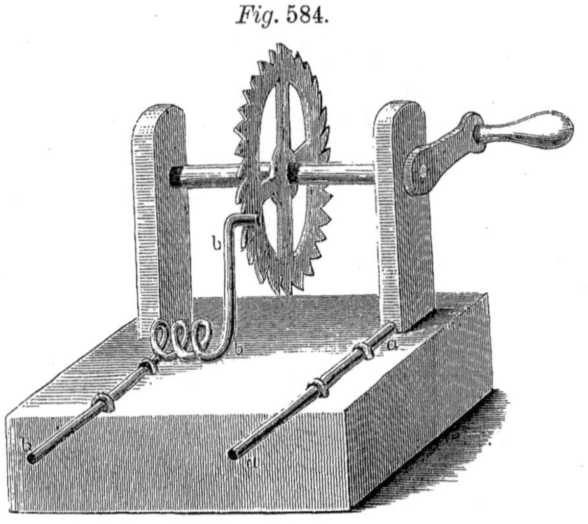

on two brass posts fixed in a board, with one of which the wire *a* is connected, while the spring *b* strikes against the teeth. As often as the spring passes a tooth the current is interrupted. The teeth may be amalgamated, to render the contact more perfect. This is done by holding a bit of paper, moistened with muriatic acid, against the teeth while the wheel is revolving, and then passing them through a cup of mercury.

Neef's apparatus for the same purpose has, instead of the notched wheel, a circle of brass, into the circumference of which pieces of boxwood or ebony are inserted at regular intervals. In this case, the spring *b* need not have so much elastic force. This apparatus is free from the clicking noise which the other makes, but the spring is apt to drag metallic particles over the surface of the wood, and thus render it a conductor.

[292] **Physical effects of the battery: sparks and calorific effects.**—Sparks of appreciable length can only be obtained with very powerful apparatus, by induction. The light evolved is generally produced by the combustion of the ends of the conducting wires, and especially of the mercury with which they are amalgamated. The sparks

produced when the current is interrupted are brilliant in proportion to the length and thickness of the conducting wires. If a part of the circuit be too thin to conduct the current, it becomes heated, even fused or volatilized. The intensity of the heat is proportioned to the thinness and shortness of the wire and the size of the plates. A single Wollaston's element of the dimensions assigned is sufficient to ignite a very fine platina wire, stretched between the poles, as shown at *p p*, in fig. 585. When the wire is very fine, and not more than $\frac{1}{4}$ to $\frac{1}{2}$ inch long, it becomes red hot as often as the element is immersed in the acid. By the method described in § 268, No. 5, it is easy to make iron wire fine enough to be ignited and melted with apparatus of moderate power. To burn a watch-spring requires a more powerful battery than a simple Wollaston's.

*Fig.* 585.

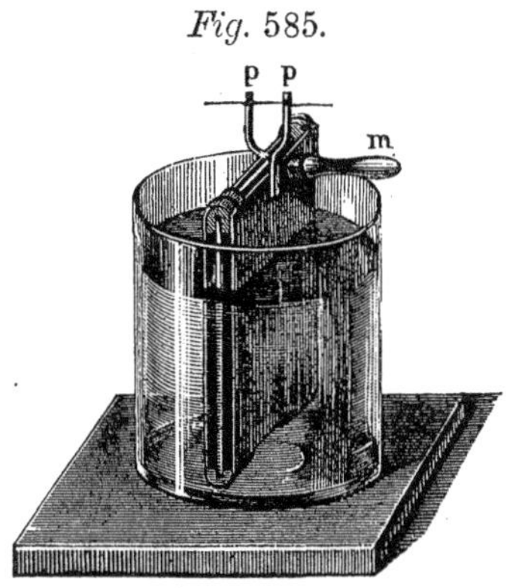


A sheet of gold or silver leaf, hung like a flag on one of the electrodes of the battery and made a part of the circuit, burns with a brilliant light; if the other electrode be passed over the foil like a knife, it burns a way for itself, so that the metal may thus be cut into strips. The same brilliant combustion is produced by connecting a file with one pole, and drawing over it a fine wire connected with the other; the wire burns with a shower of sparks. To obtain the bow of light between charcoal points requires 12 to 24 good Bunsen's elements of the size described. For this purpose, take cylindrical pieces, $\frac{1}{3}$ to $\frac{1}{2}$ an inch thick and 2 inches long, pointed at one end, and made either of the same material as the carbon elements, or of coke. Charcoal does not conduct well enough, and can, therefore, only be used with very powerful batteries.* Carbonaceous deposits are found adhering to the iron gas retorts, which are extremely dense and very suitable for this purpose. The pole wires should be wrapped around them several times to increase the surface of contact, and the points of the coal then brought together. With a battery of 40 or 50 pairs, the carbon points may be drawn apart, after they become red hot, to the distance of $\frac{1}{4}$ inch, and an arch of light of dazzling brightness obtained. Pointed wires coated with soot give a brilliant light when the circuit is closed.

* With a Bunsen's battery of 12 elements, boxwood charcoal gives very good results.—*Trans.*

As the distance between the coals must remain constant during this experiment, and the influence of the magnet on the arch should also be exhibited, a stand like fig. 586 is very convenient. Two stout metal rods fixed in a wooden base carry each a couple of sliding clamps. The conducting wires are fixed in two of these, *a a*, and the coal points, with their wires, in the other two. The wires to be melted may be held in the

*Fig.* 586.

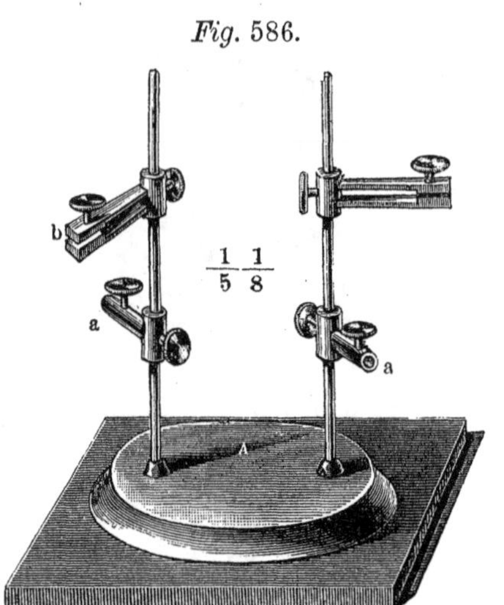


*Fig.* 587.

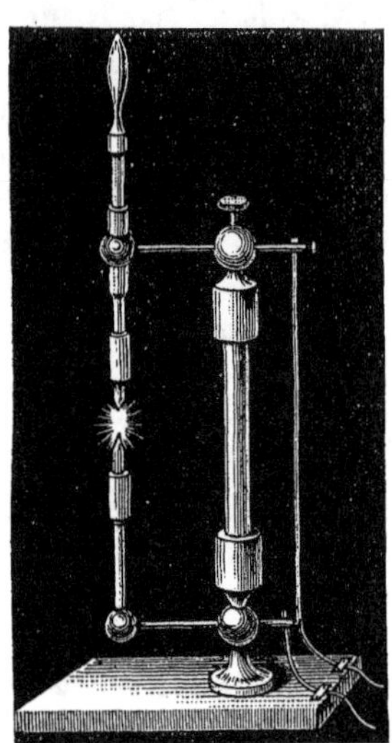

same way, the clamps being very convenient, especially for very fine wires, which are apt to slip out of the binding screws. The carbon points may be fixed at the proper distance, and a Leyden jar discharged through them to start the current, which will then continue. A special support for this experiment is shown in fig. 587. The coal is held in metallic tubes, and the upper rod slides in its socket. The middle part of the column is made of glass.

[293] **Blasting by galvanism.** — Close both ends of a small glass tube with sound corks. Through one of these thrust the bare ends of two well-insulated copper wires, so that they will project about $\frac{1}{2}$ centimeter within, and be as far apart as possible; scrape them clean and connect them by a fine, bright, iron wire. Cement the cork tightly in the tube, and, after filling it with gunpowder, cork up the other end also, and cement it. Tie the insulated wires together, wrap them again with silk as far as they are to be immersed in water, and varnish them over. These preparations must be made some weeks before the experiment, to

allow the varnish time to dry. Lay the cartridge in a tub filled with water, cover it with a board, and from a safe distance pass through the conducting wires a current sufficient to melt the small iron wire. The experiment should be made out of doors; the tub is generally burst open.

[294] **Production of cold by the galvanic current.**—Solder together a bar of bismuth and a bar of antimony-glance, each from 2 to 4 square lines, in transverse section, so as to form one continuous bar, and send a current through them; a reduction of temperature will take place at the joint when the current goes from the bismuth to the antimony-glance, and a rise when it passes the other way. The bars, with a short copper wire fastened to each, must be inserted in a glass bulb, 1 to $1\frac{1}{2}$ inches in diameter, with a long neck, like fig. 588. In soldering copper to these bars, it must first be tinned. The cementing may be done by frequent applications of thick solution of sealing-wax, as the bars cannot be made hot enough to melt sealing-wax. Plaster of Paris is too porous for the purpose. Dip the end of the tube *a* in a glass of colored water, and warm the bulb *b* with the hand, so as to expel some of the air and allow the liquid to rise in the tube. Mark the height to which it rises by a thread, and fasten the apparatus to an upright stand. Send a current, by means of a pole-changer, alternately in opposite directions

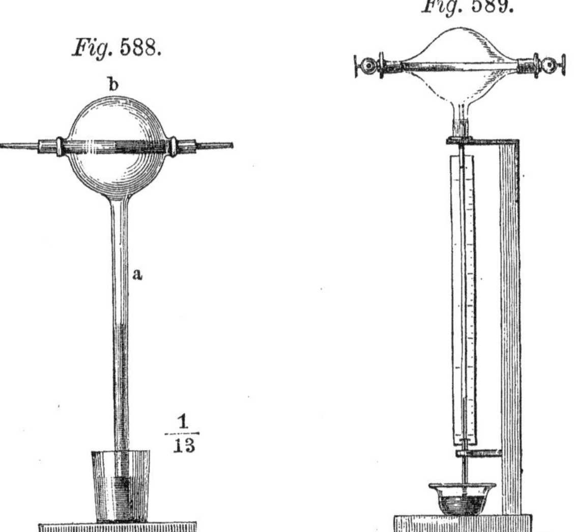


*Fig.* 588.

*Fig.* 589.

through the bars, and the thermometer will indicate rapid changes in the temperature. A strong current produces an elevation of temperature in either direction. A single Wollaston's element is very convenient for this experiment, as the quantity of liquid in the glass can be increased or diminished so as to obtain a current of suitable strength, without disturbing the apparatus. An arbitrary scale may be attached to the tube, which must be so arranged that it can be taken out and emptied of the liquid. Fig. 589 represents such an arrangement.

[295] **Chemical effects of the battery.**—*Decomposition of water.* A very simple apparatus for this purpose is represented in fig. 590. Bore two small holes in the bottom of an ordinary wineglass, with a steel drill and spirits of turpentine; close them with good corks, and thrust platina wires through the corks. Twist a wire around the stem of the glass and bend the two ends into hooks above, upon which hang two glass tubes closed at one end, at such a height that the ends of the

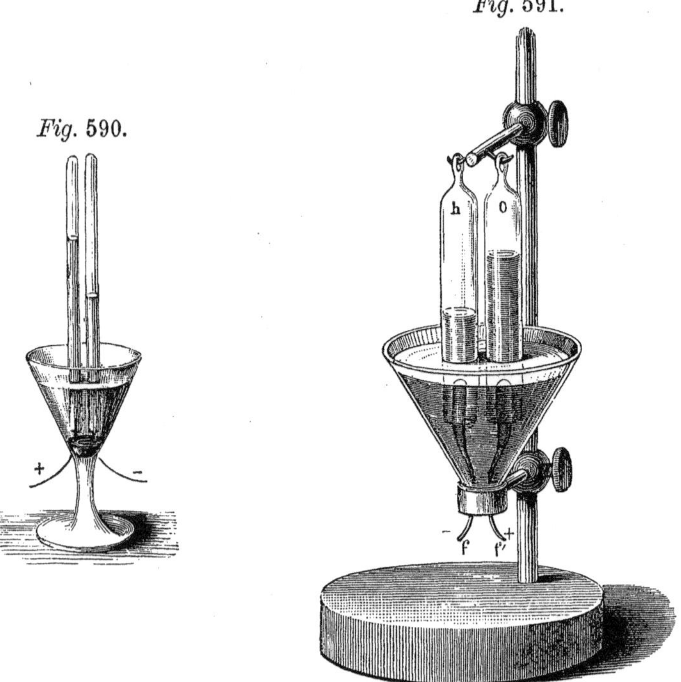


*Fig.* 590.

*Fig.* 591.

platina wires will project about 2 lines within them. The wires should be covered with sealing-wax as far as the bottom of the tubes. Only this projecting portion of the wire need be platina, the rest may be of copper. Fig. 591 represents another form of the apparatus. Take a glass funnel and cut off the tube, except about an inch; close the lower end with a good cork (inserted from above) through which two platina wires terminating in slips of foil are passed, and cover the cork with melted wax. Solder copper wires to the platina on the outside, and mount the funnel and tubes on a wooden stand.

Put slightly acidulated water in the glass, and fill the tubes with the same. The tubes should be of the same diameter, to show the difference in the quantity of the gases evolved.

The gases may be obtained in greater quantity by means of the simple apparatus, fig. 592. The glass tube *m n p* is 4 to 6 lines in diameter. The corks, the platina wires *c d*, and the tubes *a b* must be inserted air tight. The wires end in slips of platina foil *e f*, (which are needed for other experiments,) reaching nearly to the bend of the tube. The tubes *a b* are connected by india-rubber with tubes leading to a small pneumatic trough, where the gases may be collected. The platina foil by increasing the surface makes the current stronger. Fig. 593 represents a convenient support for the apparatus.

*Fig.* 592.

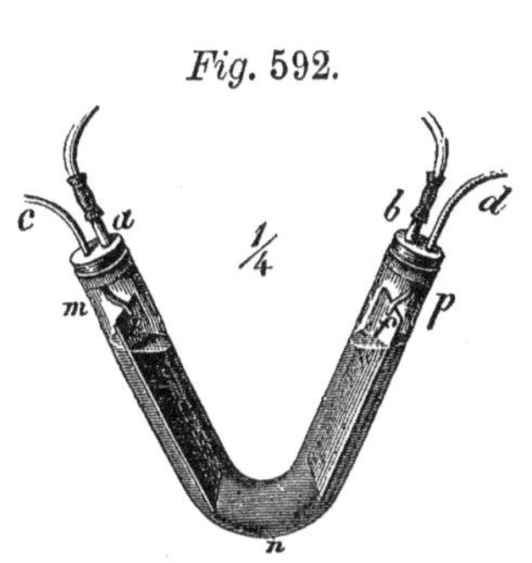


*Fig.* 593.

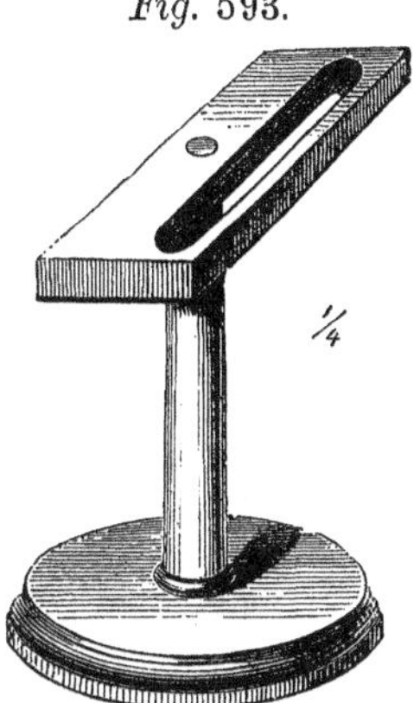


*Fig.* 594.

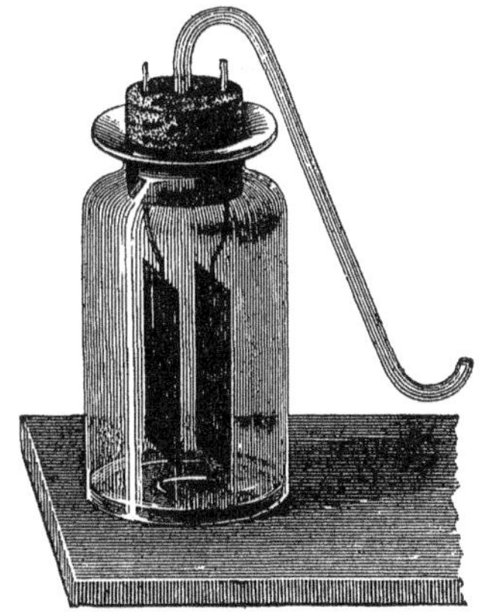

The apparatus in fig. 594 delivers both gases together. It consists of a wide-mouthed glass jar closed with a cork, through which two insulated copper wires are inserted. The wires may either be coated with sealing-wax or inclosed in glass tubes. Platina plates are soldered to these wires, and the gas is delivered by a bent glass tube. It is more convenient when this tube has a flexible joint of india-rubber. The lower side of the cork and the copper wires are coated with sealing-wax, and the plates placed very near each other. The glass is filled with slightly acidulated water. It is better to have the wires of platina soldered to the plates with gold, and connected with thick copper wires on the outside; for the latter are attacked by the acid in spite of the coating of sealing-wax. In these experiments, from 6 to 12 elements are required to obtain the gas in abundance, for the water opposes a great resistance, although decomposition begins with even three or four elements. The gas may be led directly into soap-suds, and the bubbles ignited as soon as they have detached themselves from the tube. The explosion is always sharper than that of the artificial mixture, because the gases are pure and in exactly right proportions. When the quantity of gas is to be measured, the platina plates must be constantly beneath the surface of the liquid, for they would otherwise cause a partial reunion of the two gases. For this purpose, the apparatus may be made in the form of fig. 595, in which the conducting wires pass through a

*Fig.* 595.

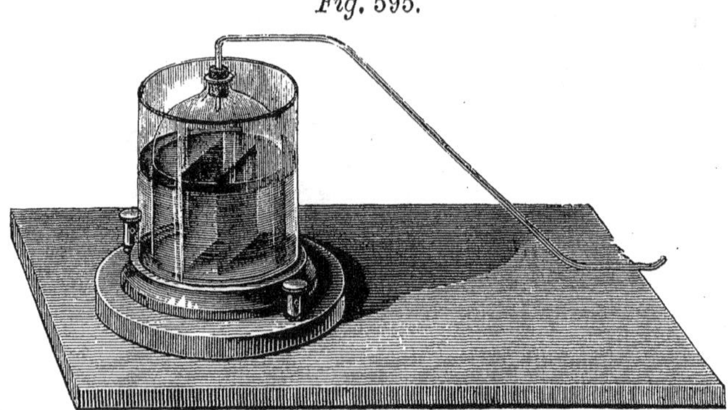

board coated well with sealing-wax, over which stands a little bell-glass, with a gas tube passing through its tubulus. If no exactly suitable vessel is at hand, it is better to bore two holes in the bottom of a glass cylinder, and pass the wires through these holes and the upper part of the board to the binding screws. The bubbles must never be ignited when standing immediately over the mouth of the tube, else the explosion will be communicated to the interior of the generator; they should always be first drawn to the side of the vessel.

Iron plates may be substituted for platina, and placed in a solution of 1 part by weight of caustic potash to 9 parts of water. The plates may be coiled up into a spiral, so as to increase their surface very much, and thereby increase also the quantity of water decomposed. Spirits of turpentine poured over the solution of potash will prevent spattering. This arrangement is not preferable to platina for measurements. With platina electrodes, the acidulated water must be free from salts of zinc, otherwise the negative pole will become coated with black metallic zinc, and the evolution of gas will cease at this pole.

[296] **Explosion of chloride of nitrogen.**—When the poles of a Bunsen's battery of 6 to 8 pairs are connected with platina plates immersed in a very strong solution of sal ammoniac, chloride of nitrogen is formed at the positive pole. If the solution be covered with spirits of turpentine, the little yellow drops of chloride of nitrogen explode very sharply as soon as they come in contact with it. The experiment succeeds better when the liquid is warmed.

[297] **Decomposition of salts.**—The simplest mode of performing these experiments is to take a bent glass tube, such as is represented in fig. 596, and fill it with a solution of some neutral salt, colored with litmus. Insert a platina electrode of a battery, of 3 to 6 pairs, in each end of the tube. The solution will be colored red at the positive pole, and blue at the negative; by reversing the poles the colors will also be reversed.

*Fig.* 596.

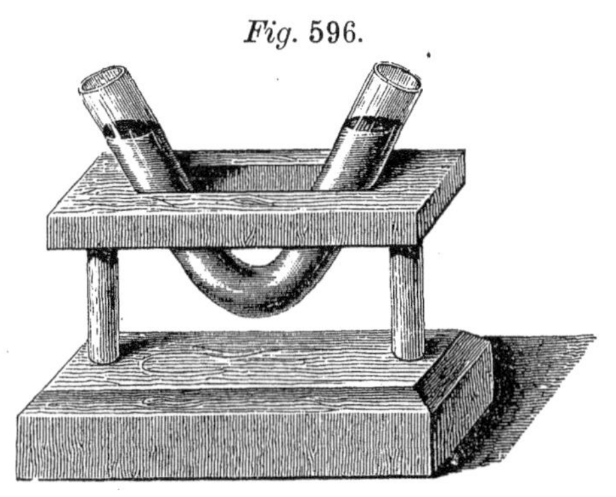

[298] **Electro-metallurgy.**—A modification of Becquerel's battery, fig. 597, is very convenient for small galvanoplastic operations: *c* is a candy jar with the bottom cut off; the mouth is closed by a piece of beef's or hog's bladder, and a sheet of zinc so arranged as to rest on the shoulder just over the bladder, but not in contact with it. It is advisable to wrap the zinc in a cloth, otherwise the falling particles of zinc will induce a deposit of copper on the bladder beneath them. Support the glass in the outer vessel on a wire tripod, leaving a space of 1 or 2 inches below the

*Fig.* 597.

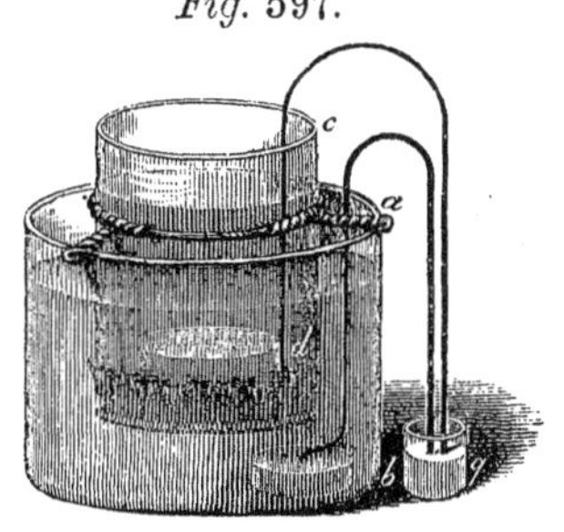

bladder. This support is made by simply twisting two wires together, as in fig. 598. A tripod of glass, fig. 599, placed beneath the jar, is, in some respects, better. A stout copper wire or ribbon is soldered to the zinc, and dips into a cup of mercury $q$, and from this a second wire goes to the mould, which lies on the bottom of the vessel and forms the second element of the battery. Those parts of the wire which dip into the liquid are varnished over. The insertion of the mercury cup makes it

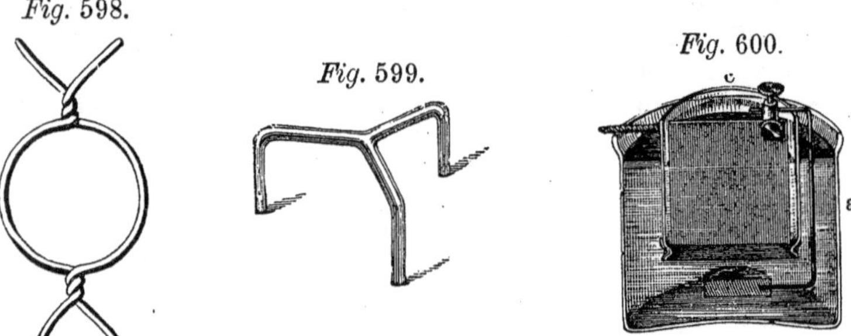

*Fig.* 598. *Fig.* 599. *Fig.* 600.

easier to take the apparatus apart, which is frequently necessary. When the mould is hard, the connecting wire is merely laid on it; if the mould be soft, it is thrust into its side. As the copper is deposited most rapidly where the wire rests, its position should be changed frequently when the mould exceeds 1 inch in diameter, or two wires be used. The zinc may also be placed vertically in its cell, and the conducting wire fastened to it by a binding screw, as seen in section in fig. 600.

Fusible metal, made of 8 parts of bismuth, 8 of lead, and 3 of tin, may be used for taking the casts; though they are generally made of wax mixed with $\frac{1}{4}$ of plaster, or of stearin. The original must be well cleaned, and a rim of paper, $\frac{1}{2}$ inch high, tied or pasted around it. Breathing upon the mould just before pouring in the wax, which must not be too hot, facilitates the removal of the cast. A thin film of grease answers the same purpose. The mould should be slightly warmed, so that the mass will not harden at once. Plaster casts are very hard to copy; they should be laid in a plate with water, and the materials poured over them as soon as the water penetrates to the surface.

To render a wax mould conducting, brush it over with precipitated metallic silver, or finely levigated graphite, with a hair pencil, making a streak also down the side, where the conducting wire is inserted. Lay a rim of shoemaker's-wax around the mould, so that the deposit of copper may not creep over the edge, which makes it difficult to detach. A mixture of 20 parts of glue and 2 parts sugar, boiled with very little water, forms an elastic mass, like that used for printer's rollers; under-wrought objects can be copied with this material. An addition of a little linseed oil is said to render it durable in the galvanic bath. Gutta-percha, pressed over the heated model in a vice, also makes good moulds.

The vessel *a*, fig. 597, is filled with a solution of sulphate of copper, which is kept saturated, and *b* with very dilute sulphuric acid ($\frac{1}{30}$ to $\frac{1}{20}$.) In order to be sure of success, the solution of copper must be free from zinc, which it will not be if it has once been used for Daniell's battery. To render the copper tough, add 5 to 10 per cent. of oil of vitriol to the solution. The copper will be thick enough to be removed in 12 to 24 hours. Remove the shoemaker's-wax first, and raise the copy straight up with a knife blade, while the mould is held in the other hand.

When any other battery is used for galvanoplastic operations, connect the mould with the negative pole, and the positive pole with a copper plate, of the same size as the mould, placed parallel, half an inch from it. (The positive current passes through the liquid element from the zinc to the copper, and out of the liquid from copper to zinc; the copper end is, therefore, the positive, and the zinc the negative pole.) For larger objects the battery is used directly, and a stronger animal membrane substituted for bladder.

[299] **Gilding and silvering.** — *General remarks.** Among the many methods proposed for the precipitation of gold, etc., by galvanism, the decomposition of the cyanides is the best for our purposes. The best battery is Daniell's, because it can be used in any situation, and yields a more constant current than any other. It can be used without interruption for weeks, requiring only the occasional addition of a little fresh acid, and an entire renewal of the same every 4 or 5 days. With Bunsen's or Grove's battery, sulphuric and nitric acids, diluted with ten times their weight of water, will produce a sufficiently strong current. For technical purposes it is advisable to insert a galvanometer, consisting of a simple strip of copper, bent at right angles, in the middle of which a needle is suspended. A slit is made in the upper part of the strip, in order to judge of the proper position of the instrument before closing the circuit. The needle may be a piece of stout knitting-needle suspended by a silk fiber (3 to 5 fold) in the manner indicated in figs. 601 and 602. An astatic needle arranged as in fig. 603 would be better, because the current is feeble. A few experiments will determine how much gold is precipitated in an hour, with various deflections of the needle, and the

---

* This subject is treated here only so far as it is useful in physical instruction, or may be required in gilding parts of apparatus, without regard to whether the process be the cheapest or the best on a large scale, provided it be simple and certain.

quantity of the other metals can be calculated from this. This furnishes an approximate measure of the quantity of gold deposited on any object,

*Fig.* 601. *Fig.* 602. *Fig.* 603.

which is very convenient even for the amateur. Fig. 604 represents a galvanometer of the simplest construction; the copper is separated at *a* by varnished wood.

*Fig.* 604.

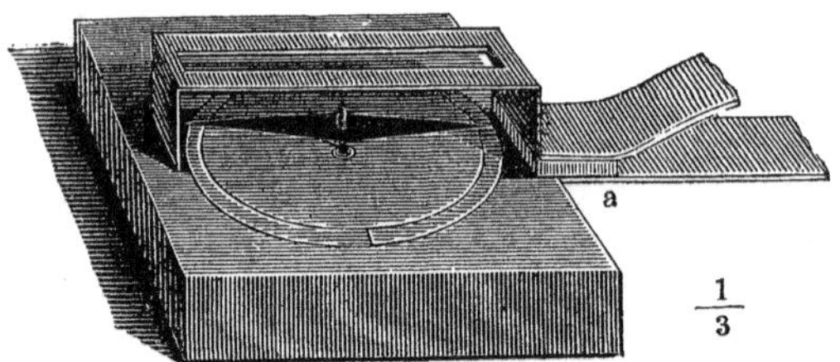


The solutions of the metals, except platinum, are prepared by galvanic agency, by soldering to the positive pole a plate of pure gold, silver, or copper; and to the negative, a slip of platina foil. The ordinary gold and silver coins are not pure enough for the purpose. Pure copper can be obtained by galvanic precipitation, for the copper of commerce is not always pure enough. Fuse the precipitated copper, under borax, in a Hessian crucible. Immerse the electrodes in a solution of cyanide of potassium in distilled water, until the metal dissolved from the positive electrode begins to be deposited at the negative pole. In preparing the silver solution, the negative electrode must be gold. The solutions of the metallic cyanides can be kept unchanged in closely stoppered bottles, but the cyanide of potassium decomposes.

All the articles to be plated must be cleaned, first with lye, then with dilute sulphuric acid, or soot and dilute nitric acid, and finally with cream of tartar and a stiff brush. They should then be washed off with pure water, and placed immediately in the bath, and the circuit closed. The positive pole is inserted in the liquid first, and then the negative pole with the object to be plated. The wires should be bent so that one will be in contact with the object, and the plate connected with the other parallel to the same.

Too much care cannot be taken to prevent the amalgamation of the articles by mercury. The least trace of it on the hands will spread over the clean metallic surfaces. The mercury can be driven off by heat, but it is apt to leave spots. In arranging the battery, the operator should wear gloves, on account of the amalgamated zinc, or cleanse the hands thoroughly afterwards.

The decomposing cell should, if possible, be made of glass or porcelain; but, to avoid the necessity of too much liquid, one will often have to use stoneware.

The temperature of the bath is not indifferent; the operation does best at a temperature of 75° to 140° F., though it succeeds between 60° and 70°.

A part of the metal is always deposited in the form of powder or of acicular crystals; as soon, therefore, as the articles cease to exhibit a bright metallic luster, they should be taken out, and washed with water, and cleaned with cream of tartar. The poisonous nature of the solution of cyanides must never be forgotten while handling the articles. This pulverulent deposit may be partially prevented by occasional additions of cyanide of potassium, of which the solution should always contain an excess.

The current must never be strong enough to cause the formation of bubbles of gas on the surface of the electrode. Feeble currents produce more uniform deposits, and require less frequent cleaning. The strength of the current can be controlled completely, within the limits required, by immersing the positive electrode to a greater or less depth, or changing its distance from the object to be plated.

[300] **Special management of the solutions.** — *Gilding and silvering.* In addition to what has already been said, it should be remarked that these operations succeed best with a single element; and that it must not be supposed that an article is gilded as soon as it becomes yellow. This error has contributed much to bring electro-plating into discredit. It is, therefore, very desirable to know how much gold is precipitated in an hour with a current of given strength; for, according to the investigations of Dumas, fire-gilt articles have 28 to 130 milligrammes of gold on 50 square centimeters. This does not amount to 2 dollars on a square foot, although gilders speak of 3 dollars to the square foot; but by the galvanic process, it is easy to deposit 4 dollars to the foot. The articles are, however, dull and must be burnished.

An object to be thickly gilded must be taken out as soon as it begins to be dull or brown, and rubbed with a wire brush or a stiff tooth-brush, cleaned with cream of tartar, and again immersed. The same process must be repeated until the gold becomes thick enough. When the gilding does not proceed well, or becomes spotted or pale yellow, the article must be cleaned, and a lump of cyanide of potassium added to the solution; a deep-yellow color cannot be produced without an excess of cyanide of potassium. When it is desirable to gild more rapidly, a current strong enough to produce a copious evolution of gas may be used, and the object allowed to become very brown. After removal from the bath it is rubbed first with the finger, or with fine linen, and then with the wire brush. If the browning be not carried too far, the linen will show no trace of gold, and even the scratching brush will detach very little. In one case in which I precipitated 1 milligramme of gold upon 1 square centimeter very rapidly, I found, after using the brush, scarcely $\frac{1}{12}$ less gold; but with twice as thick a coat the linen was colored, and, after brushing, I had lost $\frac{5}{12}$ to $\frac{6}{12}$. The precipitation is very much aided by the addition of a very small quantity, not over $\frac{1}{6000}$, of sulphide of carbon to the metallic bath as often as it is used.

A very simple apparatus for gilding and silvering is made by cutting off the bottom of a wide-mouthed phial, and tying a bladder over the mouth. In this is placed a piece of zinc, to which is soldered a strip of copper, which is afterwards bent as in fig. 605, so that the spiral will stand $\frac{1}{4}$ to $\frac{1}{2}$ inch from the bladder. The strip is coated with sealing-wax up to the spiral. By supporting the phial on a wire, as seen in fig. 598, in a small tumbler, we obtain a battery for electro-gilding. This battery is charged with fresh solution of cyanide of potassium in the zinc cell, and the gold or silver solution in the outer cell. The objects to be gilt are laid on the spiral, which is also coated, but they must be turned once. This battery works very powerfully. Gold must be added to the solution when it becomes too poor.

*Fig.* 605.

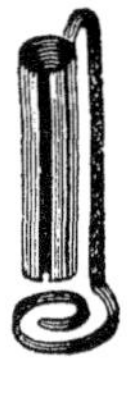

*Frankenstein's process.* In this process a brass wire, to which a bit of zinc is fastened, is wound loosely around the object, or, instead of the wire, a thin, narrow strip of zinc, and the whole placed in a gold or silver bath, warmed to about 167° F. The bath consists, according to Frankenstein, of 1 part chloride of gold, 4 carbonate of potash, 6 ferrocyanide of potassium, and 10 water, boiled together for $\frac{1}{2}$ to $\frac{3}{4}$ of an hour and decanted. Instead of this solution, the ordinary solution in cyanide of potassium, with the addition of potash or soda, may be used. For silvering, Frankenstein recommends 1 part chloride of silver, 5 ferrocyanide of potassium, 5 carbonate of potash, 2 common salt, 5 liquid ammonia; the liquid is boiled and decanted as before. The articles must be taken out and cleaned every 10 or 20 minutes, and the zinc wound over other places. The quantity of gold precipitated in a minute depends on the strength of the solution, and this must be determined for a given strength by the balance, until some experience is acquired, or by analysis of the solution remaining. When the solution is very weak the operation proceeds very slowly, which can be ascertained by the time which elapses before the article becomes yellow or brown. It is asserted, that with the proportions given, and with repeated washing, 1 dollar per square foot can be deposited in 40 minutes. The gilders usually take more dilute solutions. Iron can be gilded in this way without previous coppering.

*Coppering.* At least three elements are required for this operation, the decomposition of the cyanide of copper being more difficult than that of the combinations of gold and silver. In coating iron with copper as a preparation for subsequent gilding, it must be cleaned with great care, attached to the negative pole, the positive pole immersed first in the solution, and then the negative pole. When they are introduced in reverse order, the iron and the copper wire form with the liquid a closed circuit; and the iron being attacked, hinders the firm adhesion of the copper, so that it soon falls off, or at least the iron rusts. The article is only thinly covered with copper, and then connected with the negative pole and immersed in the solution of gold. The gold should be laid on thickly. Unfortunately, the gilding cannot always be done so perfectly as to prevent the subsequent rusting of the iron. Where the operation succeeds well, the articles can be laid in dilute nitric acid without injury.

*Brazing.* A solution of brass is prepared by placing a strip of copper connected with the positive pole in a solution of cyanide of potassium, with a platina negative electrode, until copper begins to be deposited on the latter; zinc is then used for the positive pole, until the deposit on the platina assumes a brass-yellow color. In this bath, iron can be coated with brass. The articles must first be lightly coppered, and then placed in the brass bath, using zinc and copper together as positive pole, and immersing the one or the other deeper, as the color of the deposit requires.*

*Platinizing.* A solution of ammonio-chloride of platinum in water is used for this; but the operation has never yet been brought to the perfection necessary for practical purposes. According to Jewreinoff, an

* The author has never succeeded in plating iron with brass, by using brass as the positive pole, as directed by Jacobi; the article was always covered with zinc, and the brass wire became black.

adherent coat is obtained in the following way: 1 part of platina is dissolved in aqua-regia, the solution evaporated to dryness in a water bath, redissolved in water, and 1 part of caustic potash dissolved in water added to the solution. The precipitate, together with the supernatant liquid, is boiled with a solution of 2 parts of oxalic acid, until the precipitate is redissolved, and a solution of 3 parts of caustic potash added to it. The articles must be treated in this solution with feeble currents several times, being cleaned and burnished after each treatment.

[301] **Nobili's colored rings.**—These beautiful rings are produced on a small silver coin beaten out thin, or on a daguerreotype plate. If the coin be selected, it must be made quite smooth and ground off bright; the coin is better, because the rings can easily be rubbed off and renewed at pleasure, which is not so easy with the daguerreotype plate. The experiment is made by placing on the silver a few drops of acetate of copper, and touching the plate through the liquid with a pointed piece of zinc. The contact is continued as long as the rings continue visibly to increase. The rings are finer when produced by a battery of 3 to 6 elements. The silver plate is connected with the negative pole, and the pointed wire from the positive pole dipped into the liquid without touching the plate. The plate must be held horizontal, so as to hold a considerable quantity of liquid. German-silver may be used instead of silver, but the rings will not be so handsome, for this alloy reduces copper of itself.

[302] The beautiful colors frequently seen on bells and other ornamental articles are also produced by galvanism. They were discovered by Becquerel, and are produced in the following way: Dissolve 1 part caustic potash in 5 or 6 parts water, and add an excess of finely pulverized litharge; boil the mixture half an hour in an earthen pot, stirring it continually; filter the liquid and preserve it in close vessels. Pour the liquid in a brass or leaden vessel large enough to hold the article to be colored, connect the latter with the positive pole of a battery, of 3 or more elements, immerse it in the lead bath, and connect the vessel with the negative pole. The precipitation takes place so rapidly that when the current is strong, it is not always easy to stop at the desired color; it is best, therefore, to use a feeble current. With brass, the wire must be placed in contact with the cell, and the article introduced slowly, edgewise. It is advantageous to gild it lightly beforehand.

To produce the rings distincly, so as to recognize the succession of colors, immerse in the liquid a plate of German-silver laid on a spiral wire, connected with the positive pole, but not in contact with the metallic vessel. Hold a round platinum plate, connected with the negative pole, opposite the center of this plate.

[303] **Magnetic effects of the current.**—To illustrate the action of the galvanic current on the magnetic needle, fasten to boards two copper wires bent as in fig. 606, with the plane of one vertical and of the other horizontal. Place them with the vertical plane of the one and the sides of the other in the plane of the magnetic meridian. Hold a

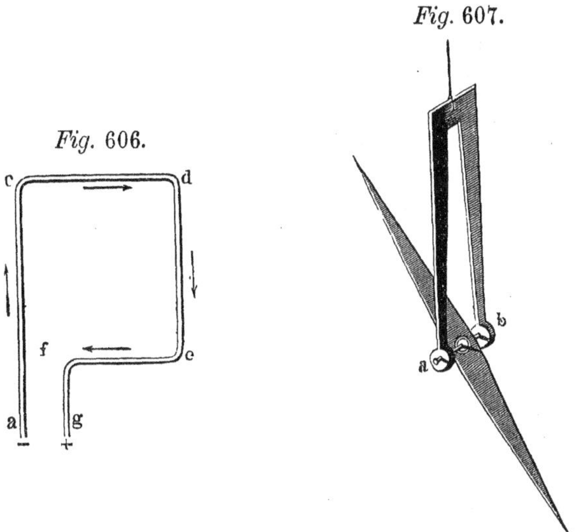

*Fig.* 606.

*Fig.* 607.

magnetic needle above and below the horizontal, and beside the vertical parts of this current. Hold a short dipping needle near them in the same manner. The needle represented in fig. 607 is very convenient for this purpose. To render Ampere's expression of the law of deflection more distinct, little jointed dolls with the left arm extended may be fastened in the direction of the current, so that they can be turned about. Turn the face of the doll each time toward the needle.

[304] **The galvanometer.**—Besides the simple compass described in § 299, accurate measuring instruments should be described: these are the multiplier, the tangent compass, and the compass of sines. All these instruments require a firm support, such as a bracket on the wall, not connected with the floor. Short conducting wires terminating in binding screws should be fastened to the wall, and the movable conductors fastened to these.

*The multiplier.* There should be at least three multipliers: one with 20 to 30 turns of wire, $\frac{1}{2}$ to $\frac{2}{3}$ millimeter thick; one with 100 to 200; and another with 1000 turns of very fine copper wire, such as is used for wrapping strings of catgut. The construction is in other respects the same in all; they are all furnished with astatic needles, the inner of which can easily be removed if necessary. For some purposes multipliers of several thousand turns are needed, but 1000 is sufficient for physical experiments.

The wire, well wrapped with silk, is wound on a square wooden frame, like fig. 608, about 2 inches long in the clear. Two opposite sides are cut away above and below to the thickness of 3 to 5 lines, so that the space within the coils will be that wide. The wire is wound in layers between projections left standing above and below, so as to leave a slit at the top. The ends of the wire must project on the same side, and be tied so that the wire cannot uncoil, which annealed wire will have no tendency to do. The frame is screwed down upon a square board, a little longer than itself, furnished with leveling screws. The needle will come to rest soon if the interior of the coil be coated with foil, so that it will vibrate as if between two metallic plates.

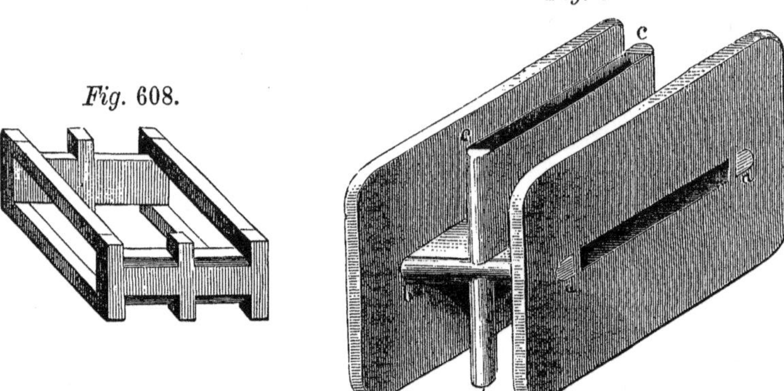

*Fig.* 608. *Fig.* 609.

On the upper bars of the frame is fastened a plate of ivory or of wood, covered with paper, with a graduated circle equal in diameter to the length of the needle. A slit is made in this circle from 0° to 180°, corresponding to the slit in the coil, and a line drawn at right angles to it through the center.

To hold 1000 or more turns of wire the space for the needle must be smaller, and the flanges which confine the wire larger. Fig. 609 represents the frame for this. As the spaces for the introduction and vibration of the needle must be made very small, they should be inclosed by thin boards, to prevent the intrusion of the wire. When the wire has several thousand turns, it is advisable to attach binding screws to the ends of each thousand separately; the operator has it then at his option to connect the various lengths so as to form a continuous wire, or a shorter and thicker wire, opposing less resistance, and less sensitive. It is convenient to make the several lengths of different colors. Figs. 610, 611, and 612 show the

*Fig.* 610. *Fig.* 611. *Fig.* 612.

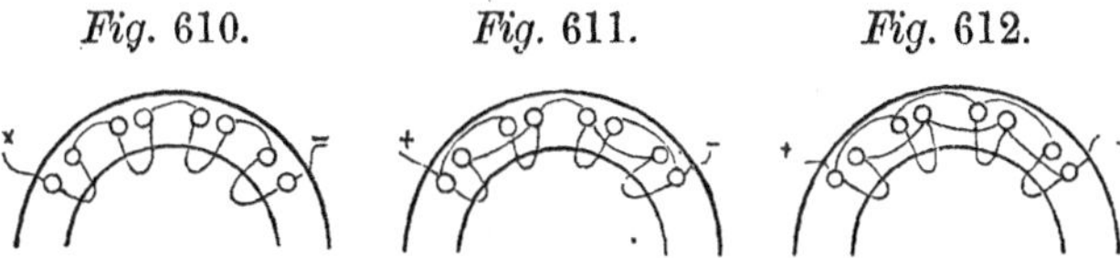


various combinations for 4000 turns. In very sensitive multipliers it often occurs that the needle recedes from the zero line, and when it is twisted steadily toward it, springs suddenly to the other side. This is caused by the presence of iron in the copper wire, and must be corrected, if necessary, by the addition or subtraction of a constant error, the determination of which is beyond the limits of this work.

The needles are made of bits of thin steel knitting-needle, 2 or 3 lines shorter than the space within the frame will admit. They are suspended in the manner indicated in fig. 613, or fig. 614, by a single cocoon fiber

*Fig* 613. *Fig.* 614.

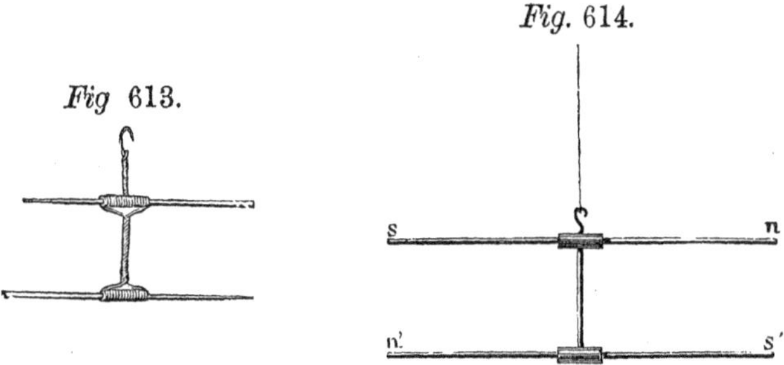


wound around a screw head, inserted in an arm which is fastened to the plate, fig. 615. The fiber rests in a smooth notch in the end of this arm, which must be exactly over the center of the circle. The needles can be raised or lowered by turning this screw. The upper needle serves as an index. There should be some means of fastening the upper needle to the plate, to prevent the breaking of the thread in moving the apparatus. Fig. 616 represents a simple way of accomplishing this: *a* is a brass slide which can be fastened over the end of the needle by the screw *b*.

*Fig.* 615.

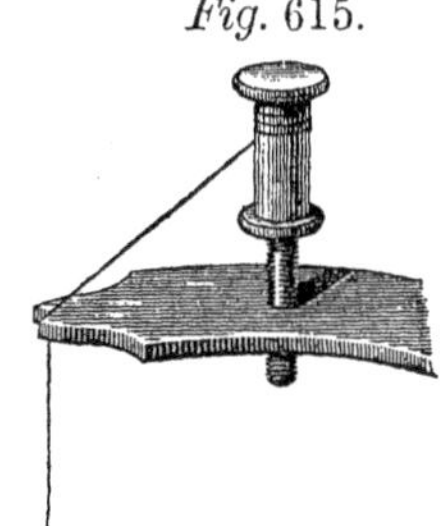

The ends of the wires should be fastened or soldered to binding screws, like fig. 549. The instrument is covered with a glass case, leaving the binding screws, of course, on the outside. In using the instrument, the zero point must be turned toward the north.

*Fig.* 616.

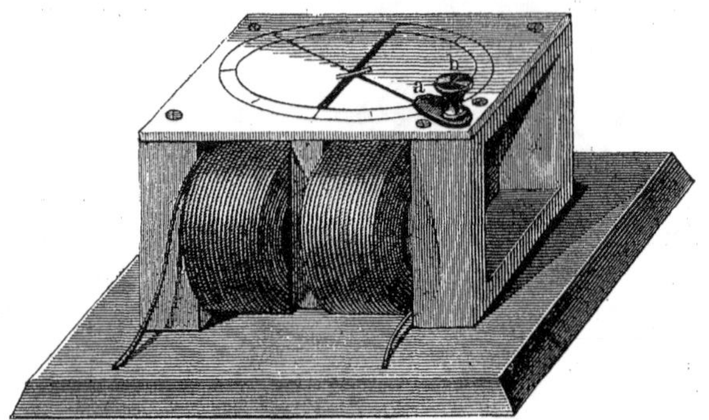


The upper needle should be a little the stronger, so that the visible north pole may point to the north. The deflection of the needle, not its first oscillation, must be taken as a measure of the force of the current. In

*Fig.* 617.

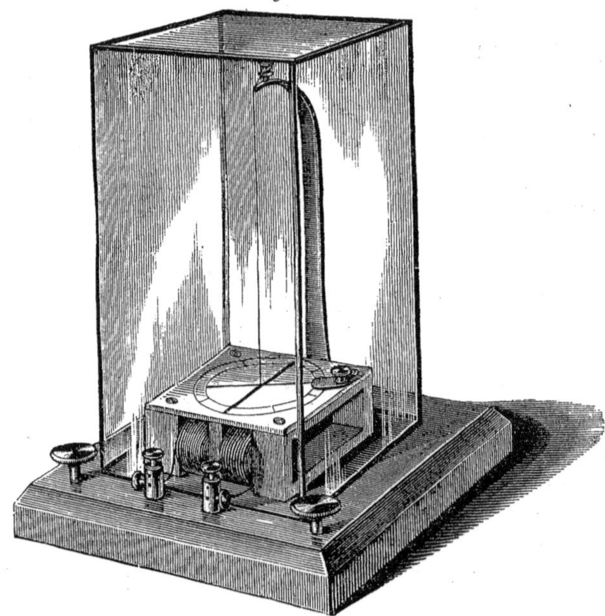

the multiplier, however, and in the simple compass, the force of the current bears no simple ratio to the deflection, as it does in the tangent compass and compass of sines. Fig. 617 represents the multiplier complete.

Astatic needles are apt to assume, unless one of them is considerably stronger than the other, some other direction than that of the magnetic

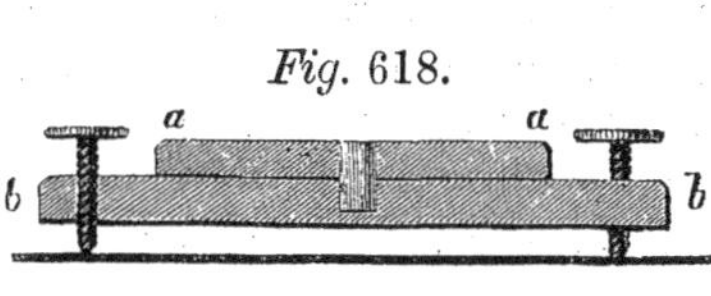


*Fig.* 618.

*Fig.* 619.

meridian. This does no harm in the galvanometer, provided the zero of the scale be placed in the direction of the needle. This difficulty arises

*Fig.* 620.

from the needles not being in the same vertical plane, and is difficult to overcome; it becomes more perceptible as the needle is more perfectly astatic. Perfectly astatic needles, not situated in the same vertical plane, ought to place themselves at right angles to the meridian.

When the upper plate is made of copper, it serves as a check upon the vibrations of the needle, bringing it to rest sooner.

It is very convenient to have the instrument movable on a pivot in the base, as represented in fig. 618, so that the position of the needle can easily be adjusted by turning it. It is well also to have brass pegs fixed in the circle at 90° and 270°, to prevent the needle from being thrown completely round.

In very delicate galvanometers, the needle should be capable of being raised and lowered without removing the glass case. Bore a hole through the top of the cover, and attach to the piece which supports the cocoon fiber a brass cap, represented of natural size in fig. 619. There is a square hole in the center of this cap, and above a ring with two grooves, in which a screw nut is inserted. The fiber hangs on a hook at the end of a square pin, which has a thread cut on the other end. By turning the nut, this pin is screwed up or down. Fig. 620 represents a galvanometer with all these conveniences. The frame for the wire is merely slipped between two wooden cleats on the upper board.

[305] **Wrapping the wire.**—Wire less than 1 millimeter thick is always wrapped with silk. Thicker wires are usually covered with wool, or with silk ribbons made by cutting up old silk and sewing the strips together. A fringe-maker can wrap almost any length of fine wire; but the thicker wire cannot be wrapped on small spools, and the length which can be wrapped in one piece depends on the room. A machine for wrapping àny length of wire is described in the next section.

To splice the wires, scrape the ends and twist them together with the pliers. When the wires are thick, this disturbs the regularity of the coils, and it is better to solder them together with soft silver-solder, and file down the joint. The bare places are wrapped by hand.

[306] **The wire-wrapping machine.**—Figures 621, 622, 623, 624, and 625 represent a simple machine for wrapping endless wires. The same parts are marked with the same letters in all the figures, and figs. 624 and 625 are half the real size.

The base *A* rests on feet, and upon it the board *B B* is fastened by a wooden screw and nut *C*, which works in a groove in the base, so that the board *B* and the wheel it supports can be slid back in order to tighten the band. The iron axle of the wheel *E* works in brass bearings. The circumference of this wheel is grooved for a band, and five grooves of varying radius are cut on the axle. The cord which carries the spool around the wire works in the large groove, and that which pushes forward the wire in one of the smaller. We will follow the latter first. It passes from the wheel over two rollers *a a*, sliding on smooth iron rods, which are fixed obliquely in the base, and fastened above by a brass band to the brace *T.* As this band is designed to be movable, it is slit so as to slip

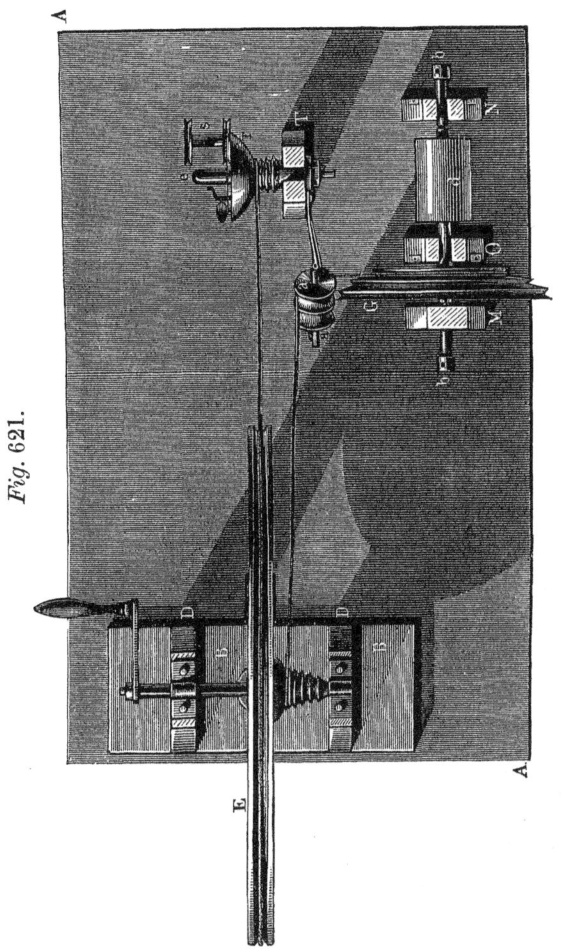

*Fig.* 621.

under the head of a screw. From the rollers *a a*, the band passes at right angles to its former course to a wheel *G*, with three grooves. The iron axle of this wheel turns on screw points *b b*, the nuts for which are sunk in the uprights *M N.* Besides this wheel, the same axle carries an iron roller covered with leather, $c$, and the third upright $O$ has a deep slit from the top to admit the axle. The roller $d$ is of lead covered with leather; its axle lies loose in the slit in $O$. It serves to press the wire on the roller $c$, so that it must be pushed forward when the wheel turns

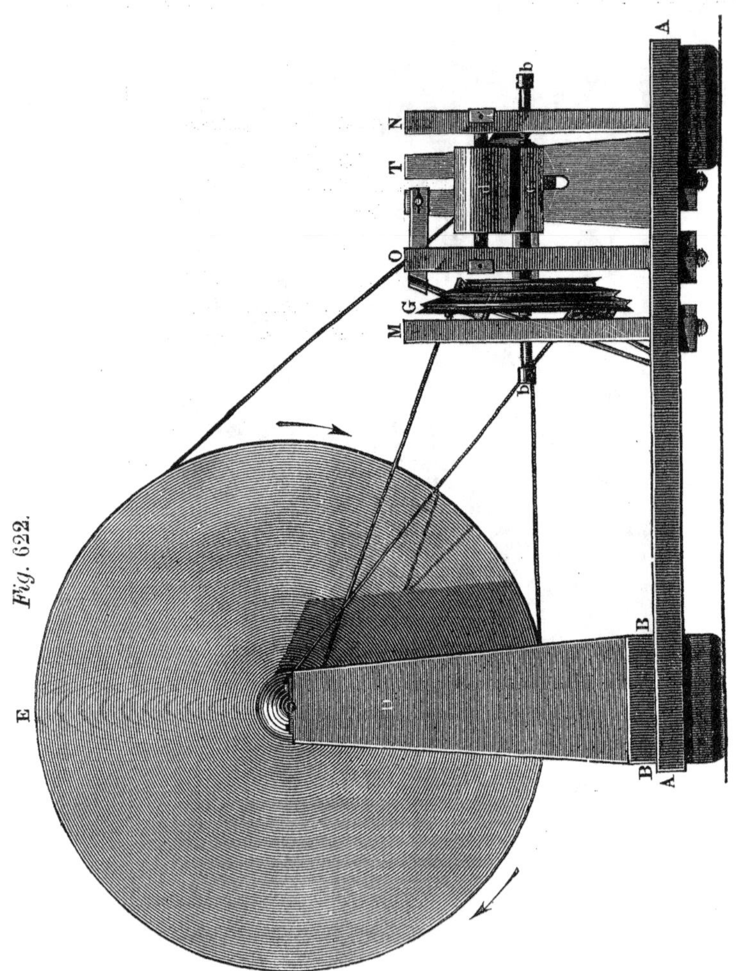


*Fig.* 622.

in the direction of the arrow. The wire is thus carried to the upright $T$, and pushed through the tube $x$, fig. 624. This tube $x$ is thrust through a slit in the upright, and held by the shoulder $y$ and the nut $Z$ in such a position that the middle of the wire always coincides with the axis of the tube. The upright *T* can also be slid in the slit *e*. On the other side of the upright the tube forms the axle of the spindle *r*, in one of the

*Fig.* 623.

grooves of which the cord from the large wheel works. The box of the spindle is of brass, and runs very easily on the tube, being kept on by the

*Fig.* 624.

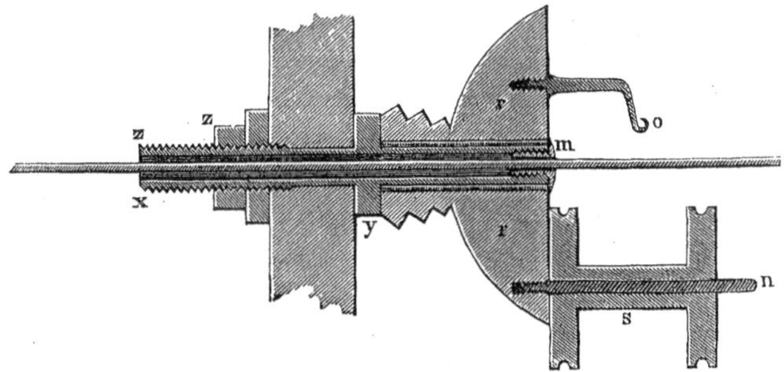


tap *m*. Several of these taps must be provided, for the hole in it must be very little larger than the wire, which passes through it. The spool *s* is stuck on a smooth pin *n* driven into *r*. The silk is easily wound on these spools on the lathe. From the spool the thread is carried over the hook *o*, before being attached to the wire. The spool is kept on the pin by a little catch *p*. A fine brass wire *q q* is passed around one end of the spool, to prevent its revolving faster than the silk is wound on the wire; by turning the pin *i* the friction of the wire can be increased at pleasure. The ratio between the revolution of the spool and the advance of the wire can be varied by changing the drums upon which the bands run, and the thickness of the wrapping varied correspondingly. The wire should be wound on a roller behind the machine, so as to run smoothly into it. The wrapped wire should be wound into coils as it comes from the machine. Before being wrapped, the wire must be well annealed.

*Fig.* 625.

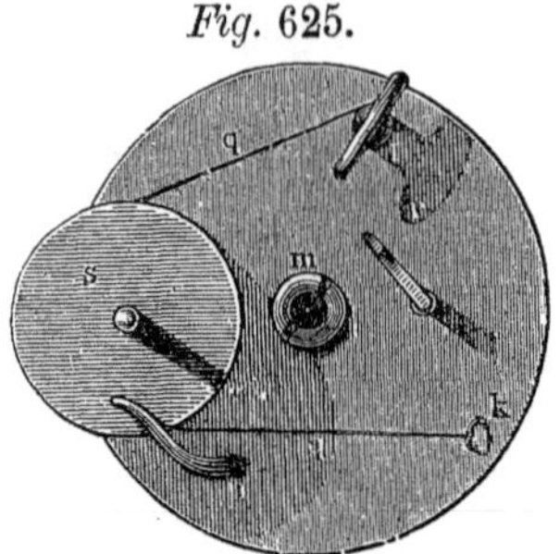


*Fig.* 626.

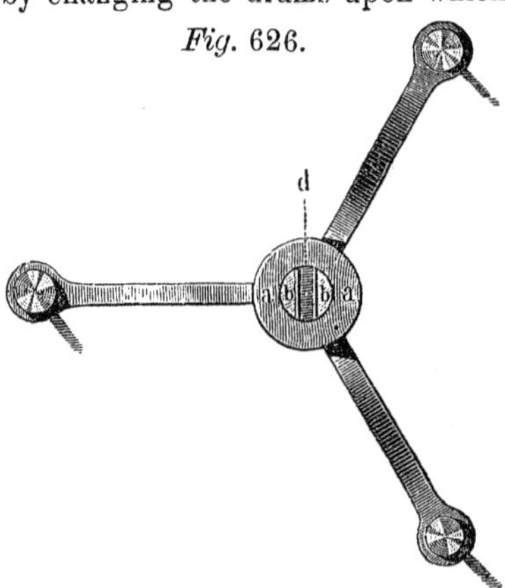


[307] **The tangent compass.**—It consists of a stout copper ring, (2 to 3 millimeters thick and 1 to 2 centimeters broad,) at least 1 foot in diameter, and bent at right angles at the ends, as seen in fig. 627, instead of being soldered together. The two extremities are

*Fig.* 627.

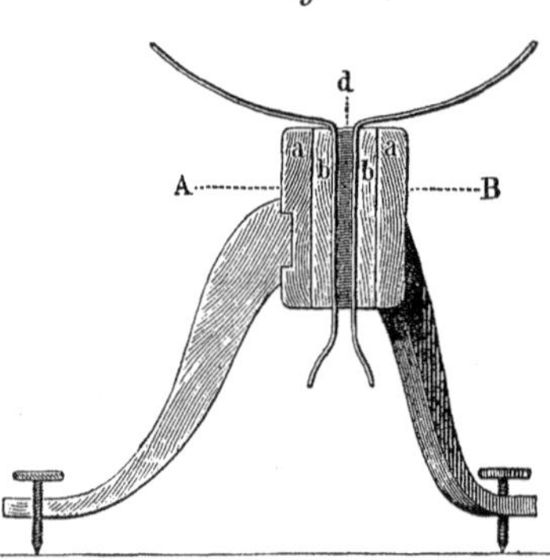


*Fig.* 628.

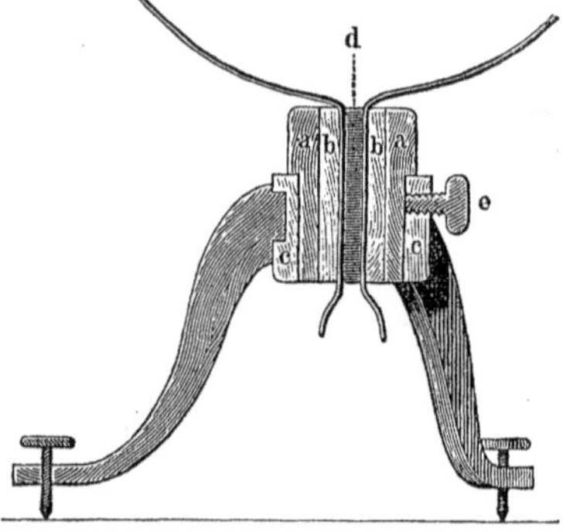

separated by a varnished board *d*, inserted in the hole in the cylinder *a a*, figs. 626 and 627, and the space filled up with blocks of wood *b b*. The cylinder *a a* is supported on three feet with leveling screws, and the ends of the copper, with binding screws attached, project between these. The conducting wires are of copper wrapped with silk, 1 meter long and 2 to 3 millimeters thick, and twisted together so as to neutralize their influence on the needle.

Instead of attaching the feet directly to the cylinder *a a*, this may be made to turn within a second cylinder *c c*, fig. 628, and fastened by a thumb-screw *e*; the ring can then be turned independently of the feet.

*Fig.* 629.

In the copper ring is fixed a wooden frame with a groove, into which the ring fits. This frame is designed to support a compass, with a needle only 1 inch in length, and must be of such a height that the center of the needle will coincide with the center of the ring. Fig. 629 represents the whole apparatus, with the simple modification of carrying off the wires parallel to each other. A needle so short as 1 inch does not admit of a more minute graduation than to 2 degrees; and the error of parallax will be considered when the needle does not swing much beyond the plane of the ring. In fig. 629 the needle is prolonged by threads of black glass, and the circle made larger. It is cheaper to attach to the cylinder *a* two arms *c c* supporting a table *d d*, fig. 630, about 6 inches

*Fig.* 630.

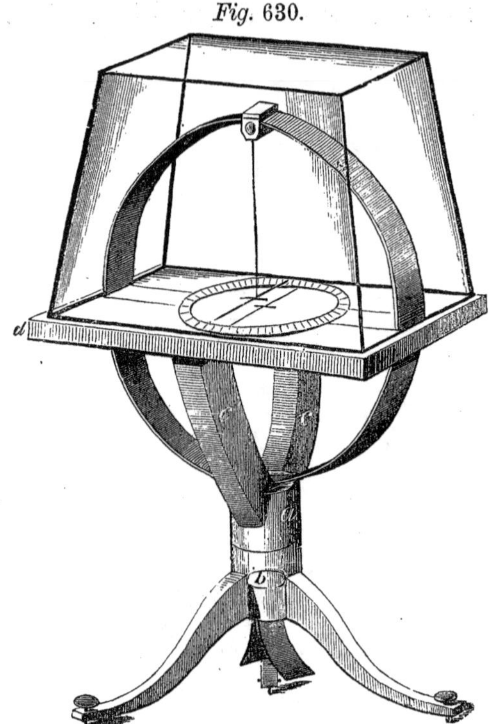


wide, and 2 inches longer than the diameter of the ring. The ring is inserted through a slit in the middle, which is afterwards filled up again. A large compass, such as is described in § 219, may be placed on this table, or a graduated circle of paper pasted on it, with the zero points at right angles to the plane of the ring. The needle consists in this case of a bit of steel knitting-needle an inch long, with a socket like fig. 601. To this socket are soldered two wires at right angles to the needle, extending to the graduation. The needle is suspended by a cocoon fiber from the top of the ring, by the contrivance represented in figs. 631 and 632. This consists of a block of wood with a groove *a* in the top to admit the ring. The depth of this groove is not quite equal to the thickness of the ring, so that when the bar *c* is screwed over it the block will be held fast on the ring. The block is also hollowed out below, and the pin *e* thrust through it, turning with considerable friction. The cocoon fiber is wound on this pin, and passes through a smooth notch in the brass plate *f*, which is screwed to the bottom of the block. The apex of this notch must be exactly over the center of the graduated circle. The whole is covered with a glass case, to protect it from currents of air. The needle may be fastened by a similar contrivance to that described for the multiplier, or it may be hung on a little hook at the end of the thread. The instrument is thus rendered more sensitive, while it may still be used with the strongest currents, is more convenient to read, and can be graduated to at most one degree. If the index wires should be slightly bent, the error may be noted and corrected before the needle is deflected.

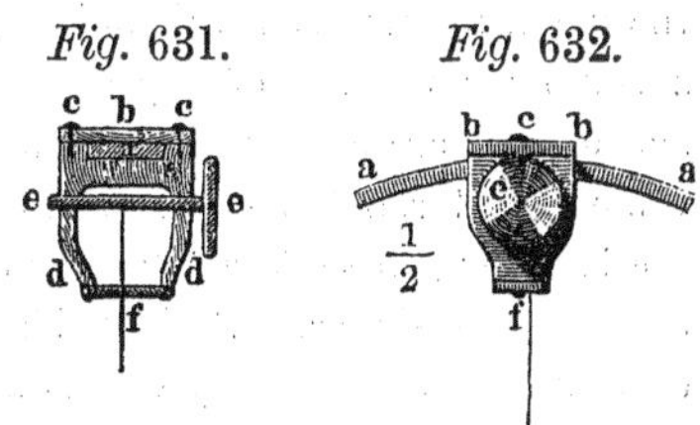

*Fig.* 631. *Fig.* 632.

The instrument can be read more accurately by inserting a mirror within the graduated circle. In reading off, place the eye in such a position that the index wire will cover its image in the mirror; error of parallax is thus entirely avoided. The same contrivance can be applied to many other instruments, where accurate reading is necessary.

With a tangent compass constructed as described, a compass of sines may be dispensed with. It differs from the tangent compass only in being more sensitive; but it is more complex, and, therefore, costlier. In using the tangent compass, place the ring parallel to the magnetic meridian, and level the instrument. It is level when the center of the needle coincides with the center of the circle, provided that the stirrup on which the thread hangs be properly adjusted, which must be done at first with the aid of a spirit-level. To fix the plane of the ring accurately in the magnetic meridian, the median line of the graduation must be produced—with a diamond on a mirror—and must coincide with the middle of the ring. Two slits are made in the ring for the purpose of observing whether the direction of the needle coincides accurately with this line.

The delicacy of this instrument, *i.e.* the ratio between the tangent of the deflection and the real strength of the current, depends upon its dimensions, and should be determined by a series of experiments, by inserting at the same time a decomposing apparatus like fig. 594 or 595, and collecting the gas in a graduated tube. Calculate from each experiment the quantity of detonating gas corresponding to the tangent 1, and take the mean of all; from this it is easy to calculate the quantity of gas produced by a current of any strength. In this way the data of various instruments are comparable. The tangent compass is a much more convenient and reliable galvanometer than the decomposing apparatus, for the latter engrosses the attention of one observer, whereas the former can be read at a glance. Besides this, the decomposition of water retards the current, and the quantity of gas must always be reduced to 32° F., and a pressure of 76 centimeters.

The delicacy of the tangent compass can be increased by making the ring of thick wire, wound several times around a wooden hoop. By attaching binding-screws to each turn, they may be combined at pleasure. The diameter of the ring need never be more than 10 to 15 times the length of the needle.

[308] **The compass of sines.**—In this instrument the current is carried through a copper wire wound once or oftener around a wooden circle, 6 or 8 inches in diameter. A magnetic needle is attached to this ring, so that its center coincides with the center of the ring, fig. 633. The ring is supported on the axis of a graduated circle, the index of which must be at zero when the coils of the wire are parallel to the magnetic meridian. When the needle is deflected by the current, the ring is turned until it again coincides with the direction of the needle; the strength of the current is then proportional to the sine of the deflection indicated on the horizontal circle. The delicacy of the instrument increases with the number of turns of the wire. It is, however, not available for currents beyond a certain strength, because the sines do not increase indefinitely as the tangents do. This instrument is too complicated for domestic manufacture, and has gone, in a great measure, out of use.

*Fig.* 633

[309] **Conducting power and Ohm's law.** — Constant batteries of considerable power are required to illustrate these subjects. The conducting wires may be wound on wooden spools, 1 to 1½ inches in diameter and 2 to 3 inches long, with screw threads, about 1 line apart, cut on them. Well-seasoned pear-wood is the best for the purpose. The ends of the wires are thrust through holes bored at right angles to the axis, and fastened by a plug. Thick copper wires, 3 or 4 centimeters long, are soldered to the ends of the wires, to protect them. The copper wires with which the comparisons are made should be of precipitated copper melted down and drawn out. The difference in the conducting power of the metals is made most evident by winding upon the spools just described equal lengths of copper, brass, iron, and German-silver wire of the same thickness. A copper wire, of the same length and double the thickness, is also required. The thickness of the wires should be about ⅓ millimeter, and the length 3 meters.

It will generally be sufficient to show the effect produced upon the tangent compass by inserting the wires successively in the circuit. If more than this be required, the resistance of the element, together with the compass and its conducting wires, must be determined by inserting the compass, first alone, then with various lengths of copper wire, and calculating the resistance of the element and the compass from the tangent of each inflection, according to Ohm's law, combining each observation with the first.* But this process is too complicated for class instruction. The laws can be illustrated much more simply by means of the rheostat.

[310] **The rheostat.** — This is a contrivance by which varying lengths of the same wire, accurately measured by the instrument itself, can be inserted in the circuit without interrupting the current. Bunsen's apparatus will serve to illustrate the use of all similar instruments.

Bunsen's rheostat consists of a wooden cylinder, about 4 inches in diameter and a foot long. A screw, with threads about 1 to 1½ lines apart, is cut on the surface, and brass or German-silver wire, ½ to 1 millimeter thick, wound in the threads. The cylinder is turned by a crank on an iron axle, to which a wire is attached leading to the battery; the beginning of the brass wire is also connected with the axle. From the wire, the current is carried through a brass slide which fits into the thread of the screw, and presses constantly on the wire. When the crank is turned, this piece slides back and forth, so that more or less of the wire can be inserted at pleasure

* By combining all together we obtain numbers greater according to the length of wire inserted, because the polarization changes.

Figs. 634 and 635 represent this apparatus, $\frac{1}{7}$ the real size, except the threads, which, for the sake of distinctness, are made 4 times too large.

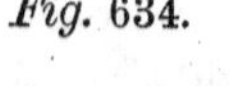

*Fig.* 634.

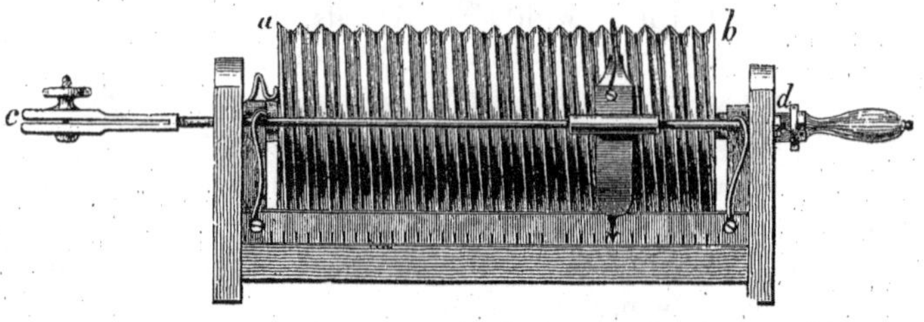


*Fig.* 635.

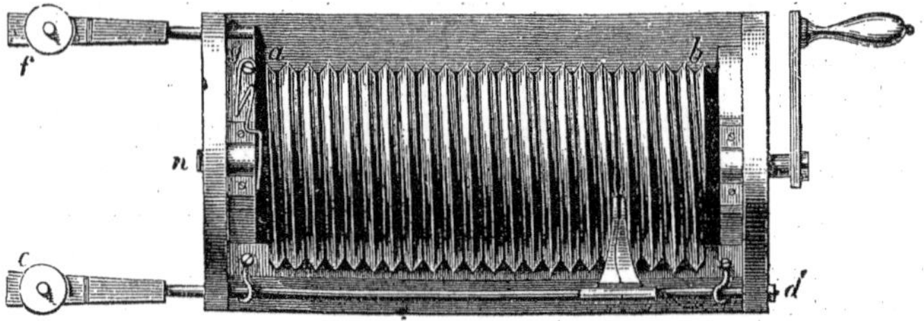


The current is introduced, for instance, through the clamp *f*, passes through the iron axle, and from this to the wire at *b*. From the wire it passes to the elastic piece of brass which slides upon the brass rod *d c*. An index fixed to one end of this slide indicates the number of turns between it and the end of the wire at *b*. The slide is pushed forward by a small block of wood with two or three threads fitting those of the cylinder. Fig. 636 represents this slide. The rod *c d* is loose in the frame, and is pressed against the cylinder by two springs. The tube of slide is slit, to make it more elastic.

*Fig.* 636.

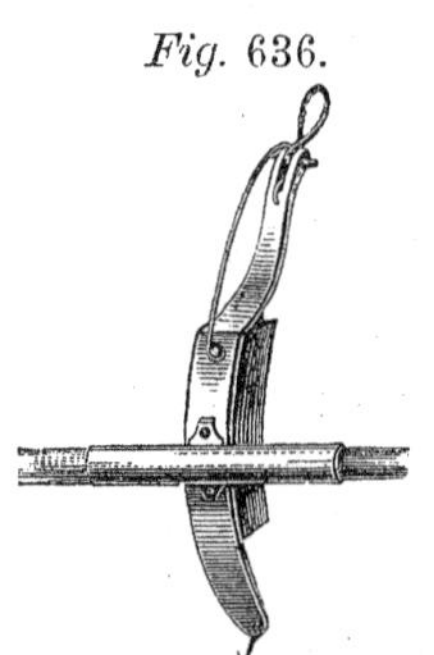

The primary design of the rheostat is to render the force of the current constant, when the resistance or the power of the element changes.

Any other resistance may be expressed in terms of the length of rheostat wire required to produce the same effect, by placing the index of the rheostat at zero, inserting the resistance, and noting its effect on the compass, and then, after removing the resistance, turning the rheostat until the needle is deflected to the same degree. The conducting power of different kinds of wire of the same length and thickness will be inversely proportioned to the length of the wire of the rheostat required to produce the same effect, thus obviating the uncertain determination and tedious calculation of the resistance of the element. Ohm's law may be illustrated in the same way, by inserting successively wires of varying length and thickness.

The conducting power of good copper has been shown by these experiments to be almost four times that of the best brass; and, therefore, as the price of copper is not more than one-half greater than that of brass, it is cheaper to use the former for all galvanic apparatus. An advantage is also gained by being able to use finer wire, which is easier to handle, and occupies less space. The size of the wire is of still greater importance when it is wrapped with silk, a wire of double thickness requiring twice as much silk. Wires of from 2 to 3, 1 and $\frac{1}{4}$ millimeters thickness, will be sufficient for almost every purpose.

[311] **Measure of resistance.** — The standard now generally assumed, for the measure of resistance to the galvanic current, is the resistance of a copper wire 1 meter long and 1 millimeter thick. The copper must be chemically pure, and not stirred with an iron rod during fusion. It is not necessary that the wire should be exactly 1 millimeter thick, if its size be accurately measured. This can be done with most certainty by means of a micrometer screw under the microscope. If the length and thickness of the wire coiled on the cylinder of the rheostat be known, it is easy to calculate its resistance in terms of the given unit, which should be written on the cylinder.

*Fig.* 637.

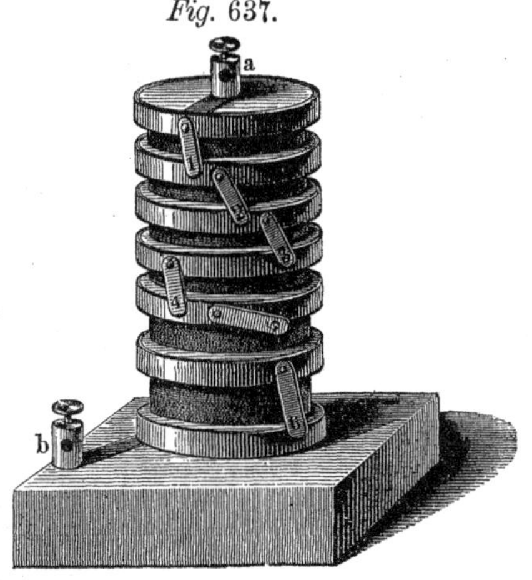


In many experiments it is necessary to insert quickly a given resistance in the circuit; for this purpose, Eisenlohr's column of resistance, fig. 637, is very convenient. It consists of a cylinder of mahogany, or other wood saturated with oil varnish, in which about nine rings are cut. The spaces between these rings are bound with brass bands, and a little brass bar, turning on a screw, made to extend from one to the other, as represented in the figure; these bars must be slightly bent, so as to press with some force on the bands. Wrapped wire, whose resistance is known, is wound in these grooves, the shortest length containing the assumed unit once, or an even number of times, which is marked on the column. The length of the coils of wire in the successive grooves increases from 1 to 9; the ends of each wire are soldered to the two nearest bands, the upper band being connected with the screw *a*, and the lowest with *b*. When this column is inserted, the current passes through the bars, whose resistance is inconsiderable; but if one of the bars be wanting, as in fig. 637, the current must pass through the intervening wire.

[312] **Conducting power of liquids.**—To show the great resistance which liquids oppose to the galvanic current, we need only insert in the circuit a voltameter, and, noting the effect on the tangent compass, it will be evident enough, even when the electrodes are very large; in this case, however, the battery should consist of fewer and larger elements.

*Fig.* 638.

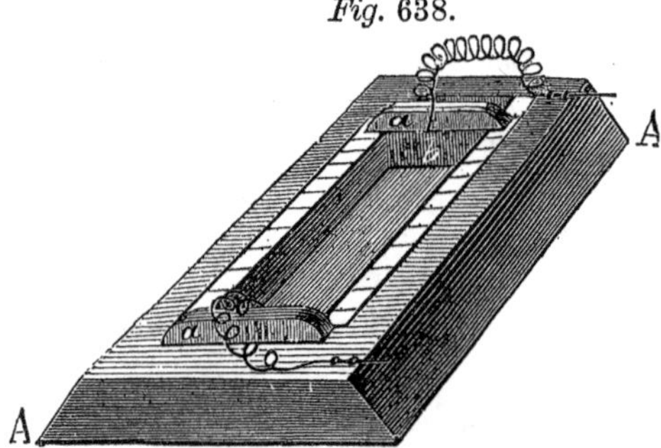


To measure actually the resistance, make a wooden trough, 4 or 5 inches long, *A A*, fig. 638, and cement it with sealing-wax. In this trough place two movable boards *a a*, the edges of which, extending over the sides, will serve as indices to the scale. To each of these boards is attached a platina plate, soldered to a spiral copper wire, the ends of which are fastened to the trough. The liquid is placed in the trough, and the plates placed at any convenient distance from each other; after observing the compass, the wire of the rheostat is substituted for the liquid until the same deflection is produced. Since Ohm's law holds for liquids as well as solids, the resistance of the stratum of liquid can be calculated from the length, breadth, and thickness, when the resistance for the unit of section and length is known.

## (*e.*) EXPERIMENTS ON ELECTRO-MAGNETISM.

[313] **Electro-magnets.**—These can be made at home at a comparatively moderate cost. Soft, round, charcoal iron is best for the purpose. Have a horseshoe bar made of this, about the thickness of a finger and 10 or 12 inches long; after coating it with clay, heat it and

leave it to cool among the embers, in order to show the almost total loss of magnetism when the current ceases. Wind copper wire, wrapped with silk or wool, closely around it, leaving the bend free, and pass on from one side to the other, making the turns on both sides in the same direction. A single layer of wire, 1 line thick, will impart, with a strong battery, a very considerable attractive power. The armature is made like those of ordinary magnets, figs. 639 and 640.

*Fig.* 639. *Fig.* 640.

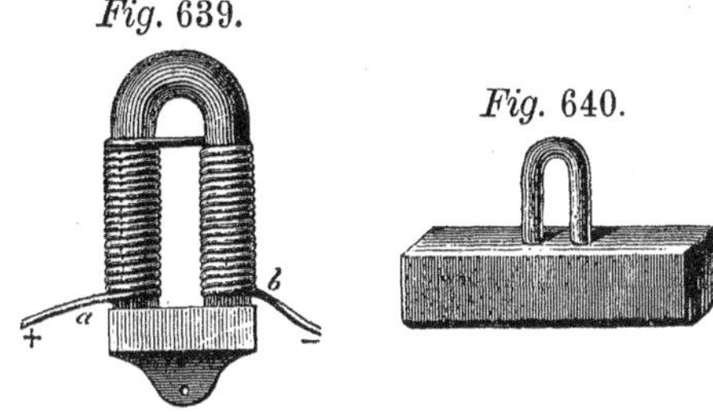


To make a powerful electro-magnet for magnetizing steel, etc., prepare a bar, 2 inches in diameter and 3 feet long, in the manner just described. The armature should be provided with a hook,

*Fig.* 641.

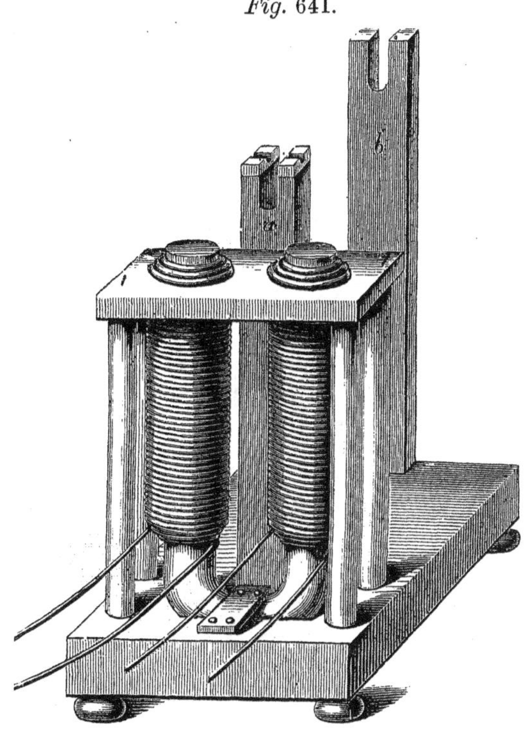


and its upper surface made somewhat convex, as in fig. 640. Fasten the magnet on a stout wooden frame, like fig. 641. The post *a* has two steel bearings for the axis of an iron lever. The longer arm of the lever is graduated, and carries a sliding weight of 25 to 50 pounds. The post *b* is to catch the lever when the armature is pulled off. The bend of the magnet is let into the base, and held down by a stout bar screwed over it.

Several layers of copper wire, about 3 millimeters thick, must be used, in order to impart to the magnet a power equal to several hundredweight. The ends of each layer may be kept separate, and combined at pleasure. It is advisable not to wind the wire directly on the limbs of the magnet, but on thin wooden or pasteboard spools. The spools must be slipped over wooden cylinders, to prevent their being crushed during the winding. Each layer of wire is separated by a sheet of paper. The ends of the spools are finally cut down to within 3 lines of the wire. These spirals can be used separately for many experiments.

As it is difficult to bend iron 2 or 3 inches in diameter without altering its thickness, two straight pieces may be connected by a short bar and two strong screws.

If there be a machine-shop near by, it is best to have the armature and the ends of the magnet planed; it is then easy to grind them perfectly even. The ends of the magnet should not be left round, but squared

*Fig.* 642.

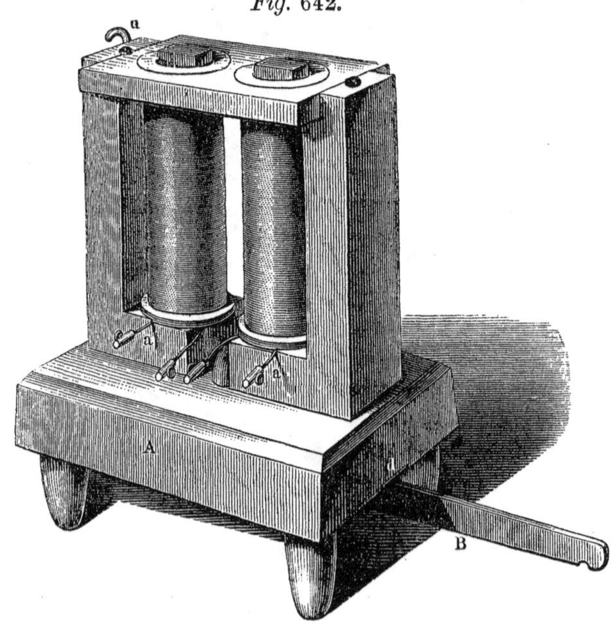


off to one-half the diameter, as seen in fig. 642. The surface of the armature is made just the same width.

Fig. 642 represents an electro-magnet with a frame designed for great weights. Its poles are directed upward; the spirals are even with the top of the table, and can be easily taken out.

*Fig.* 643.

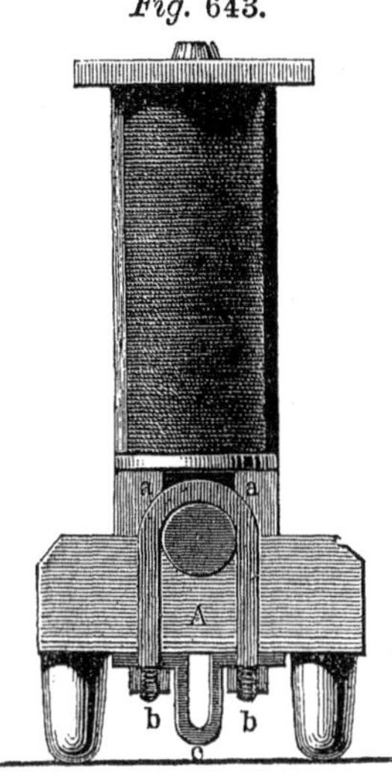


Fig. 643 represents a section through the fastenings of the bar, which lies between the posts *a a*, which support the spirals also. The nuts *b b*, which hold the bar to the base, support also the stirrup *c*, fig. 643, into which the end of the bar *B*, fig. 642, is inserted; *d*, fig. 642, is an iron plate let into the base, against which *B* presses.

To exhibit the power of the magnet, lay the armature, without the stirrup *E*, fig. 644, on the poles within the iron arch *D*, which rests on the frame; fasten the stirrup to the armature by its bolt, and insert the lever, as shown in the figure. The lever lies in the same plane with the bar *B*. The longer arm is ten times as long as the shorter. Hang a spring balance on the longer arm, and connect it by a chain and screw-coupling with the bar *B*. The armature is torn loose by turning the screw, the balance indicating $\frac{1}{10}$ of the strain. The armature is prevented from moving any great distance, by the projections *x y* on the arch, which is itself held down by the hook *u*, fig. 642. The apparatus remains in complete order after the parting of the armature.

The poles of the magnet are made to project above a little table, in order to use it in experiments on diamagnetism.

In order to obtain the maximum effect, the resistance of the battery must be equal to that of the conductors; knowing, therefore, the resistance of one element, and the length and thickness of the wire, it is a matter of simple calculation whether the elements should be combined singly, or by twos or threes, or used as a single element. Let $N$ be the number of elements, $w$ the resistance of one of them, $l$ the resistance apart from the battery, and $n$ the number of pairs to produce the maximum effect; then $n = \sqrt{\frac{N\,l}{w}}$, and $\frac{N}{n}$ pairs are to be combined in one, therefore $n$ must have such a value as to make $\frac{N}{n}$ a whole number.

It is well to mark upon every such spiral the length and thickness of the wire, even though there be no present necessity for knowing its resistance; it may afterwards be required for other purposes.

*Fig.* 644.

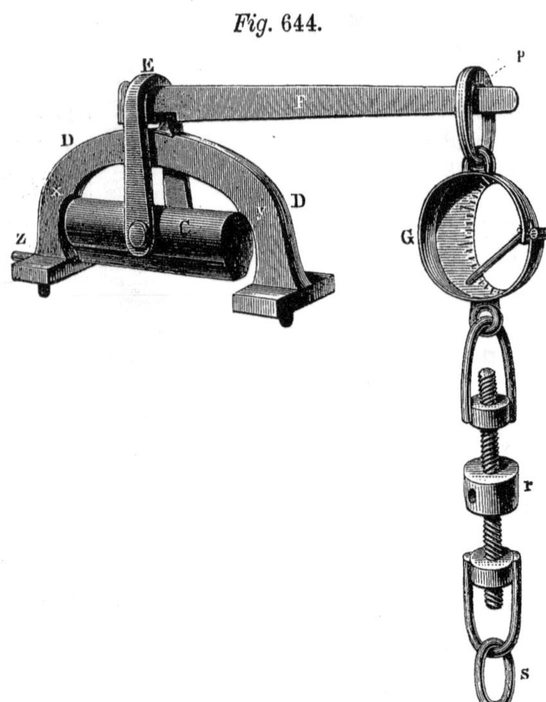


Fig. 645 represents an armature for testing the strength of the poles separately: *a b* and *a′ b′* are iron screws, *c c* a thick brass plate, and *d d* a bar of beech-wood. The hook *e f* is made of brass. As the strength of the poles depends on the mass of the armature, the plates *b b′* must be several inches thick.

*Fig.* 645.

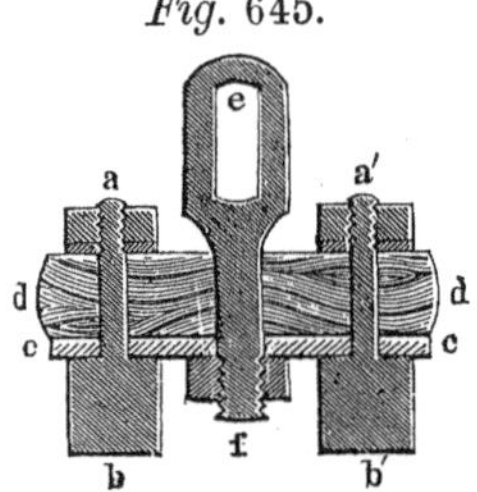


[314] **Magnetizing hard steel.**—In order to impart permanent magnetism to hard steel by the galvanic current, wind 7 or 8 meters of silk-wrapped wire, 3 millimeters thick, into a coil, like fig. 646. The size of the inner space depends on the size of the steel; the length of the axis of the coil should not exceed 3 centimeters. Place the steel within the coil, and pass through the latter a powerful current, from Grove's battery, with about 40 square inches of platina, moving the bar 6 to 20 times

back and forward. Arrest the motion at the middle of the bar, taking care that each half has been passed through the coil an equal number of times, and break the circuit. If the bar be a horseshoe, apply the armature, and do not remove it to take out the bar until the circuit is broken.

*Fig.* 646.

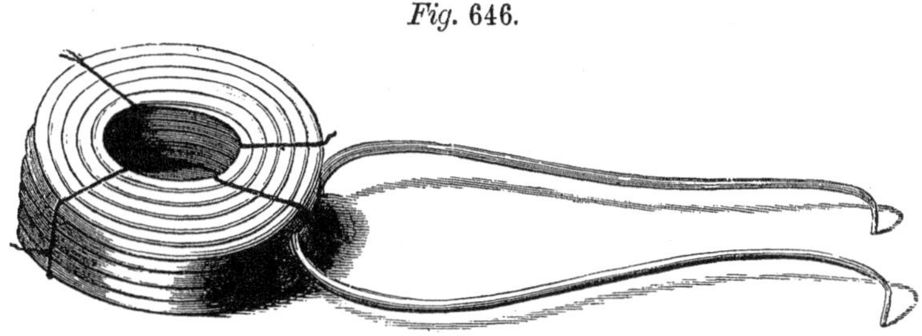

This method is not advantageous, though less troublesome, unless a powerful current can be obtained. It is much better to employ the same wire and the same current to magnetize a horseshoe bar of iron, 4 or 5 pounds in weight, and use this to magnetize the steel, with the double or even the single touch. The poles of the electro-magnet must be close together for the double touch, else it is apt to produce succession points. With very powerful currents the double touch cannot be used, because it is sure to produce such points.

[315] **Electro-magnetism as a motive power.**—Fig. 647 represents a very simple contrivance for producing a continuous rotary motion, by means of an electro-magnet. A circular cavity *a b* is made in the board *M N*, fig. 648, and divided into two equal parts by a partition

*Fig.* 647. *Fig.* 648.

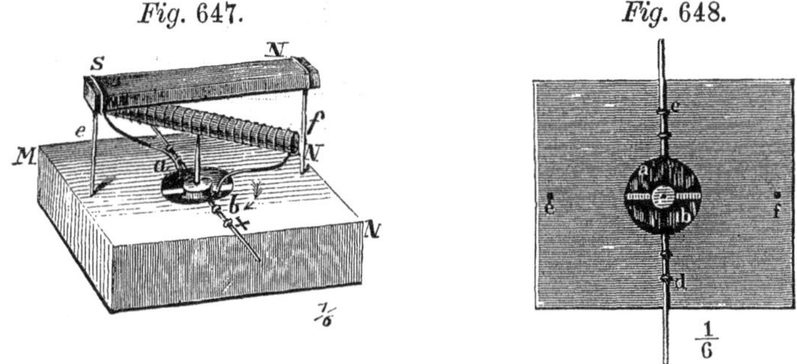


of some insulating substance, 1 or two lines lower than the depth of the cavity, and 2 lines thick. A pointed steel wire is driven into the projection left standing in the center of the cavity, (by seizing it with a hand-vice and striking this.) An iron rod, wrapped with insulated copper wire, is balanced on this pivot, so as to turn freely like a magnetic needle. The bar may be covered with silk, and wire coiled around it, keeping the turns a little way apart. The ends of the wire are bent downward, and dipped into the mercury in the cavity just deep enough to clear the partition. A copper wire *c d* connects the mercury in each half of the cavity with a battery. Mercury is poured into each side until the convex surface rises above the partition but does not flow over. A bar-magnet, about the same length as the electro-magnet, is fixed horizontally a little way above it. The action of the apparatus is evident.

Another more complicated apparatus for the same purpose is represented in fig. 649. A horseshoe magnet is fixed upright on a board, and

*Fig.* 649.

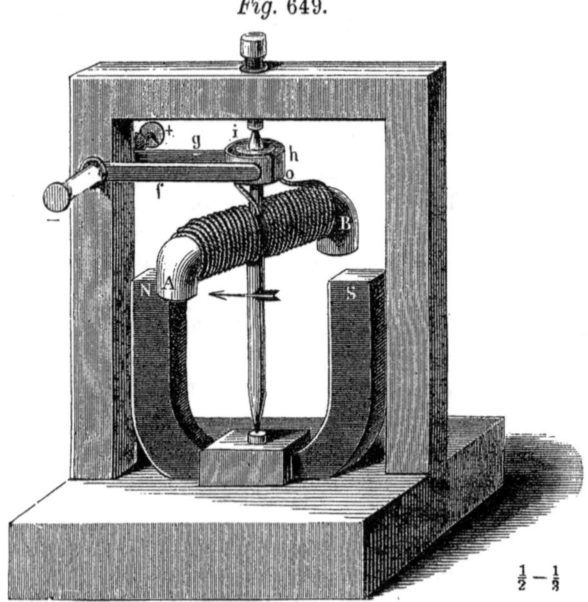


the electro-magnet *A B* made to revolve above it. The axle of this carries a wooden drum, to the circumference of which two brass plates are fastened, with a space of about 1 line between them; one end of the spiral is soldered to each of these plates. The vacant spaces are at the ends of a diameter, at right angles to the length of the electro-magnet. By means of the metallic springs *f g* a communication is made with the battery, until the electro-magnet passes the poles of the steel magnet, when it is broken, and the current immediately passed in the opposite direction. The friction of the springs offers more resistance than the wires meet in passing through the mercury.

A cheap and simple working apparatus is represented in fig. 650. The

*Fig.* 650.

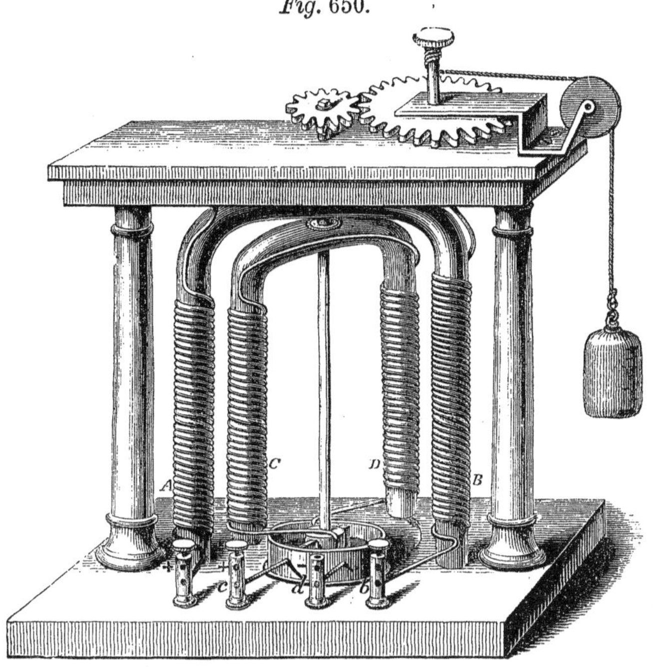


electro-magnet *A B* is substitued for the steel magnet. The construction of the apparatus is evident from the drawing. The binding-screws *a b* are connected with the wire of the fixed magnet, and *c d* with the opposite sides of the mercury cup. It is better to employ two separate currents than to send the same current through both spirals. In the latter case, the current must enter at *a* and pass out at *c*, *b* and *d* being connected. The figure is on a scale of 1 to 5, and it would require a strong current to raise the weight represented.

[316] **The electric telegraph.**—If a well-made working model cannot be obtained, the apparatus fig. 651 will serve to show how telegraphic signals can be made by electricity. Two boards, *M N* and *O P*, are fixed at right angles to each other, and to one of them is attached a little electro-magnet of soft iron. A small clock bell is fastened near it. The ends of the spiral are fastened to *M N*, and connected with the poles of a distant battery. The wire *b*, which serves as a hammer, is elastic,

and has opposite the magnet a plate of soft iron. The position of this wire and the length of the cross piece should be such that when free, the further knob will be 0·1 line from the side of the bell, and the plate 0·5 line from the magnet; and when attracted by the magnet, the nearer knob will be 0·1 line from the bell. To prevent the magnetism remaining in the iron after the current ceases from holding the hammer fast, they are kept from actual contact by a piece of paper pasted on the iron plate. By opening and closing the circuit, the hammer will be made to strike the bell by the alternate attraction of the magnet and the elasticity of the spring. A considerable number of conventional signs can thus be made.

*Fig.* 651.

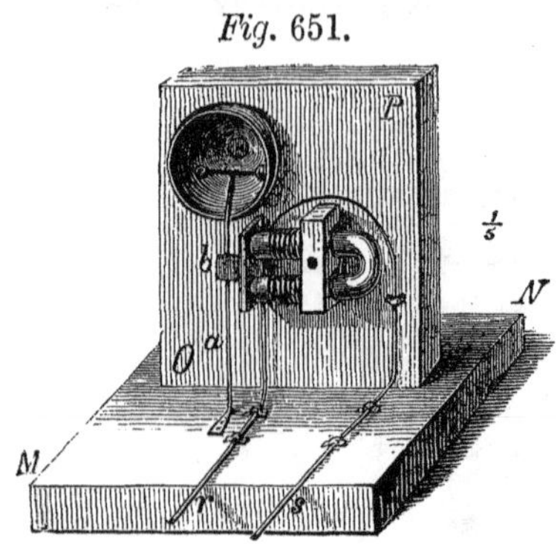


The principle of the English needle telegraph can be shown with the galvanometer, by inserting a pole-changer, and using a very feeble current.

[317] **Ampere's stand.** — A simple form of this indispensable apparatus, which is tolerably easy to make, is represented in figs. 652 and 653; *a b c* are three brass plates let into a piece of hard wood, their

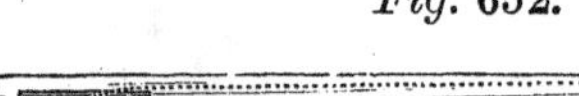

*Fig.* 652.

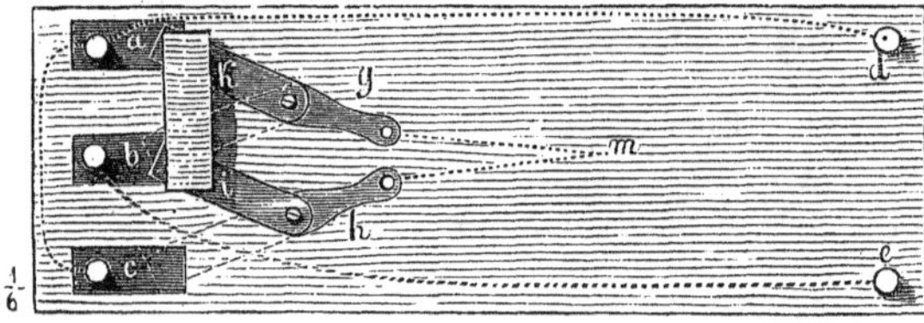


slightly convex surfaces rising in the middle, a very little above the surface, binding-screws or wires being attached to all; and *a* and *c* are connected with each other and with the screw *d* by a wire sunk in the wood. The middle plate is connected with the screw *e*. Two other plates, *g* and *k*, are fastened to the board, and upon these two others, *i* and *k*, turning on screws passing through them and the two former; *k* and *i* are connected by a block of wood fastened by a single screw from below, so as to turn freely. This block may be loaded with lead. It is evident when the plates *i k* are in the position shown, and *a* or *e* connected with the positive pole and *b* with the negative, the plate *g* will be connected with the positive; but if they be placed as indicated by the dotted lines, *g* will be connected with the negative and *h* with the positive. In an intermediate position they will not form part of the circuit.

Two thick brass wires, *r s* and *r′ s′*, fig. 653, are inserted in the ends of *g* and *h*, and fastened by nuts and screws below. These wires are bent at a right angle and terminate in mercury cups, which stand vertically over each other and about 2 inches apart. The wires are separated by bits of wood, and the whole wrapped with silk. The mercury cups are very shallow, and a bit of watch-glass is cemented with sealing-wax to the botom of the upper one. The cups should be both riveted and soldered to the wires. Before use they should be scraped bright in spots on the inside.

*Fig.* 653.

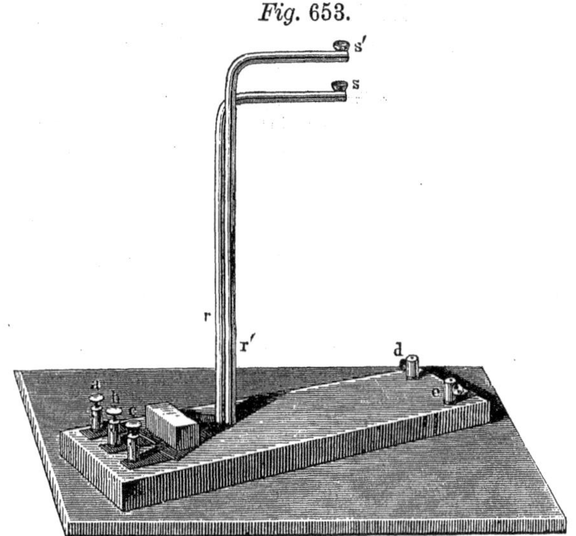


[318] **The pole-changer.**—Among the numerous contrivances for this purpose, none seems to the author so easily understood as the one just described. Those pole-changers are preferable which do not require the use of mercury.

It is convenient to have a pole-changer independent of the rest of Ampere's stand, which can be made by putting binding-screws in the place of the wires *a a′*, *d* and *e* (fig. 652) being then superfluous. In using the pole-changer it is important, not only that the course of the current should be perfectly clear, but that the instrument should occupy little space, and be easily inserted in any circuit. The pole-changer devised by Prof. J. Müller, and represented in fig. 654, on a scale of $\frac{1}{2}$, is admirable in all these respects. It consists in a wooden cylinder, turning on an iron

*Fig.* 654.

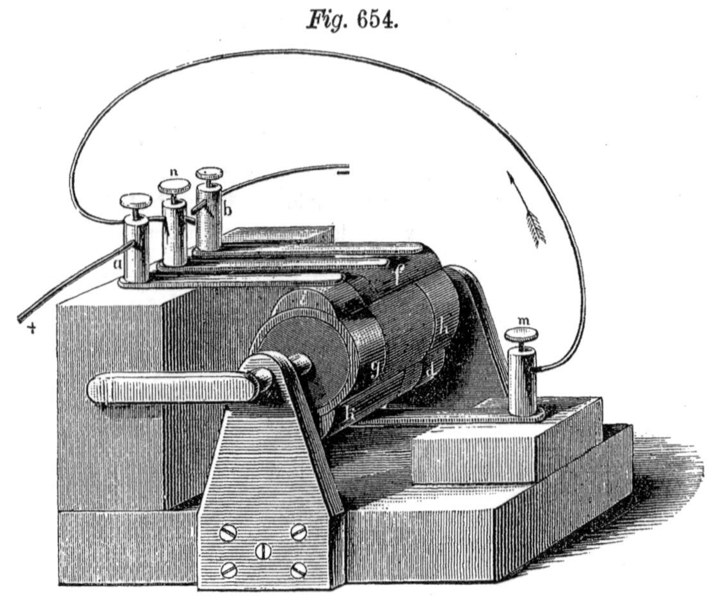


axis by a crank. The ends of this roller are bound with metallic rings *g h*, to which the metallic plates *i k* and *d f* are soldered on opposite sides. Two of these, *d* and *i*, are of the same breadth as the rings; but the other two, *k* and *f*, extend beyond the middle of the roller, not, however, to the other ring. The four binding-screws, *a b m n*, rest upon metallic springs, which rub against the roller. The screws *a* and *b* are connected with the poles of the battery, and *m* and *n* with the rest of the circuit. In the position shown in the figure, the current goes from *a* to the plate *i*, thence through the ring and the plate *k* to *m*, from which it passes through the required circuit to *n*, and thence through *f*, *h*, and *b* back to the battery. By making a half turn of the handle, the spring *a* will fall on *k*, and the current will pass from *k* to *n*, and thence through *m*, *f*, *h*, and *d*, in a direction the reverse of the former. When the arm *l* is in a vertical position, the springs *m n* rest on wood, and the circuit is, therefore, broken. Care must be taken that the springs be quite smooth, so as not to abrade the plates, otherwise the particles of metal will render the wooden cylinder a conductor. When the plates are made of the thickness indicated in the figure, the springs need not touch the wood.

By using mercury, a very convenient pole-changer can be made at a trifling expense. Take a board, *A B C*, fig. 655, 3 inches wide and 4 in length, and bore in it 6 holes, half an inch in diameter. Fasten copper wires by staples in the positions shown in the figure, keeping the two which cross each other apart. Fasten two bent wires in a bit of wood as long as the space between the holes 5 and 6, and across these fasten two others, as seen in the figure, just long enough to clear the board when the first two stand upright in the holes 1 4. The least impulse will throw this cradle from one side to the other. Fill the holes with mercury, and amalgamate the ends of the wires, and the apparatus is complete. The current is admitted at 1 and 4, and goes out at 5 and 6. The arrows indicate the course of the current for the position given.

*Fig.* 655.

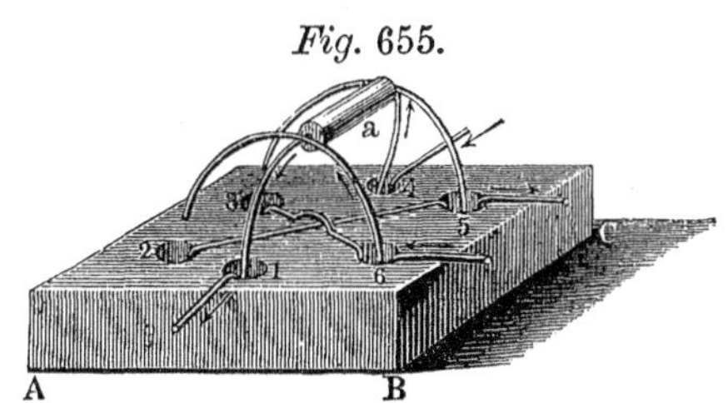


[319] **Rotating conductors.**—The conductors to be hung in the cups *s s'*, fig. 653, are made of copper wires, well wrapped with silk at the point where they are tied together. The steel points are soldered to the copper, and must be brightened before using the apparatus, for iron in contact with copper rusts easily, especially when soldered with tin. For this reason the points should not be made of very thin wire. The copper should be flattened, and the points both riveted and soldered. The lower wire, which need not be pointed, must be exactly in the axis of rotation, otherwise the mercury will offer too much resistance to its motion. The conductors, figs. 656 and 657, are made as large as the horizontal arms of the supporting wires will admit.

*Fig.* 656. *Fig.* 657.

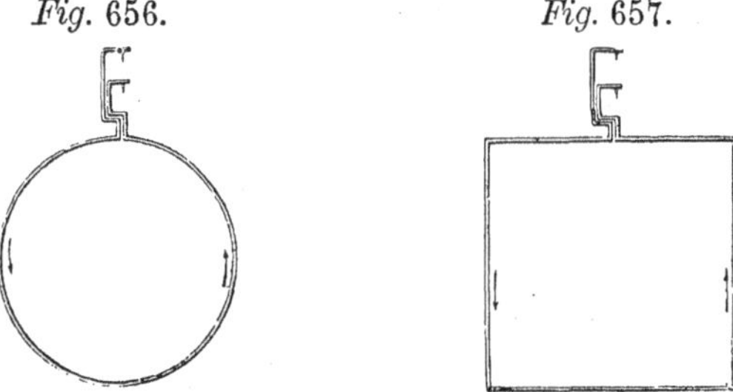

To show the effect of a second current, conductors like figs. 658 and 659 are preferable, because they are astatic in respect to the terrestrial magnetism. The first is for horizontal, the latter for vertical currents. Where the wires come together they are insulated either by slips of cork or by wrapping one with silk, and bound together with silk.

*Fig.* 658. *Fig.* 659.

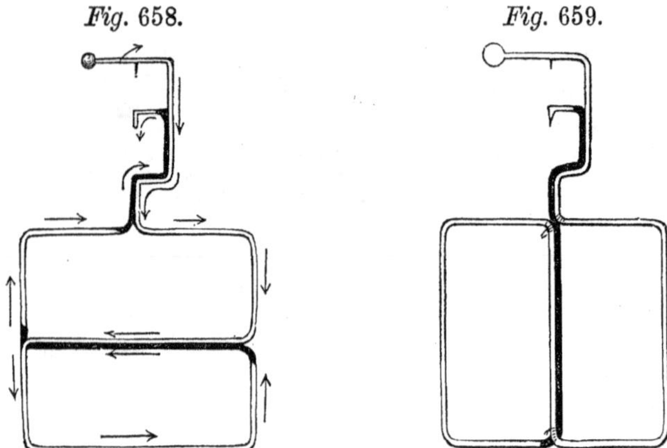

Besides the conductors figs. 656 and 657, we must have two wires of the shape of fig. 660, fitting into the binding-screws *d e*, fig. 654. The one must be bent, so that the straight portion will stand upright between the screws *d* and *e*, while the curve projects out behind the board; it serves to exhibit the attraction and repulsion between the vertical parts of the conductors, figs. 657 and 659, the two rods which support the latter not acting on them, because the current passes in opposite directions in the two. It requires a tolerably strong current for this experiment, because it has a considerable circuit to traverse. The second wire, like fig. 660, is bent so that when inserted in the screw the straight part *a b* will lie in a horizontal position just below the horizontal part of the conductor, figs. 657 and 658, in order to show the effect of rectilinear currents crossing each other.

*Fig.* 660.

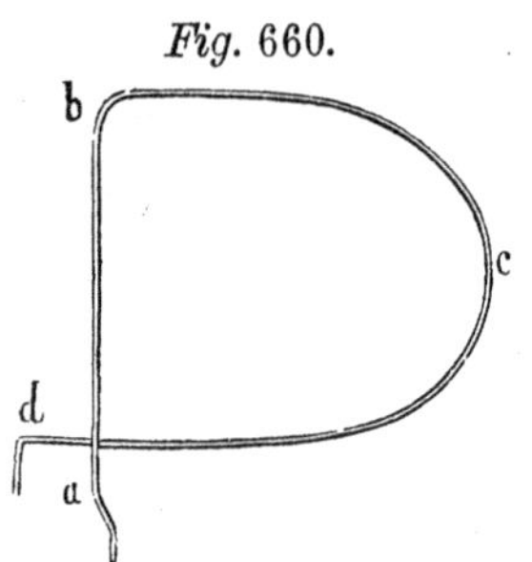


*The helix.* The ordinary form is represented in fig. 661. The wire begins in the middle, passes from one end to the other in the axis of the spiral, and makes the same number of turns in returning to the center. The rings should not be less than 2 inches in diameter, and the current must traverse them all in the same direction. Small rings give the apparatus more the form of a magnet, but they have not sufficient rotary force. The rings are tied with silk, at equal distances, to a slip of wood. In the form figured, it is difficult to make the rings neatly. It is easier to make the coil entirely first, tie it to the bar, and then bend the ends of the wire together, fig. 662, in the center.

*Fig.* 661. *Fig.* 662.

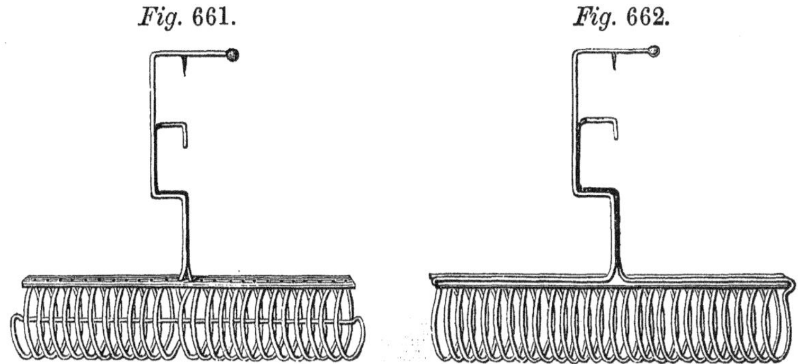

The ends of a second helix are inserted in the binding-screws *d e*, fig. 654, to show the effect of two helices on each other. Many helices do not work because the wire is too slender; it should be over 1 millimeter in thickness. There is less to fear from the increased weight than from enfeebling the current, for the resistance of the helix is considerable. The action of a steel magnet on the helix can be shown under almost any circumstances, but with terrestrial magnetism only it requires nicer construction. This action is easily seen in the conductors, figs. 656 and 657, because they are large.

It should be remarked of this apparatus, that the division of the current between two conductors makes both branches weak, and as they are unequal, the result is often doubtful, unless a strong current can be produced. It is better, in such a case, to pass separate currents through the two conductors. A Grove's battery, with 10 to 12 square inches of platina, is sufficient for any form of the experiment; or six Bunsen's elements, three being used for the fixed and three for the movable conductor.

If only limited means are at command, the stand may be made in the form of fig. 663; the two rods then attracting and repelling the rotating conductor, it must place itself in the plane of these rods. A single Wollaston's element is sufficient for this.

The attraction and repulsion may be exhibited well in two coils of copper ribbon, suspended by strips about 3 feet long, as seen in fig. 664, a pole-changer being inserted in one circuit.

*Fig.* 663. *Fig.* 664.

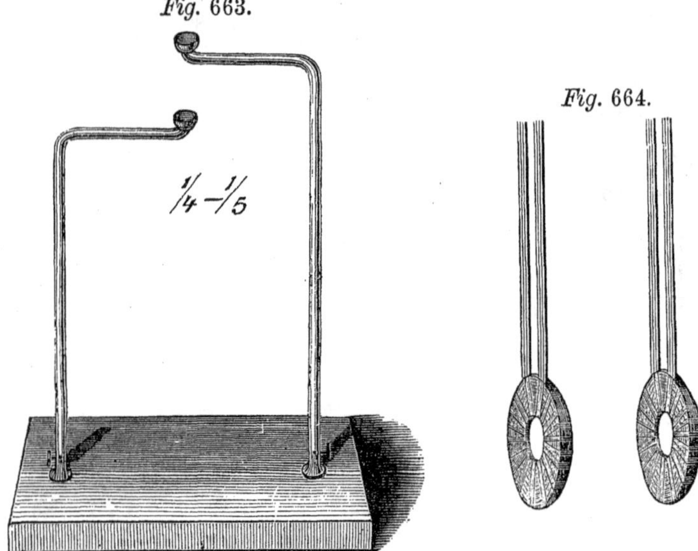


[320] **Floating currents.**—The easiest mode of showing the effect of magnets and terrestrial magnetism on a movable current, is to float the conductors. Take a Wollaston's element, made of 1 to $1\frac{1}{2}$ square inches of thin zinc. Solder copper wires, 1 to 2 millimeters thick, to both metals, and tie the plates together with waxed thread. The element must be made as light as possible, that the wire may be larger. After the plates are put together thrust the wires through a cork, bend them into a ring, and solder them together. The ring may be made of several turns and the ends soldered to the plates after being thrust through the cork, fig. 665. If this element be floated on acidulated water, the ring will take a position at right angles to the magnetic meridian, or parallel to a straight current passing over it. The effect of a magnet is most striking when it is but little longer than the diameter of the ring, and is held horizontally near the wire.

*Fig.* 665.

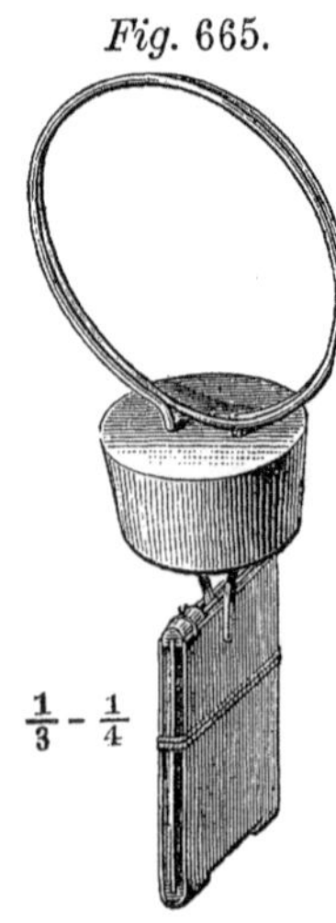


[321] **Rotary apparatus.**—Among the various contrivances for

producing the rotation of a current around a magnet, or *vice versa*, the following are worthy of notice for their simplicity and certainty.

(*a*) Have a tray made of zinc, like fig. 666, with a broad strip of copper soldered across the middle of it, into which screw a stout copper wire, terminating in a mercury cup. Fill the tray with acidulated water, set it on a little perforated stand, fig. 667, and place the steel pivot of the copper conductor, fig. 668, in the mercury cup, and the ring in the liquid. If the pole of a magnet be held in the aperture in the center of

*Fig.* 666.

*Fig.* 667.

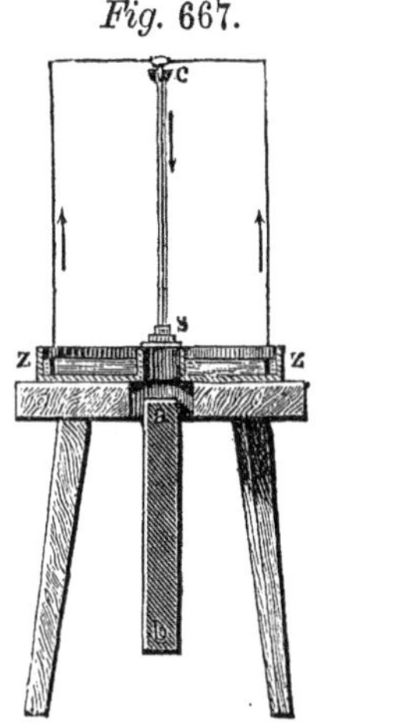


*Fig.* 668.

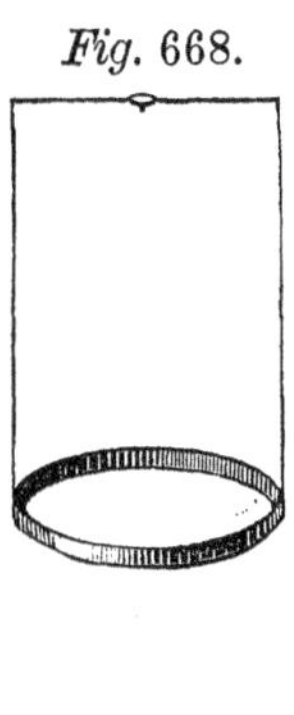

the disk, as represented in fig. 667, the ring will begin to rotate. Instead of the magnet, a coil of insulated copper ribbon, about 30 feet long and ½ inch wide, may be laid around the outside of the zinc disk. The left side of fig. 669 represents a modification of the same apparatus. The copper band is bent into a stirrup, by which the whole apparatus can be hung on one limb of a horseshoe magnet.

(*b*) The right side of fig. 669 represents another rotating battery. A wooden cup *a a* is slipped over the end of a horseshoe magnet, and kept there by a spring. This cup contains mercury, and a little cavity is made in the pole of the magnet in which a pivot rests, supporting several wires which dip into the mercury. A pointed wire from the rod in the middle dips into a mercury cup on the top of the pivot. A rotary motion begins as soon as this wire and the wire *b*, communicating with the mercury in *a a*, are connected with the battery.

(*c*) Fig. 670 represents the rotation of a helix around the pole of a magnet, the poles being changed, as explained in § 315. A horseshoe magnet is supported on a brass tripod, which can be used for the foregoing apparatus, and in the center of it a brass rod *b* sliding in the tube *a*. This rod supports a disk of wood, with a circular basin for mercury. A low rim of glass or pasteboard is fixed around it, to prevent the spattering of the mercury. Each side of the basin is connected by an amalgamated wire with a binding-screw, one of which is visible at *c*. The rod *b* passes through the block without communication with the mercury, and has a hole in the top, in which the pivot of the copper bar *k k* turns freely. Two coils of wire *s s*, consisting of 12 to 18 turns, arranged in 2 or 3 layers, are fastened to the ends of this bar. One end of each coil is tied with silk to the bar, and the other carried, without contact, along the bar to the center, where it dips into the mercury. The current traverses both coils in the same direction, and they rotate like an electro-magnet when the binding-screws are connected with a battery.

*Fig.* 669.

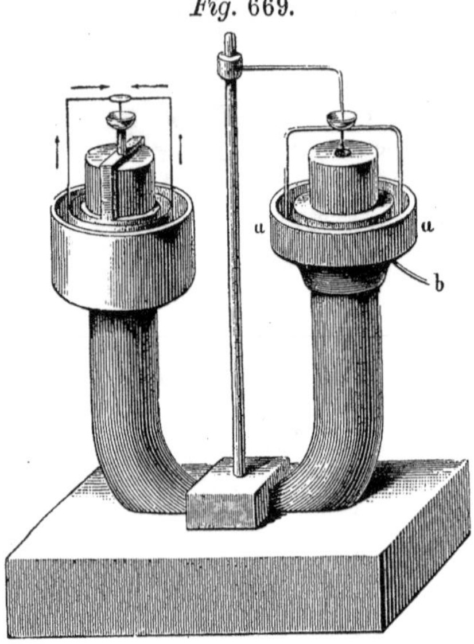


*Fig.* 670.

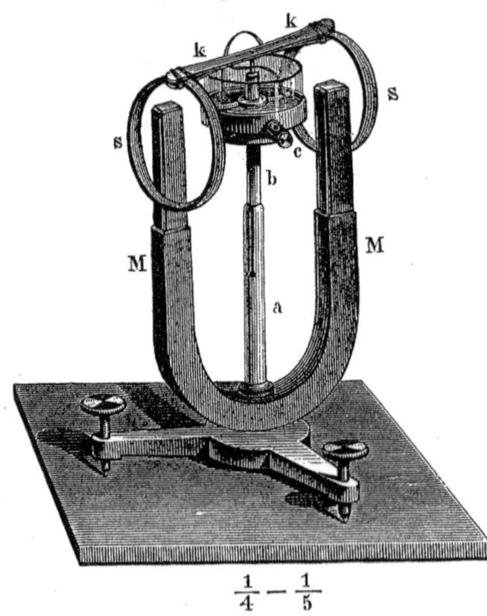


$\frac{1}{4} - \frac{1}{5}$

(*d*) The following apparatus produces rotary motion without the aid of a magnet. *A*, fig. 671, is a mercury cup divided into two parts, around which is placed a ring *B B*, formed of three layers of insulated copper wire with 6 turns in each, laid in a groove *a a*, as seen in section in fig. 672. The ends of the wire are amalgamated and thrust obliquely through the bottom into the mercury, one on each side. A second ring *C C*, exactly similar, turns on ivory bearings around a steel rod, fig. 673, fixed in the center of the mercury cup. The lower bearing rests on a brass shoulder on the rod, and the ends of the wire dip into the mercury. The mercury cup is fixed on a base furnished with leveling-screws, and with binding-screws connected with the mercury.

*Fig.* 671.

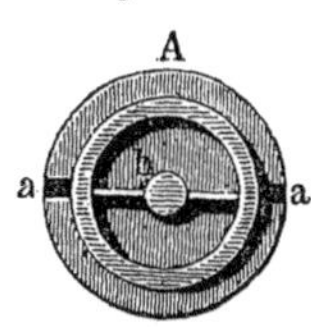


*Fig.* 672. *Fig.* 673.

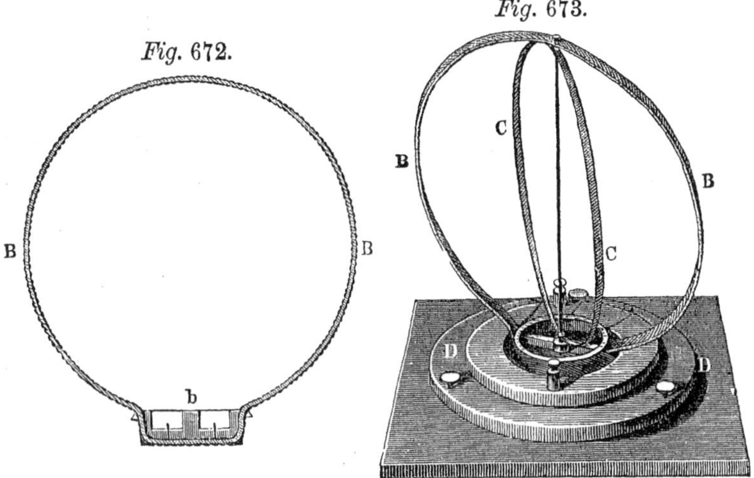


The current divides itself pretty equally between the two rings, and is reversed in the movable one. The axis of the inner ring might be made to turn with it.

(*e*) The easiest method of making a magnet rotate by a galvanic current is represented in section in figs. 674 and 675. A tumbler is set in the board *M M*, upon the bottom of which is cemented a bit of wood with a steel point in the center. For the magnet, take a bit of round steel, 3 inches long and 2 to 3 lines thick, turn it accurately on the lathe, then harden, polish with emery, and magnetize it. Fit a cap of hard wood or bone *a a* around the pole *N*, and set the magnet between the steel pin below, and a screw *b* fixed in the brass brace *c d*, allowing it a little play. The wire *e* is bent into an elastic ring, so as to fit closely against the inside of the glass, and amalgamated. Pour some mercury in the cup *a a*, and enough in the tumbler to buoy up the magnet so that it will just float between its two bearings, and turn very easily. The screw

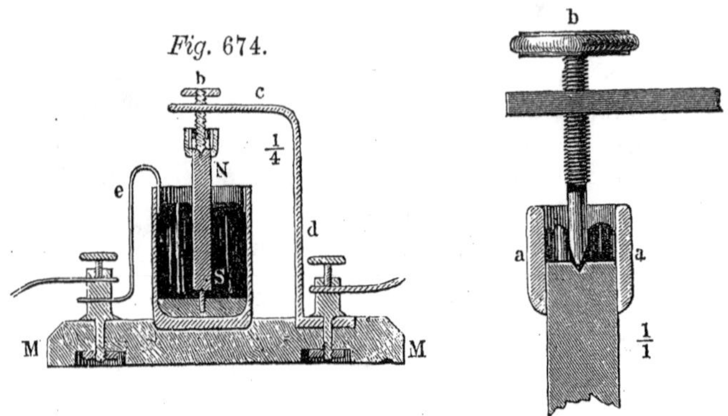


*Fig.* 674. *Fig.* 675.

*b* must work rather stiffly. The battery is connected with the wire *e* and the brace *c d*. If the cap be placed on the pole *S*, or the current reversed, the magnet will rotate in the contrary direction.

Some such rotary apparatus must be exhibited, and the explanation of others will serve as a good task for practice in the theory.

## (*f*.) EXPERIMENTS ON INDUCTION.

[322] **Winding insulated wires.**—When insulated wires are wound on spools, they should be either wooden spools with very thin walls, or tubes of pasteboard with wooden disks, about 3 lines thick, glued to the ends. In metal spools, the ends should be 2 millimeters and the shaft $\frac{1}{2}$ millimeter thick; but they must be slit their whole length and the space filled with some non-conducting substance, otherwise secondary currents will be induced, which disturb the result. In winding, a wooden plug must be fitted tightly into the tube, to prevent its being compressed. Varnished paper is laid between the several layers, so that if there is any imperfection in the insulation the current cannot pass from one layer to the other. Imperfections of insulation should be mended during the winding, by wrapping varnished paper or silk around the spot.

For experiments on tension, the wire must be varnished before winding. To do this without gluing the whole together is a tedious task. It can only be done rapidly in the open air, by standing at such a distance from the spool that the varnish will dry before it reaches it. For many purposes it is recommended to varnish, moreover, each layer; in this case no paper is required, but each layer must dry thoroughly.

Thick wires should be wound by hand, and require the aid of another person. Thin wires can be wound very rapidly and smoothly by fixing the spool in a lathe.

The ends of the wires are generally thrust through the ends of the spools, but they ought in addition to be tied with silk.

[323] **Ribbon-coils.**—Flat coils of copper ribbon are needed in many experiments. The ribbon should be of uniform breadth, and the joints filed down to the thickness of the rest. A ribbon of silk or wool a trifle wider is laid between the several turns, and the whole bound fast with silk. In most cases the connecting strip must be soldered on at right angles to the coil. A strip is soldered to the inside, well wrapped, and brought to the outer edge at *a*; here a strip is soldered at right angles, so as to lie flat against the spiral, and another piece at right angles to this, which is bent out at *b*. The spiral is generally closely wrapped with ribbon, 12 turns of which are represented in fig. 676.

*Fig.* 676.

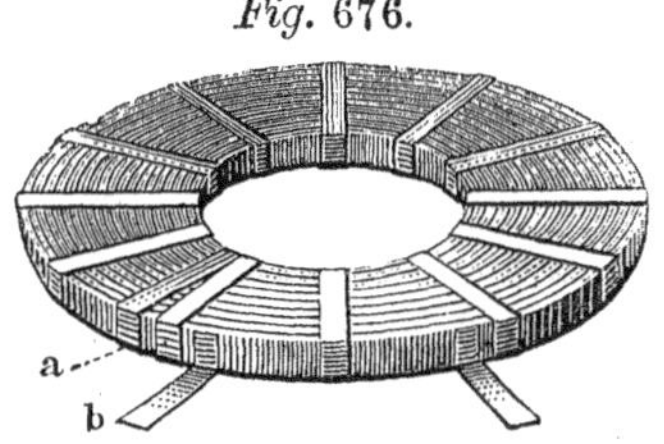


*Fig.* 677.

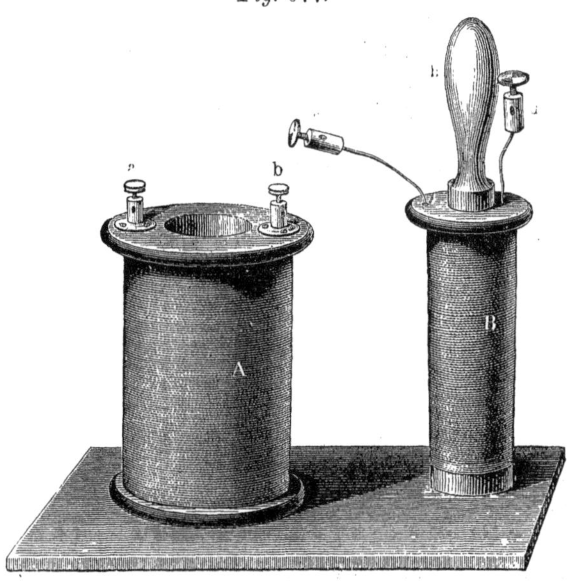


[324] The fundamental laws of galvanic induction can be illustrated very simply by two spirals, *A* and *B*, fig. 677. Each contains about 100

feet of well-insulated wire; and the smaller can be placed inside the larger, whose walls are very thin; for which purpose the ends of the wire are thrust through the handle. The cylinder *B* is also hollow, to admit an iron rod about 2 lines thick. To use the apparatus, connect the ends of the coil *A* with a galvanometer, and *B* with the battery, without inserting the rod. If *B* be quickly placed inside *A*, the galvanometer needle will be deflected, and again on removing *B* after the needle has come to rest. When *B* is left in *A*, the needle is deflected whenever the circuit is opened or closed. The same result, and even stronger, is produced by using a magnet instead of *B*. The effect of the coil is increased by inserting the rod, for the hypothetical current produced in it by the current in *B*, rotates in the same direction as that in *B*. For these experiments, the galvanometer ought not to have over 200 to 300 turns. If handles be fastened to *A*, and the current in *B* interrupted by a spur wheel or a file, a Wollaston's element, with about 12 square inches of zinc, will give perceptible shocks, even with this short wire.

[325] To obtain a very powerful inductive action, wind 1200 to 1800 feet of well-insulated copper wire, $\frac{1}{2}$ to $\frac{2}{3}$ of a millimeter in diameter, on a cylinder like *A*, fig. 677. Over this, wrap about 100 feet of wire, 2 to 3 millimeters thick. Attach handles like fig. 583 to the ends of the small wire, which are thrust through the end of the spool. Connect one end of the thick wire with the pole of a feeble galvanic battery, (a Wollaston's element, with 2 or 3 square inches of zinc,) and the other end with a file. When the other pole-wire is drawn over this file, any one taking hold of the handles, with moist hands, will perceive very sensible shocks, which will be increased by inserting a rod of iron or a bundle of wires inside the cylinder. This bundle should be coated with sealing-wax. Fig. 678 represents the apparatus complete. The ends of

*Fig.* 678.

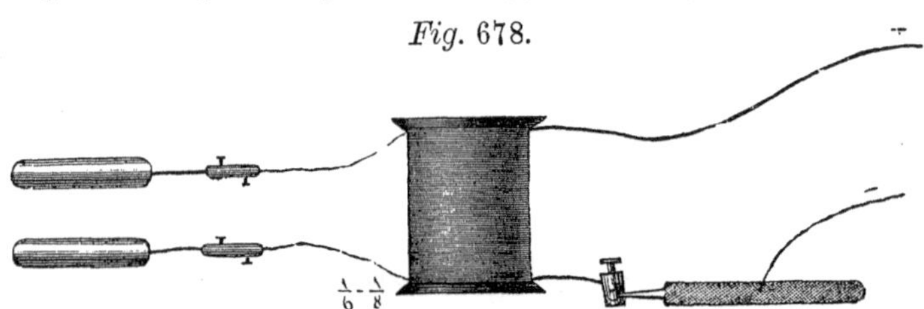

the wire to which the handles are attached should be about 2 feet long, for many persons are very sensitive and injure the apparatus by throwing the arms apart. It is well to direct the person who receives the shock to let the handles drop when the shock is too severe; though this is often impossible, on account of the spasmodic contraction of the hands.

[326] **The extra current.**—After uniting tne two wires, attach to one end a handle and a file by binding-screws, connect the other end, to which a handle is attached in the same way, with the battery; a person taking hold of the handles, while another passes the pole-wire of the battery over the file will receive shocks from the extra current. Fig. 679 represents the arrangement. The apparatus may be made much more convenient by placing the cylinder in a little frame, as in fig. 680, with two little grooves in the base to receive the ends. The ends of the wires are connected with four binding-screws, like fig. 549, fastened to the top of the frame. The interruptions of the current can be regulated better by using a spur wheel, fig. 584, instead of the file. This wheel is usually fastened on the top of the frame, the free end of the spring being in connection with the binding-screw No. 1, through which the current is introduced; and the wire connected with the axis of the wheel is one end of the thick spiral, the other end being connected with the binding-screw No. 2. The ends of the fine wire are connected with the screws 3 and 4, to which the handles are attached for the experiments on induction. To obtain the extra current, connect two binding-screws by a wire, so that the current will traverse the whole coil in the same direction; these two screws must be marked directly after making the coil, for an error may arise afterwards. The handles are then attached to the thick wire connected with the axis of the wheel and to the end of the fine wire, which passes to the battery.

*Fig.* 679.

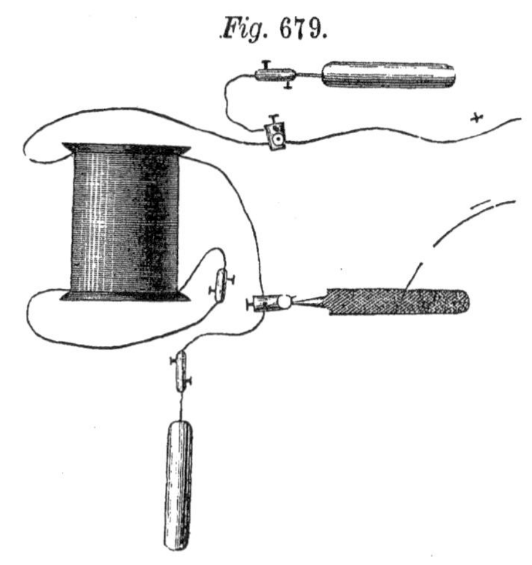

*Fig.* 680.

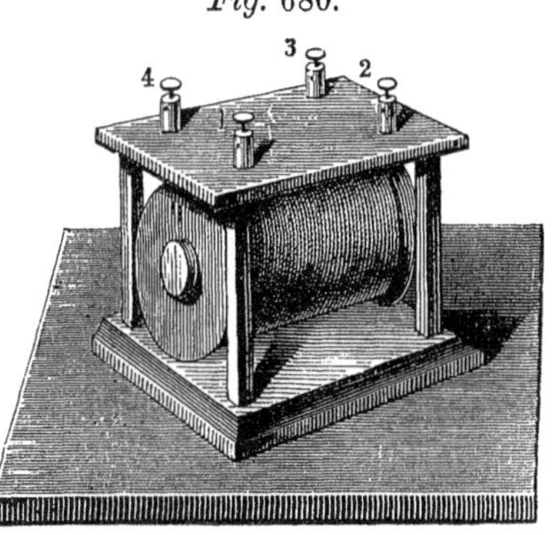

[327] A very simple arrangement, called the magnetic hammer, may be substituted for the spur wheel. Fasten the frame, fig. 680, between two cleats upon a board, fig. 681, and place the iron rod in the cylinder. Screw a piece of soft iron *b*, which has a smooth slightly convex surface, to a bit of brass wire *a*, about a line thick, and pointed at one end. The point is amalgamated and a hole bored in the center, by which it is riveted to a copper wire *c d e*, beaten flat at the end to serve as a spring. The hammer is fastened to the board so that the iron will strike about the center of the cylinder. The wire *c d e* is made long enough to reach the binding-screw 1. A second wire *f g*, terminating in a little amalgamated disk, is also fastened to the board. The spring of the hammer is so arranged that the point just touches the disk *h*. It is better to solder a little plate of platina to the disk *h*; the piece *a b* may then be made entirely of iron, and a little platina tip soldered to it, which renders amalgamation unnecessary. The magnetic force is also greater in a long piece of iron, and the apparatus consequently acts better and with a feebler current. A current introduced at *f* will pursue the course *f g h a c d e* 1, and pass from thence to the thick wire, and out at 2. The iron core is thereby magnetized, and attracts the hammer *a b*; this breaks the current, and the iron, losing its magnetism, the spring carries the hammer back to *h*.

*Fig.* 681.

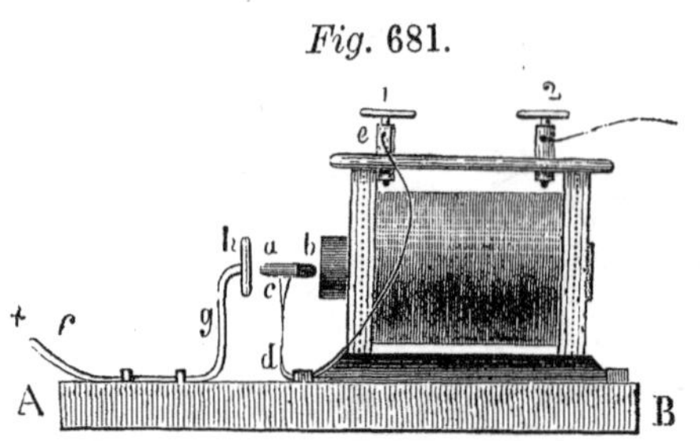


The same arrangement can be used for the extra current. It is done by attaching a handle to *c d e* by the screw 1, connecting the two wires, and attaching to the fourth screw the second handle and the wire leading to the battery.

[328] Reinsch has devised an induction apparatus very well adapted for instruction, which is represented about $\frac{1}{3}$ to $\frac{1}{4}$ the real size in fig. 682. It consists of two cylinders fitting inside of each other, each wrapped with about 100 feet of wire of different colors. The inner one is hollow also, to admit a bundle of wires. The board *A* is screwed to the base, after fastening the binding-screws from below, and has an aperture just large enough to admit the larger spiral. The ends of the outer or of the inner spiral can be connected with the screws 2 and 3, and those of the other, together with the wires of the galvanometer or the handles, with 4 and 5. The circuit may be broken by the magnetic hammer. The

hammer and the bundle of wires may also be removed and a magnetic bar introduced into their place, or the outer spiral connected with the screws

*Fig.* 682.

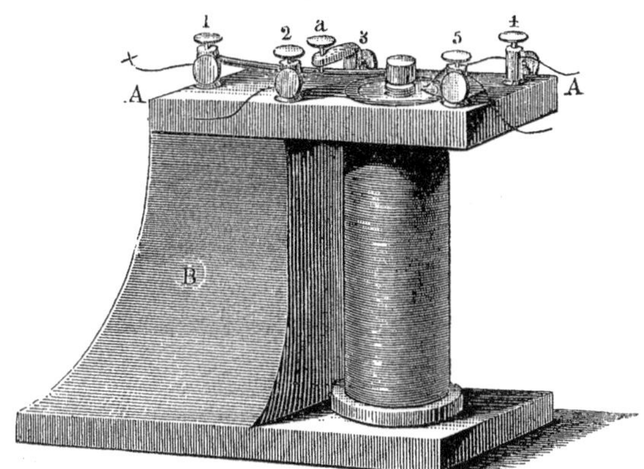


4 and 5 and the galvanometer, while the inner spiral is connected by long wires with the battery, and alternately inserted in and drawn out of the outer. The spirals may be united so that the current shall traverse both in the same direction, and used for obtaining the side current. Where the means at command are only moderate, this apparatus is very good.

[329] When an apparatus is required merely for the purpose of giving shocks, one like fig. 683 will answer, the current being broken by means

*Fig.* 683.

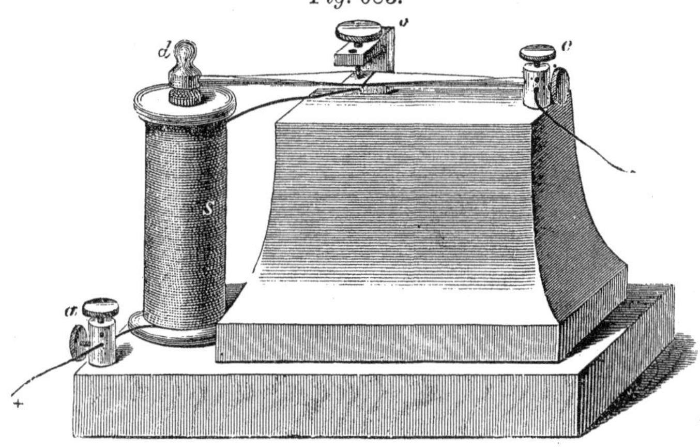


of the magnetic hammer. The ends of the spiral are connected with the screws *a* and *b*, and the poles of the battery with *a* and *c*.

[330] **Electrical tension on the induction wire.**—If the ordinary handles be attached to the ends of the fine wire of the induction apparatus, fig. 681, and, without being brought into actual contact, laid very close to each other, a constant succession of sparks will be seen in the dark to pass between them while the apparatus is in action. A higher tension cannot, of course, be obtained in an apparatus whose parts are only separated by wood. The iron rod within the spiral also becomes electrical, and little sparks pass constantly from it to the hammer. If bits of wire be screwed to the binding-screws of the fine wire, and touched with the knob of a gold-leaf electrometer, the leaves will diverge, and a permanent divergence may be produced by lightly touching the vibrating wire, which cannot be produced by a continuous contact. Upon investigation the electricity will be found to originate from the opening current; and the iron core manifests the same electricity. Electricity is developed at the end of the outer wire only when the inner is touched with a conductor, and then it is of the opposite kind. In fig. 681 the inducing wire is on the outside.

[331] **Ruhmkorf's apparatus.**—As this apparatus has been much talked of lately, and I know of no accurate description of it, I will describe one which I made for myself. The form chosen was based upon one made in Paris for the University of Tübingen. The changes which I introduced are chiefly in the arrangement of the hammer, and the course of the current on the base. My design was to enable me to apply at pleasure the mode of fastening the hammer described by Sinsteden, and also to connect the condenser with the inducing wire in two different ways, or to omit it entirely, or substitute for it a Leyden jar.

The spiral of the apparatus is wound on a very thin wooden spool, with a perforation 3 centimeters in diameter and octagonal end pieces of glass. Holes for the passage of the wires are bored in one of the end pieces, and the direction of the coil from these holes marked by white arrows painted on the glass. Upon this spool are wound, first, four layers consisting each of 80 turns of wire, 2 millimeters thick; and then about 30 layers of 500 turns each, of wire not quite $\frac{1}{3}$ millimeter thick. The former is wrapped with wool, and the latter with silk, and varnished. Varnished paper is placed between each layer, and the two wires are separated by a double fold of the same, thickly varnished, and cemented at the ends to the glass. The separate layers of fine wire were not varnished, in order that this expensive material might be available for other purposes.

This spiral rests on a board an inch thick, between four triangular pieces, which are glued down and screwed fast from below, the hypothenuse of each being notched to receive the glass plates. The spiral is thus held firmly and can be easily lifted up when the ends of the wires are detached from the binding-screws. Fig. 684 and 685 represent the

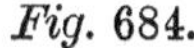

*Fig.* 684.

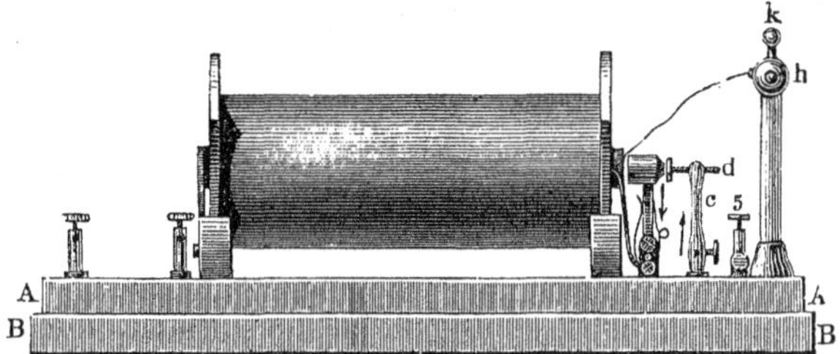


*Fig.* 685.

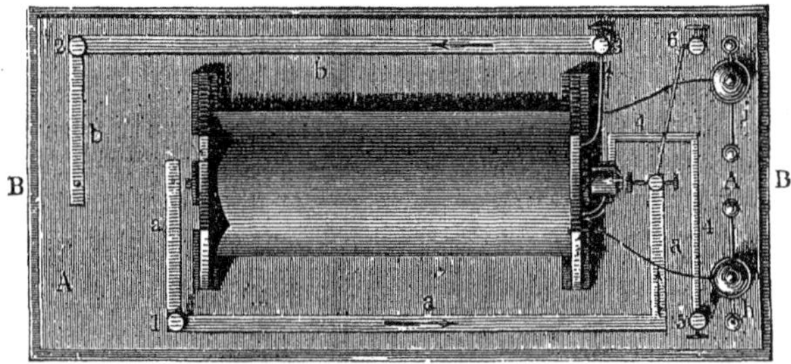


side view and plan of the apparatus. The board *A A* can be separated from the base *B B*, (which will be mentioned presently,) but is ordinarily fastened to it by four screws from below. The course of the current is as follows:—

Two strips of copper *a a a*, *b b*, are inserted in the board *A A*, which are bent at right angles on the left, so that a pole-changer can be laid on them there; they are also furnished with binding-screws, 1 2, to make a direct communication. The current entering at *a* passes into the upright *c*, and thence to the horizontal screw *d*, the head of which is plated with platina. From this it passes into the hammer, which is also platina plated, and descends through the stem. The stem of the hammer is hinged by a screw to the iron pin *e*, fig. 686; the end of the thick wire is bent into a ring and screwed to the same pin, and into this the current passes. In order to insure a perfect contact between the stem of the hammer and *e*, a spring *x*, made of hard copper wire, is fastened between them by two little screws seen in fig. 686; it serves at the same time to press the hammer against the screw *d*, and to draw it away from the bundle of wires in the spiral. This spring not being strong enough with powerful currents, a steel spring *f* is fastened to *e*, the pressure of which is regulated by the screw *g*. The pin *e* ends with a square head and a screw, by which it can be fastened to a plate of iron *m m*, an inch wide, let into the base. A copper wire, (mentioned subsequently,) bent into a ring at the end, is screwed in between the plate *m m* and the nut *n*. This plate is longer than the spiral, and is bent up at a right angle at the other end, until it comes opposite the projecting ends of the spiral wire. Here, however, it consists of two pieces, and the upper part can be taken off after removing the screw *o*. If the stem of the hammer be of iron, the hammer is now in magnetic communication with the bundle of wires in the spiral. This can be prevented by making the stem of the hammer of brass, but the iron *m m* is not entirely without effect until the other end is removed by the screw *o*.

*Fig.* 686.

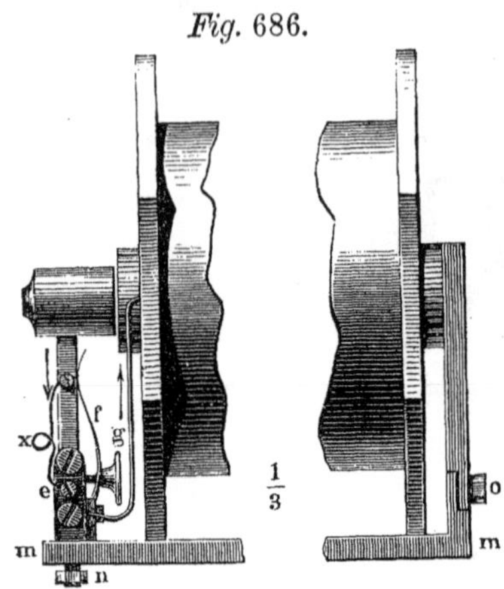


The current having traversed the thick wire passes through the binding-screws 2 and 3 back to the battery.

Pieces of somewhat thicker wire are soldered to the ends of the fine wire, and, after making one turn around the spool, thrust through holes in the glass plate, and fastened under little screws on the insulated balls *h i*, figs. 684 and 685. One of the knobs of the cross wires can be unscrewed and screwed to a pointed copper wire, so that either the pointed end of one wire can be directed toward the knob of the other, or two points or two knobs turned toward each other.

A wire was mentioned above, which was screwed to the end of the pin *e*, fig. 686. This wire is bent twice at right angles, and lies in a covered cavity 4 4 in the board *A A*, passing to the binding-screw 5, under which it is bent to a ring and pressed down by the screw. There is also a binding-screw at the bottom of the pillar *c*, and the binding-screw 6 may be connected by a wire either with this pillar or with the screw No. 3.

In order to obtain a shock by the secondary current, the handles must evidently be attached to the binding-screws 5 and 6, which are made double for this purpose. If *c* be connected with 6, the secondary current on opening is strengthened by the current from the battery; but if 3 be connected with 6, the secondary current alone is felt. The binding-screws 5 and 6 are connected with a condenser placed between *A A* and *B B*, by means of spiral springs of hard-hammered copper wire, fig. 687, the ends of which are fastened under the nuts which hold the screws. *B B* is a frame with a thin bottom and a projecting rim, as shown in section in fig. 688. *A A* rests on the frame, and is held in place

*Fig.* 687.

*Fig.* 688.

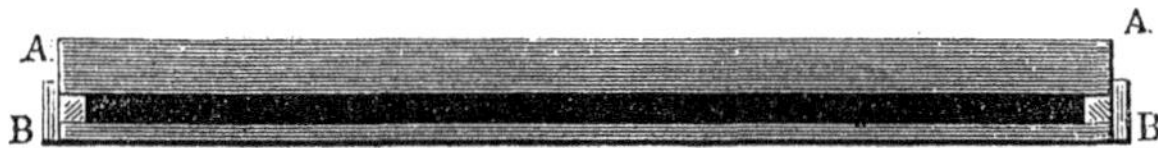


by the projecting rim. The condenser is placed in the space between. It consists of a strip of oiled silk, 3 meters long, made of several pieces cemented together with shellac. The strip is 19 centimeters wide, and coated on both sides to within 1 centimeter of the edge with tin-foil fastened on with shellac. It is folded up several times and paper laid between the folds. The upper end is cut out, as represented in fig. 689, so that the upper sheet of foil is visible at *a*, and a part of the lower at *b*, the top sheet of varnished paper extending only to *m n*. The spiral springs from the screws 5 and 6 press upon brass plates laid on *a* and *b*, and the condenser is thus connected with the secondary current, by which it is charged like any other condenser, and discharged when the main circuit is closed.

*Fig.* 689.

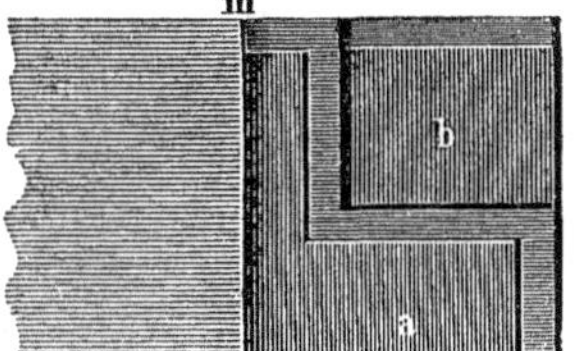


A large battery of Leyden jars, inserted in place of the condenser by connecting one coating with the pillar *c* and the other with screw No. 5, acts in the same manner as the condenser, though less powerfully, on account of the thickness of the glass.

The battery employed should consist of not more than two pairs, otherwise the insulating layer between the turns of the fine wire may be broken through, and the apparatus permanently weakened.

Fig. 690 is a perspective view of the whole apparatus from another side. The pole-changer is left out for the sake of simplifying the figure.

Any pole-changer will answer the same purpose, but for permanent attachment to any apparatus, that which comes with the Paris instru-

*Fig.* 690.

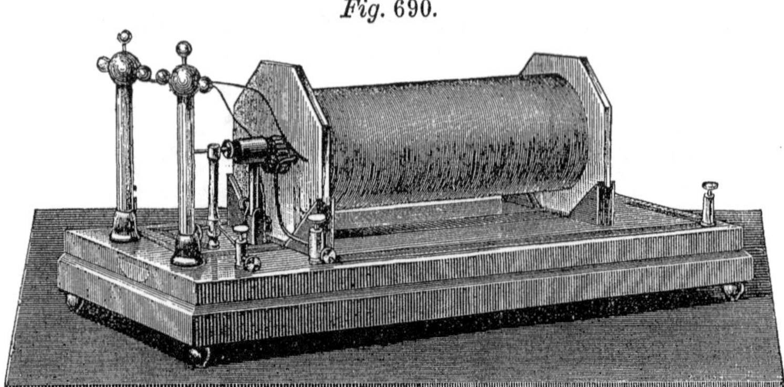

ments excels in convenience and compactness all others. Fig. 691 represents a section, and fig. 692 a view of the side turned toward the spiral, but in a different position from fig. 691. It is very simple. On the ivory cylinder *a* are screwed two copper bars, the screws *e* and *h* coming in contact with the axis, *f* and *g* not going so deep. The ends of the axles turn in pillars *m* and *n*, and the bearings can be contracted by screws so as to make the contact with the axle perfect. The screws which touch the axis are indicated by a mark. An insulating handle *o o* is attached to the axle. Two binding-screws are connected by springs *p q* with the copper bars, the pillars which support the axle being screwed down over two other copper bars *x y* let into the base of the apparatus.

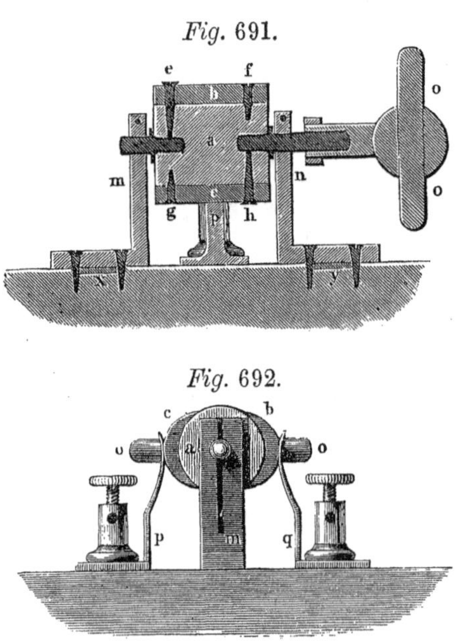


*Fig.* 691.

*Fig.* 692.

[332] When the apparatus is in action, frequent sparks, from 3 to 5 millimeters or more in length, pass between the knobs or points which are connected with the ends of the fine spiral, but they become much less frequent, and are reduced to less than 1 millimeter, as soon as the condenser is taken out. An examination with the electrometer proves the presence of free electricity only at the ends of the outer wire, and all parts of the same excite a burning sensation when touched. The contact with the electrometer must be only momentary, otherwise the leaves immediately collapse; this does not occur, however, when a wire is connected with the electrometer, one end of which is wrapped with silk and bent into a ring, and this ring brought in contact with the end of the spiral. From the character of the electricity and the known direction of the current and the coils, it is evidently due to the current produced when the circuit is opened. The inner end of the wire shows free electricity only by the aid of a condenser, and it is of the opposite kind to that at the outer end. But strong electricity is developed here also when the other end is placed in communication with the earth. When this is done, no part of the apparatus which is in communication with either wire can be touched without receiving violent shocks, unless the person be himself insulated. In order to charge jars, their knobs must also be provided with a wire, the end of which is wrapped with silk.

*Fig.* 693.

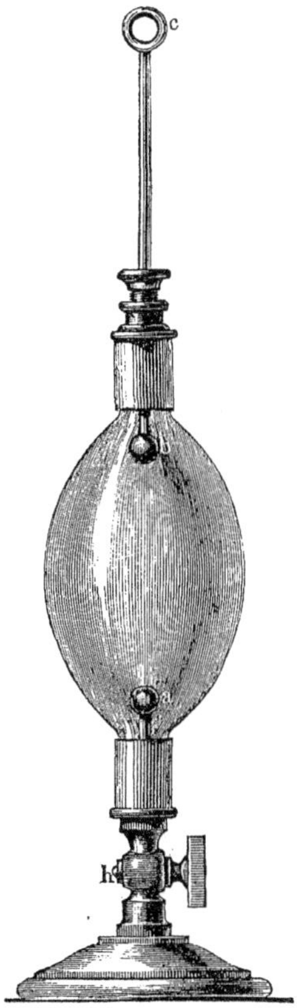

[333] But the most beautiful experiments with Ruhmkorff's apparatus, are those made in a vacuum. For this purpose the receiver, fig. 693, is exhausted of air as completely as possible. The barometer gauge must fall to 1 or 2 meters. One end of the wire is connected with the lower cap, and the other with the sliding rod, the connecting wires being fastened in insulated knobs in which the ends of the spiral terminate. In order to make the experiment very successful, the varnish must be removed from both knobs of the receiver. A splendid stream of reddish light is then seen to pass from the positive to the negative knob, while the latter is entirely enveloped in blue light, and

sparks are visible all over its surface, similar to those which are seen under the microscope in Neef's apparatus. If a few drops of ether or oil of turpentine be introduced into the receiver before the exhaustion, the stream of light is interrupted by dark bands, which seem to advance toward the other knob. If the current be reversed the knobs also change color.

[334] In order to show that the magnetism developed in a piece of soft iron by induction also produces a current of electricity in a spiral wire, take a bit of soft iron, a little longer than the armature of your horseshoe magnet, round it off in the middle, and coil round it 10 or 20 feet of copper wire, so as to form a spiral which will pass between the poles of the magnet without the least contact. The board *A B*, figs. 694 and 695, is cut out so as to leave the end *A* thicker than the rest by

*Fig.* 694.

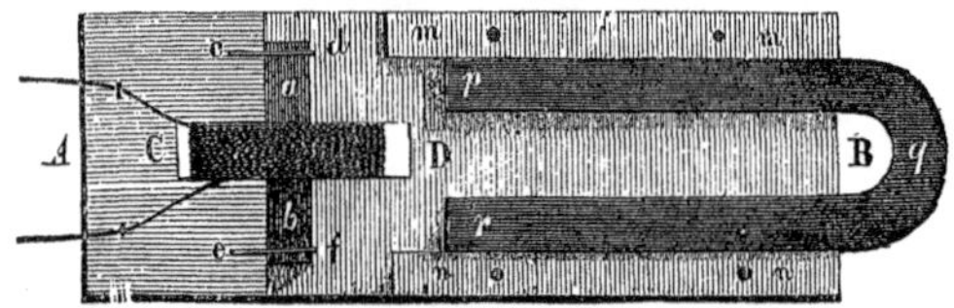

*Fig.* 695.

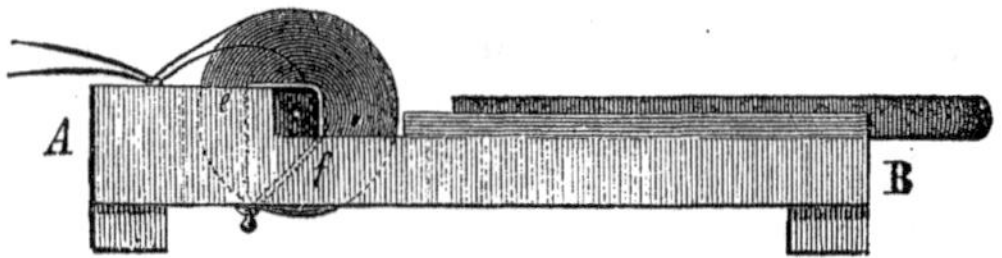

the thickness of the iron, and a slit *C D* made in the middle to receive the coil, the axis of which is fastened down to the board. The ends of the spiral are fastened to the board, so that they can be connected with the galvanometer. Two small bars, *m m* and *n n*, are nailed to the other end of the board, between which the horseshoe magnet, *p q r*, can be rapidly pushed against the armature and torn loose from it. At each separation the galvanometer will indicate a current in the direction previously determined.

[335] **Magneto-electric machines.**—The foregoing experiment is the basis of the explanation of the magneto-electrical machines, the details of which are very various. Those which are provided with steel magnets have many advantages over those which depend upon electro-magnets, but they are more costly. The machines of this kind made by Stöhrer, in Leipsic, are peculiar in their arrangement, and are

convenient, although they have not steel magnets, because the magnet can be made to rotate by the current, and thus afford an illustration of a machine driven by galvanism. The machines with steel magnets made by the same maker are worthy of notice for their powerful effect and convenient arrangement.

It is difficult to give general directions for the management of such machines, on account of the differences in their construction. The interruption of the induced current must occur nearly at the instant when the surface of the armature has passed the pole of the magnet. When not in use, an armature of soft iron must be laid on the magnet. In ordering a machine of this kind, directions should be given to arrange the wire coiled around the revolving armature so that it can be used either as a single long wire or a short, thick one, or as it is usually expressed, so as to increase either the quantity or the intensity of the current. Separate armatures may be used for these purposes.

[336] **Magnetic phenomena produced by rotating disks.**—The experiments on this subject are performed with the whirling tables. To a thick copper disk *a b*, fig. 696, solder a piece of brass *c*, in which a screw is cut to fit the axis of the whirling table, and with this screw as a center turn the disk perfectly round and even. Upon the stirrup of the machine described in § 117, screw a circular board *A B*, fig. 697, about six inches in diameter, with a flange cut on the edge.

*Fig.* 696. *Fig.* 697.

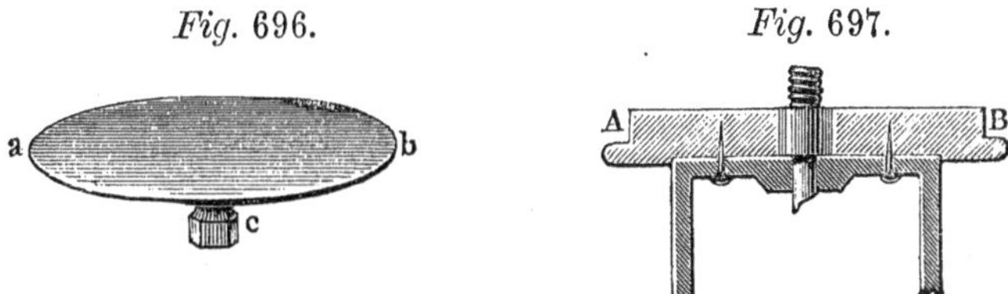


On this fit a cylinder of pasteboard, fig. 698, in which a disk of glass is fastened by rings in such a position that when the copper disk is screwed to the axis of the machine and the cylinder fitted over it, the copper will be close below the glass without touching it. On the glass place a low stand, fig. 699, supporting a magnetic needle nearly as long as the diameter of the copper. A rapid rotation of the copper in its case will cause the needle to rotate in the same

*Fig.* 698. *Fig.* 699.

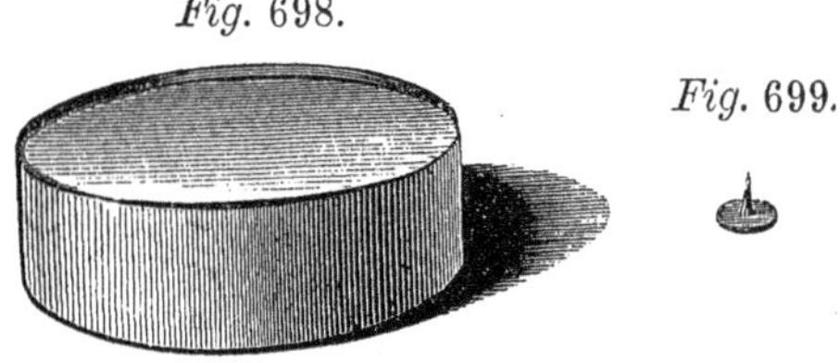

direction. The effect of different disks is very unequal, even when cut near each other from the same piece of copper. Saw out radiating notches in another copper disk, to show the well-known effect of this.

### (*g.*) EXPERIMENTS ON DIAMAGNETISM.

[337] For the experiments with bismuth and other solid bodies, attach to the poles of a strong electro-magnet pieces of iron, such as represented in fig. 700. They are pierced at exactly the same height to admit iron cylinders, pointed at one end, about 2 centimeters in diameter, which slide easily in the holes and can be fastened by screws. By this means the poles of the magnet can be brought within any distance of each other. The poles may be covered with a case made by pasting together pieces of window glass, fig. 701, and resting on a strip of velvet pasted to the table which surrounds the poles, figs. 641 and 642. A hole is bored in the top of the case, through which a glass tube, 6 to 8 inches high, is fitted by a

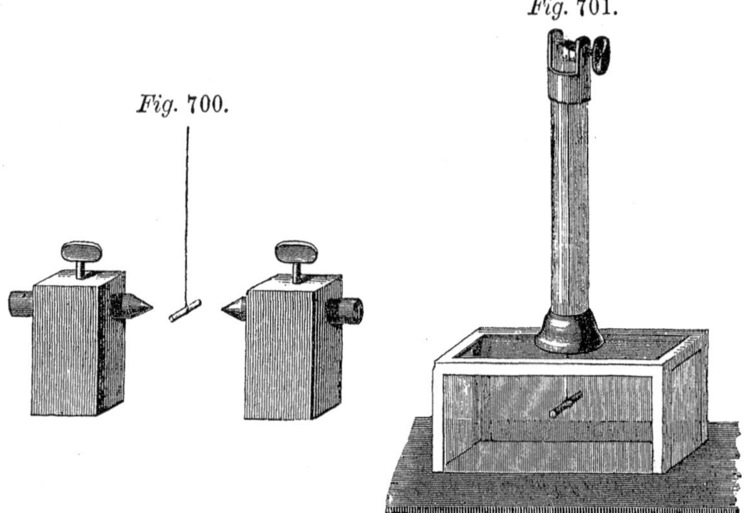

*Fig.* 700. *Fig.* 701.

wooden cap. This tube is not cemented in, but both the upper and lower caps lined with velvet. The upper cap has two projections, through which a pin is thrust. On this pin is wound a fiber of silk, with a little hook at the end.

[338] The substances to be submitted to the action of the magnet are attached to cocoon fibers, of about $\frac{1}{2}$ to $\frac{2}{3}$ the length of the tube. The hook is let down inside of the box, the cocoon fiber hung on it by a loop

and drawn up again. By moving the box a little, the object can easily be brought in a straight line between the two points, and its position of equilibrium may be changed at pleasure, by turning the upper cap. The points must be far enough apart to admit the axis of the object between them, without touching. The substances experimented on should be in the form of bars, about 2 centimeters long, tied to the cocoon fiber. There should be bars of copper, iron, platina, wood, etc., a little rhombohedron of Iceland spar, a prism of tourmalin, and especially a bar and a little ball of bismuth, as free as possible from iron, though even the ordinary commercial bismuth can be used. Bars of bismuth are easily made by dipping a glass tube, $\frac{1}{2}$ to 1 line in diameter, in the melted metal and sucking it in quickly. The glass tube is then broken, if the expansion of the bismuth has not already broken it. A bit of one of these cylinders may be filed into a ball.

These experiments require a powerful current, but the repulsion of the ball of bismuth from the line of the poles, and the equatorial position of the bar can be shown with an electro-magnet which will support only a few hundred-weight. For the experiment with the ball of bismuth, the poles must be brought so near each other as just to admit the ball freely.

## (*h.*) EXPERIMENTS ON THERMO-ELECTRICITY.

[339] **Electrical currents with one metal.** — *a.* Connect with the galvanometer two copper wires, one of which ends in a little plate; a current will be produced whenever this plate is warmed and touched with the cold end of the other wire.

*b.* Take a platina wire, several inches long and $\frac{1}{2}$ to 1 millimeter thick, bend it near one end into three or four spiral turns about a line in diameter, and connect it with the conducting wires of a galvanometer; by warming the spiral with a spirit-lamp a current will be produced which will pass from the spiral toward the cold part of the wire.

*c.* Heat one end of a steel knitting-needle to redness, connect it with the galvanometer, and heat the spot where the soft and hard steel join; a galvanic current will pass from the soft to the hard part. The current may be maintained by moving the lamp toward the hard portion, so as to keep that part which is in the flame always red hot. A current may be produced with a needle which has been entirely softened, by moving the flame along the needle in the same way; the current passes in the direction in which the flame moves.

[340] **Currents produced by different metals.** — For this purpose two elements are especially needful: one of copper and

bismuth, the other of copper and antimony-glance. Cast two bars of these metals, about as thick as your finger and 3 inches long, in paper moulds, and file them bright; the mould for the antimony-glance must be thickly wrapped. Then bend two bits of copper wire or ribbon twice at right angles, and solder each with tin, as shown in fig. 702, to one of the bars. The strip intended for the bismuth must be previously tinned, for the bismuth will not bear heat sufficient to cause the solder to run on the copper. It is convenient to fasten beforehand a needle point in each strip and each bar, to support little magnetic needles.

*Fig.* 702.

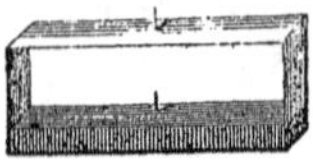

In experimenting, set magnetic needles on the elements and place them parallel to each other and to the magnetic meridian, on a wooden stand, with one joint of each projecting. On applying heat to the joints with a spirit-lamp, a considerable deflection will be seen, in opposite directions in the two elements.

[341] **The thermo-pile.**—As this is ordinarily used only in connection with the galvanometer as a thermoscope, it must be made on a small scale, so as the more rapidly to assume the temperature to which it is exposed. The elements are made of bars of antimony-glance and bismuth, cast in iron moulds, and shaped with a file as represented of natural size in fig. 703. The bars of antimony-glance must be tinned on both heads with very soft tin solder, because they, like copper, do not take the tin without considerable heat. The soldering must be done with an iron, on account of the exceeding fusibility of bismuth. The bars may be held between spring forceps made of wire, like fig. 704, and the spaces

*Fig.* 703. *Fig.* 704.

between the bars filled with bits of wood, which may be left there to impart greater solidity to the pile, but they must not extend over the joints. The vertical rows of 5 pairs each are first soldered, and these united when all the layers are complete. The end pieces of each row

*Fig.* 705. *Fig.* 706.

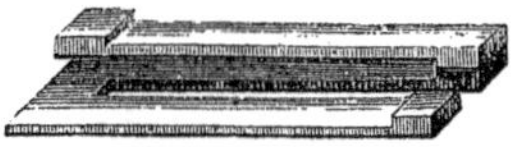

must have an offset at right angles to the bar, like fig. 705. Fig. 706 represents the combination of the end pieces of two vertical rows. When the pack of 20 or 25 pairs is completed, lay it in a round or square case of brass, having first soldered to the middle of the first and last bars short copper wires, which pass through two holes in the case. These holes should be mounted with ivory for better insulation, and the ends of the wires provided with permanent binding-screws. The vacant spaces are filled up with plaster of Paris, which is afterwards scraped away so as to leave the ends of the bars free, and these are then blackened. Fig. 707 represents the pile complete and with a double case, because for certain purposes a funnel-shaped cap must be fitted on, sometimes to exclude extraneous rays, sometimes to condense the rays of heat. This conical cap must be polished on the inside, not blackened, as is sometimes done. It must have a loose fitting cover. A tube of the same width, blackened within and also fitted with a cover, is slid over the other end of the case. As all this is easier to make when the case which incloses the pile is round, it is usually made in this form, and also provided with a rod fitting in a stand so that it can be adjusted to any

*Fig.* 707.

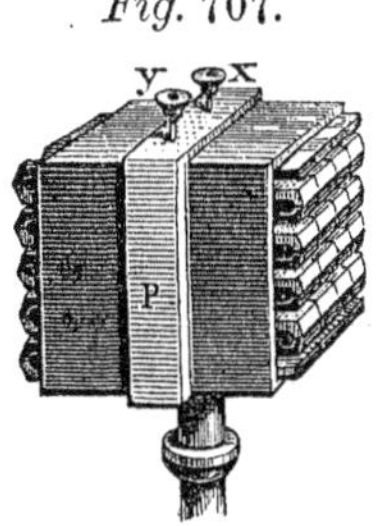


*Fig.* 708.

*Fig.* 709.

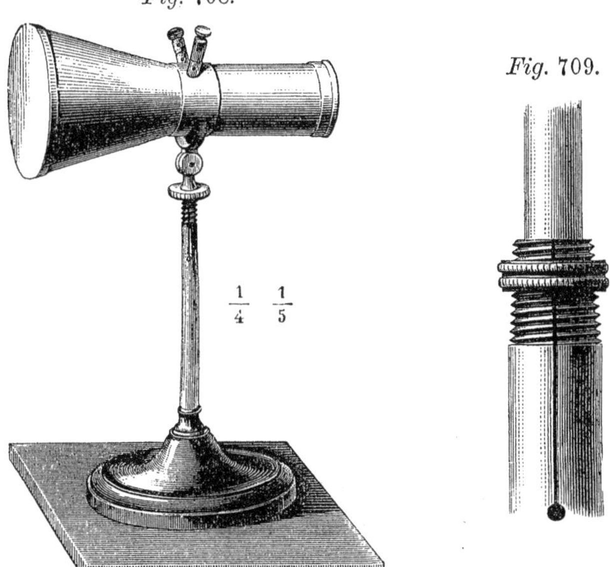


height. Fig. 708 represents the apparatus complete. In fig. 708 the tube of the stand is slit, and is furnished with a slightly conical screw and a nut, by which the rod can be clamped tightly. Fig. 709 is an enlarged view of the arrangement. The conductors should be spiral copper wires, which are yielding enough to prevent the apparatus from being overset by an accidental jerk. Connecting wires of this kind are useful in many other cases. This work requires a great deal of patience, for soldering these thin pieces is a delicate matter; if you have them soldered you must watch closely that the proper order is observed. It must be expected that many bars, especially of antimony-glance, will be broken in the operation; a good supply must therefore be provided. If great delicacy be not aimed at, the work may be rendered very easy by taking bars about 4 square lines in cross section. A galvanometer, with few turns of thick wire, should always be used. The use of the apparatus comes under the subject of heat.

# CHAPTER VII.

## EXPERIMENTS ON HEAT.

### (*a*.) EXPERIMENTS ON EXPANSION.

[342] The thermometer is based upon the expansion of matter by heat. As this instrument is so constantly used, and with all care in handling one will be broken occasionally, there is the more inducement to acquaint one's self with the construction of them, good thermometers being rather expensive.

[343] **Construction of thermometers.**—(*a*) *Selection of the tubes.* These are had from the glass-house, and only 5 or 6 feet taken from the middle of the tubes, which are drawn 15 to 20 feet long, although this enhances the cost. As soon as they are made they are cut into lengths of $1\frac{1}{2}$ to 2 feet, and sealed at both ends. For ordinary purposes, glass tubes with flat bores have obvious advantages; but for making very accurate thermometers I would always prefer round tubes, because they can be tested more exactly. This operation is performed by sucking into the tube a column of mercury about an inch long, and moving it by gentle agitation along the tube, carefully comparing its length in different parts with the original length. Having in this way found a part of the tube of uniform size, if it be only 3 inches long, mark it by tying waxed thread around it and break it, so that the glass wasted in blowing the bulb may be taken from the uneven part. The operator will soon learn by unpleasant experience that uniform thermometer tubes are very rare. For ordinary thermometers this is of little importance; the inferior tubes may be used for house-thermometers, which need not range above 100° F., and may be graduated to this point by a standard thermometer. Thermometers for ordinary physical purposes may also be made from the inferior tubes, and the limits within which they are correct marked on them.

In sucking up the mercury, moisture is very apt to be introduced into the tube, and it is almost impossible to get rid of it. The only means likely to be at all successful is to heat the whole of the tube except the bulb, and then heat the bulb, the air from the latter driving out the vapor. If any other impurity enter the tube, it may as well be thrown away at once. Instead of sucking up the mercury with the mouth, it may be done by tying the tube in the mouth of a thick india-rubber bottle, squeezing the bottle together and inserting the other end of the tube in mercury; the elasticity of the india-rubber will draw up enough.

(*b*) *Preparation of the tube.* After sealing and thickening the tube at one end, a bulb is blown on it according to the directions given under that head. Care must be taken not to introduce moisture into the tube. As this will sometimes happen with all the pains that can be taken, some recommend fastening an india-rubber bottle to the end and blowing the bulb by pressing this. It is much better to close both ends, heat the whole length of the tube, and then insert one end in the flame and allow the elasticity of the confined air to raise a small bulb. The other end is then opened, closed again after cooling, and the operation repeated until the bulb is about 2 lines in diameter. The flame is then directed against the end so as to burst the bulb there; the orifice is widened with an iron rod, and strengthened by melting down the rim. A glass tube, 1 to 1½ lines in diameter, with thin walls, is then welded on and the joint blown out somewhat. A bulb can now be blown on the other end without introducing any moisture in the tube. The tube when completed has the form of fig. 710, and is prepared in the best possible way for filling. The size of the bulb may be determined by comparison with another thermometer.

*Fig.* 710.

If the tube requires a bulb more than 4 to 5 lines in diameter, a cylinder, 3 or 4 inches long, should be welded on, as directed above, and the joint blown out but little, or not at all. The superfluous glass is then melted off. Thermometers with very stout tubes often have a cylinder of the same size, so that the instrument can be thrust through a cork, the scale being etched on the tube.

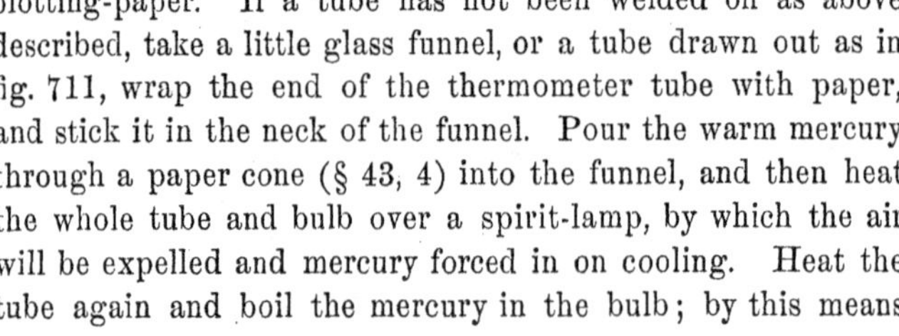

(*c*) *Filling.* The mercury is first freed from air by boiling, then agitated with dilute nitric acid, washed with water, and dried with blotting-paper. If a tube has not been welded on as above described, take a little glass funnel, or a tube drawn out as in fig. 711, wrap the end of the thermometer tube with paper, and stick it in the neck of the funnel. Pour the warm mercury through a paper cone (§ 43, 4) into the funnel, and then heat the whole tube and bulb over a spirit-lamp, by which the air will be expelled and mercury forced in on cooling. Heat the tube again and boil the mercury in the bulb; by this means

*Fig.* 711.

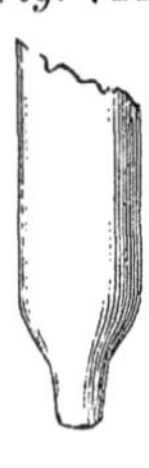

nearly all the air will be expelled. The bulb must not be removed suddenly from the flame, for the sudden influx of mercury might break the tube. A little bubble of air usually adheres to the bulb at its junction with the tube; but the thermometer is allowed to cool, so as to force in as much mercury as possible. Now heat the thermometer in a vertical position, and the expansion of the mercury will drive the air bubble up the tube. As soon as you think that the mercury below the bubble has reached the end of the tube, remove the lamp, and watch whether the bubble draws back into the tube. If it does so, heat it still higher and stir about the end of the tube with a knitting-needle. All this is much easier with the tube shown in fig. 710, in which the mercury can be boiled and the bubble is never out of sight. If, after all, a minute bubble of air should remain, it will do no harm, provided it be not large enough to be visible before the tube is closed, for the removal of the external pressure of the air will cause it to expand.

When the tube is perfectly cool, pour the mercury out of the funnel and expel a drop or two from the tube, in order to determine, by a preliminary experiment, the length of ten degrees. You may then make up your mind how you will arrange the scale to suit the length of the tube. Ordinary thermometers should indicate from 20° to 40° F. below zero, but in order to reach or exceed the boiling point, one must often be content with 10° below zero. The thermometer may be made to range as high as 648° F. Taking these things into consideration, make a mark on the tube which shall indicate the present temperature of the air, and expel or add mercury until the top of the column stands at its height.

Thermometers which are to be exposed to a temperature below the freezing point of mercury are made with spirits of wine, but with this exception mercury alone is used for the purpose. The tubes of spirit thermometers are made as much as ½ millimeter in diameter, to render the spirit more visible, and for the same reason it is generally colored red with cochineal. To prepare the tincture of cochineal, rub the cochineal fine in a mortar, make it into a paste with spirit of wine, and rub this up for some time. Then add more spirit and a little sulphuric acid, stir it well, let it settle, and decant through a filter. Wash the residue with spirit, and remove the lead-gray scales.* The process is repeated with what remains on the filter until the solution becomes too pale. A drachm of cochineal will make a pound of tincure. The aqueous infusion is prepared in the same way. As cochineal is apt to deposit coloring matter, a tincture of Brazil-wood is preferable for coloring spirit. The

* Cochineal is usually adulterated to the extent of 25 per cent. with metallic lead.

tube is filled in the same way as with mercury, or else drawn out to a point and immersed in a capsule containing spirit of 60 or 70 per cent., as shown in fig. 712.

*Fig.* 712.

(*d*) *Closing the tube.* When the quantity of mercury is regulated, melt off the funnel and draw out the tube to a fine point. Now heat the thermometer until a drop of mercury appears at the capillary aperture, then withdraw the bulb from the flame and insert the point. It melts quickly, especially with the aid of the blow-pipe. The mercury contracts and leaves a vacuum above it. The end of the tube should then be rounded in the flame, and, if it is to be attached to a flat scale, bent over, as seen in fig. 713. If the column of mercury should be separated by an air bubble after closing, fasten the thermometer to a string 3 or 4 feet long and swing it in a circle; if the tube be not too narrow the centrifugal force will cause the mercury to unite, and the little bubble of air above it will do no harm.

*Fig.* 713.

To leave air in the thermometer with or without an expansion of the upper end is very objectionable; for just as the pressure of the external air in exhausted thermometers causes an elevation of the freezing point by compressing the bulb so as to make it necessary to wait 2 or 3 months after closing the tube before graduating the instrument, in the same way the inclosed air becoming compressed when the thermometer is heated, lowers the boiling point.

Spirit thermometers are also freed from air. They will still, however, bear the temperature of boiling water, the pressure of the vapor preventing the fluid from boiling, unless the bulb is very thin. The size of the tube renders it very easy to remedy the separation of the column of spirit by swinging.

(*e*) *Determination of the fixed points.* The freezing point is fixed by immersing the bulb and tube as far as the top of the mercury in a mixture of ice or snow with a little water. Tie a waxed silk thread around the tube, and, when the mercury remains stationary, slide the thread along until its lower edge is tangent to the top of the column. This point may afterwards be marked with a scratching diamond. It is well to do this when the temperature of the air is but little above freezing. The boiling point is determined in a metallic vessel like fig. 714, whose long neck has two apertures for the escape of the steam. The thermometer should be thrust through a cork so as to slide easily, and the cork fitted in the neck of the vessel, in which the water has been boiling for some time. The bulb must nearly touch the boiling water. The lower part of the vessel is made of copper, and the upper part, which need not be narrow, is soldered in with zinc and set by a projecting flange on the retort stand, fig. 715, which is needed for so many other purposes.*

*Fig.* 715.

*Fig.* 714.

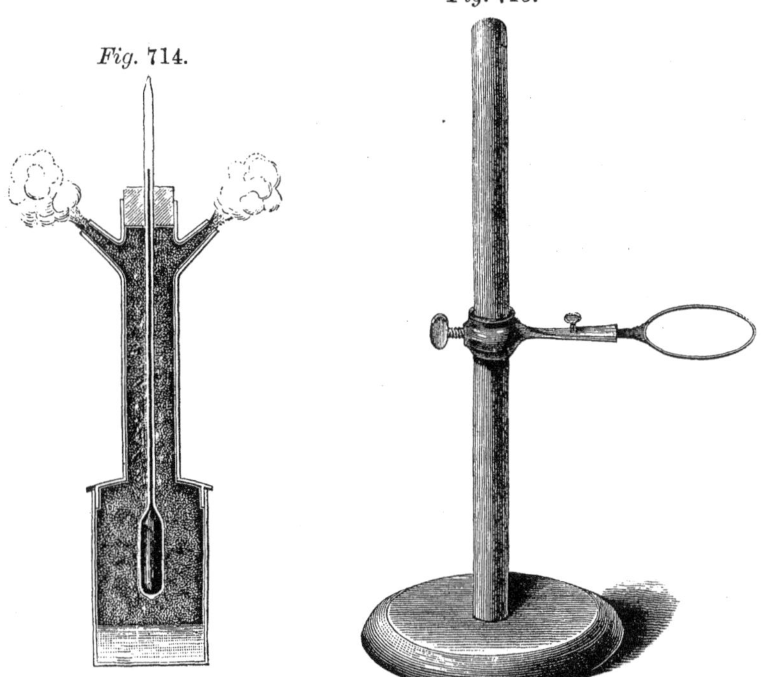

Heat may be applied either by a gas or spirit flame. The tube is frequently drawn a little way out of the vessel, and the thread, previously

* The slide may be triangular although the rod is round, but the screw should always act against a brass plate which presses on the rod.

tied around it, moved by a rod to the top of the column as before. When the point is permanently fixed, mark it with a diamond. The boiling

*Fig.* 716.

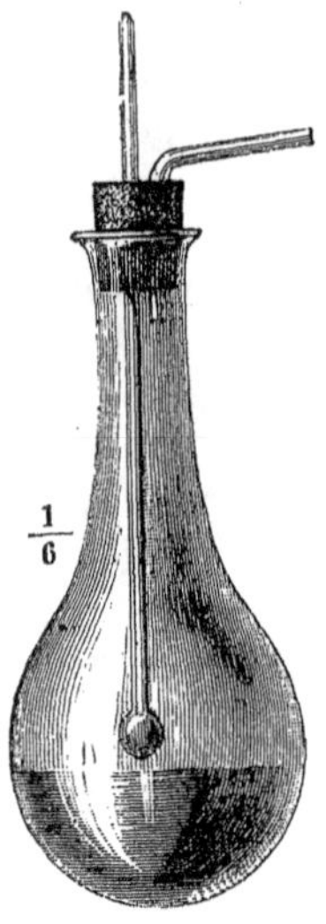


point can be determined in a wide-necked glass flask, by placing a spiral coil of wire in the water and closing the neck of the flask with a cork, which has an aperture for the thermometer and a bent tube for the escape of the steam, fig. 716.

In fixing the boiling point, the pressure of the air must be observed, the boiling point of thermometers being reduced to a pressure of ·76 centimeters, or 336·9 Paris lines. This correction must be made by a table of the pressure of aqueous vapor at different temperatures; but as the height of the barometer is not so very variable, it is enough to know that when the barometer stands at 70·7 centimeters, water boils at 98° C., and at 73·3 at 99° C., the intermediate pressures being proportional to the temperature. For instance, if the barometer stood, during the determination of the boiling point, at 72·5 centimeters, the point found would correspond not to 100 but to 98·7, and the space between it and the freezing point must be divided accordingly. A shorter (and for most purposes sufficiently accurate) method is to deduct from 100 0·0378° Centigrade for every millimeter which the barometer stands below 76 centimeters, and for every Paris line below 336·9 deduct 0·0881° C. In these reductions the height of the barometer is supposed to be reduced to the freezing point.

*Fig.* 717.

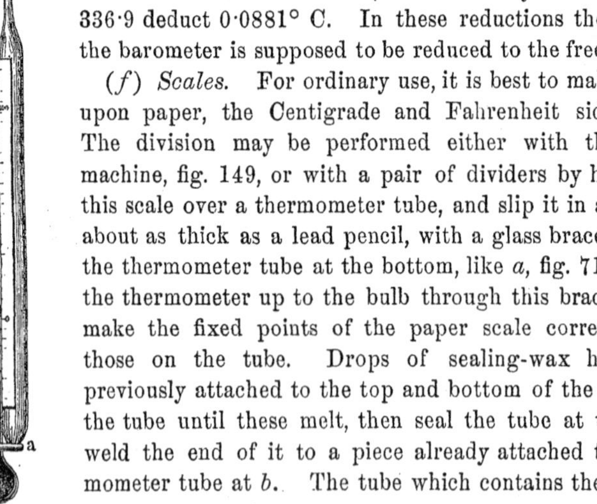


(*f*) *Scales.* For ordinary use, it is best to make the scale upon paper, the Centigrade and Fahrenheit side by side. The division may be performed either with the dividing machine, fig. 149, or with a pair of dividers by hand. Roll this scale over a thermometer tube, and slip it in a glass tube about as thick as a lead pencil, with a glass brace as wide as the thermometer tube at the bottom, like *a*, fig. 717. Thrust the thermometer up to the bulb through this brace, and then make the fixed points of the paper scale correspond with those on the tube. Drops of sealing-wax having been previously attached to the top and bottom of the paper, heat the tube until these melt, then seal the tube at the top and weld the end of it to a piece already attached to the thermometer tube at *b*. The tube which contains the scale must not be too large, otherwise it renders the instrument too bulky when it is to be immersed in small quantities of fluid; for the same reason, small thermometer tubes are preferable. The thermometer and scale are often inlosed in another tube so as to leave only the bulb projecting, as in fig. 718. This is a difficult operation, and though very useful for some technical purposes, does not admit of great accuracy.

*Fig.* 718.

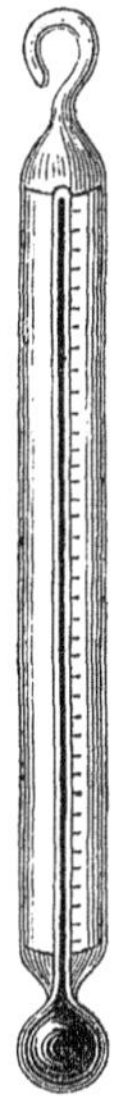

When the scale is made of wood, brass, or ivory, a hole must be made in the top to receive the hook, fig. 713. On ivory the lines may be blackened with indian-ink, and the scale scraped over with glass. Brass is warmed and coated with black sealing-wax; when cold it is again ground with water and then varnished with shellac, or silvered.

The scales of spirit thermometers accord well enough with those of mercury below the freezing point, but above this point the observations must be corrected by a special table, unless the scale has been regulated for every ten degrees by a mercurial thermometer.

(*g*) *Scales etched on glass.* It is often desirable to have the scale upon the thermometer tube itself. Scratching the lines with a diamond renders the tube liable to break; it is better to etch them with fluoric acid. The scale is first drawn on paper, transferred to the glass in the manner described in § 32, and then etched.

[344] **Leslie's differential thermometer.**—Take two thermometer tubes, about 1 to $1\frac{1}{2}$ millimeters wide, and blow upon them two nearly equal bulbs, 2 to 3 centimeters in diameter. One of the tubes should be calibred (§ 343 *a.*) The calibred tube should be long enough to form one vertical limb of fig. 719, and the other tube to form the other limb and the cross piece. Weld the tubes together, and with a pointed flame and a little glass rod draw out some point of the horizontal part into a fine tube. Bend the tubes as shown in the figure, and warm the bulbs so as to draw in enough sulphuric acid, colored with cochineal, to fill one vertical limb and the horizontal part. Seal the opening, and mount the tube on a stand with a scale divided into lines for the calibred

*Fig.* 719.

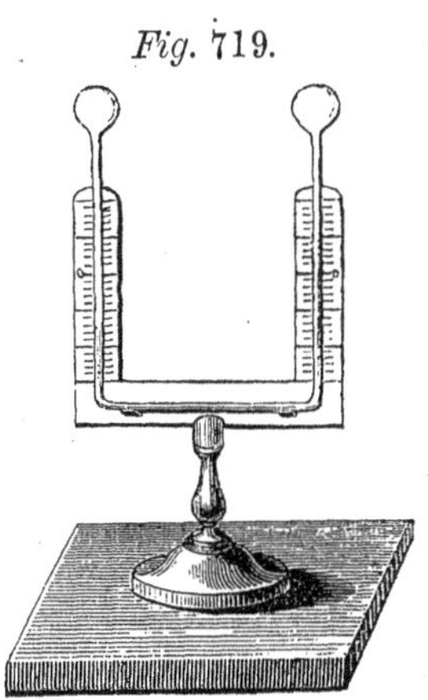

limb. If the tube be not calibred, there must be a scale for each limb. Distribute the air by heating one bulb or the other, so that the liquid will stand at the same height in both limbs when the bulbs have the same temperature. Blacken one bulb with indian-ink when it is to be exposed to radiant heat. This instrument is only used as a thermoscope, although the scale can, of course, be made comparable with that of the mercurial thermometer. The instrument is more delicate when filled with colored spirit of wine and nearly exhausted of air by boiling.

[345] **Rutherford's maximum and minimum thermometers.**—Wider tubes are taken both for the mercurial and spirit thermometers than is usual. The former should have a diameter $\frac{1}{2}$ to $\frac{2}{3}$ millimeter, the latter 1 to $1\frac{1}{2}$ millimeter. The scale, which is graduated by a standard thermometer, ranges usually from — 35° + 100° F. A bit of wood is perhaps preferable to the steel index, because the steel after long use sometimes adheres to the mercury, and of course does not fulfill its design. The glass index in the spirit thermometer consists of a slender tube of black glass, swelled at both ends, and so adjusted as to sink slowly in the spirit. Fig. 720 represents this index, magnified twice.

*Fig.* 720.

[346] **Expansion of solids.**—This is frequently measured by instruments called pyrometers, which consist of a metallic rod, one end of which is fixed and the other acts against a system of levers or cogged

*Fig.* 721.

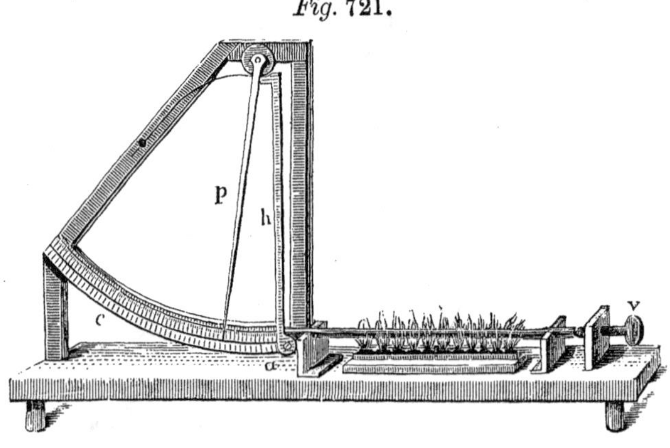


wheels. The former is preferable. Fig. 721 represents an instrument of this kind, and the mode of heating the bar by a spirit-lamp. Such a contrivance evidently cannot be used to measure the heat, but only to exhibit the fact of the expansion.* This is usually shown by a metallic ball, which when cold fits accurately in a metal ring, but when heated will not pass through it. Such a ball is easily turned, as it need only be round in one direction. The ring is mounted on three feet, fig. 722.

*Fig.* 722.

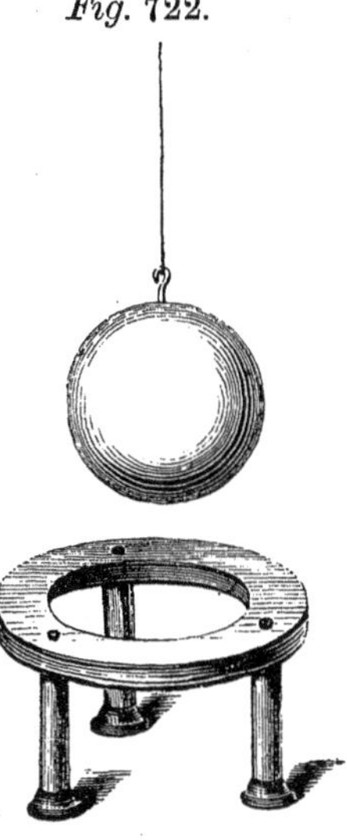

To show the unequal expansion of different metals, fasten together a slip of iron and one of brass, about a foot long, $\frac{1}{2}$ inch wide, and $\frac{1}{2}$ line thick, by rivets $\frac{1}{2}$ an inch apart, fig. 723. The compound bar, though straight when cold, bends when heated. Soldering the bars together leaves them perceptibly crooked even when cold.

In a gridiron pendulum of zinc and iron, the zinc rods must be $\frac{7}{11}$, or, according to others, 0·75 of the length of the pendulum from the point of suspension to the center of the ball. A couple of zinc rods is enough. The rods may all be riveted in the cross bars, except the middle one, which supports the ball, which must be screwed into the last upper cross bar; the work thus becomes very simple. The ball is made of two spherical segments of brass soldered

*Fig.* 723.

together with silver, and the space between filled with lead. It may afterwards be turned on a lathe. It is most convenient to fasten it to the rod in a horizontal position, by a hole bored through the centers of the two disks.

The construction of metallic thermometers is very difficult, and too often those obtained from good instrument-makers are inaccurate. They are usually too dull for almost any use, because they are inclosed in glass.

[347] **Expansion of liquids.**—If the experimenter does not wish to confine himself to the exhibition of the fact, with a mercurial or spirit thermometer, but to show the expansion of other liquids, the vessel

* Experimental measurements of the expansion of bodies, whether solid or liquid, are not suited for instruction, since, apart from the loss of time, all the necessary precautions cannot be observed.

represented in $\frac{1}{2}$ or $\frac{2}{3}$ the natural size in fig. 724, is very well adapted for the experiment. It is made of glass, and has a mark on the neck at *a*. It is filled with the liquid by warming it, and left to cool; the liquid is then removed from the funnel by means of blotting-paper until the surface coincides with the mark on the neck. The application of heat causes it to rise in the funnel again. The apparent expansion of the liquid can be easily determined by filling the vessel up to the mark at the freezing

*Fig.* 724.

*Fig.* 725.

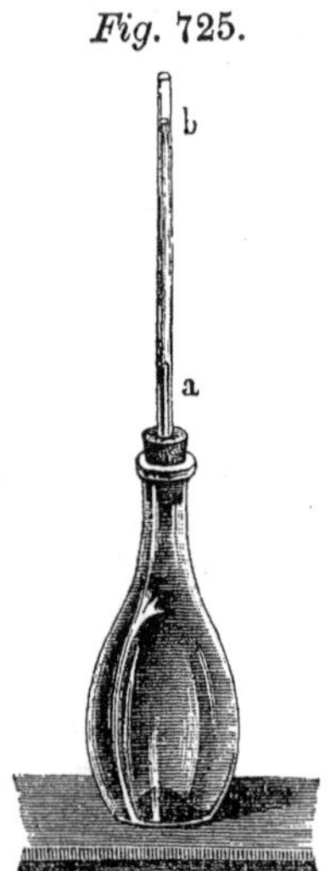


point, weighing it, and after heating it to a certain degree in water and removing the excess above the mark, weighing it again. The tare of the vessel must be deducted from both weights. To ascertain the actual expansion, the expansion of the glass must be taken into consideration. If a vessel of this kind is not at hand, it is easy to make one out of a glass bottle and a tube, as shown in fig. 725.

The expansion of water may be shown independently of the glass, by the apparatus fig. 726. It consists of two tolerably wide glass tubes, closed with a cork at one end and connected by a small bent tube. The whole system is fastened upon a board, with a scale graduated to inches and lines. The tubes are first half filled with cold water, and hot water then poured into one of them so as to run down the side, until both tubes are nearly full. The warm water will always stand 3 to 4 centimeters higher than the cold. The experiment succeeds better when the small tube is provided with a stop-cock, which is not opened until both tubes are filled with water. The small tube can easily be welded to the large ones, instead of being fastened with corks. The small tube may also be long enough to pass through the corks nearly to the top of the larger ones. The smaller is then filled with colored water, one of the large ones with pure hot, and the other with cold water.

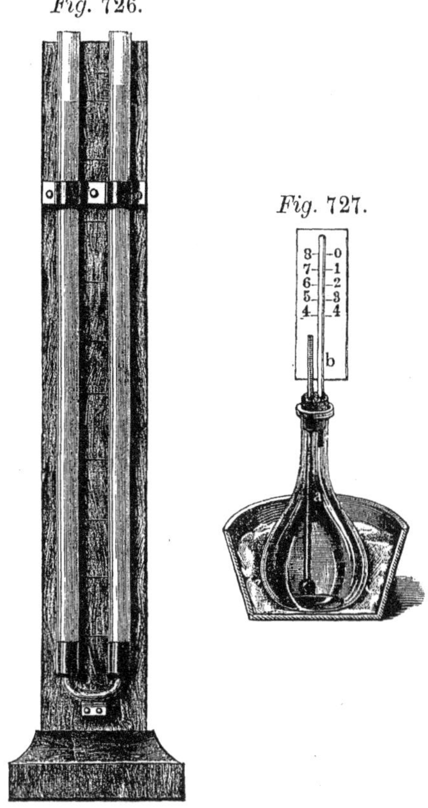

*Fig.* 726. *Fig.* 727.

[348] **Maximum density of water.**—To demonstrate the fact, take a stout thermometer tube, about 1 millimeter in bore, and blow on the end of it a bulb 4 centimeters in diameter. Fill this thermometer with distilled water, and seal it when free from air. Any bubbles which show themselves may be easily brought by swinging into the upper part of the tube. Next, tie two waxed threads around the tube, plunge it in melting ice, and move the lower thread after the water as it sinks, until it begins to rise again; when it ceases to rise, place the second thread at the top of the column. The ball may be removed from the ice to the hand, in order to show to several persons at the same time that the column of water sinks, rapidly at first and then more slowly, until it reaches the lower thread and then rises again. When the water has risen above the upper thread, the bulb may again be plunged in ice in order to show this important fact to all the audience without unnecessary loss of time. The expansion of glass need not be taken into consideration. If the apparatus were made larger, in order to render the column of water more apparent, the cooling and warming would proceed too slowly. With the dimensions given above, the distance between the two threads will be about 4 to 5 millimeters. The temperature of the maximum

density may be easily measured by having a little bottle, in the cork of which is inserted a thermometer, *a*, fig. 727, and a glass tube *b*, 1 to 2 millimeters in diameter, provided with an arbitrary scale.

[349] **Expansion of gases.**—To show this, requires only a simple air thermometer, which serves by comparison with a mercurial thermometer to show the comparative expansion of gases and liquids.

## (*b*.) EXPERIMENTS ON CHANGES OF THE PHYSICAL STATE.

[350] **Latent heat of water.**—Weigh out in a tared glass vessel a pound of crushed ice or snow, and add to it quickly a pound of water at 174° F. and stir them together; the temperature of the mixture will be 32° F. To make the experiment successful, the temperature of the room must be but little above 0° C. It is better to take 2 or 3 pounds of water at a temperature of 85° to 100°, because this loses less heat while pouring in. The quantity of latent heat is determined by a simple calculation, when the temperature of the mixture is above 32°. In summer, however, the experiment will always give bad results, on account of the water already mingled with the ice. It is desirable in summer to choose the temperature of the water, so that the mixture will have about the temperature of the room. This experiment, even when the figures are not very accurate, is very well adapted to illustrate the important law of latent heat.

[351] **Freezing mixtures.**—The mixture of three parts of pounded ice or snow, and one part of common salt, rapidly stirred with an iron spoon, yields a semi-fluid mass, the temperature of which falls even in summer to 14° or 10° F. A temperature as low as 3° F. may be obtained by previously cooling the salt with ice, and surrounding the vessel in which the mixture is made with the same. The quantities employed should be enough to yield at least 2 or 3 pounds of the mixture. Other freezing mixtures cannot be recommended for class illustration; they show nothing more than can be shown with ice and salt, although some of them produce a lower temperature. In the absence of ice, a sufficiently low temperature to freeze water in thin tubes can be produced by a mixture of crystallized Glauber salts and commercial fuming muriatic acid. The following mixture is valuable, because the salts can be recovered by crystallization, and it may therefore be used with advantage on a large scale—for example, by confectioners. It consists of 5 parts of sal ammoniac, 5 parts of saltpeter, and 10 parts of water at a temperature of 50° F.; the mixture falls to 10° F. Grubaud's useful apparatus

for this purpose is represented in figs. 728, 729, and 730. The ice is formed in 9 conical tin tubes, opening above into a common cylindrical

*Fig.* 728.

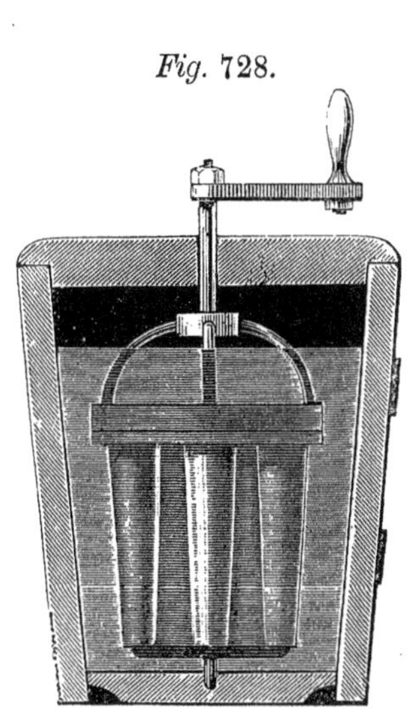

*Fig.* 729.

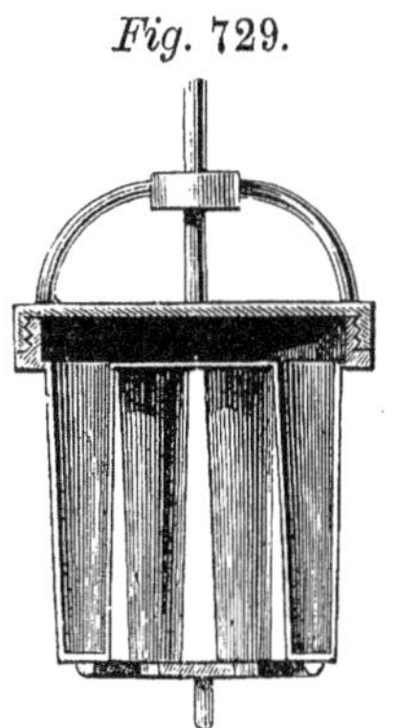

*Fig.* 730.

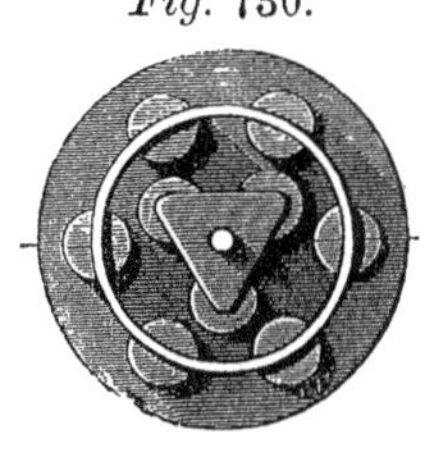

space, which is closed by a cover as represented in fig. 729. This figure is a section of the freezing apparatus proper, in the line *a b*, fig. 730, which is a plan of the same apparatus. The substance to be frozen is placed in the conical tubes, the cover screwed on, and the lower pivot of the apparatus set in a hole made for it in the bottom of a wooden tub, fig. 728. The tub is either filled with the above mixture, or else half filled with very cold water, and an equal weight of fine crystallized nitrate of ammonia added to it; the cover is then fastened on the top, and the freezing apparatus rapidly turned by the handle. In a short time, the water in the conical tubes will be frozen. The following are also freezing mixtures: 3 parts of crystallized chloride of calcium and 2 parts of snow yield a temperature of 32° to —18° F. Equal parts of snow and dilute sulphuric acid yield from 23° to —60° F.

[352] Fusible alloys are obtained by the following mixtures:—

| Bismuth. | Tin. | Lead. | Fusing Point. |
|---|---|---|---|
| 4 | 1 | 1 | 201·2 F. |
| 4 | $1\frac{1}{2}$ | $2\frac{1}{2}$ | 212 |
| 4 | $1\frac{1}{2}$ | 4 | 225·5 |
| 4 | 2 | 4 | 254 |
| 4 | $3\frac{1}{5}$ | $\frac{4}{5}$ | 246 |
| 4 | 4 | 4 | 254 |
| 4 | 4 | 5 | 266·6 |
| 4 | 4 | 6 | 269·6 |
| 4 | 4 | | 286·2 |
| 4 | 7 | 8 | 290 |
| 4 | 6 | 8 | 293·8 |
| 4 | 12 | 11 | 309 |
| 4 | 18 | 16 | 320·4 |
| 4 | 24 | 16 | 331·8 |
| 4 | 8 | | 334 |
| 4 | 12 | 15 | 341·6 |
| | 5 | 1 | 381·2 |
| 4 | 32 | | 392 |
| | 4 | 7 | 419 |
| | 4 | $7\frac{1}{2}$ | 429·8 |
| | 4 | 8 | 442·4 |
| | 4 | $8\frac{1}{2}$ | 449·6 |
| | 4 | 10 | 469·4 |
| | 4 | 14 | 489·2 |
| | 4 | 19 | 509 |
| | 4 | 30 | 530·6 |
| | 4 | 48 | 550·4 |

The last seven alloys are used as a bath for imparting an uniform temper to cutting instruments, which it is difficult to do when they are long or vary in thickness. The seven alloys named impart respectively the following tints to steel: pale yellow, gold yellow, dark yellow, purple, violet, dark blue. Light blue is obtained in boiling linseed oil at 599° F. Instruments heated in these metallic baths at the fusing point need no further tempering.

[353] The liberation of caloric at the moment of solidification is best exhibited in the following way:—Place in a phial 2 parts of crystallized Glauber salts and 1 part water, heat it to the boiling point, and after inserting a thermometer in it, let the mixture cool down to 65·7° F. In summer it must be set in cool water. If a grain of Glauber salts be then introduced, the temperature rises rapidly to 81° or 86° F., while the

greater part of the salt crystallizes. If the vessel be standing in water, this must be removed by a siphon just before the crystallization takes place, else the temperature will not rise so high. When the glass with its contents is to be preserved, the water which evaporates must be replaced from time to time, a mark on the glass serving to indicate the original quantity. The phial should be only half filled.

Hyposulphite of soda may be used for this experiment, and does not require the addition of water, since the salt dissolves in its own water of crystallization. The solution is allowed in this case to cool to 65° or 77°; a slight agitation of the glass will cause the salt to crystallize, and the thermometer will rise 72°. The result is more striking in this experiment than in the last, but hyposulphite of soda is not always to be had, and is rather expensive.

[354] **Crystallization of bismuth.** — This is effected most easily by melting 2 to 4 lbs. of the metal in a hemispherical iron ladle, allowing it to cool slowly until a crust is formed on the surface, then breaking this with a wire and pouring out quickly the still fluid metal from within. This yields, if not always large crystals, at least faces, from which project the corners of numberless cubes. Fine large crystals with beautiful stair-like arrangement can be obtained only by making the bismuth chemically pure, which is a rather tedious operation.

*Fig.* 731.

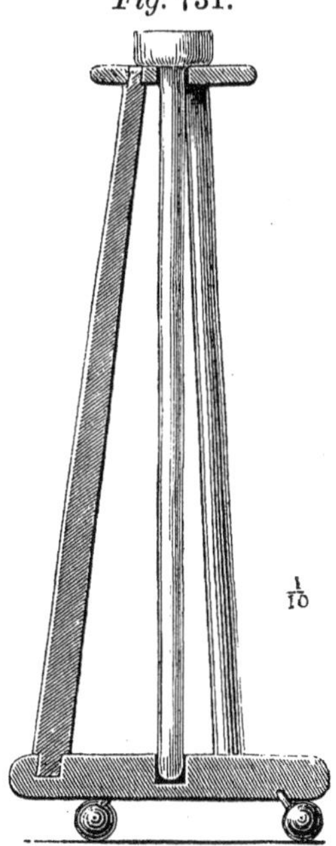


[355] **Laws of vaporization.** — To illustrate this subject, we use the vessel fig. 731, and a wide glass tube, a meter or more in length, well cleaned and closed at one end. This tube, previously warmed, is filled through a paper funnel with pure boiled mercury, to within ½ or 1 centimeter of the top, and the air bubbles carefully removed in the manner described in § 96. This may also be effected by placing the filled tube under the receiver of an air-pump, and exhausting the air as completely as possible; the escape of the bubbles being aided by gentle agitation. After the air is again admitted, invert the tube in mercury, replace it under the receiver, and repeat the exhaustion; then incline the tube so that the mercury may touch the top, close it with the finger, and remove it from the mercury, invert it, and replace it under the air-pump with the

aperture upward. Fill it at last, completely with mercury. Without these precautions, the results would vary too much from the law, on account of the air inclosed in the space above the mercury. Fill the rest of the tube with boiled sulphuric ether, close it with the finger, and invert it in the vessel. The ether may be introduced when the inverted tube is entirely filled with mercury, in order to show the immediate depression of the latter; for this purpose, take a crooked tube, like fig. 732, 3 to 6 inches long, pour into it enough sulphuric ether to fill it up to the point, immerse it in quicksilver, with the point under the barometer tube, and add a little more ether from a pipette; then close the upper end of the tube with the finger, and expel as much ether as necessary by the warmth of the hand.

*Fig.* 732.

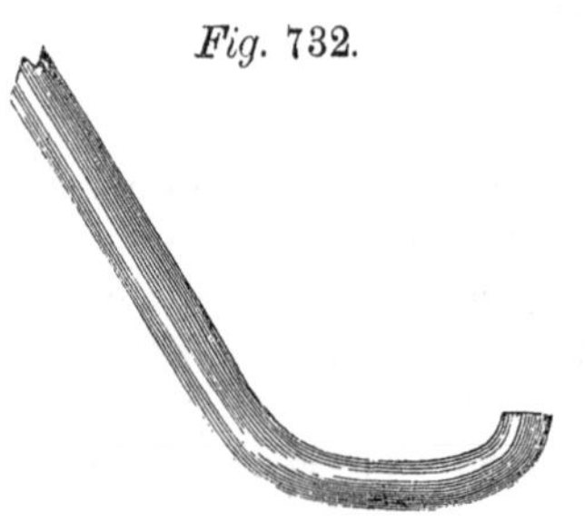

*Fig.* 733.

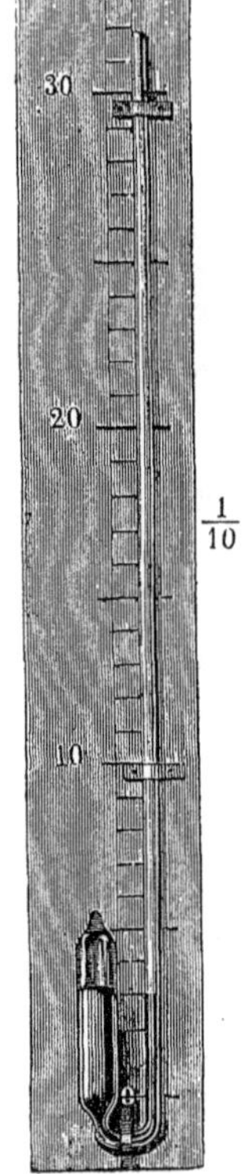


The laws of vaporization may be very well illustrated with the apparatus thus arranged, care being taken not to introduce into the tube enough ether to balance the pressure of the air, when the tube is drawn out of the vessel as far as possible. Five millimeters of ether are enough for a tube not over 1 meter in length. The condensation and expansion of the vapor of ether may also be exhibited when it does not fill the vacuum, and likewise the increase and decrease of its elasticity, as the space is made smaller or larger. Tubes, closed at one end with an iron screw, are very convenient. They are first inserted, entirely open, in the vessel fig. 731, drawn out as far as possible, immersed again, opened, and the remaining space filled to overflowing with boiled ether; the aperture is then closed, and the apparatus is ready for experiment.

For these experiments, it is very convenient to have a scale fastened to the frame, the zero point of which indicates how far the vessel should be filled with mercury; the scale must be graduated to correspond with the barometer used. The scale might be marked with indian-ink on the tube itself.

To show the effect of a rise of temperature in increasing the elasticity of the vapor of ether, you need only hold a piece of heated metal near the tube.

To measure the elasticity of the vapor of water below 212° F., take a siphon barometer, like fig. 733, with a tolerably large bulb, and draw out the end of the bulb to a point. Introduce water into the bulb above the mercury, boil it until the air is entirely expelled, and then seal the aperture. The difference of level of the mercury in the tube and in the bulb indicates directly the elastic force of the vapor of water at the observed temperature. Ether may be employed in the same way, and boiled by being plunged in hot water. The tube is attached like a thermometer to a board, on which a scale is drawn. The mercury in the tube must not be more than sufficient to fill $\frac{2}{3}$ of the bulb, and the water, after boiling, must occupy less than the remaining third. A boiled barometer is not necessary. One filled with boiled mercury and placed under the air-pump, as directed above, yields sufficiently accurate results for temperatures below the boiling point. In filling, it is convenient to have the point bent outward in the plane of the two tubes. The tube is then placed in a safe position, with the bulb over a large dish, and the upper end lowest; the mercury introduced through a glass tube drawn out to a fine point, and gradually worked round the bend by holding the tube nearly horizontal. In small tubes, the successive portions of mercury which pass the bend do not unite readily, and recourse must be had to the centrifugal force, by swinging the tube with a short string tied around the bend. By taking a tube similar to fig. 733, and sealing up water boiled in the bulb over the mercury, we have an apparatus by which the tension of vapors over 212° F. may be determined, so far as the height of the tube allows. The water is heated in a mercury bath.

*Fig.* 734.

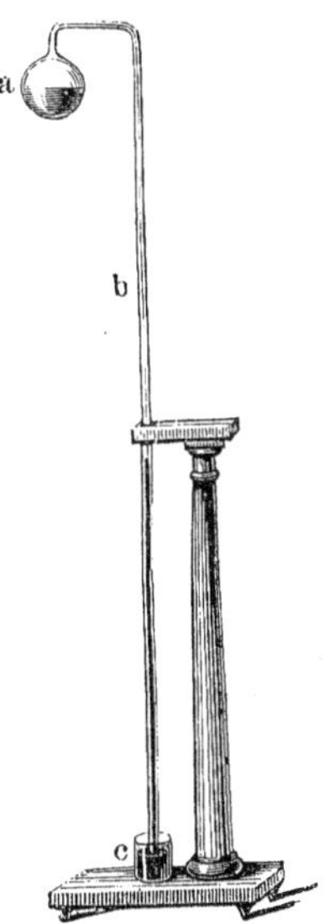


[356] The following experiment serves to illustrate the law, so important for the condensing steam-engine, that vapor in a closed space always has the tension which corresponds to the coldest part of the apparatus. Blow a bulb, 1 inch in diameter, upon a barometer tube, 30 inches long; bend the tube twice at right angles and fasten it to a support, as shown in fig. 734. Introduce sulphuric ether into the bulb. Boil the ether either by a spirit-lamp or hot water, and keep it boiling until it is certain that all the air is expelled from the bulb and the tube; if the bulb was half full of ether, the half of it may be rapidly evaporated. While the ether is still boiling, plunge the end of the tube into a glass containing warm, boiled mercury. On cooling, the

mercury rises gradually in the tube, to a height corresponding to the temperature of the air, a little ether condensing on its surface, especially when cold mercury is used. When the mercury ceases to rise, plunge the bulb into a glass of ice water; the mercury will again begin to rise, and soon reaches the height corresponding to this new temperature, although only the bulb and not the tube was cooled. During this, the ether condensed on the surface of the mercury evaporates. If the mercury has not been boiled, air bubbles will rise through it on cooling; this affects the main result only in so far as it lessens somewhat the height of the mercury in the tube.

*Fig.* 735.

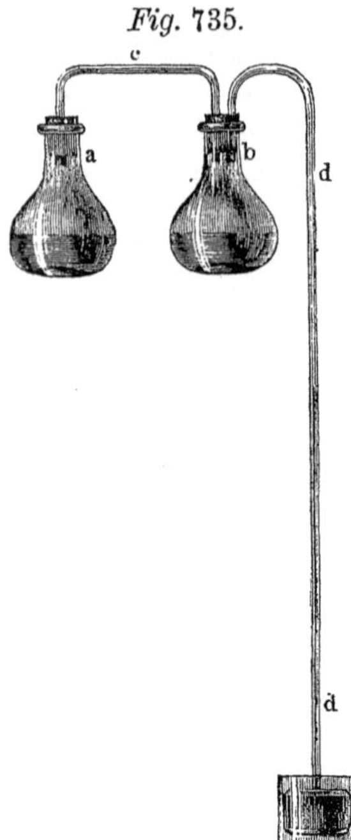


It is not very easy to blow a large bulb on so long a tube; if this cannot be done, fit a little thin glass phial to the tube by means of a cork. It is well, in this case, to prepare the experiment, as far as cooling with ice, before the lecture, because the cooling to the temperature of the air proceeds slowly, on account of the thickness of the glass. This experiment is usually made with two small flasks of the capacity of 1 to 2 cubic inches, as in fig. 735. The ether is boiled in both, and when the whole has assumed the temperature of the air, one of the flasks is cooled.

[357] The appartus represented in fig. 736, $\frac{1}{6}$ to $\frac{1}{5}$ of the real size, is designed to show that the same laws which apply to vapors of ether and water hold good also for other gases. It consists of a strong glass cylinder *C*, with brass caps on each end; into the upper one is screwed a forcing-pump, and a water reservoir *b*. With the dimensions given, the glass must be from 7 to 8 millimeters thick, (this, at least, is sufficient,) and the joints well secured by washers. Within the cylinder is placed an iron vessel with a long handle *a*, containing mercury, and in this, four small tubes, one of which contains atmospheric air, and is graduated by atmospheres; the others contain gaseous sulphurous acid, ammonia, and cyanogen. The mercury cistern need not be more than half as high as shown in the figure, and the tubes must be fastened in some way to the handle. For this purpose, two disks are fastened to the handle, each

*Fig.* 736.

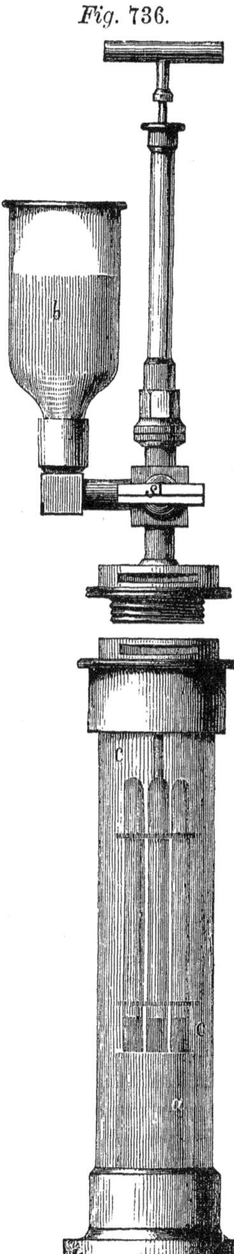


provided with four split rings, which can be tightened by a screw. The handle which supports these disks must be unscrewed from the bottom of the vessel, in order to pass the rings over the tubes; but since the rings must, at all events, be made elastic and furnished with a screw, as in fig. 737, No. 1, it would, perhaps, be better to adopt one of

*Fig.* 737.

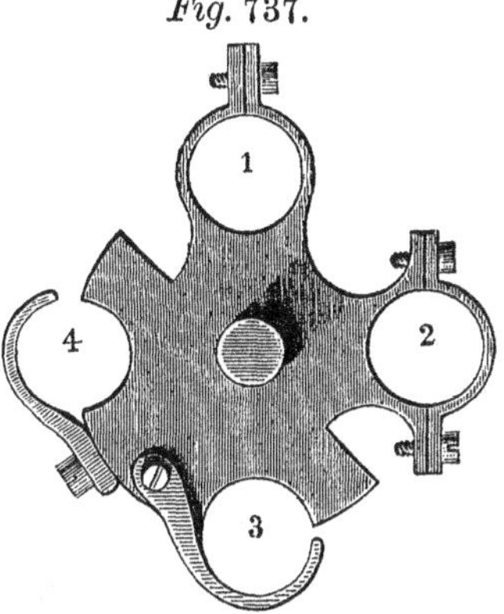


the three other modes of fastening indicated in fig. 737, so that the tubes need only be set in their places. I would prefer No. 3. The tubes must be long enough to reach nearly to the upper cap, so that the stratum of fluid condensed above the mercury will be apparent enough. When the vessel *a* is in its place, fill *C* with water, screw on the pump, fill *b* also with water, and turn the stop-cock *s* so as to make a communication between the pump and the reservoir *b*, as shown by the marks on the handle. Then draw up the piston so as to fill the pump with water, turn the cock a quarter of a revolution so as to close the communication between the pump and *b*, and open that with *C*, then depress the piston. After a few strokes of the piston, turn the cock carefully so as to make a communication between *C* and *b*, in order to allow the air left in *C* to escape. This is repeated until no more air escapes. The compression is now continued, the ascent of the mercury in the tube filled with air indicating at each moment the pressure; the other gases, one after the other, become liquid, and as soon as one of them begins to liquefy, the mercury rises only in this tube, until the gas is entirely converted to liquid. By opening the cock *s* a little toward *b*, the water gradually escapes, and when the pressure is sufficiently diminished to allow one of the liquids to become gaseous, it assumes this condition as rapidly as the water can escape. In opening the apparatus to empty it, the communication between *b* and the cylinder *C* must be left opened; otherwise a rarefaction is produced in *C*, when the pump is unscrewed, and part of the gases escape from the tubes. It is well, before opening the apparatus, to remove the water from *b* by a siphon. The apparatus must be emptied each time, because one of the gases decomposes, and all are gradually absorbed.

With this apparatus, a vessel is usually furnished to show the compressibility of water, according to the method of Colladon and Sturm.

[358] A similar experiment can be made with the air-pump and ether. Take a perfectly clean glass tube, about 24 centimeters long, somewhat contracted and closed at one end, and fill it to within 5 millimeters of the top, with hot boiled mercury, poured through a long, pointed tube, (§ 96.) Fill the rest with boiled ether, and insert it in mercury. Beside this tube place one of the same length, inclosing about 3 centimeters of air. If the air is slowly exhausted, the mercury at first sinks in the air tube until the tension of the vapor of ether is equal to the pressure of the air which remains, when vaporization begins and the mercury in the air tube remains stationary; on the readmission of the air, the reverse takes place. The pump must be worked very slowly, else the mercury will not remain stationary in the tube. The experiment should be tried before exhibiting it, for upon the admission of the air, an air bubble usually appears above the ether; the ether tube must then be closed with the finger, removed from the cistern, and filled up with boiled ether. If too much ether has been introduced at first, exhaust the air, until a portion of it escapes by evaporation. Fig. 738 represents the apparatus complete. A wooden cap is fitted tightly upon the glass cup *a*; this supports a board with a cross-piece having two holes for the tubes. When the tubes are plunged in the mercury the cistern must be nearly full, and should,

*Fig.* 738.

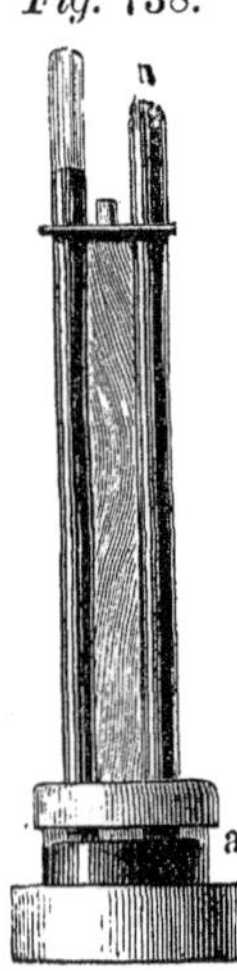

therefore, be set in a larger vessel; the excess must be poured out before the wooden cap is slipped over the tubes. The experiment is still more instructive, if a boiled barometer tube can be introduced under the receiver at the same time, but it must be, at least, 15 inches high. Instead of filling the tubes hot, it can be done as described in § 357.

## (*c.*) EXPERIMENTS ON THE DIFFUSION OF VAPORS THROUGH GASES.

[359] In order to demonstrate the law that gases, which, in a state of saturation fill any space, when mingled with other gases have the same tension as in a vacuum, make the same arrangement as for Mariotte's law, § 98, and then introduce into the barometer tube a few drops of sulphuric ether. The mercury sinks immediately. If the air be reduced to its former volume, by depressing the tube, the tension will be increased by an amount equal to the tension of the vapor of ether in vacuo, at the given temperature. This may also be shown very simply in the following way: Take a flat-bottomed flask, fig. 739, fit in it a good cork with two glass tubes, one of which is bent to serve as a manometer. Dry the flask thoroughly by connecting one of the tubes with the air-pump and the other with the chloride of calcium tube, fig. 739, by means of caoutchouc connectors, and making several strokes with the pump, at intervals of 15 minutes. When the vessel is filled with perfectly dry air,

*Fig.* 739.

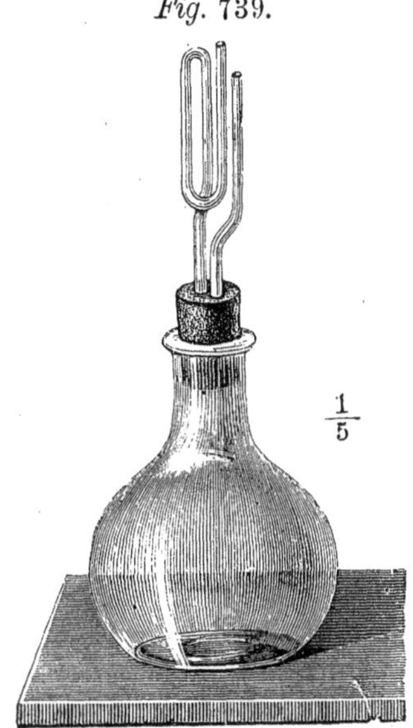


*Fig.* 740.

disconnect it, and pour some mercury into the manometer tube. Connect a small funnel, fig. 741, to the other tube by means of india-rubber, which is squeezed together by a clamp, (§ 104.) Put water in the funnel, and let a few drops fall into the flask. Before long, the mercury in the manometer will indicate an increase in the tension of the air within. There are several other pieces of apparatus for similar experiments, which are, however, expensive and not easy to handle.

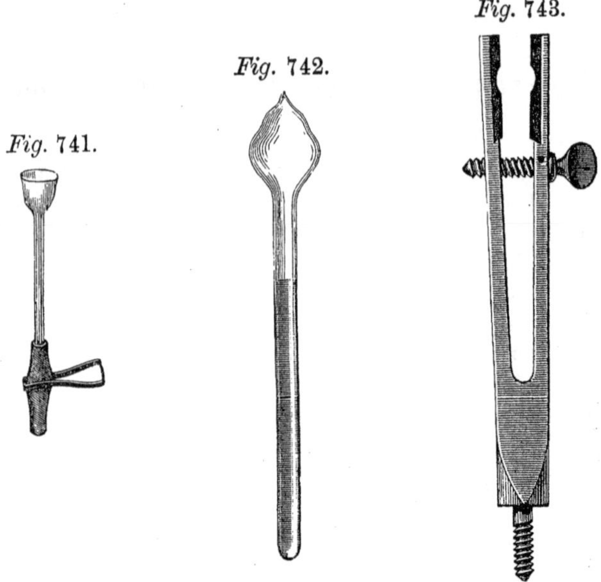
*Fig.* 741. *Fig.* 742. *Fig.* 743.

[360] The following experiments, in addition to those enumerated in § 102, 9, serve to show the influence of pressure upon the boiling point:

(*a*) *The water-hammer.* Blow a bulb upon a stout glass tube, fig. 742, 1 foot long, and draw out the tube beyond this to a point. Fill the tube with water up to the bulb, boil away enough to leave it half full, and, while boiling, melt the fine point. In such cases, as the tube becomes very hot, it may be held with a wooden clamp, fig. 743, taken from a stand to be described hereafter. As the tube is now vacuous, water will boil in it by the heat of the hand, when it is held by the bulb in a nearly horizontal position. If quickly turned up, so as to let the water run in the other end of the tube, it will strike with a sound like mercury in the barometer, without breaking the tube, if well annealed. To make this requires a hot fire and a steady hand.

(*b*) *The pulse-glass.* Blow two bulbs on a glass tube, bend the tube twice at right angles, as seen in fig. 744, and draw out the end to a point. Fill the bulb *a* and the tube with 70 to 80 per cent. alcohol colored with

Brazil-wood, expel the air by boiling, and seal the tube, with about half a bulb full of alcohol in it. When held horizontally by one bulb with the bulbs uppermost, the liquid will be driven with lively ebullition from one bulb to the other, and a sensible degree of cold produced by the evaporization.

*Fig.* 744.

(*c*) A similar experiment, well known as the culinary paradox, may be made as represented in fig. 745. The flask *a* is closely corked, while the water within is in rapid ebullition. The application of cold to the exterior causes the ebullition to commence again, until the temperature of the whole falls below the point at which water boils in vacuo.

*Fig.* 745.

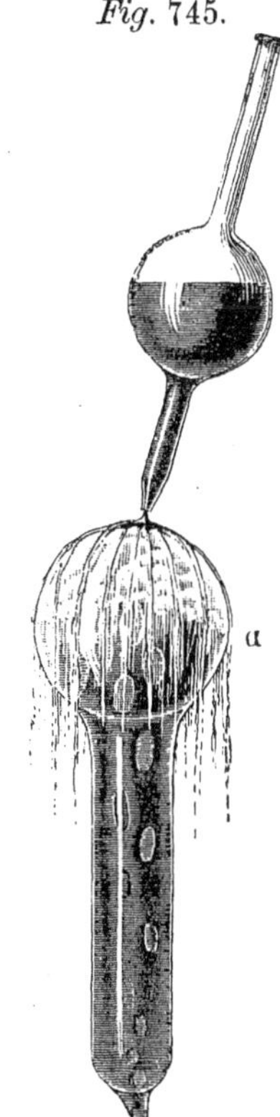


(*d*) *Papin's digester.* For illustration, it is sufficient to show the rise of the boiling point with the increase of pressure up to a few atmospheres. For this purpose the boiler should be in a cylindrical form, 5 inches wide and 12 to 15 inches long; the metal must be 3 millimeters thick, to bear with safety an over-pressure of four atmospheres. The boiler must have a safety-valve, and a gauge-cock to test the height of the water, which should not sink below the level of the surface exposed to the fire, a steam-cock, and a thermometer. The boiling point of the latter must be without the vessel; its scale need not be graduated above 400° F. It is cemented in with litharge cement. If the fire-pan be insulated by four glass legs, as in figs. 746 and 747, the boiler may be used as a source of electricity. The fire-pan is made simply of sheet iron, the posterior edge *a a*, fig. 747, being bent slightly inward. It terminates at the back in a short pipe, which should be connected with a good chimney. It should have a grate and door. The boiler may be heated with spirit contained in a box fitting into this pan, and provided with 12 to 20 wicks, stuck through holes in the top. A pint of spirit is sufficient for a series of experiments with a boiler of the dimensions given, especially when filled at first with hot water. Fig. 748 represents the lamp. The precaution must never be neglected not to allow the water to fall so low as to leave any part of the heated surface uncovered. With such small dimensions a gauge-tube cannot be applied, and when the pressure is high the water-cock always

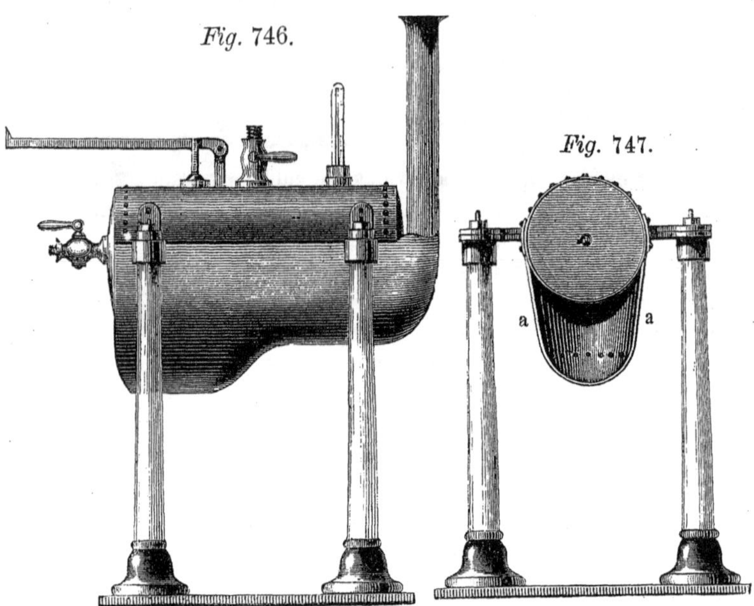


*Fig.* 746.

*Fig.* 747.

seems to emit steam, because the very hot water evaporates in the open air; but if the hand be passed through the jet, the copious moisture deposited will furnish pretty safe evidence whether the steam contains liquid water.

*Fig.* 748.

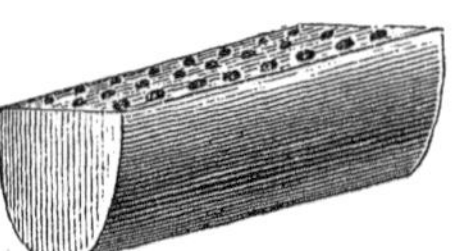

*Fig.* 749.

If it be desired to open the boiler, either to clean it or to introduce something into it, an iron ring should be riveted to the end, and the plate with the gauge-cock screwed to this, as seen in fig. 749.

For these experiments, the weight required by the safety-valve for each atmosphere should first be calculated, the boiler placed under the balance, the lever suspended by its fulcrum, and the slide moved until it counterbalances the weight in the other pan. The position of the weight corresponding to each half atmosphere is then marked on the lever. The valve must close the aperture exactly, and the size of the aperture must be very accurately measured. The thermometer is then inserted, and the boiler heated with varying loads upon the valve, until the steam not only escapes here and there, but really opens the valve, and the load of the latter is thus controlled, before the marks are permanently cut into the lever. When the apparatus is designed for no other use, the form shown in fig. 750 is the simplest. In this figure, the thermometer is inserted in mercury contained in a short iron tube screwed into the top.

*Fig.* 750.

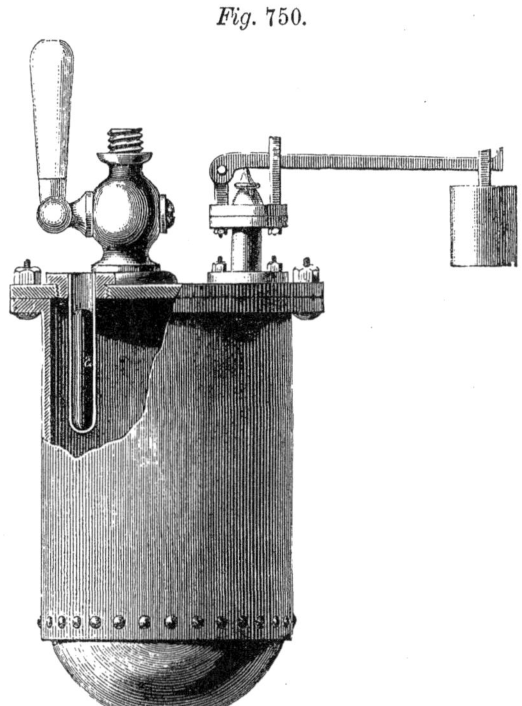

[361] **Leidenfrost's experiment.**—Take a small silver or platina capsule, fig. 751, heated to redness over a spirit-lamp, and let a drop of water fall upon it from a glass tube. When the spirit flame is good, several drops may be let fall in quick succession. While the capsule remains over the fire, the water evaporates very slowly, with a constant, tremulous, rotary motion; large drops assuming often an almost angular form.

*Fig.* 751.

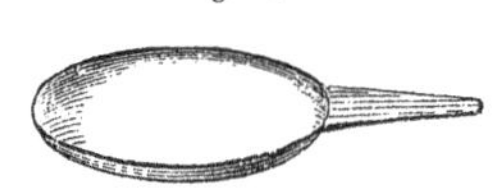

The last minute drop explodes suddenly. When removed from the fire the water begins, as soon as the capsule is cooled below red heat, to adhere to the surface, and evaporates very rapidly. Other metals, and even glass, exhibit the same phenomena. The experiment can be made with a silver teaspoon, if the lamp burns well. It succeeds beautifully in a large platina dish, but it must be heated over burning coals, or a gas flame. In a platina crucible, water can be accumulated sufficient to immerse in it a thermometer, which will indicate 203° to 209° F. To see the tremulous motion, and the angular figures which the drops assume, the spoon should be fastened to some support.

The experiment of passing the hand through molten metals belongs in this place. If a sufficient quantity of lead (10 to 20 lbs.) be melted in an iron pan, it may be stirred with the hand without the least injury, and, indeed, a sensation of coolness is felt, when the hand is moistened with ether.

[362] **Hygrometers.**—Under this name are included all instruments used to indicate the amount of moisture in the atmosphere. They are of two classes. To the first belong two instruments which give the degree of moisture directly: that is, whether the air is more or less saturated with moisture; the actual quantity of vapor of water being determined by the aid of a thermometer. The instruments of the other class give directly the quantity of water, and the fraction of saturation is determined by the thermometer. To the first class belong those toys which have a catgut, 2 or 3 inches long, fastened at one end, with an index at the other. Usually a human figure is chosen, upon the back of which a quill is fixed; the catgut is contained in this, and is fastened at the further end of the quill, with the other end of the string passing through the figure to the arm, which serves as an index. These things have no scale.

Besides these, there are hygrometers of wood, whalebone, and hair. They all have a centigrade scale. The zero point is obtained by placing the instrument, with chloride of calcium or Nordhausen sulphuric acid, under a bell-glass, and marking the position of the index, when stationary, which will be in about six hours. The point of saturation is 100°, and is obtained by placing the instrument with wet blotting-paper under a bell-glass. It is incorrect to plunge the whole instrument in water, as directed by De Luc, in order to find the point of greatest saturation of the air.

(*a*) *The wood hygrometer.* Figs. 752 and 753 are a plan and view of this. Two wooden rods are fastened upright in the board *a a*; these are grooved on the inside, and in these grooves is inserted a slip of fir *c c*, not over one line in thickness. It must be made of fine grained wood, free from knots, and must slide easily in the grooves. To one of the bars is fastened a plate *d*, upon the back of which is the pin *e*, fig.

*Fig.* 752.

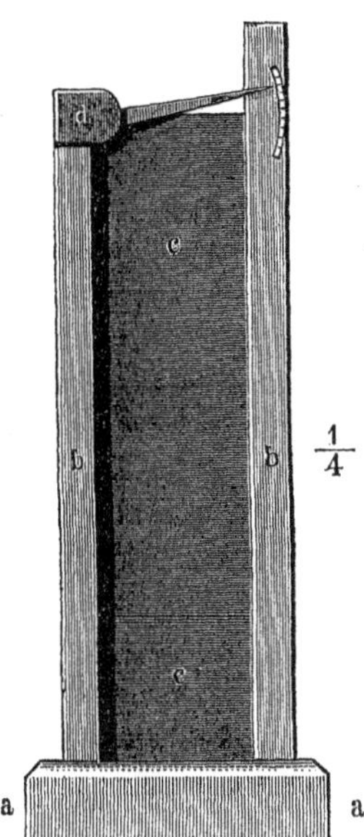


*Fig.* 753.

*Fig.* 754.

754. An index is fixed on the first offset of this pin, and held down by a rim *f*, riveted on the second offset. The index moves freely, and rests by a pin on the slip *c*, which consequently raises the index when expanded by the moisture. When the wood contracts, the index sinks by its own weight. The scale is marked on the other bar.

(*b*) *The hair hygrometer* is the most useful of all, because the hygroscopic body, which is a smooth, straight, human hair boiled in weak alkali, is easily obtained. The hair expands from its condition of greatest dryness to that of greatest moisture, about $\frac{1}{50}$ of its length. Figs. 755 and 756 represent the instrument: *a a* is a brass frame, on one side of which is the dial-plate *b*, and on the other an arm *c*, supporting between them an axle, whose circumference is a little more than $\frac{1}{50}$ the length of the hair. The axle is made to move an index over the dial-plate. A pin is inserted in the other end of the frame, and held down by a clamp, as seen in fig. 757. The hair may be fastened directly to this nail, the other end carried over the axle, and a weight of about 15 centigrammes fastened to it. The instrument is then complete except the graduation. Instead of the weight, which is inconvenient in carrying the instrument, a fine spiral spring may be used, as seen in the figure. It must be made rather long, so as to yield a tolerably uniform resistance for the whole extent of the motion, not exceeding 15 centigrammes. Instead of carrying the hair itself around the axle, it is usually fastened by clamps, fig. 758, to silk threads, and these fastened to the lower pin and the axle. To prevent any slipping on the axle, make a double cut on it, as in fig. 756, and carry a separate string to the spring. The ends of the strings can be easily fastened to the axle, by a hole bored through it. This gives the additional advantage, that when a new hair is stretched, its length may be altered until it corresponds to the fixed points.

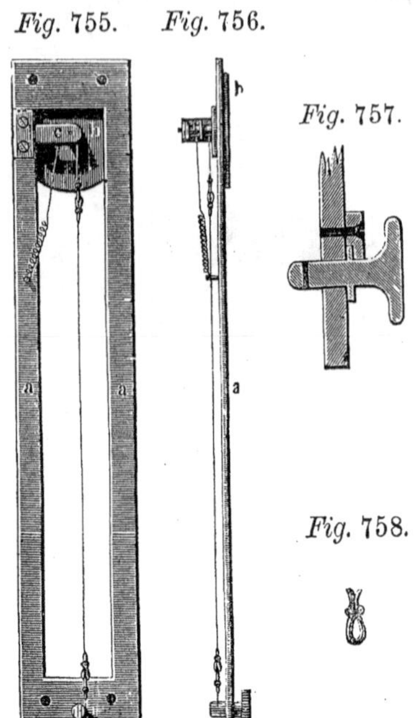

*Fig.* 755. *Fig.* 756. *Fig.* 757. *Fig.* 758.

(*c*) *The whalebone hygrometer* is arranged essentially like the hair hygrometer, but the strip of whalebone is made only $\frac{2}{3}$ as long as the hair; it must not be over 2 millimeters wide, and cut perpendicular to the fibers. Uncut whalebone is difficult to obtain, and therefore not very suitable for a hygrometer. Much pains have been taken to make these instruments comparable with each other; but when one reflects that they all gradually lose their hygroscopic properties, and even hair hygrometers, whose fixed points agree, vary in the intermediate parts of the scale, it seems useless to attempt further perfection.

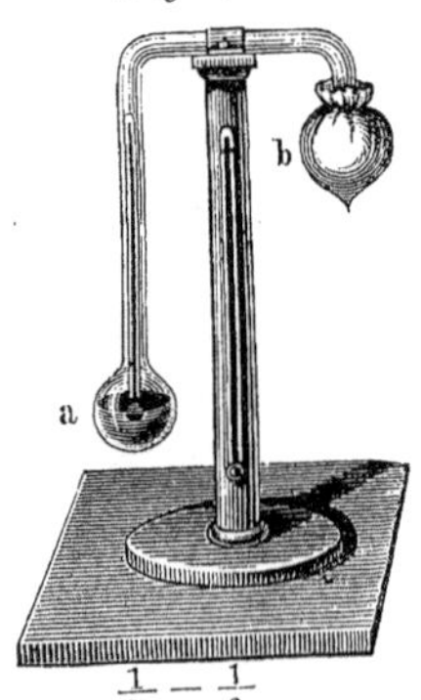


*Fig.* 759.

(*d*) *Daniel's hygrometer*. This instrument is represented in fig. 759, and must be bought from a reliable maker. The scale of the thermometer inclosed in the longer limb is now usually made of glass, and united with the upper end of the thermometer. It must be immovable in the tube, which is, therefore, narrowed toward the bulb. The bulb of the thermometer must dip into the ether, without touching the bottom of the bulb. A ring, 2 or 3 lines wide, around the middle of the bulb *a* must be gilded, and the bulb must contain enough ether to reach the middle of the gilt ring. The bulb *b* is covered with muslin or fine linen. Another thermometer is fixed on the support, and both have centigrade scales. To use the instrument, pour on the bulb *b* enough sulphuric ether to moisten it thoroughly, but not to hang in drops. The bottle, fig. 760, is convenient for this purpose, the heat of the hand being sufficient to drive out the ether. Not only the temperature at which the dew forms on the gold ring should be observed, but also that at which it disappears; the mean is the dew point. After the observation, the ether condensed in *b* must be allowed to run back into *a*. Very exact observations can be made only with the telescope, the heat of the body rendering the data of the instrument incorrect. In the following table gives the amount of vapor and its tension, for temperatures for — 20° to + 40° C.

*Fig.* 760.

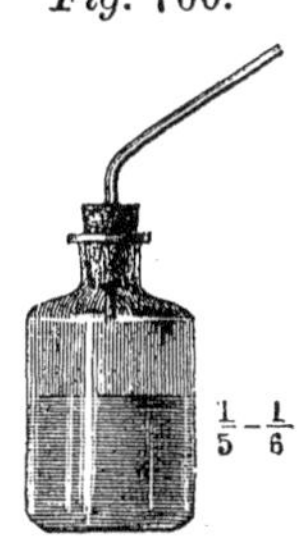


| Temperature of the dew point. | Tension of the aqueous vapor. | Weight of aqueous vapor in 1 cubic centimeter of air. | Dew point. | Tension of the aqueous vapor. | Weight of aqueous vapor in 1 cubic centimeter of air. |
|---|---|---|---|---|---|
| | mm. | gr. | | mm. | gr. |
| — 20° | 1·3 | 1·5 | 19° | 16·3 | 16·2 |
| — 15 | 1·9 | 2·1 | 20 | 17·3 | 17·1 |
| — 10 | 2·6 | 2·9 | 21 | 18·3 | 18·1 |
| — 5 | 3·7 | 4·0 | 22 | 19·4 | 19·1 |
| 0 | 5·0 | 5·4 | 23 | 20·6 | 20·2 |
| 1 | 5·4 | 5·7 | 24 | 21·8 | 21·3 |
| 2 | 5·7 | 6·1 | 25 | 23·1 | 22·5 |
| 3 | 6·1 | 6·5 | 26 | 24·4 | 23·8 |
| 4 | 6·5 | 6·9 | 27 | 25·9 | 25·1 |
| 5 | 6·9 | 7·3 | 28 | 27·4 | 26·4 |
| 6 | 7·4 | 7·7 | 29 | 29·0 | 27·9 |
| 7 | 7·9 | 8·2 | 30 | 30·6 | 29·4 |
| 8 | 8·4 | 8·7 | 31 | 32·4 | 31·0 |
| 9 | 8·9 | 9·2 | 32 | 34·3 | 32·6 |
| 10 | 9·5 | 9·7 | 33 | 36·2 | 34·3 |
| 11 | 10·1 | 10·3 | 34 | 38·3 | 36·2 |
| 12 | 10·7 | 10·9 | 35 | 40·4 | 38·1 |
| 13 | 11·4 | 11·6 | 36 | 42·7 | 40·2 |
| 14 | 12·1 | 12·2 | 37 | 45·0 | 42·2 |
| 15 | 12·8 | 13·0 | 38 | 47·6 | 44·4 |
| 16 | 13·6 | 13·7 | 39 | 50·1 | 46·7 |
| 17 | 14·5 | 14·5 | 40 | 53·0 | 49·2 |
| 18 | 15·4 | 15·3 | | | |

From the temperature, which the exterior thermometer indicates, is derived the highest tension which the vapor could assume, and the quotient of this into the actual tension according to the interior thermometer, gives the moisture of the atmosphere.

(*e*) *Regnault's hygrometer.* Since glass is a bad conductor of heat, the gilding of Daniel's thermometer has not exactly the temperature indicated by the internal thermometer, for which reason Regnault, and before him Döebereiner, has proposed the following modifications: A vessel of thin, polished silver, about 2 centimeters wide and 4 centimeters high, is fitted tightly on a glass tube *a*, fig. 761, and filled with sulphuric

*Fig.* 761.

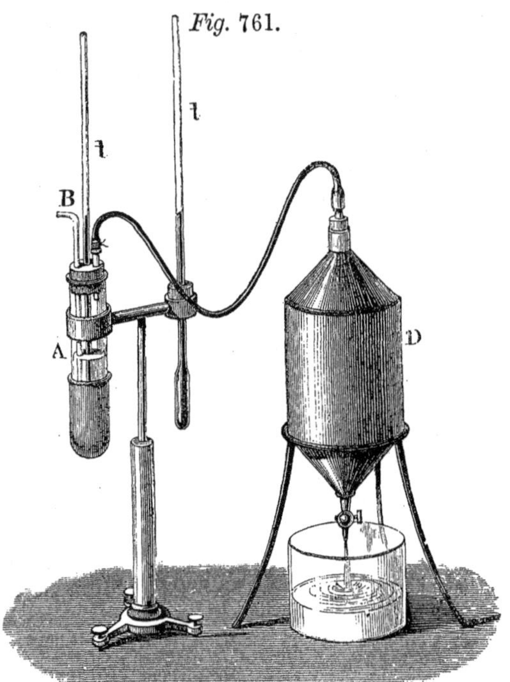


ether. The other end of the tube is closed by a cork, through which pass a thermometer *t*, and a tube *B* dipping into the ether, and a second short glass tube. By blowing through *B* with the mouth the ether evaporates and thereby cools the silver, so that the temperature at which dew is deposited on it can be observed; a second thermometer indicating the temperature of the air.

It is better to fix the apparatus on a stand, and draw the air through an aspirator, as seen in the figure. The aspirator is simply a tin vessel filled with water. The lower aperture has a stop-cock, by which the flow of the water can be regulated. The aperture at the top is connected by a caoutchouc tube with the short tube of the hygrometer. The glass bottle fitted up with a siphon, according to § 162, fig. 296, may be used instead of a separate aspirator. The siphon serves for the exit tube, the air-tube is connected with the hygrometer, and serves at the same time to regulate the flow. In this way the whole apparatus, except the silver cup, can be made at home at small expense.

(*f*) *The psychrometer.* The thermometers in this instrument must be more delicate than those of the previous ones, and must be graduated, at least, to half degrees. The correctness of the fixed points, the correspondence of both thermometers, and the accuracy of the graduation, must be tested by careful observations. Both thermometers must be fastened on a common stand, so that the bulbs shall be free. The covered bulb should be moist, without drops of water hanging from it, and it is probably best to connect it with a cotton wick, dipping into a vessel of water placed just below it. Sufficient water is thus drawn up by capillarity. The wick must be renewed occasionally, which is no great trouble; fig. 762 represents this instrument. The instrument must be suspended in the open air, and protected from strong currents. Under these circumstances, the tension $x$ of the aqueous vapor may be calculated by the following formula: $x = f - 0{\cdot}00778\ (t - t')\ b$, $f$ being the tension of the saturated vapor, at the temperature $t'$, in millimeters; $t'$, the temperature of the wet; $t$, that of the dry thermometer, according to Celsius; $b$, the height of the barometer reduced to the freezing point. The effect of $b$ is seldom greater than 0·1 of a millimeter, and, therefore, where the greatest accuracy is not required, the mean height of the barometer may be taken, and the factor $b$ combined with the other constants; if $b$ be taken $= 740$ *mm*, then $x = f - 0{\cdot}574\ (t - t'.)$ To obtain the quantity $y$ of moisture contained in the air, let $q$ be the quantity at the temperature $t'$, then $y = q - 0{\cdot}65\ (t - t'.)$ The following table contains these numbers calculated for the psychrometeric differences from 0° to 12°, and the temperature of the air from — 20° to + 35° C.

*Fig.* 762.

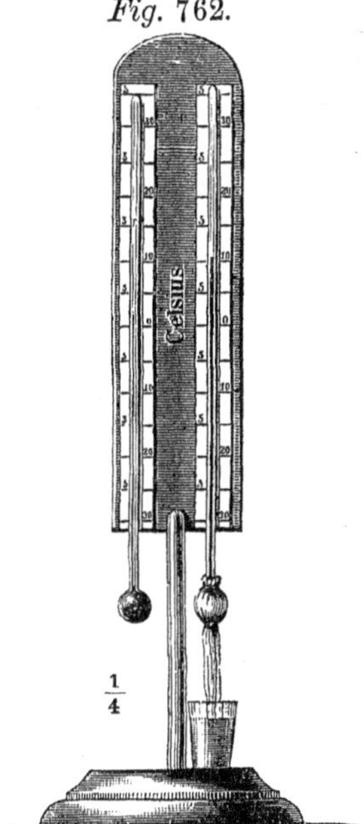

| Temperature of the air. | Difference between the wet and dry bulbs. | | | | | | | | | | | | |
|---|---|---|---|---|---|---|---|---|---|---|---|---|---|
| Degrees Cent. | 0 | 1 | 2 | 3 | 4 | 5 | 6 | 7 | 8 | 9 | 10 | 11 | 12 |
| — 20 | 1·5 | 0·8 | 0·1 | ...... | ...... | ...... | ...... | ...... | ...... | ...... | ...... | ...... | ...... |
| — 19 | 1·6 | 0·9 | 0·2 | ...... | ...... | ...... | ...... | ...... | ...... | ...... | ...... | ...... | ...... |
| — 18 | 1·8 | 1·0 | 0·3 | ...... | ...... | ...... | ...... | ...... | ...... | ...... | ...... | ...... | ...... |
| — 17 | 1·9 | 1·1 | 0·4 | ...... | ...... | ...... | ...... | ...... | ...... | ...... | ...... | ...... | ...... |
| — 16 | 2·0 | 1·2 | 0·5 | ...... | ...... | ...... | ...... | ...... | ...... | ...... | ...... | ...... | ...... |
| — 15 | 2·1 | 1·4 | 0·6 | ...... | ...... | ...... | ...... | ...... | ...... | ...... | ...... | ...... | ...... |
| — 14 | 2·3 | 1·5 | 0·8 | ...... | ...... | ...... | ...... | ...... | ...... | ...... | ...... | ...... | ...... |
| — 13 | 2·4 | 1·6 | 0·9 | 0·1 | ...... | ...... | ...... | ...... | ...... | ...... | ...... | ...... | ...... |
| — 12 | 2·6 | 1·8 | 1·0 | 0·3 | ...... | ...... | ...... | ...... | ...... | ...... | ...... | ...... | ...... |
| — 11 | 2·7 | 2·0 | 1·2 | 0·4 | ...... | ...... | ...... | ...... | ...... | ...... | ...... | ...... | ...... |
| — 10 | 2·9 | 2·1 | 1·3 | 0·6 | ...... | ...... | ...... | ...... | ...... | ...... | ...... | ...... | ...... |
| — 9 | 3·1 | 2·3 | 1·5 | 0·7 | ...... | ...... | ...... | ...... | ...... | ...... | ...... | ...... | ...... |
| — 8 | 3·3 | 2·5 | 1·7 | 0·9 | 0·1 | ...... | ...... | ...... | ...... | ...... | ...... | ...... | ...... |
| — 7 | 3·5 | 2·7 | 1·9 | 1·1 | 0·3 | ...... | ...... | ...... | ...... | ...... | ...... | ...... | ...... |
| — 6 | 3·7 | 2·9 | 2·1 | 1·3 | 0·5 | ...... | ...... | ...... | ...... | ...... | ...... | ...... | ...... |
| — 5 | 4·0 | 3·1 | 2·3 | 1·5 | 0·7 | ...... | ...... | ...... | ...... | ...... | ...... | ...... | ...... |
| — 4 | 4·2 | 3·4 | 2·5 | 1·7 | 0·9 | 0·1 | ...... | ...... | ...... | ...... | ...... | ...... | ...... |
| — 3 | 4·5 | 3·6 | 2·8 | 1·9 | 1·1 | 0·3 | ...... | ...... | ...... | ...... | ...... | ...... | ...... |
| — 2 | 4·8 | 3·9 | 3·0 | 2·2 | 1·4 | 0·5 | ...... | ...... | ...... | ...... | ...... | ...... | ...... |
| — 1 | 5·1 | 4·2 | 3·3 | 2·4 | 1·6 | 0·8 | ...... | ...... | ...... | ...... | ...... | ...... | ...... |
| 0 | 5·4 | 4·5 | 3·6 | 2·7 | 1·9 | 1·0 | 0·2 | ...... | ...... | ...... | ...... | ...... | ...... |
| 1 | 5·7 | 4·7 | 3·8 | 2·9 | 2·1 | 1·2 | 0·4 | ...... | ...... | ...... | ...... | ...... | ...... |
| + 2 | 6·1 | 5·1 | 4·1 | 3·2 | 2·3 | 1·4 | 0·5 | ...... | ...... | ...... | ...... | ...... | ...... |
| + 3 | 6·5 | 5·4 | 4·4 | 3·4 | 2·5 | 1·6 | 0·7 | ...... | ...... | ...... | ...... | ...... | ...... |
| + 4 | 6·9 | 5·8 | 4·8 | 3·7 | 2·7 | 1·8 | 1·0 | ...... | ...... | ...... | ...... | ...... | ...... |
| + 5 | 7·3 | 6·2 | 5·1 | 4·1 | 3·1 | 2·1 | 1·2 | 0·3 | ...... | ...... | ...... | ...... | ...... |
| + 6 | 7·7 | 6·6 | 5·5 | 4·5 | 3·4 | 2·4 | 1·4 | 0·5 | ...... | ...... | ...... | ...... | ...... |
| + 7 | 8·2 | 7·0 | 5·9 | 4·9 | 3·8 | 2·8 | 1·8 | 0·8 | ...... | ...... | ...... | ...... | ...... |
| + 8 | 8·7 | 7·5 | 6·4 | 5·3 | 4·2 | 3·2 | 2·1 | 1·1 | 0·2 | ...... | ...... | ...... | ...... |
| + 9 | 9·2 | 8·0 | 6·9 | 5·7 | 4·6 | 3·6 | 2·5 | 1·5 | 0·5 | ...... | ...... | ...... | ...... |
| + 10 | 9·7 | 8·5 | 7·3 | 6·2 | 5·1 | 4·0 | 2·9 | 1·9 | 0·9 | ...... | ...... | ...... | ...... |
| + 11 | 10·3 | 9·1 | 7·9 | 6·7 | 5·6 | 4·4 | 3·3 | 2·3 | 1·2 | 0·2 | ...... | ...... | ...... |
| + 12 | 10·9 | 9·7 | 8·4 | 7·2 | 6·0 | 4·9 | 3·8 | 2·7 | 1·7 | 0·6 | ...... | ...... | ...... |
| + 13 | 11·6 | 10·3 | 9·0 | 7·8 | 6·6 | 5·4 | 4·3 | 3·1 | 2·1 | 1·0 | ...... | ...... | ...... |
| + 14 | 12·2 | 10·9 | 9·6 | 8·3 | 7·1 | 5·9 | 4·8 | 3·6 | 2·5 | 1·4 | 0·4 | ...... | ...... |
| + 15 | 13·0 | 11·6 | 10·3 | 9·0 | 7·7 | 6·5 | 5·3 | 4·1 | 3·0 | 1·9 | 0·8 | ...... | ...... |
| + 16 | 13·7 | 12·3 | 10·9 | 9·6 | 8·3 | 7·0 | 5·8 | 4·6 | 3·5 | 2·4 | 1·3 | 0·2 | ...... |
| + 17 | 14·5 | 13·1 | 11·6 | 10·3 | 9·0 | 7·7 | 6·4 | 5·2 | 4·0 | 2·9 | 1·7 | 0·7 | ...... |
| + 18 | 15·3 | 13·8 | 12·4 | 11·0 | 9·6 | 8·3 | 7·0 | 5·8 | 4·6 | 3·4 | 2·2 | 1·1 | ...... |
| + 19 | 16·2 | 14·7 | 13·2 | 11·7 | 10·3 | 9·0 | 7·7 | 6·4 | 5·1 | 3·9 | 2·8 | 1·6 | ...... |
| + 20 | 17·1 | 15·5 | 14·0 | 12·5 | 11·1 | 9·7 | 8.3 | 7·0 | 5·8 | 4·5 | 3·3 | 2·2 | ...... |
| + 21 | 18·1 | 16·5 | 14·9 | 13·4 | 11·9 | 10·5 | 9·1 | 7·7 | 6·4 | 5·1 | 3·9 | 2·7 | ...... |
| + 22 | 19·1 | 17·4 | 15·8 | 14·2 | 12·7 | 11·2 | 9·8 | 8·4 | 7·1 | 5·8 | 4·5 | 3·3 | ...... |
| + 23 | 20·2 | 18·5 | 16·8 | 15·2 | 13·6 | 12·1 | 10·6 | 9·2 | 7·8 | 6·4 | 5·2 | 3·9 | 2·5 |
| + 24 | 21·3 | 19·5 | 17·8 | 16·1 | 14·5 | 12·9 | 11·4 | 10·0 | 8·5 | 7·2 | 5·8 | 4·5 | 3·1 |
| + 25 | 22·5 | 20·6 | 18·9 | 17·1 | 15·5 | 13·8 | 12·3 | 10·8 | 9·3 | 7·9 | 6·5 | 5·2 | 3·9 |
| + 26 | 23·8 | 21·8 | 20·0 | 18·2 | 16·5 | 14·8 | 13·2 | 11·6 | 10·1 | 8·7 | 7·3 | 5·9 | 4·6 |
| + 27 | 25·1 | 23·1 | 21·2 | 19·3 | 17·5 | 15·8 | 14·2 | 12·6 | 11·0 | 9·5 | 8·1 | 6·7 | 5·3 |
| + 28 | 26·4 | 24·4 | 22·4 | 20·5 | 18·7 | 16·9 | 15·2 | 13·5 | 11·9 | 10·4 | 8·9 | 7·5 | 6·1 |
| + 29 | 27·9 | 25·8 | 23·7 | 21·7 | 19·8 | 18·0 | 16·3 | 14·6 | 12·9 | 11·3 | 9·8 | 8·3 | 6·8 |
| + 30 | 29·4 | 27·2 | 25·1 | 23·0 | 21·1 | 19·2 | 17·4 | 15·6 | 13·9 | 12·3 | 10·7 | 9·1 | 7·7 |
| + 31 | 31·0 | 28·7 | 26·5 | 24·4 | 22·4 | 20·4 | 18·5 | 16·7 | 15·0 | 13·3 | 11·6 | 10·1 | 8·5 |
| + 32 | 32·6 | 30·3 | 28·0 | 25·8 | 23·8 | 21·7 | 19·8 | 17·9 | 16·1 | 14·3 | 12·7 | 11·0 | 9·4 |
| + 33 | 34·4 | 31·9 | 29·6 | 27·3 | 25·2 | 23·1 | 21·1 | 19·1 | 17·3 | 15·4 | 13·7 | 12·0 | 10·4 |
| + 34 | 36·2 | 33·7 | 31·2 | 28·9 | 26·7 | 24·5 | 22·4 | 20·4 | 18.5 | 16·6 | 14·8 | 13·1 | 11·4 |
| + 35 | 38·1 | 35·5 | 33·0 | 30·6 | 28·2 | 26·0 | 23·8 | 21·8 | 19·8 | 17·8 | 16·0 | 14·2 | 12·5 |

[363] **Latent heat of steam.**—The great quantity of latent heat in steam may be shown by the apparatus fig. 763. Pour into the cylinder *c* a certain weight of water at a known temperature, and cause the water in *a* to boil before inserting the tube *b* into the cylinder. The boiling is kept up until all the air is driven out of the apparatus, which,

*Fig.* 763.

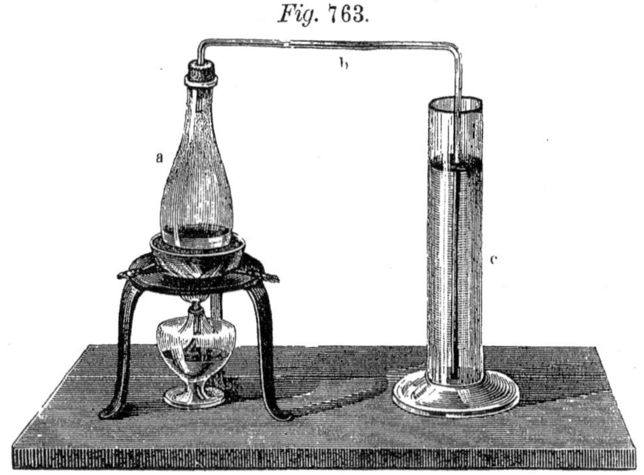


for an apparatus 5 times the size of the figure, will require 2 or 3 minutes. The tube *b* is then introduced into the cylinder, without interrupting the boiling. The steam condenses, and the water rises in the cylinder. When it has increased about $\frac{1}{10}$, withdraw the tube *b*, and weigh the water again, after noting its temperature. A thermometer with a very small bulb must be used, and only the bulb immersed, as the immersion of the whole instrument with its scale would cause a greater error by cooling the water than would be avoided by sinking the mercury to the top of the column.

The calculation is simple, as the following example shows: Water weighed before the experiment 60 grammes, and afterwards 66·1. The temperature before the experiment was 12° C., afterwards 68·5°; the 60 grammes of water gained, therefore, 56·5°, that is, 60 . 56·5 = 3390 units of heat; 6·1 grammes of steam lost 31·5° of heat, imparted therefore 31·5 . 6·1 = 192 units of heat to the other water; 3198 units of heat are due to the latent heat of 6·1 grammes of steam, 1 gramme has, therefore, yielded 524 units. Although in this experiment the latent heat of steam always results too low, on account of the loss from the glass and the air, together with the water which condenses in the tube *b*, it is very well suited to show the fact, and to give an approximate idea of the quantity of latent heat in steam.

[364] **The condenser.**—A very simple, and for small operations very useful condenser, is represented in fig. 764, which is particularly suitable for the illustration of the apparatus. It consists of a glass tube, 1 to $1\frac{1}{2}$ feet long and 1 inch in diameter, through the center of which passes

*Fig.* 764.

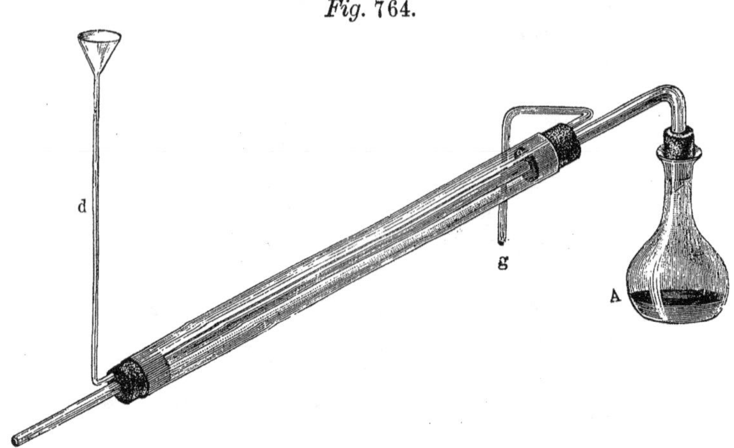


another tube, $\frac{1}{4}$ inch in diameter, projecting on both sides and fitted tightly through corks. One end of the inner tube, which projects but little, is somewhat enlarged; the other end, which projects 3 to 4 inches, is contracted. At this end a small tube, 5 or 6 inches long, bent at an acute angle and terminating in a funnel, passes through the cork. A similar tube, bent twice at right angles, passes through the other cork. The condenser is fixed in a sloping position, the funnel tube perpendicular, and rising higher than the upper end of the condenser. The neck of a little retort is inserted in the large end of the inner tube. Water is poured through the funnel so as to fill the large tube, and as it gradually becomes heated, it runs off again through *b*, condensing the vapors in the inner tube. The water may be drawn by a siphon from a higher vessel, and the flow thus regulated as occasion requires, or a Mariotte's vase may be used instead of the siphon.

*Fig.* 765.

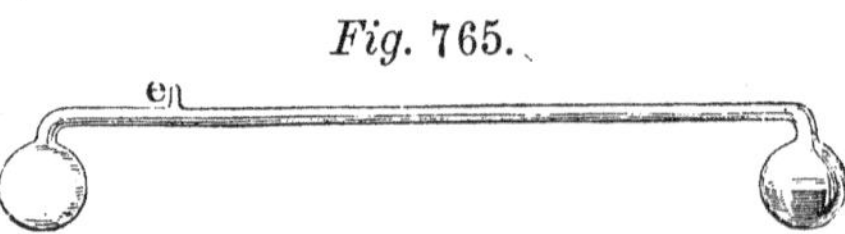


[365] **The cryophorus.**—This is represented in fig. 765, about $\frac{1}{8}$ the real size, but it is convenient to have the two arms, on which

the bulbs are blown, somewhat longer than in the figure, and the cross-piece proportionally shorter. The glass is constructed like the pulse-glass. It is filled with water, and the boiling must be done over a wide charcoal fire, so as to heat the tube throughout its whole extent. When the water has boiled long enough, and one bulb is not quite half full of water, seal the opening with a blow-pipe. To make the experiment, all the water is got into one bulb, and the other plunged into a freezing mixture of pounded ice and salt, so as to cover the entire bulb and part of the tube. When the vessel containing the freezing mixture is small, it may, previously, be surrounded with ice. It is advisable to interrupt the experiment as soon as a film of ice forms in the free bulb, because the entire freezing of the water is apt to burst the bulb. When the experiment is made in summer, a draught must be created, so that the air may be as little charged with aqueous vapor as possible, which is apt to occur with a numerous audience, but the bulb itself must not be exposed to the draught.

## (*d.*) EXPERIMENTS TO ILLUSTRATE THE STEAM-ENGINE.

[366] **Heros' rotating ball.**—Blow a large glass bulb with two points, pass these through a wire handle and bend them as shown in fig. 766. Warm the bulb, and, closing one point with the finger, draw in a little water. If the water be heated to boiling over a spirit-lamp, while the bulb is slowly turned, the pressure of the escaping steam will cause it to rotate rapidly. The handle itself may be made of glass, in which case the jets are first bent, and then the handle, made of two slender glass rods, which are attached to the bulb, and afterwards joined together.

*Fig.* 766.

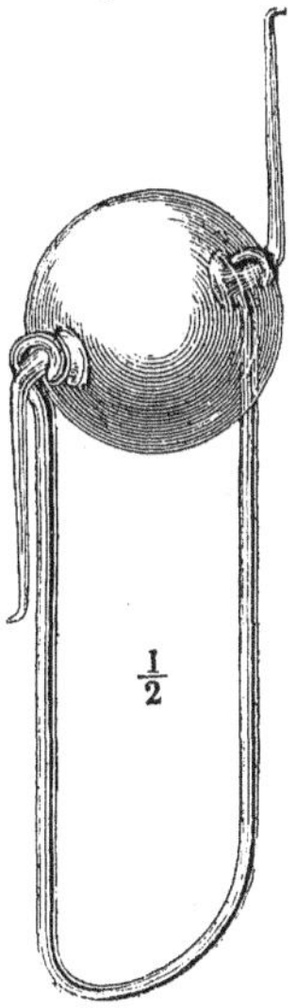

$\frac{1}{2}$

[367] **The steam-engine.**—In illustrating the steam-engine, it is of the greatest importance to give a clear idea of the mode of working the valves, and of the action of the condenser in the low-pressure engine. For the latter, you may revert to the experiment described in § 364.

In respect to the working of the valves, it is enough to explain any of the various arrangements by which the new steam is introduced, alternately above and below the piston, and the spent steam withdrawn; that is, to consider the action of a double-acting engine, no others being now in use. The best for this purpose are the sliding-valves, and a sectional model should be made. This can be done very simply in the following way: First, bend an iron wire, fig. 767, so as to obtain two cranks *a* and *b*, at right angles to each other, and the third *c*, *d*, whose position is indifferent. Make for this a wooden frame, fig. 768, consisting of two rectangular divisions *A* and *B*; the first of which represents the cylinder, the other the steam-chest. The channels 1 2 3 are painted on the partition between them, with white paint on black ground: 1 and 3 leading from the steam-chest to the cylinder, and 2 into the open air, or to the condenser. The size of the apertures in the steam-chest is equal to the length of the crank *a*, and the flanges of the slide are made of the same width. The slide is made either of brass, or wood painted yellow, and pressed against the partition between the two compartments by a slight spring *e e* acting against the side of the frame. The bars *n* and *o* are grooved up as far as the length of the box *A*; a cross-bar *p* is fixed between *m* and *n* to serve as a guide to the valve-rod *l l*. Before the frame is put together, the cross-bar, fig. 769, is put into the grooves between *n* and *o*, so as to slide freely. The piston *C* consists of a bit of wood, and the piston-rod *D* is passed through holes in the cross-bar *q* and the top of the cylinder at *r*, and glued fast to *q* and *C*. Finally, make the rods *t* and *u* of stout wire; the latter appears foreshortened in the figure because its crank is horizontal; lay the axis in notches in the bars *m n o*, and fasten it down with little wire staples. The valve-rod *l* is made of wire, with a screw on the end, and the slide is fastened to it by two nuts, so that when the crank *a* stands horizontal, the slide would be in the middle of

*Fig.* 767.

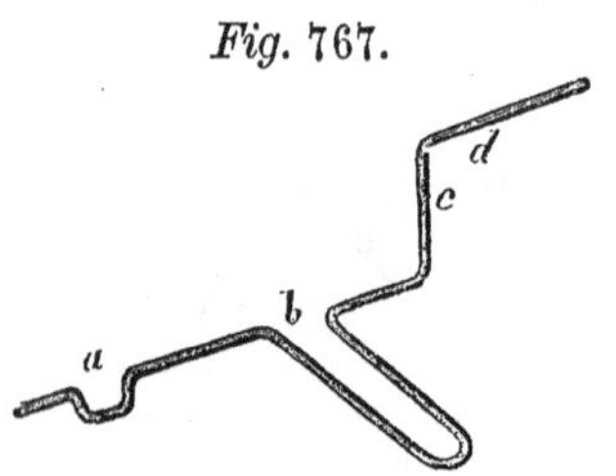


*Fig.* 768.

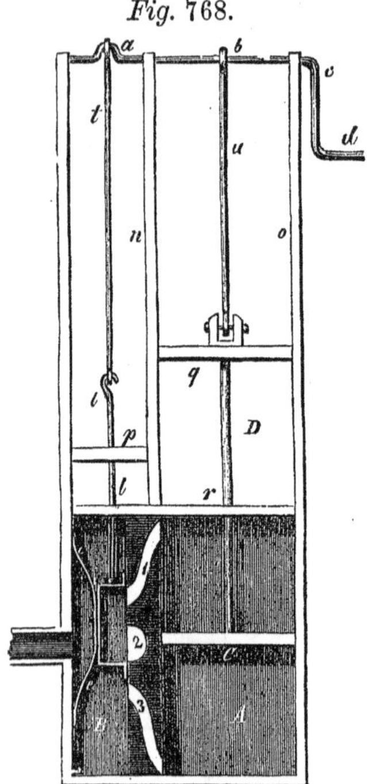


*Fig.* 769.

the partition between *A* and *B*. It is best not to mark the channels 1 2 3 until this is done, and to fix their position according to the actual motion of the slide. A little tube *x* represents a steam-pipe. By turning the crank *c d*, the piston and the slide will make the corresponding motions, so as to illustrate fully this important part of the steam-engine.

An eccentric wheel, fastened at *a*, instead of the crank, would serve to illustrate this much-used contrivance. In this case, the support outside of the slide could be omitted, and the axis made a little stronger. Large diagrams may be used for further illustrations, but, better than all, a working model.

[368] **Models of steam-engines.**—That the lecturer must have some means of giving a correct conception of the working of a steam-engine, is beyond doubt; but whether a model is necessary, is quite another question. If the means can be commanded, it will be well to provide a model; but certainly, there are many things much more valuable and necessary. The following points are important in constructing a model:—

(*a*) It must not be executed in what is called the model style, *i. e.*, it must not be merely a little machine, which turns a wheel by steam, omitting everything which is not necessary to this object; but it must be accurately copied from some steam-engine in common use.

(*b*) The construction of the model-machine must be plain and easily understood, and must admit of the steam-chest being covered with glass. It is not necessary to have the cylinder and the pumps made of glass.

(*c*) The model must not be on too small a scale.

It is more important to show the action of the separate parts by taking them to pieces, than to set the machine actually in motion. It is better to work the model by a force-pump, than by the boiler, because it injures it to be left standing a year after using it with steam without a thorough cleaning, which consumes considerable time.

[369] **The locomotive.**—One of the most valuable special forms of the steam-engine is the locomotive, and the principal points of its construction cannot be well omitted in a course of physical instruction. The chief peculiarities of the locomotive engine, are the construction of the boiler and the arrangement for running forward or backward at pleasure. Both of these may be well illustrated by the following sectional model:—

Have a frame, *a a b b*, made of hard wood, figs. 770 and 771, of such dimensions that the model will be $\frac{1}{10}$ of the real size; *c c* are only pieces which serve to fill up. At the front a block of wood, *m*, is inserted in the frame, so as to be even at the top at *e e*. Fashion the half of this block into the section of the cylinder *f*, fig. 772, and of half of the steam-chest $g$, which lies between the two cylinders, as shown in section in the figure, leaving the front and rear walls standing. Excavate the upper

*Fig.* 770.

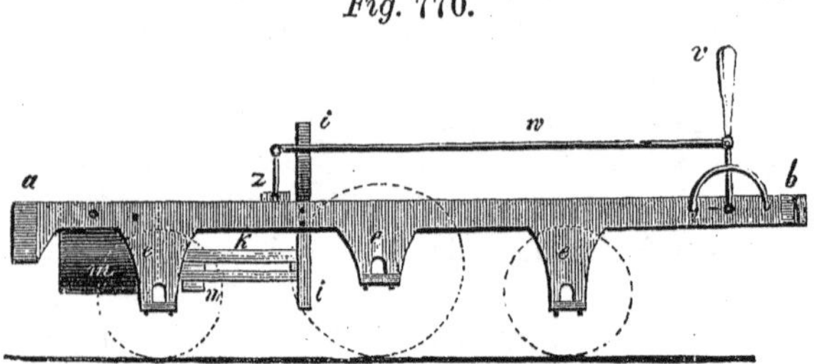


*Fig.* 771.

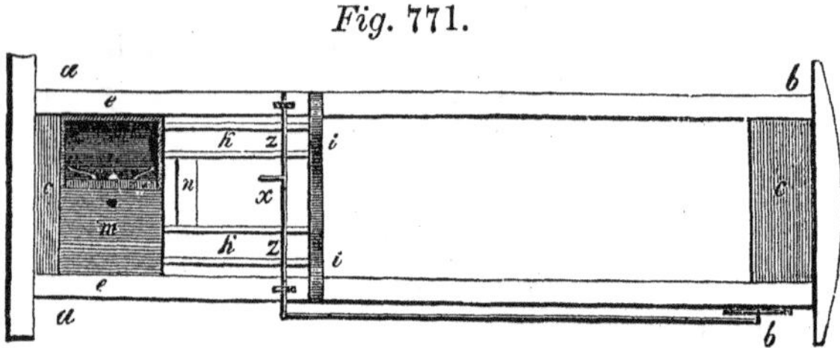


surface, as in the figure, and screw the block in its place. Bore holes for the driving-rod and the valve-rods. On the partition $h$, fig. 772, between the cylinder and the steam-chest, mark the apertures for the valves, just

*Fig.* 772. *Fig.* 773.

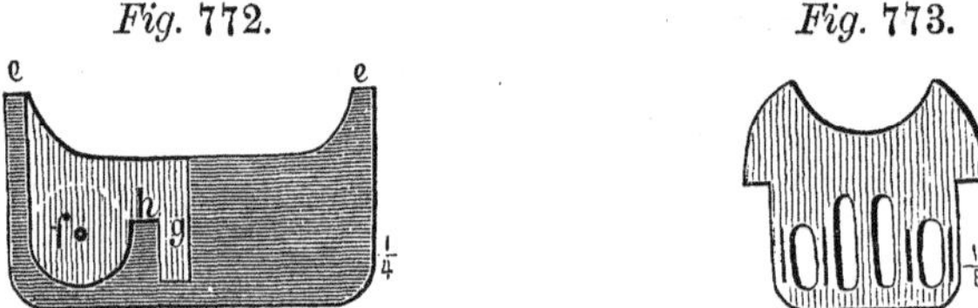


as the model, fig. 768. In both cases this is the last thing to be done; $i$ $i$ is a board, like fig. 773, set upright in the frame, with the necessary holes for the rods connected with the piston and eccentric. Between this board and the block $m$ are fastened four grooved bars $k$, two of which serve as guides for the piston-rods, only half the machine being worked up, and the others, therefore, remaining unemployed; $n$ is a cross-piece screwed between two bars, and serves as a guide for the valve-rod; it is, for this reason, made of two halves. Notches are made in the cleats $e$ $e$ $e$, fig. 770, for the wooden axles of the wheels, which are only disks of wood. The wheels run outside of the frame, and are glued to the axles; the driving-wheels are made rather smaller, so that they can be turned freely by a handle, when the model stands on the four running wheels. The driving-axle is made of hard wood, in one piece, not glued together, the length of the cranks, which are placed at right angles to each other, being dependent upon the stroke of the piston. The piston consists of a thin disk of wood. The eccentrics are also made of wood, and must, therefore, be rather larger than the proper proportion, to insure the requisite strength. Their edges are grooved, to receive rings made of wire. The hole for the axle is so arranged that the fibers of the wood may run perpendicular to the line drawn through the center. The disk is split in the line *a b*, fig. 774, and glued upon the axle in such a position that the prolongation of the valve-rod will pass between the two disks, which lie close together. The eccentricity of both is placed in a straight line, at right angles to the axis of the crank, no account being taken here of the effect of expansion.

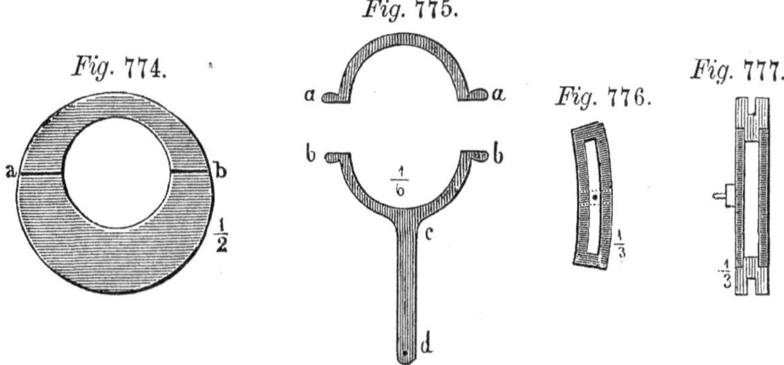

*Fig.* 774. *Fig.* 775. *Fig.* 776. *Fig.* 777.

The rings *a a*, *b b*, fig. 775, are made of flattened wire, and a thicker, flattened wire, *c d*, soldered to one of each pair. The projections *a a* are filed out a little close to the ring, so that the two sections of the ring can be tied together with binding-wire. The two rods must be bent toward each other so that their ends *d* may lie in the same vertical plane. The links are made of two plates, fig. 776, riveted to two strips of brass, fig. 777, in which the rods of the eccentrics run on pins; the complete imitation of the ordinary construction on such a small scale being difficult, and quite unnecessary for the present purpose. On one side a cross-piece is riveted, with a pin ending in a screw, with which the arm *x*, fig. 771, of the bar *z*, is connected, so that the links can be raised and lowered by means of the lever *v* and the rod *w*.

*Fig.* 778.

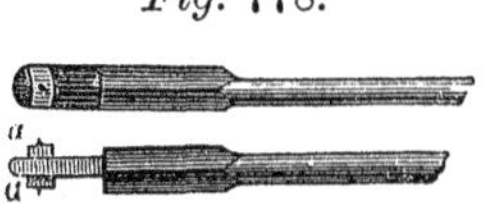


The valve-rod, fig. 778, which has two rollers

*a a*, is inserted in the link, in which the rollers must move smoothly up and down; they must, therefore, be ground with emery. The portion of the valve-rod which runs in the guide *n*, fig. 771, is square. The valve itself consists of a strip of brass, as already explained in fig. 768; the spring is omitted. Figs. 779 and 780 represent the parts with the valve *k* and the crank *n* in two positions, the form of the links being somewhat changed.

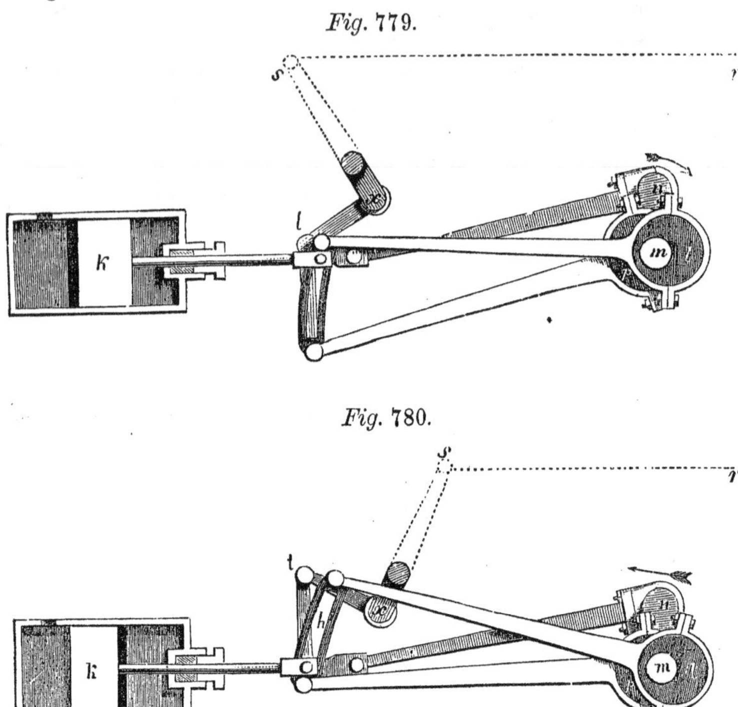


*Fig.* 779.

*Fig.* 780.

The piston-rods and connecting-rods are made as described in § 368, the guide of the piston-rod being shorter as the bars *k k* come much closer together in this case than in that. Feed-pumps are not necessary in the model.

Set on the frame a half-round wooden model of the boiler, with its furnace, smoke-chamber, chimney, and dome, a longitudinal section of the internal arrangement of these parts being painted on the block. Fig. 781 represents the surface which is painted, and figs. 782 and 783 are sections of the block in the lines indicated in fig. 781. It will be seen that this model has a rabbet on the round side of the furnace, which fits on the frame *a a*, *b b*, fig. 771, the half-round part of the boiler fitting in the concavity of the board, fig. 773. If the model is made to be taken apart, as indicated by the dotted lines, these surfaces may also be covered with paper, and the respective sections painted on them, the pieces being held together by pegs.

*Fig.* 781. *Fig.* 782. *Fig.* 783.

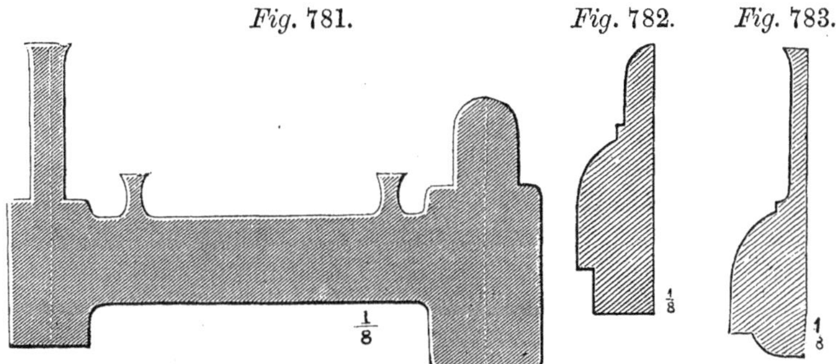


The foregoing paragraphs contain hints for the construction of a sectional model, but the execution presupposes an acquaintance with all the parts of a locomotive, and some accurate drawing must be taken as a guide.

Large diagrams may be used for further illustration of the separate parts, when a railroad station is not accessible. Even in this case, a sectional model is useful for the previous explanation of these parts which cannot be seen at all, or at least not in action.*

What has been said respecting a working model of a steam-engine applies to the locomotive, and the latter is much less indispensable.

## (*e.*) EXPERIMENTS ON SPECIFIC HEAT.

[370] Of the various methods of showing the specific heat of bodies, that by mixture is the only one adapted to class illustration. The most instructive experiments are: (1) The demonstration of Richmann's law, by mixing two weighed quantities of water at different temperatures, in order to show that the capacity of water is constant at different temperatures; and (2) the mixture of mercury and water, to show the low specific heat of the former. In the first experiment, the temperature of the mixtures being first calculated, the vessel is warmed with water to

* Models of both these machines, well executed, and in many respects stronger than here described, with internal or external cylinders, may be had of Widman, in Freiburg, for from 8 to 10 dollars.

this degree; this is emptied out, and the hot and cold water immediately poured in, and stirred with a thermometer. In the second experiment, this precaution is not necessary; the mercury being added to water at the temperature of the air, and rapidly stirred with an iron rod. It is simplest to take 1 lb. of water at the temperature of the surrounding air $= t$, and heat 1 lb. of mercury to 108·5° F. $+ t$. The mixture will have a temperature $t + 34{\cdot}2$. The precautions given will insure results as nearly accordant as the purposes of instruction require.

The change of the specific heat of gaseous bodies, by expansion and compression, can be shown by the air-pump. For this purpose, suspend a sensitive thermometer (one with fine tube and large bulb) in a small receiver, and exhaust the air rapidly; the thermometer will sink from 2 to 4 degrees. If the communication with the air be now cut off, and the apparatus allowed to stand until the temperature is again equalized, and the air then admitted, the thermometer will rise. The reverse effect is produced by condensation. The result is certain, but can only be seen by a few at a time. Spirit thermometers, although more easily observed, are too dull for this experiment. The diminished capacity of gases, for heat, by condensation, may also be shown by the fire-syringe. This consists of a glass or metallic cylinder, *A B*, fig. 784, with a rounded base, which is laid in the hollow of the hand, the cylinder itself being held between the fore and middle finger. A piston, *c*, fig. 785, with a cavity in the end, fits air tight into this cylinder. This cylinder has also a rounded handle, and a little hook is usually placed in the cavity, to hold a bit of tinder. If the piston be forcibly driven into the cylinder and quickly withdrawn, the tinder will take fire. This apparatus can be very well made in the following way: Take a piece of brass of the requisite thickness—it is convenient but not necessary to have it cast; when soldered together it renders boring difficult—and bore out a block of wood on the lathe, so that the cylinder can be driven into it. Turn off the top of the brass, and bore it through with a cannon drill. It is better to use two or three drills successively. Next, take the block with the brass from the lathe, close it at one end, insert in it a stout iron wire, and fill it with melted lead. This wire, with the lead, is fixed on the lathe, smeared with emery and oil, and the tube ground out with it, being constantly turned and slid back and forth.

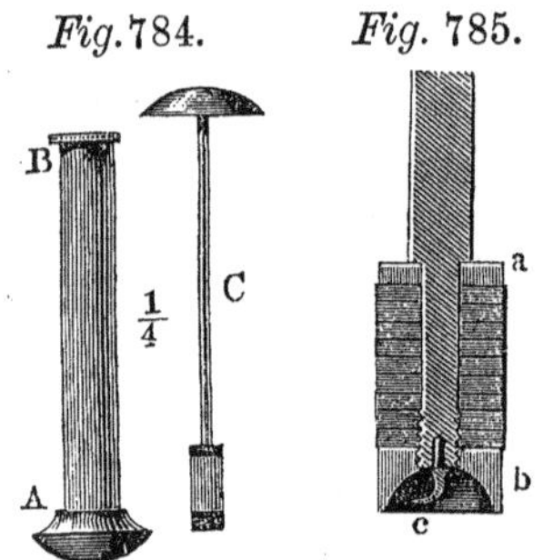

*Fig.* 784. *Fig.* 785.

Several cylinders of lead must generally be used, the last with fine pumice powder and oil. The piston-rod is made of a steel wire, turned off to make a shoulder for the plate *a*. A screw is cut on the end for the piece *b*, and a little hole made for the hook *c*. Disks of leather soaked with tallow are pressed between *a* and *b*, and turned on the lathe. This piston must move very easily, and yet air tight; it is, also, the best test of the uniformity of the tube. Finally, the handle of the tube is soldered to it with tin, the wooden case having been previously split off, and the tube is then turned off smooth and polished. The apparatus sold in shops is not always made with so much care, and my chief reason for being thus minute, is to describe the mode of boring a tube.

## (*f*.) EXPERIMENTS ON THE DIFFUSION OF HEAT.

[371] The radiation from heated bodies is easily shown by the following experiment: Take a sheet of pasteboard and cut in it a round hole, a little larger than the blackened bulb of the differential thermometer, or the aperture of the funnel of the thermo-multiplier. Place this screen upright on the table, and behind it the thermoscope, so as to be entirely protected, and support a mass of heated iron on a level with the aperture, at the distance of 1 or 2 feet. The thermoscope is not affected by the iron, until it is placed before the aperture in the screen in a straight line with the heated mass. The thermoscope must not be more than 1 foot from the screen.

[372] For the remaining experiments on radiant heat, a pair of spherical or parabolic metallic reflectors, 14 to 20 inches in diameter, are required. They need not be very highly polished. Spherical mirrors are generally preferable, because they can be more accurately ground than parabolic; the latter are not easily ground on account of their irregular curvature; but spherical mirrors can easily be made accurate enough to be used for optical purposes. They may be made at home in the following way:—

Make a model of sheet iron, with a radius of 1½ to 2 feet, the chord of which measures 14 to 20 inches. Have a mirror made after this model, of stout brass, (yellow metal,) with a strengthened rim. A good workman will give the mirrors very nearly the correct form, with only slight inequalities of the surface. Lay these disks in a tub filled with saw-dust, and grind them with a large piece of pumice-stone and water, until the whole surface appears uniform, moving the pumice-stone in epicycloidal curves. The surface will very soon become spherical. If, after grinding awhile, considerable inequalities still appear, they can be remedied by hammering. The tedious operation of grinding can be done by any person, under proper supervision. When the surface is uniformly ground with pumice-stone, take a thick piece of beech-wood charcoal, cut upon it a face inclined to the axis at an angle of 45°, round it off into the shape of the mirror, and with it grind away the marks of the pumice-stone, by the aid of oil. The final strokes must be made from the margin toward the center. The final polish is imparted with whiting.

The finished mirrors are screwed to a strong plate of brass, fig. 786, by holes drilled before the grinding. The plate is attached below the center of the mirror, and hinged to the end of a bar of wood, about 1½ inches square, on which it may be elevated or depressed by a wooden screw, *b*, fig. 787. The bar must be, at least, 3 feet long, and be fastened in a triangular or square base, so that it can be placed immediately upon the

*Fig.* 786. *Fig.* 787.

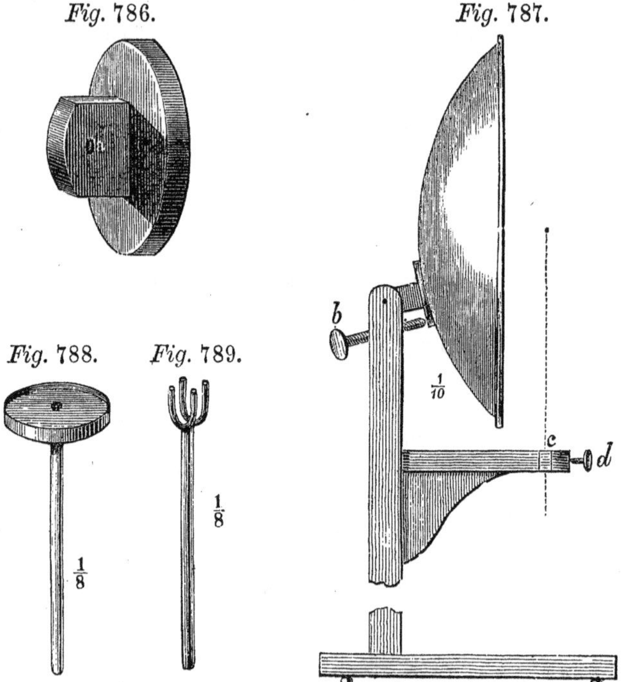


*Fig.* 788. *Fig.* 789.

floor. No other adjusting-screws are required. It is convenient to attach to the vertical support a horizontal arm, of such a length that the hole *c*, bored in the end of it, will be directly under the focus of the mirror. In this hole can be fastened by a binding-screw *d*, either a little table, fig. 788, or an iron fork, fig. 789, or a pointed wire. The table must be broad enough to support the differential thermometer witn one of its bulbs in the focus. Separate stands may be used for many purposes, as in fig. 790.

A cover of pastebord should be made to protect the mirror from injury.

[373] **Experiments with the reflectors.** — Place the two mirrors parallel to each other, fig. 790, at a distance of 10 or 20 feet, with their axes in the same straight line. The proper position is most easily found, by placing a light in the focus of one mirror and moving both until the image of the light is seen in the focus of the other.

*Fig.* 790.

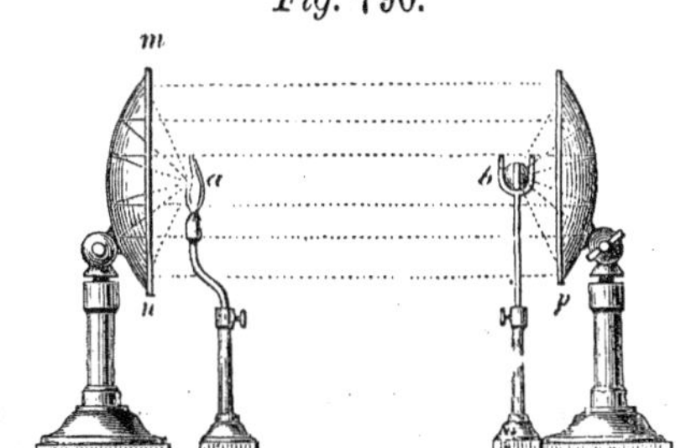


(*a*) Place in the focus of one mirror the blackened bulb of the differential thermometer, and in the other the iron fork, fig. 789, supporting a mass of 1 or 2 pounds of iron, heated nearly to redness. The thermometer will immediately indicate considerable heat, which will cease when a screen of wood, pasteboard, or glass is placed between the two mirrors—glass being, however, athermanous only for rays of heat radiating from bodies below 212° F.

(*b*) Place a live coal in the iron fork, and a bit of tinder in the focus of the other mirror; the tinder will soon be ignited, and the effect is not prevented by a pane of glass held between the two mirrors. In this case, however, it is advisable to place the mirrors nearer together, especially when they are not accurately spherical, because the heat is somewhat diminished by the glass, and the rays not being reflected entirely parallel, that portion which strikes the second mirror may not be sufficient to ignite the tinder. The coal must be briskly blown by a hand-bellows.

(*c*) A lump of ice placed in the focus of one mirror will cause the thermometer to sink in the focus of the other.

(*d*) Take a cubical brass box, polish one side, roughen another, paint a third white with lead and gum water, and blacken the fourth with lamp-black. When this is filled with hot water and placed in the focus of one of the mirrors, the difference in the radiating power of the four sides may be shown by the differential thermometer. The mirrors must be placed only a few feet apart, and the side of the cube next the thermometer covered with white paper to prevent radiation in that direction. This experiment can be made with one mirror, fig. 791, the focus corresponding to the position of the cube being found by the aid of a light. The cube must not be placed far from the mirror, and it is better to use two mirrors.

Although the effect in the ordinary differential thermometer is very perceptible, a thermo-multiplier, § 341, is much better for the purpose,

*Fig.* 791.

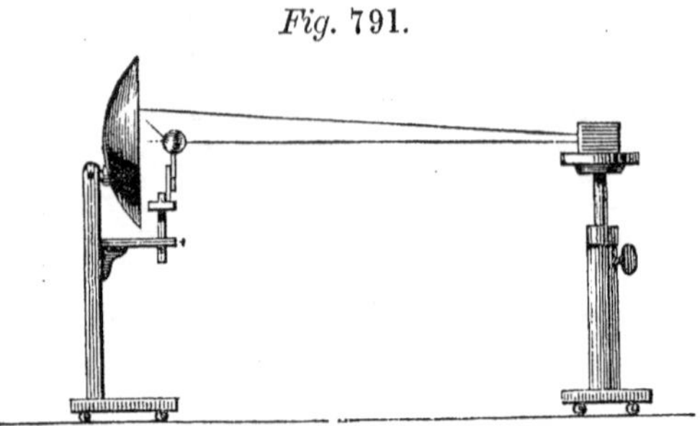

for, although the differential thermometer may be free from air and filled with alcohol, some rays of heat always fall upon the other bulb, although they may not be concentrated, and, what is of more importance, the instrument partially covers the mirror. Besides this, it requires much longer for the differential thermometer to rest than for the needle of the galvanometer. It is not even necessary to wait for this, for the final deflection of the needle may be estimated before it comes to rest. The great difference in the effect of the various sides of the cube makes this easier. The arrangement of the experiment is the same as in fig. 791, except that the thermo-pile, with its funnel turned toward the mirror, is put in the place of the differential thermometer. The pile is kept covered until the poles are connected with a galvanometer. The water in the cube is, in the mean while, heated to boiling by a lamp placed under a stand, fig. 792. After removing the lamp, uncover the pile, observe the deflection, and replace the cover. While the needle is coming to rest, the lamp may be replaced under the cube, and another face turned toward the mirror. When the thermo-pile is used, the mirror is not necessary; the funnel of the pile has only to be placed 1 or 2 inches from the side of the cube, even when the galvanometer has only a few hundred turns.

*Fig.* 792.

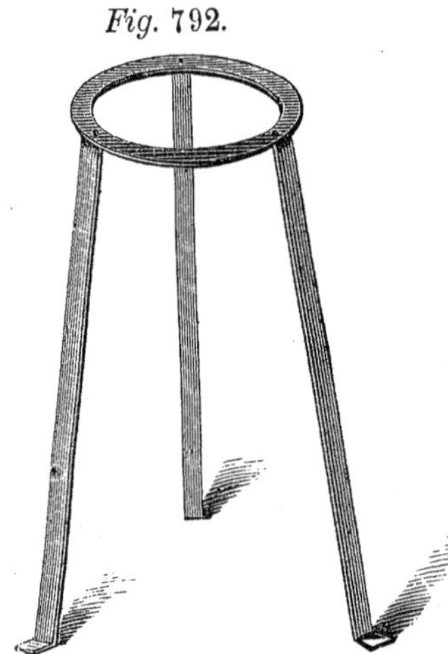

[374] **Conducting power of solids.**—In the side of a brass vessel, fig. 793, solder from 4 to 6 similar tubes, 2 or 3 lines long, and in these insert bars of various substances, such as brass, iron, lead, glass, wood, etc., of equal length and thickness. Coat these bars lightly with melted wax, and pour heated water or oil in the vessel; the wax will be melted on the bars to unequal distances from the vessel.

*Fig.* 793.

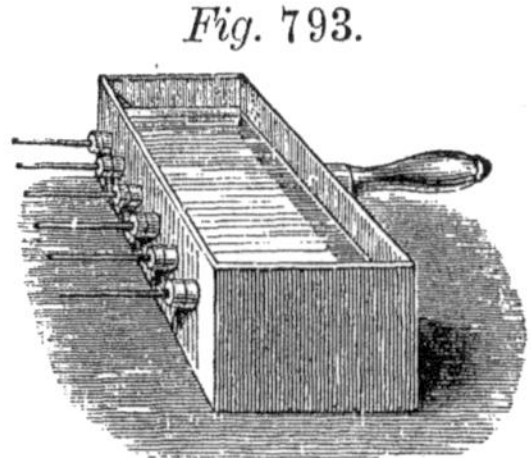

*Fig.* 794.

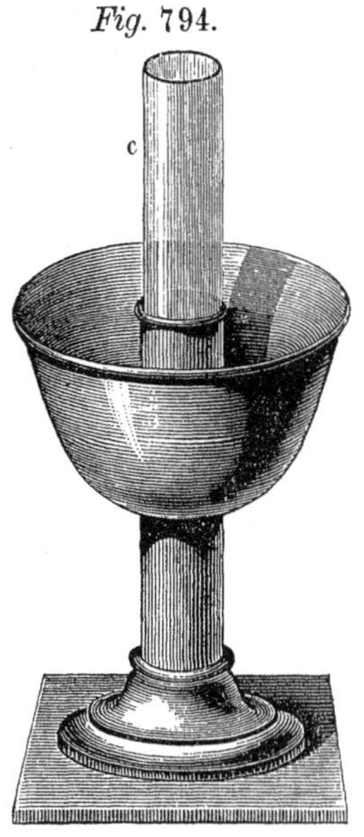


[375] **Conducting power of liquids.**—Surround a glass cylinder, fig. 794, with a metallic vessel; fill the cylinder with cold water, holding chalk or amber in suspension, then place a thermometer, fig. 717, on the bottom of the glass cylinder, and hang a second in the water at the top. When this is done pour hot water into the metallic vessel. The motion of the particles of chalk will prove that the heated water ascends at the sides of the cylinder, while the colder portions sink in the middle. The upper thermometer rises rapidly, while the lower one remains unaffected.

The low conducting power of liquids may be shown by suspending a thermometer in a vessel of cold water, and then carefully pouring heated oil upon the surface, or setting a capsule filled with burning alcohol upon it. The thermometer remains a long time unaffected. An air thermometer may be inserted through a cork, fitted in a hole in the side of the vessel, the bulb being placed near the surface of the liquid.

The simplest mode of showing this experiment is to fill a test-tube with water, and hold it by the bottom in a slanting position over the lamp until the water boils.

The following experiment answers the same purpose. Fasten a glass tube, 10 or 12 inches long and 1 inch in diameter, closed at one end, on a support, fig. 795. Fill the tube with water, and sink a lump of ice to the bottom of it by a leaden weight; the water may be boiled in the upper part of the tube without melting the ice.

*Fig.* 795. *Fig.* 796.

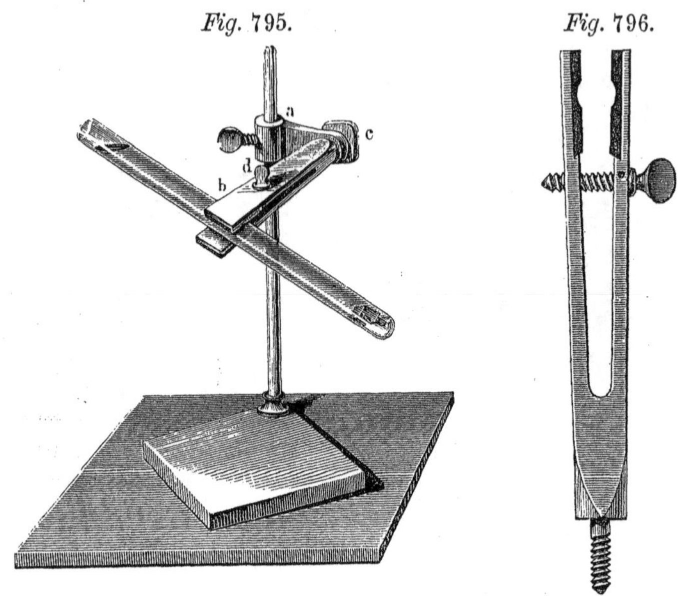


The support fig. 795 is useful for many purposes. The sliding arm *a* supports a clamp *b*, seen on a large scale in fig. 796. This can be fixed in any position by means of the screw and nut *c*. The apparatus may be had of all philosophical instrument makers.

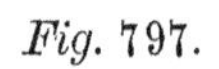
*Fig.* 797.

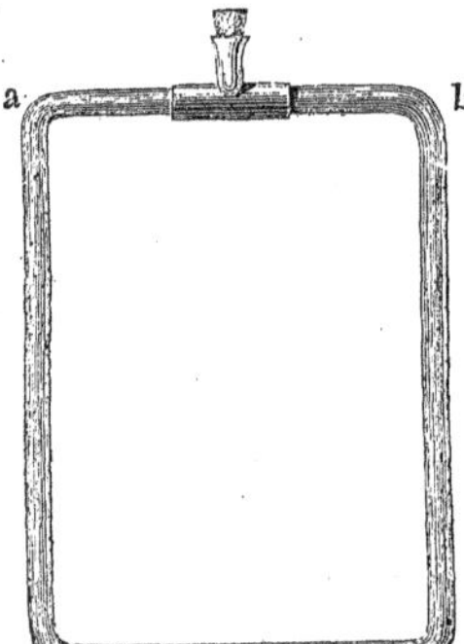


The circulation of heated water is well shown by the apparatus fig. 797. A glass tube is bent into a rectangle, and both ends cemented into a tubulated brass collar. The tube is filled with water, rendered turbid by powdered amber, supported by a clamp, and heated at one corner. The two corners, *a* and *b*, should be bent first, one of them cemented in and the other coated with sealing-wax; and then the other two bends made, after marking their positions with chalk. The tube might be made of two pieces connected by vulcanized india-rubber.

# APPENDIX.

## TABLES FOR REDUCING FRENCH MEASURES TO ENGLISH.

### TABLE I.

COMPARISON OF THE BAROMETER SCALES IN MILLIMETERS AND INCHES.

| Mms. | Inches. | Mms. | Inches. | Mms. | Inches. | Mms. | Inches. |
|---|---|---|---|---|---|---|---|
| 700 | = 27·560 | 725 | = 28·543 | 750 | = 29·528 | 775 | = 30·512 |
| 701 | = 27·590 | 726 | = 28·583 | 751 | = 29·567 | 776 | = 30·552 |
| 702 | = 27·638 | 727 | = 28·622 | 752 | = 29·607 | 777 | = 30·591 |
| 703 | = 27·678 | 728 | = 28·661 | 753 | = 29·646 | 778 | = 30·631 |
| 704 | = 27·717 | 729 | = 28·701 | 754 | = 29·685 | 779 | = 30·670 |
| 705 | = 27·756 | 730 | = 28·741 | 755 | = 29·725 | 780 | = 30·709 |
| 706 | = 27·795 | 731 | = 28·780 | 756 | = 29·764 | 781 | = 30·749 |
| 707 | = 27·835 | 732 | = 28·819 | 757 | = 29·804 | 782 | = 30·788 |
| 708 | = 27·876 | 733 | = 28·859 | 758 | = 29·843 | 783 | = 30·827 |
| 709 | = 27·914 | 734 | = 28·898 | 759 | = 29·882 | 784 | = 30·867 |
| 710 | = 27·953 | 735 | = 28·938 | 760 | = 29·922 | 785 | = 30·906 |
| 711 | = 27·992 | 736 | = 28·977 | 761 | = 29·961 | 786 | = 30·945 |
| 712 | = 28·032 | 737 | = 29·016 | 762 | = 30·000 | 787 | = 30·985 |
| 713 | = 28·071 | 738 | = 29·056 | 763 | = 30·040 | 788 | = 31·024 |
| 714 | = 28·111 | 739 | = 29·095 | 764 | = 30·079 | 789 | = 31·063 |
| 715 | = 28·150 | 740 | = 29·134 | 765 | = 30·119 | | |
| 716 | = 28·189 | 741 | = 29·174 | 766 | = 30·158 | | |
| 717 | = 28·229 | 742 | = 29·213 | 767 | = 30·197 | | |
| 718 | = 28·268 | 743 | = 29·252 | 768 | = 30·237 | | |
| 719 | = 28·308 | 744 | = 29·292 | 769 | = 30·276 | | |
| 720 | = 28·347 | 745 | = 29·331 | 770 | = 30·315 | | |
| 721 | = 28·386 | 746 | = 29·371 | 771 | = 30·355 | | |
| 722 | = 28·426 | 747 | = 29·410 | 772 | = 30·384 | | |
| 723 | = 28·465 | 748 | = 29·449 | 773 | = 30·434 | | |
| 724 | = 28·504 | 749 | = 29·489 | 774 | = 30·473 | | |

| Inches. | Mms. |
|---|---|
| 28 | = 711·187 |
| 29 | = 735·587 |
| 30 | = 761·986 |
| 31 | = 787·386 |

# TABLE II.

## MEASURES OF LENGTH.

| | 1 | 2 | 3 | 4 | 5 | 6 | 7 | 8 | 9 |
|---|---|---|---|---|---|---|---|---|---|
| *Meter.* | | | | | | | | | |
| Yards. | 1·09363 | 2·18727 | 3·28090 | 4·37453 | 5·46816 | 6·56180 | 7·65543 | 8·74906 | 9·84270 |
| Feet ... | 3·28090 | 6·56180 | 9·84270 | 13·12360 | 16·40450 | 19·68539 | 22·96629 | 26·24719 | 29·52809 |
| Inches. | 39·37080 | 78·74158 | 118·11236 | 157·48315 | 196·85394 | 236·22473 | 275·59552 | 314·96630 | 354·33709 |
| *Decimeter.* | | | | | | | | | |
| Feet ... | 0·32809 | 0·65618 | 0·98427 | 1·31236 | 1·64045 | 1·96854 | 2·29663 | 2·62472 | 2·95281 |
| Inches. | 3·93708 | 7·87416 | 11·81124 | 15·74832 | 19·68539 | 23·62247 | 27·55955 | 31·49663 | 35·43371 |
| *Centimeter.* | | | | | | | | | |
| Inches. | 0·39371 | 0·78742 | 1·11812 | 1·57483 | 1·96854 | 2·36225 | 2·75596 | 3·14966 | 3·54337 |
| *Millimeter.* | | | | | | | | | |
| Inches. | 0·03937 | 0·07874 | 0·11811 | 0·15748 | 0·19685 | 0·23623 | 0·27560 | 0·31497 | 0·35434 |

## TABLE III.

### MEASURES OF WEIGHT.

| | 1 | 2 | 3 | 4 | 5 | 6 | 7 | 8 | 9 |
|---|---|---|---|---|---|---|---|---|---|
| *Kilogramme.* | | | | | | | | | |
| Pounds (avoir.) | 2·20486 | 4·40971 | 6·61457 | 8·81943 | 11·02428 | 13·22914 | 15·43400 | 17·63886 | 19·84371 |
| Pounds (Troy) | 2·67951 | 5·35903 | 8·03854 | 10·71805 | 13·39757 | 16·07708 | 18·75659 | 21·43610 | 24·11562 |
| *Gramme.* | | | | | | | | | |
| Grains........... | 15·44242 | 30·88484 | 46·32726 | 61·76968 | 77·21210 | 92·65352 | 108·09694 | 123·53936 | 138·98178 |
| *Decigramme.* | | | | | | | | | |
| Grains........... | 1·54424 | 3·08848 | 4·63273 | 6·17697 | 7·72121 | 9·26535 | 10·80969 | 12·35394 | 13·89818 |
| *Centigramme.* | | | | | | | | | |
| Grains........... | 0·15442 | 0·30885 | 0·46327 | 0·61770 | 0·77212 | 0·92654 | 1·08097 | 1·23539 | 1·38982 |
| *Milligramme.* | | | | | | | | | |
| Grains........... | 0·01544 | 0·03089 | 0·04633 | 0·06177 | 0·07721 | 0·09265 | 0·10810 | 0·12354 | 0·13898 |

Example of the use of these tables: Let it be required to find how many grains are equal to 87·435 grammes:—

| Grammes. | | Grains. |
|---|---|---|
| 80 | = | 1235·3936 |
| 7 | = | 108·0969 |
| 0·4 | = | 6·1770 |
| 0·03 | = | 0·4633 |
| 0·005 | = | 0·0772 |
| 87·435 | = | 1350·2080 |

# LIST OF APPARATUS

## FOR A COMPLETE COURSE ON PHYSICS.

---

### *A.*—ON EQUILIBRIUM.

A NUMBER of weights of any unit, or, instead of these, tin boxes for coins, or sand, § 49.

Some small scale-pans, § 49.

Arrangement for the parallelogram of forces, § 50.

Model of the knee-press, § 51.

Inclined plane, § 53, 54.

Apparatus to illustrate the screw, § 55.

Archimedes' screw, § 55.

Model of a screw-press, § 55.

Apparatus to illustrate the wedge, § 56.

Several pulleys, and gangs of pulleys, § 57, 58.

Arrangement of levers, § 59.

Wheel and axle, § 60.

Jack for hoisting, and models of other complex machines, § 60.

Set of regular figures of wood to illustrate the center of gravity, § 61.

A figure balancing on a point, § 62.

Double cone, § 63.

A roller of homogeneous wood.

A roller loaded on one side with lead, § 63.

Chinese tumblers, § 64.

Arrangement to illustrate stable equilibrium, § 65.

Apparatus to show the position of the center of gravity in the balance, § 66.

A fine balance, § 67.

A common balance, § 69.

Steelyard, § 71.

Platform, or counter scales, § 72.

Spring balance.

Iron weights up to 25 lbs., § 70.

Small brass weights, made to fit into each other, § 70.

Apothecary's weights, § 70.

A set of French weights, § 70.

Marble plate and ivory ball, § 73.

A glass trumpet, a tube cut spirally, and a bunch of glass threads, § 73.

Bologna flasks, Rupert's drops, § 73.

Arrangement to compare the absolute cohesion of various substances, § 75.

Cohesion plates of glass and metal, § 76.

Arrangment for using the cohesion plates in vacuo, § 76.

Plates for the adhesion of liquids, § 76.

Apparatus to exhibit the uniform transmission of pressure, § 77.

Hydraulic press, § 77.

Labels for boxes and bottles.

Hydrostatic bellows, § 79.

Apparatus to show the downward pressure of water, § 80.

Apparatus to show the upward pressure, § 81.

Communicating tubes, § 82.

Communicating tubes for different liquids, § 82.

Several cylinders of wood to illustrate the subject of floating bodies, § 83.

Arrangement to show the loss of weight in water, § 84.

Cartesian devil, § 85.

Specific gravity bottle, § 86.

Glass bulb loaded with mercury, § 86.

Nicholson's areometer, § 87.

Volumeter, § 88.

Tralle's hydrometer, § 91.

Tall glass cylinder for the areometers to float in, § 91.

Areometers of Beck, Beaumé, and Cartier, § 92.

A common saccharimeter, § 93.

Hydrometer indicating the specific gravity directly, § 93.

Capillary tubes, connecting with wider tubes, § 94.

A lot of simple capillary tubes, § 94.

Conical tubes, § 94.

A number of pieces of plate glass, § 94.

A number of small glass bulbs, some empty, some loaded with mercury, and balls of wood, cork, wax, etc., § 94.

Some plates of mica, § 94.

Endosmometer, § 95.

Torricellian tubes, § 96.

A cistern barometer, § 97.

A siphon barometer, § 97.

Huyghen's double barometer, § 97.

A wheel barometer, § 97.

Arrangement to demonstrate Mariotte's law, § 98.

An air-pump, § 99.

Receivers for the air-pump, including one with a sliding rod.

Magdeburg hemispheres, § 102.

A cylinder for bursting bladders, § 102.

Siphon barometer inclosed in receiver, § 102.

Bell to ring in vacuo, § 102.

Apparutus for freezing water, § 102.

Flint lock to strike in vacuo, § 102.

Siphon to be placed under a receiver, § 102.

Arrangement to illustrate the theory of suction, § 102.

Arrangement for shower of mercury, § 102.

Balance manometer, § 102.

Receiver for condensed air, § 102.

Fountain with compressed air, § 102.

Glass balloon for weighing air, § 102.

Tube for falling bodies in vacuo, § 102.

Air-gun, § 103.

Siphons, § 104.

Tantalus' cup, § 104.

Pipettes, § 104.

Magic funnel and cans, § 105.

Vestal's sieve, § 105.

Hiero's fountain, § 106.

Magic cask, § 106.

Models of suction and force pumps, § 107.

Model of a fire-engine, § 107.

Intermitting fountain, § 108.

Manometers of various kinds, § 109.

Hydrogen lamp, § 110.

Gas generator, § 110.

Balloon of gold-beater's skin, or collodion, § 111.

Glass vessels communicating by a cock, for the diffusion of gases, § 113.

## *B.*—ON MOTION.

Atwood's machine, § 115.

Apparatus to show the curve of projectiles, § 116.

Whirling machine, § 117.

Centrifugal railway, § 118.

Gyroscope, or Bohnenberger's machine, § 119.

Apparatus for Foucault's experiment, § 120.

Arrangement for the fall of bodies through chords and arcs, § 121.

A pendulum moving on knife edges, § 121.

A reversion pendulum, § 121.

Pendulum to beat audibly, § 122.

Models of the common clock and watch escapements, § 122.

Apparatus to show the laws of impact, § 123.

Apparatus to demonstrate the law of incidence and reflection, § 123.
Tribometer, § 124.
Spouting fluid machine, § 125.
Mariotte's vase, § 126.
Barker's mill, § 128.
Water ram, § 129.
A gasometer, § 130.
Plate paradox for Clement's experiment, § 132.
Apparatus to illustrate the draft of the locomotive, § 132.

## *C.*—ON ACOUSTICS.

Wheatstone's undulation apparatus, § 133.
Troughs of glass and wood to hold water and mercury, § 133.
A rope to show undulations, § 134.
Violincello bow.
A tube to conduct sound, § 138.
Closed and opened organ pipes, § 141.
Lip of an organ pipe with tubes of various materials to fit it, § 142.
Clamp with glass and metal disks for sound-figures, § 137.
Extensible tube to sound in unison with a bell, § 142.
Arrangement to produce sounds with a jet of hydrogen, § 143.
Apparatus to show nodes in tubes, § 144.
A siren, § 145.
A monochord, § 146.
Apparatus for longitudinal vibrations, § 147.
Two tuning-forks, § 148.
Apparatus for interference of sound-waves, § 149, 150.
Apparatus for communicating vibrations, § 151.
Model to illustrate vocal organs, § 152.
Model of the ear, § 153.

## *D.*—ON OPTICS.

A theodolite which may be used as a goniometer.
A sheet of gold-leaf laid between glass, § 154.
A long graduated bar, § 161.
Stands for one and four candles, § 155.
Photometer, § 155.
Arrangement for the camera clara, § 157.
Plane mirrors, common and black, § 158.
Hinged mirrors, § 158.
Kaleidoscope, § 158.
Concave and convex mirrors, § 161, 162.
Cylindrical and conical mirrors, § 162.
Apparatus to show the law of reflection, § 159.
Heliostat, § 160.
Apparatus to show the law of refraction, § 163.
Camera lucida.
The six principal forms of lens, § 165.
A large lens of 2 to 4 feet focal distance, and several of shorter focus, § 165.
A pocket lens.
Several screens covered with white paper, 1 to 10 square feet, and a small one covered with straw paper.
Prisms of flint and crown glass, solid and hollow, § 166.
Screens with apertures of various shapes and sizes, § 167.
An oscillating prism, § 167.
Apparatus for measuring refraction, § 167.
Large glass bulb to illustrate the rainbow, 168.
Reusch's apparatus for the same purpose, § 168.
Apparatus to illustrate chromatic aberration, § 169.
Achromatic prism.
Achromatic lens.
Apparatus to illustrate spherical aberration, § 170.
Arrangement to exhibit fluorescence, § 171.
Stand for demonstrating the anatomy of the eye, § 172.

Stampfer's optometer, § 174.
Rainbow disk, § 175.
Thaumatrope, § 176.
Phantascope, § 177.
Figures and colored paper for subjective colors, § 178.
Nörremberg's apparatus for subjective colors, § 178.
Arrangement for making colored shadows, § 179.
A stereoscope, § 180.
Arrangement to make, pictures stereoscopic without lenses or mirrors, § 180.
Camera obscura, § 181.
Solar microscope, § 182.
Compound microscope, with prepared objects, § 183.
Simple microscope with reflector.
Achromatic terrestrial telescope, § 184.
Astronomical telescope, with an arrangement for equatorial mounting, § 184.
Lenses mounted so as to illustrate the theory of the telescope, § 184.
A reflecting telescope, § 185.
Mirror and prism for producing interference, § 187, 188.
Glasses for producing Newton's rings, § 189.
Arrangement for projecting the rings on a screen, § 189.
Apparatus for Grimaldi's experiment on interference, § 190.
Magic lantern and slides, § 186.
Arrangements for experiments on diffraction, § 192.
Apparatus to illustrate the vibrations of polarized light, § 194.
Polarizing apparatus, § 195.
Tourmalin apparatus, § 197.
Various crystals cut to show polarization.
A crystal of Iceland spar, and a model to show the sections, § 200.
Nichol's prism, § 202.
Thin plates of selenite and mica.
Polarizing apparatus for liquids, § 211.
A piece of unannealed glass.
A press for glass, § 212.
Daguerreotype apparatus.

## *E.*—ON MAGNETISM.

Magnetic bars.
Magnetic bar with succession points.
Horseshoe magnet.
A piece of load-stone, § 217.
Magnetic needles, § 218.
A compass, 219.
Dipping needle, § 228.
Bars of soft iron and iron-filings.

## *F.*—ON ELECTRICITY.

Pith balls on linen and silk threads.
Amalgam—Chains and hooks, § 234, 235.
Rods of glass and sealing-wax.
An electrical needle, § 237.
Electrometers of different sorts.
Torsion balance, § 244.
Proof-plane, § 246.
Electrical machine, § 247.
Steam electrical machine, § 255.
Insulating stool, § 256.
Electrical spider—Chime of bells, § 257.
Electric float—Gold fish, § 257.
Arrangement for hail storm and puppet dance, § 257.
Electric pistol, § 257.
Insulated globe and insulated cylinder, § 258.
Arrangement to show the effect of points, § 259.
Insulated ball with hemispherical caps, § 260.
A roll of gilt ribbon on an insulated roller, § 260.
Conductors to exhibit electrical induction, § 261.
Franklin's plate with detached coatings, § 262.
An assortment of Leyden jars, § 263.

Discharger — Henley's universal discharger, § 264, 264.
Lanne's measuring jar, § 266.
Powder cup, § 268.
Press for glass, § 268.
Thunder house, § 268.
Electric mortar, § 268.
Kinnersley's thermometer.
Disks for producing currents by induction, § 268.
Electrophorus, § 269.
Condensers of brass, zinc, and copper, § 271.
Lightning tube, § 272.
Spotted plate, § 272.
Luminous letters, § 272.
Electrical egg—Aurora tube, § 272.

## *G.*—ON GALVANISM.

Copper and zinc plates for Volta's fundamental experiment, § 276.
Voltaic pile of 50 pairs, § 278.
An assortment of binding-screws, § 279.
A single Wollaston's couple, § 280.
Hare's calorimotor, § 281.
Daniell's battery, § 285.
Grove's battery, § 286.
Bunsen's battery, § 287.
Zamboni's dry pile, § 289.
Handles for receiving shocks, § 290.
A spur wheel, § 291.
Stand for supporting charcoal points, § 292.
Apparatus for production of cold, § 294.
Voltameter, § 295.
Arrangements for electro-metallurgy, § 298.
Apparatus to show the effect of the current on the magnetic needle, § 303.
Several galvanometers, § 304.
A tangent compass, § 307.
Rheostat, § 310.
Several coils of wire of known resistance.
Apparatus to show the resistance of fluids, § 312.
A small electro-magnet, § 313.
An electro-magnet for diamagnetism, § 313, 337.
A spiral for making artificial magnets, § 314.
A model of some machine driven by electro-magnetism, § 315.
Electric telegraph, § 316.
Ampere's stand, § 317.
A pole-changer, § 318.
A floating ring, § 320.
Apparatus to show the rotation of magnets or currents, § 321.
Helices for fundamental experiments on induction, § 322.
Ruhmkorf's apparatus, § 331.
Arrangement for the extra current, § 326.
Electro-magnetic machine, § 335.
Apparatus to show the effect of rotating disks, § 336.
Melloni's apparatus, § 341.

## *H.*—ON HEAT.

An assortment of thermometers.
Differential thermometer, § 344.
Maximum and minimum thermometers, § 345.
Apparatus to show the expansion of bodies by heat, § 346.
Model of a gridiron pendulum, § 346.
Model of chronometer compensation, § 346.
Expansion apparatus for liquids, § 347.
Apparatus to determine the maximum density of water, § 348.
Barometer tubes arranged to show the tension of vapors in vacuo, § 355.
Apparatus to illustrate the condenser in steam-engines, § 356.
Condensing apparatus for gases, § 357.
Apparatus to show the tension of aqueous vapor in a space filled with air, § 359.

Water-hammer, § 360.
Pulse-glass, § 360.
Papin's digester, § 360.
Platina capsule for Siedenfrost's experiment, § 361.
Various hygrometers, § 362.
Psychrometer, § 362.
Liebig's condenser, of glass, § 364.
Cryophorus, § 365.
Hero's rotating ball, § 366.
Model of sliding valves, § 367.
Fire-syringe, § 370.
Concave reflectors, § 372.
Cube for radiation, 373.
Conductometer, § 374.
Apparatus to show the low conducting power of liquids, § 375.
Safety-lamp.
Model of a steam-engine, § 367, 368.

# INDEX.

## Nott and Gliddon's Types of Mankind.

Types of Mankind, or Ethnological Researches, based upon

the Ancient Paintings, Monuments, Sculptures, and Crania of Races, and upon Natural, Geographical, Philological, and Biblical History. By J. C. NOTT, M.D., of Mobile, Alabama, and GEORGE R. GLIDDON, formerly United States Consul at Cairo. One vol. 8vo. $5.00.

Truly, "Types of Mankind" is a work in all respects worthy of our age and honorable to our country. It is unique in conception, remarkable in execution, and in the thoroughgoing and outspoken manner in which it strikes at the root of historical falsehoods, clerical ignorance, and philanthropical humbugs. At the same time it places the study of mankind upon philosophical foundations, new as they are noble, and durable as they are simple. More useful scientific knowledge, and more varied erudition, were never before compressed into the same space.—*New York Herald.*

## Nott and Gliddon's Indigenous Races.

Indigenous Races of the Earth; or, New Chapters of Ethno-

Buddha. Mahmood Adil Shah.

Shah Jehan. Musa Khan.

logical Inquiry. Contributed by ALFRED MAURY, FRANCIS PULSZKY, J. AITKEN MEIGS, M.D.; presenting fresh investigations, documents, and materials. By the Editors, J. C. NOTT, M.D., and GEORGE R. GLIDDON, authors of "Types of Mankind." One vol. royal 8vo. $5.00.

Whatever may be the reader's view of the main doctrine advanced, he cannot fail to be instructed and delighted by much contained in the pages now before us. One involuntarily feels an expansion and elevation of the soul as he approaches the great mysteries of life and being, and tracks back again the distant ages until they are lost in obscurity, to interrogate the monuments of humanity that mark their course.—*North American.*

## Smith's (R. C.) Geographies.

**Smith's New Geography:** Containing a concise text and explanatory notes. With over 100 Maps. For the use of schools in the United States and Canada. By ROSWELL C. SMITH, A.M. 4to. $1.00.

**Smith's New Geography** has been in course of preparation for many years, and is the crowning production of the distinguished author. No pains have been spared to combine in it all that is essental to a complete and comprehensive school geography, and great care has been taken to render it of the greatest practical usefulness in the school-room and family.

Prof. F. A. ALLEN, Principal of Chester County Normal School, speaks of this work in the following terms:—

"It is without doubt the most comprehensive work of its size now published, and sufficiently comprehensive to answer all the purposes of an advanced work for higher institutions. . . . The test to which we are daily submitting it proves highly satisfactory to all concerned; and though somewhat prejudiced against it at first, we are now fully convinced of its merits, an evidence of which is found in the fact that we have adopted it as our STANDARD TEXT-BOOK in geography."

One can see at a glance that a vast amount of labor has been expended in its preparation, and the result is the most complete text-book, both as regards method, and lucidity, and variety, and range of details, that we have seen on this subject.—*Evening Press.*

**Primary Geography.** An introductory geography, designed for children. Illustrated with 126 engravings and 20 maps. Twenty-ninth edition, revised. By ROSWELL C. SMITH, A.M. 16mo. 38 cents.

**Modern and Ancient Geography.** Geography on the productive system, for schools, academies, and families. Latest revised and improved edition, containing the addition of ancient geography. Accompanied by a large and valuable **Atlas of Modern and Ancient Maps.** By ROSWELL C. SMITH, A.M. 12mo. $1.25.

**Smith's Quarto Geography.** A concise and practical system of geography, for schools, academies, and families. Designed as a sequel to the First Book. Illustrated with 32 steel maps and numerous engravings. By ROSWELL C. SMITH, A.M. 12mo. 75 cents.

## Theism and Skepticism.

A Treatise on Theism and on the Modern Skeptical Theories. By Professor FRANCIS WHARTON, of Kenyon College. One vol. 12mo. $1.25.

It is a very interesting volume even to us upon this side of the Atlantic. It may be placed with great advantage in the hands of thoughtful and inquiring young persons; for it conducts the several lines of argument it takes up to sound conclusions, while the path is made pleasant by anecdote and illustration. . . . A more interesting book, or one likely to be more useful to young and ardent minds passing through that anxious state which often intervenes to such between the simple happy acquiescence of childhood, and that firm faith and undisturbed repose which is the fruit of many a bitter conflict, we have not lately met with, and we shall be glad to contribute anything to its success and wider circulation.—*London Christian Observer.*

He illustrates his positions from law books and from Gœthe; from Agassiz, with his fossil fishes, and from Shakspeare; from lectures on coal, and from Hugh Miller's Word-Pictures on the Rocks. . . . We are certain no one can rise from a perusal of Mr. Wharton's book without being both interested and instructed.—*Southern Churchman.*

## The Human Body, etc.

The Human Body and its Connection with Man. Illustrated by the principal organs. By J. J. GARTH WILKINSON, Member of the Royal College of Surgeons of England. One vol. 12mo. $1.00.

A very curious and original work. It considers the organism of man in connection with his moral and spiritual nature, and ignores those maxims of natural science which regard him merely as an animal. The author is said to favor the doctrines of Swedenborg, and he has drawn largely on the writings of that remarkable author in relation to the human body. No work on this subject has ever attracted so much attention from the religious world as this volume on "The Human Body and its Connection with Man."

## The Two Lights.

The Two Lights. By the author of "Struggles for Life." One vol. 12mo. $1 00.

A spirited novel, written with a religious purpose, and designed to rebuke the skeptical spirit so prevalent in Christendom. The incidents are striking, the characters well drawn, and the style lively and readable.

www.ingramcontent.com/pod-product-compliance
Lightning Source LLC
LaVergne TN
LVHW020932110826
845150LV00004B/844

* 9 7 8 1 4 2 5 5 5 2 7 4 9 *